# Lecture Notes in Control and Information Sciences

Edited by A. V. Balakrishnan and M. Thoma

Vol. 22: Optimization Techniques
Proceedings of the 9th IFIP Conference on Optimization Techniques,
Warsaw, September 4–8, 1979
Part 1
Edited by K. Iracki, K. Malanowski, S. Walukiewicz
XVI, 569 pages. 1980

Vol. 23: Optimization Techniques
Proceedings of the 9th IFIP Conference on Optimization Techniques,
Warsaw, September 4-8, 1979
Part 2
Edited by K. Iracki, K. Malanowski, S. Walukiewicz
XV, 621 pages. 1980

Vol. 24: Methods and Applications in Adaptive Control
Proceedings of an International Symposium
Bochum, 1980
Edited by H. Unbehauen
VI, 309 pages. 1980

Vol. 25: Stochastic Differential Systems – Filtering and Control
Proceedings of the IFIP-WG7/1 Working Conference
Vilnius, Lithuania, USSR, Aug. 28 – Sept. 2, 1978
Edited by B. Grigelionis
X, 362 pages. 1980

Vol. 26: D. L. Iglehart, G. S. Shedler
Regenerative Simulation of Response Times in Networks of Queues
XII, 204 pages. 1980

Vol. 27: D. H. Jacobson, D. H. Martin, M. Pachter, T. Geveci
Extensions of Linear-Quadratic Control Theory
XI, 288 pages. 1980

Vol. 28: Analysis and Optimization of Systems
Proceedings of the Fourth International Conference on Analysis and Optimization of Systems
Versailles, December 16–19, 1980
Edited by A. Bensoussan and J. L. Lions
XIV, 999 pages. 1980

Vol. 29: M. Vidyasagar,
Input-Output Analysis of Large-Scale Interconnected Systems – Decomposition, Well-Posedness and Stability
VI, 221 pages. 1981

Vol. 30: Optimization and Optimal Control
Proceedings of a Conference Held at Oberwolfach, March 16–22, 1980
Edited by A. Auslender, W. Oettli, and J. Stoer
VIII, 254 pages. 1981

Vol. 31: Berc Rustem
Projection Methods in Constrained Optimisation and Applications to Optimal Policy Decisions
XV, 315 pages. 1981

Vol. 32: Tsuyoshi Matsuo,
Realization Theory of Continuous-Time Dynamical Systems
VI, 329 pages, 1981

Vol. 33: Peter Dransfield
Hydraulic Control Systems – Design and Analysis of Their Dynamics
VII, 227 pages, 1981

Vol. 34: H.W. Knobloch
Higher Order Necessary Conditions in Optimal Control Theory
V, 173 pages, 1981

Vol. 35: Global Modelling
Proceedings of the IFIP-WG 7/1 Working Conference Dubrovnik, Yugoslavia,
Sept. 1–5, 1980
Edited by S. Krčevinac
VIII, 232 pages, 1981

Vol. 36: Stochastic Differential Systems
Proceedings of the 3rd IFIP-WG 7/1 Working Conference
Visegrád, Hungary, Sept. 15–20, 1980
Edited by M. Arató, D. Vermes, A.V. Balakrishnan
VI, 238 pages, 1981

Vol. 37: Rüdiger Schmidt
Advances in Nonlinear Parameter Optimization
VI, 159 pages, 1982

Vol. 38: System Modeling and Optimization
Proceedings of the 10th IFIP Conference
New York City, USA, Aug. 31 – Sept. 4, 1981
Edited by R.F. Drenick and F. Kozin
XI, 894 pages. 1982

Vol. 39: Feedback Control of Linear and Nonlinear Systems
Proceedings of the Joint Workshop on Feedback and Synthesis of Linear and Nonlinear Systems
Bielefeld/Rom
XIII, 284 pages. 1982

Vol. 40: Y.S. Hung, A.G.J. MacFarlane
Multivariable Feedback: A Quasi-Classical Approach
X, 182 pages. 1982

Vol. 41: M. Gössel
Nonlinear Time-Discrete Systems – A General Approach by Nonlinear Superposition
VIII, 112 pages. 1982

Vol. 42: Advances in Filtering and Optimal Stochastic Control
Proceedings of the IFIP-WG 7/1 Working Conference
Cocoyoc, Mexico, February 1–6, 1982
VIII, 391 pages. 1982

For information about Vols. 1–21 please contact your bookseller or Springer-Verlag.

# Lecture Notes in Control and Information Sciences

Edited by A.V. Balakrishnan and M.Thoma

IRIA

62

# Analysis and Optimization of Systems

Proceedings of the Sixth International Conference on Analysis and Optimization of Systems
Nice, June 19–22, 1984

## Part 1

Edited by
A. Bensoussan and J. L. Lions

Springer-Verlag
Berlin Heidelberg GmbH 1984

**Editors**
A. Bensoussan
J. L. Lions

INRIA
Institut National de Recherche en Informatique et en Automatique
Domaine de Voluceau, Rocquencourt, B.P. 105
78153 Le Chesnay/France

Library of Congress Cataloging in Publication Data
International Conference on Analysis and Optimization of Systems
(6th : 1984 : Nice, France)
Analysis and optimization of systems.
(Lecture notes in control and information sciences ; 62–63)
"Organized by the Institut national de recherche en informatique et [en] automatique"
Foreword. English and French.
1. System analysis——Congresses.
2. Mathematical optimization——Congresses.
3. Automatic control——Congresses.
4. Biotechnology——Congresses.
I. Bensoussan, Alain.
II. Lions, Jacques Louis.
III. Institut national de recherche en informatique et en automatique (France).
IV. Title.
V. Series.
QA402.I533 1984 003 84-5601

ISBN 978-3-540-13551-7 ISBN 978-3-540-39007-7 (eBook)
DOI 10.1007/978-3-540-39007-7


Originally published by Springer-Verlag Berlin Heidelberg New York Tokyo in 1984

2061/3020-543210

# FOREWORD

This volume contains most of the 94 papers presented during the Sixth International Conference on Analysis and Optimization of Systems organized by the Institut National de Recherche en I formatique et Automatique.

The audience has increased by more than 50 % in comparison with the Fifth Conference. These papers, some invited and most of them submitted, were presented by speakers coming from 26 different countries. Most of the topics of System Theory are covered.

At the theoretical level, a trend towards algebraic and geometric methods was confirmed. Signal processing which was one of the main topics of the call for papers had a favourable result : two special sessions on non stationary models and on rupture detection were organized. In the field of applications, one can notice the increasing importance of the CACSD tools. Also, the progress of the biomedical and biotechnological engineering session is remarkable. It has justified the sponsorship of INSERM, for the first time.

In order to improve the coordination with the IEEE Control and Decision Conference, the Organizing Committee has decided to shift the date of the conference which from now on will be held in June. The conference took place near the new center of INRIA at Sophia Antipolis.

We would like to express our thanks to the Organisations which have given their sponsorship to this meeting : AFCET, IEEE, IFAC and INSERM.

We also would like to extend our gratitude to :

- the authors who have shown their interest in this conference,
- the numerous referees who have accepted the difficult task of selecting papers,
- the Chairpersons for having run with energy and efficiency the different sessions,
- our colleagues of the Organisation Committee,
- the Scientific Secretaries,
- Miss Bricheteau and the staff of the Public Relations Department for the difficult but successful job they have carried out in the organization of the Conference,
- Professor Thoma who has accepted to publish these proceedings in the Lecture Notes in Control and Information Sciences, and to the Publisher SPRINGER VERLAG.

A. BENSOUSSAN J.L. LIONS

# PREFACE

Ce volume contient la presque totalité des textes des 94 communications présentées lors de la Sixième Conférence Internationale sur l'Analyse et l'Optimisation des Systèmes, organisée par l'Institut National de Recherche en Informatique et Automatique.

Cette Conférence connaît une audience grandissante puisque le nombre de communications soumises a augmenté de plus de 50 % par rapport à sa dernière édition, confirmant ainsi une tendance antérieure. Ces communications, invitées ou pour la plupart soumises, émanent de 26 pays différents. La plupart des domaines de la "Théorie des Systèmes" y sont abordés.

Sur le plan théorique, on constate la confirmation d'une évolution vers les méthodes géométriques et algébriques. Le traitement du signal qui était l'un des thèmes principaux de l'appel aux communications a connu un succès certain : deux sessions spéciales sur les modèles non stationnaires et les détections de ruptures ont été organisées.

Du point de vue des applications, les communications présentées portent plus sur des outils généraux de CAO en Automatique que sur des applications spécifiques. Il faut cependant noter les progrès de la session présentant des applications au domaine du génie biomédical et des biotechnologies. Pour la première fois, la Conférence a reçu le patronage de l'INSERM.

La coordination avec la "Control and Decision Conference" de l'IEEE a conduit a déplacer les dates de la Conférence qui se tient désormais au mois de juin. La Conférence s'est déroulée à proximité du nouveau centre INRIA de Sophia-Antipolis.

Nous tenons à remercier les organismes qui ont accepté d'accorder leur patronage à cette manifestation : AFCET, IEEE, IFAC, INSERM.

Nos remerciements s'adressent également :

- aux auteurs qui ont manifesté leur intérêt pour cette conférence ;

- aux nombreux experts qui ont accepté la difficile tâche de sélectionner les communications,

- aux présidents de sessions qui ont accepté d'animer les débats,

- à nos collègues du Comité d'Organisation,

- aux Secrétaires Scientifiques,

- à Mademoiselle Bricheteau et ses collaboratrices du Service des Relations Extérieures qui ont largement participé à l'organisation de cette Conférence,

- à Monsieur le Professeur Thoma pour avoir accepté la publication de ce volume dans la série qu'il dirige, ainsi qu'à l'éditeur SPRINGER VERLAG.

A. BENSOUSSAN J.L. LIONS

**This conference is under the sponsorship of**

*Cette Conférence a reçu le patronage de*

- Association Française pour la Cybernétique Economique et Technique (AFCET)
- Institute of Electrical and Electronics Engineers, Inc (I.E.E.E.) "Co-sponsorship"
- International Federation of Automatic Control (I.F.A.C.)
- Institut National de la Santé et de la Recherche Médicale (INSERM)

## Organization Committee

*Comité d'Organisation*

| | |
|---|---|
| **K. J. ASTRÖM** | Lund Institute of Technology, (Suède) |
| **A. BENSOUSSAN** | Université Paris-Dauphine / INRIA-Rocquencourt (France) |
| **A. BENVENISTE** | INRIA-Rennes (France) |
| **P. BERNHARD** | INRIA-Sophia-Antipolis (France) |
| **P. FAURRE** | SAGEM (France) |
| **A. J. FOSSARD** | ENSAE-Toulouse (France) |
| **J. L. LIONS** | Collège de France / INRIA (France) |
| **A. G. J. Mac FARLANE** | Cambridge University (GB) |
| **M. THOMA** | Technische Universität Hannover (RFA) |
| **J. C. WILLEMS** | Groningen University (Pays Bas) |

## Scientific Secretaries

*Secrétaires Scientifiques*

| | |
|---|---|
| **F. DELEBECQUE** | INRIA-Rocquencourt (France) |
| **J. HENRY** | INRIA-Rocquencourt (France) |

## Conference Secretariat

*Secrétariat de la Conférence*

| | |
|---|---|
| **Th. BRICHETEAU** | INRIA (France) |
| **S. GOSSET** | Service des Relations Extérieures |

REFEREES

| | | |
|---|---|---|
| ABRAMATIC | J.F. | (FRANCE) |
| AEYELS | D. | (BELGIUM) |
| ALING | | (THE NETHERLANDS) |
| ALMEIDA | L.B. | (PORTUGAL) |
| ASTROM | K.J. | (SWEDEN) |
| AUBIN | J.P. | (FRANCE) |
| BABARY | J.P. | (FRANCE) |
| BAILLIEUL | John | (U.S.A.) |
| BARAS | J. | (U.S.A.) |
| BARATCHART | L. | (FRANCE) |
| BARRAUD | M. | (FRANCE) |
| BASSEVILLE | Michelle | (FRANCE) |
| BENSOUSSAN | Alain | (FRANCE) |
| BENVENISTE | A. | (FRANCE) |
| BERNHARD | P. | (FRANCE) |
| BERNUSSOU | J. | (FRANCE) |
| BERTHOMIER | C. | (FRANCE) |
| BINDER | | (FRANCE) |
| BISMUT | J.M. | (FRANCE) |
| BOIS VIEUX | J.F. | (FRANCE) |
| BONNANS | J.F. | (FRANCE) |
| BORNARD | | (FRANCE) |
| BOSGRA | O. | (THE NETHERLANDS) |
| BREMAUD | P. | (FRANCE) |
| BRILLET | J.L. | (FRANCE) |
| BROCKETT | R.W. | (U.S.A.) |
| CALLIER | F. | (BELGIUM) |
| CARPENTIER | | (FRANCE) |
| CHAPMAN | | (G.B.) |
| CHAVENT | G. | (FRANCE) |
| CHENIN | P. | (FRANCE) |
| CHERRUAULT | Y. | (FRANCE) |
| CHERUY | Arlette | (FRANCE) |
| CHEVALIER | F. | (FRANCE) |
| CHOPLIN | J. | (FRANCE) |
| CHRETIEN | | (FRANCE) |
| CLAASEN | | (THE NETHERLANDS) |
| CLARA | F. | (FRANCE) |
| CLAUDE | | (FRANCE) |
| CLERGEOT | H. | (FRANCE) |
| CLERGET | | (FRANCE) |
| COHEN | Guy | (FRANCE) |
| COLLETER | P. | (FRANCE) |
| COMMAULT | | (FRANCE) |
| COURVOISIER | J.P. | (FRANCE) |

| | | |
|---|---|---|
| DAMLAMIAN | A. | (FRANCE) |
| DAVIS | M.H.A. | (G.B.) |
| DELEBECQUE | F. | (FRANCE) |
| DELFOUR | Michel | (FRANCE) |
| DELMAS | J. | (FRANCE) |
| DENHAM | M. | (G.B.) |
| DEPEYROT | Michel | (FRANCE) |
| DESCUSSE | M. | (FRANCE) |
| DESHAYES | J. | (FRANCE) |
| DESOER | C.A. | (U.S.A.) |
| DION | J.M. | (FRANCE) |
| DODU | | (FRANCE) |
| DUBOIS | D. | (FRANCE) |
| DUPONT | | (FRANCE) |
| EKELAND | I. | (FRANCE) |
| ESPIAU | B. | (FRANCE) |
| FAUGERAS | O. | (FRANCE) |
| FAVIER | | (FRANCE) |
| FLIESS | M. | (FRANCE) |
| FORESTIER | J.P. | (FRANCE) |
| FOSSARD | A. | (FRANCE) |
| GAUTHIER | | (FRANCE) |
| GAUVRIT | | (FRANCE) |
| GERMAIN | F. | (FRANCE) |
| GLOWINSKI | Roland | (FRANCE) |
| GOMEZ | C. | (FRANCE) |
| GONDRAN | | (FRANCE) |
| GOODWIN | G.C. | (AUSTRALIE) |
| GOURSAT | M. | (FRANCE) |
| GRENIER | Y. | (FRANCE) |
| GUEGEN | C. | (FRANCE) |
| HALME | | (FINLAND) |
| HAUTUS | M.L.J. | (THE NETHERLANDS) |
| HAZEWINKEL | M. | (THE NETHERLANDS) |
| HENRY | J. | (FRANCE) |
| IRVING | E. | (FRANCE) |
| ISIDORI | | (ITALY) |
| JACOB | G. | (FRANCE) |
| KERNEVEZ | J.P. | (FRANCE) |
| KOKOTOVIC | P. | (U.S.A.) |
| KOREZLIOGLIU | H. | (FRANCE) |
| KRENER | A.J. | (U.S.A.) |
| KUCERA | W. | (TCHEKOSLOVAKIA) |
| LACOUME | J.L. | (FRANCE) |
| LANDAU | I.D. | (FRANCE) |
| LAUB | A. | (U.S.A.) |
| LE LETTY | C. | (FRANCE) |
| LEDERER | P. | (FRANCE) |
| LEMARECHAL | C. | (FRANCE) |
| LEVINE | Jean | (FRANCE) |
| LIONS | P.L. | (FRANCE) |
| LJUNG | L. | (SWEDEN) |
| LOBRY | C. | (FRANCE) |
| LORINO | H. | (FRANCE) |

| | | |
|---|---|---|
| MARMORAT | J.P. | (FRANCE) |
| MARROCCO | A. | (FRANCE) |
| MAURRAS | J.F. | (FRANCE) |
| MENALDI | J.L. | (FRANCE) |
| MICHEL | | (FRANCE) |
| MIGNOT | F. | (FRANCE) |
| MINOUX | | (FRANCE) |
| MIQUEL | | (FRANCE) |
| MOALLA | | (TUNISIA) |
| MORSE | A.S. | (U.S.A.) |
| MUNACK | A. | (F.R.G.) |
| MURON | O. | (FRANCE) |
| NAIN | Philippe | (FRANCE) |
| NEPOMIASTCHY | P. | (FRANCE) |
| NIJMEYER | H. | (THE NETHERLANDS) |
| OPPENHEIM | G. | (FRANCE) |
| ORTEGA | | (FRANCE) |
| PARDOUX | E. | (FRANCE) |
| PAVE | A. | (FRANCE) |
| PICCI | G. | (ITALY) |
| PLATEN | R. | (G.D.R.) |
| POLAK | E. | (U.S.A.) |
| PRALY | L. | (FRANCE) |
| PROTH | J.M. | (FRANCE) |
| PUN | | (FRANCE) |
| QUADRAT | J.P. | (FRANCE) |
| ROBIN | Maurice | (FRANCE) |
| ROFMAN | E. | (FRANCE) |
| ROUBELLAT | | (FRANCE) |
| ROUCHALEAU | Y. | (FRANCE) |
| RUCKEBUSH | G. | (FRANCE) |
| SAGUEZ | Christian | (FRANCE) |
| SAMSON | C. | (FRANCE) |
| SENTIS | R. | (FRANCE) |
| SERMANGE | M. | (FRANCE) |
| SORINE | M. | (FRANCE) |
| STEER | S. | (FRANCE) |
| SULEM | Agnes | (FRANCE) |
| SZPIRGLAS | Jacques | (FRANCE) |
| TEMPELAAR | D. | (THE NETHERLANDS) |
| TITLI | A. | (FRANCE) |
| VAN DER SCHAFT | A. | (THE NETHERLANDS) |
| VAN DER WEIJDEN | A. | (THE NETHERLANDS) |
| VAN DOOREN | P. | (BELGIUM) |
| VAN SCHUPPEN | | (THE NETHERLANDS) |
| VARAIYA | P. | (U.S.A.) |
| VIOT | M. | (FRANCE) |
| WEISS | | (U.S.A.) |
| WILLEMS | J.C. | (THE NETHERLANDS) |
| WILLEMS | J.L. | (BELGIUM) |
| WILLSKY | A.S. | (U.S.A.) |
| WONHAM | M.W. | (CANADA) |
| YVON | J.P. | (FRANCE) |
| ZABCZYK | J. | (POLAND) |
| ZAMES | George | (CANADA) |
| ZOLESIO | | (FRANCE) |

## TABLE OF CONTENTS / TABLE DES MATIERES

SESSION 1

### NON STATIONARY PROCESSES / PROCESSUS NON STATIONNAIRES



SESSION 2

### STABILITY I / STABILITE I



SESSION 3

### UTILITY SYSTEMS / RESEAUX DE SERVICE

SESSION 4

## DETECTION OF CHANGES IN SYSTEMS / DETECTION DE CHANGEMENTS DANS LES SYSTEMES



SESSION 5

## STABILITY II / STABILITE II

SESSION 9

**DETERMINISTIC CONTROL / CONTROLE DETERMINISTE**



SESSION 10

**FILTERING / FILTRAGE**

PART 2
(published as Lecture Notes in Control and Information Sciences, Vol. 63)

## TABLE OF CONTENTS / TABLE DES MATIERES

SESSION 11

### NUMERICAL METHODS / METHODES NUMERIQUES



SESSION 12

### STOCHASTIC CONTROL / CONTROLE STOCHASTIQUE



SESSION 13

### LINEAR SYSTEMS II / SYSTEMES LINEAIRES II

SESSION 16

**NONLINEAR SYSTEMS I / SYSTEMES NON LINEAIRES I**



SESSION 17

**BIOTECHNOLOGICAL SYSTEMS AND BIOENGINEERING**

**GENIE BIOMEDICAL ET SYSTEMES BIOTECHNOLOGIQUES**

SESSION 18

**NONLINEAR SYSTEMS II / SYSTEMES NON LINEAIRES II**



SESSION 19

**COMPUTER AIDED CONTROL SYSTEM DESIGN II / CAO EN AUTOMATIQUE II**

SESSION 20

## PRODUCTION AUTOMATION / AUTOMATISATION DE LA PRODUCTION



ADDITIONAL INFORMATION CONCERNING
SOFTWARE DEMONSTRATIONS PRESENTED DURING THE MEETING

INFORMATION SUPPLEMENTAIRE CONCERNANT
LA PRESENTATION DE LOGICIELS AU COURS DE LA CONFERENCE

Session 1

# NON STATIONARY PROCESSES

# PROCESSUS NON STATIONNAIRES

# COVARIANCE EQUIVALENT FORMS AND EVOLUTIONARY SPECTRA FOR NONSTATIONARY RANDOM PROCESSES

J.K. Hammond and R.F. Harrison
Institute of Sound and Vibration Research
University of Southampton,
England.

## ABSTRACT

Many nonstationary random processes exhibit a 'frequency modulated' structure. In this paper a method of modelling such processes as the output of a time variable filter driven by white noise is described. The basis of the method relies on producing a process that is 'covariance equivalent' to the process under consideration. This particular formulation makes it possible to predict the evolutionary (time-frequency) spectral density of the process.

The theoretical basis of the method is explained and this is followed by a detailed example which illustrates the theory. The example is concerned with the prediction of the evolutionary spectral density of the motion of a vehicle accelerating over rough terrain. The terrain is described as a spatially homogeneous random process and the nonstationarity of the response arises because of the variable velocity of the vehicle.

The covariance equivalent formulation is quite general and other applications (which include problems in acoustics) are noted.

## 1. INTRODUCTION

A spectral representation for a class of nonstationary random processes was defined by Priestley [1], resulting in a two-dimensional (time-frequency) 'evolutionary spectral density'. Using this representation a nonstationary process $x(t)$ may be expressed

$$x(t) = \int_{-\infty}^{\infty} A_t(\omega) e^{j\omega t} dZ_x(\omega) \tag{1}$$

where $Z_x(\omega)$ is an 'orthogonal' process. In words, $x(t)$ is the (weighted) sum of *amplitude modulated* sines and cosines. Many nonstationary random processes that are observed in practice exhibit forms of *frequency modulation*, e.g., perceived acoustic signals are influenced by range, directivity and Doppler effects; vehicles running over rough terrain at variable speed encounter inputs having a 'changing spectrum'. The objective of this paper is to show how nonstationary processes having a frequency modulated form may be modelled using the representation (1) which is essentially an amplitude modulated form, and thus allow the definition of evolutionary spectra for such processes.

The basis of the approach uses the concept of 'covariance equivalence' and the theory will be illustrated with a detailed example describing the response of a (simple) vehicle as it accelerates over rough ground.

## 2. EVOLUTIONARY SPECTRA

The key feature in the representation of equation (1) above is that $Z_x(\omega)$ is orthogonal, i.e., increments $dZ_x(\omega_1)$, $dZ_x(\omega_2)$ are uncorrelated when $\omega_1 \neq \omega_2$. It follows therefore that the variance of $x(t)$ is

$$E[x^2(t)] = \int_{-\infty}^{\infty} |A_t(\omega)|^2 S_{xx}(\omega)d\omega \tag{2}$$

where $S_{xx}(\omega)d\omega = E|dZ_x(\omega)|^2$. (3)

The evolutionary spectral density is $S_{xx,t}(\omega)$, where

$$S_{xx,t}(\omega) = |A_t(\omega)|^2 S_{xx}(\omega) \tag{4}$$

which is a decomposition of the power of $x(t)$ over frequency at time $t$.

An interpretation of this class of nonstationary processes in terms of the response of time varying filters [1] is important in what follows. Consider the response of a time-varying linear filter to a stationary random excitation $s(t)$. This may be expressed

$$x(t) = \int_{-\infty}^{\infty} h(t, u)s(t - u)du \tag{5}$$

Using the usual spectral representation for $s(t)$, equation (5) may be re-written in the form (1) with

$$A_t(\omega) = \int_{-\infty}^{\infty} h(t, u)e^{-j\omega u}du \tag{6}$$

Such an interpretation of a nonstationary process is common (e.g., speech wave-forms) and long standing [3]. We shall show below how such a form may be constructed for nonstationary processes having a frequency modulated structure.

It is noted here that we shall *not* be concerned with the problem of *estimating* evolutionary spectra from samples of data. Priestley has addressed this in detail in [2] and the literature abounds with descriptions of 'short-time' spectral analyses. Furthermore we shall not allude to other spectral forms for nonstationary processes in this paper. Our objectives are specifically to show how an important class of processes may be modelled in the form (1).

## 3. COVARIANCE EQUIVALENT MODELS FOR NONSTATIONARY PROCESSES HAVING A FREQUENCY MODULATED FORM

The substance of this section has previously been reported in [4] and is summarized here.

The objective is to construct mathematical models for nonstationary processes having a frequency modulated structure. The mathematical models should be of such a form that they allow development of evolutionary spectra. To do this we shall begin by considering a process $y$ that is not time dependent but a function of another variable, say $s$ (which might be a space variable, for example), i.e., $y(s)$. We shall assume that $y(s)$ is a stochastic process that is stationary (i.e., homogeneous in the s-domain, having zero mean, variance $\sigma_y^2$, and autocovariance function (ACVF) $E[y(s_1)y(s_2)] = R_{yy}(|s_2 - s_1|)$. To create a frequency modulated process let us now regard $s$ as a (deterministic) function of another variable $t$ (time). Our aim now is to describe $y$ not as a function of $s$, but as a function of $t$, i.e., we create $\tilde{y}(t) = y[s(t)]$ where $\tilde{y}$ is regarded as a function of $t$. The functional dependence of $s$ on $t$ will be described by $\dot{s}(t)$. If $\dot{s}$ is constant then $\tilde{y}(t)$ is a stationary process, but if $\dot{s}$ is not constant $\tilde{y}(t)$ is obviously nonstationary but with the properties

$$E[\tilde{y}(t)] = E[y(s)] = \text{constant (assumed zero)}$$
$$E[y^2(t)] = E[y^2(s)] = \sigma_y^2 \text{ (constant)}$$

Even though the mean and variance are constant, it is obvious that the temporal structure of the signal varies. This is, in turn, reflected in the ACVF for $\tilde{y}(t)$, i.e.,

$$R_{\tilde{y}\tilde{y}}(t_1, t_2) = E[\tilde{y}(t_1)\tilde{y}(t_2)] = R_{yy}(s(t_2) - s(t_1)) \tag{7}$$

which is a function of $t_1$ and $t_2$ and not simply $(t_2 - t_1)$ only (unless $\dot{s}$ is constant).

### Shaping Filter Models

In order to be able to develop evolutionary spectral forms for such processes, we shall now assume that $y(s)$ admits a particular representation, namely that it can be described as the output of a shaping filter that is driven by white noise. This is a common model employed in time series analysis and whilst imposing some restrictions is of sufficient generality to be of great use. The point here is that the filter we require will be specified in a differential equation form in the s domain. It is convenient to use a state form to describe this filter. Let $y(s)$ be expressed as

$$y(s) = \underline{c}^T \underline{x}(s) \tag{8}$$

where $\underline{c}^T$ is a (constant) vector having $n$ components (superscript $T$ denotes the transpose). $\underline{x}(s)$ is an n-dimensional 'state vector' which is assumed to satisfy the shaping filter equation

$$\frac{d}{ds}\underline{x}(s) = A\underline{x}(s) + \underline{b}w(s) \tag{9}$$

$A$ is an $(n \times n)$ constant matrix, $\underline{b}$ is an $(n \times 1)$ constant vector and $w(s)$ is a scalar white noise process with

$$E\left[w(s_1)w(s_2)\right] = \delta(s_1 - s_2) \tag{10}$$

N.B. There is no need to restrict $w$ to be a scalar process; this is merely a convenience for the purposes of this paper.

To create the nonstationary process $\tilde{y}(t)$ we introduce $s(t)$ and it is argued in [4] that (for $t_2 > t_1$)

$$R_{\tilde{y}\tilde{y}}(t_1, t_2) = \underline{c}^T R_{\underline{\underline{xx}}}(0)\Phi_A{}^T(s(t_2) - s(t_1))\underline{c} \;^{\dagger} \tag{11}$$

We wish to obtain a *time variable* shaping filter description for $\tilde{y}(t)$, if we are to obtain the evolutionary spectral form. We now do this as follows. From (8) we see that

$$\tilde{y}(t) = \underline{c}^T\underline{x}[s(t)] = \underline{c}^T\tilde{\underline{x}}(t) \tag{12}$$

where $\tilde{\underline{x}}(t)$ denotes the vector $\underline{x}[s(t)]$ regarded as a function of time. To 'convert' (9) to a form amenable to describe $\tilde{\underline{x}}(t)$ we note that

$$\frac{d}{dt}\tilde{\underline{x}}(t) = \frac{d}{ds}\underline{x}(s)\Big|_t \dot{s} \tag{13}$$

$|_t$ denotes the evaluation of $\frac{d}{ds}\underline{x}(s)$ as a function of time. Using (9) in (13) gives

$$\frac{d}{dt}\tilde{\underline{x}}(t) = \dot{s}A\tilde{\underline{x}}(t) + \dot{s}\underline{b}w[s(t)]. \tag{14}$$

---

$^\dagger$ $R_{\underline{\underline{xx}}}(0)$ is the zero lag autocovariance matrix for $\underline{x}(s)$

$\Phi_A$ is the state transition matrix for matrix A.

This equation shows $\tilde{x}(t)$ to be the solution of an equation of the general form of a time variable differential equation driven by a white process $w[s(t)]$. It is necessary to obtain the excitation as a function of $t$ alone and in reference [4] it is argued that

$$E\{w[s(t_1)]w[s(t_2)]\} = \frac{\delta(t_1 - t)}{\dot{s}(t)} \tag{15}$$

where it is assumed that $\dot{s} > 0$.

An 'equivalent' covariance function would arise if we conceive of *another* white noise process, written as $w_1(t)/\sqrt{\dot{s}}(t)$ where $w_1(t)$ is stationary with

$$E[w_1(t_1)w_1(t_2)] = \delta(t_1 - t_2) \tag{16}$$

so that

$$E\left[\frac{w_1(t_1)}{\sqrt{\dot{s}}(t_1)} \quad \frac{w_1(t)}{\sqrt{\dot{s}}(t)}\right] = \frac{\delta(t_1 - t)}{\dot{s}(t)} \tag{17}$$

The process $w_1(t)/\sqrt{\dot{s}}(t)$ is nonstationary in that it is a uniformly modulated white process, having an ACVF which is indistinguishable from the required form in (15). Accordingly we shall use $w_1(t)/\sqrt{\dot{s}}(t)$ in place of $w[s(t)]$ in equation (14) and so produce a vector process which we shall call $\underline{x}_1(t)$ satisfying

$$\frac{d}{dt}\underline{x}_1(t) = \dot{s}A\underline{x}_1(t) + \sqrt{\dot{s}}\underline{b}w_1(t) \tag{18}$$

Associated with (18) we write

$$y_1(t) = \underline{c}^T\underline{x}_1(t) \tag{19}$$

We use the notation $y_1$ rather than $\tilde{y}$ since it is apparent that $y_1$ and $\tilde{y}$ must differ in some respects. But in view of the fact that the equations (14) and (18) are both driven by excitations that are 'covariance equivalent' (i.e., $w[s(t)]$ and $w_1(t)/\sqrt{\dot{s}}(t)$) then it is reasonable to expect that $\tilde{y}(t)$ and $y_1(t)$ are also covariance equivalent, i.e., $R_{\tilde{y}\tilde{y}}(t_1, t_2) = R_{y_1y_1}(t_1, t_2)$. That this is indeed so, can easily be shown. Thus we now see that (19) and (18) together show that we can consider $y_1$ as covariance equivalent to $y(t)$ and also that $y_1(t)$ is obtained as the output of a time variable filter operating on a stationary input. The general form of (18) is

$$\dot{\underline{x}}_1 = A_1(t)\underline{x}(t) + \underline{b}_1(t)w_1(t) \tag{20}$$

and is referred to below.

## 4. EVOLUTIONARY SPECTRA FOR COVARIANCE EQUIVALENT MODELS

Evolutionary spectral forms for frequency modulated processes follow directly from the results of the previous section. It is important to emphasise that we shall develop an evolutionary spectral form for $y_1(t)$, but as this is covariance equivalent to $y(t)$, the evolutionary spectral density applies to $y(t)$ also.

Express the stationary process $w_1(t)$ as

$$w_1(t) = \int_{-\infty}^{\infty} e^{j\omega t} dW(\omega) \tag{21}$$

with power spectral density for $w_1(t)$ written $S_{w_1 w_1}(\omega) = 1$, then the solution of (19) may be expressed as

$$y_1(t) = \int_{-\infty}^{\infty} e^{j\omega t} A_t(\omega) dW(\omega) \tag{22}$$

where

$$A_t(\omega) = \underline{c}^T \int_0^{\infty} \Phi_{\dot{s}A}(t, t - \tau)\sqrt{\dot{s}(t - \tau)}\, e^{-j\omega\tau} d\tau \underline{b}. \tag{23}$$

The evolutionary spectral density for $y_1(t)$, and hence (by covariance equivalence) for $\tilde{y}(t)$, is

$$S_{yy,t}(\omega) = |A_t(\omega)|^2. \tag{24}$$

N.B. $\Phi_{\dot{s}A}(t_2, t_1) = \exp A[s(t_2) - s(t_1)]$ is the state transition matrix for the system in (18) (and could also be expressed $\Phi_{A_1}(t_2, t_1)$, as $A_1 = \dot{s}A$).

From the above it follows that $A_t(\omega)$ may be computed from (23) with knowledge of $\underline{c}$, $\underline{b}$, $\dot{s}$ and $\Phi_{\dot{s}A}$. In fact it is simpler from a computational point of view to solve a set of differential equations to obtain $A_t(\omega)$. This is explained as follows.

Using (21) in (20), it follows (by linearity) that $\underline{x}_1(t)$ may be expressed as

$$\underline{x}_1(t) = \int_{-\infty}^{\infty} \underline{\alpha}(t, \omega) e^{j\omega t} dW(\omega) \tag{25}$$

where $\underline{\alpha}(t, \omega)$ is a vector satisfying

$$\underline{\dot{\alpha}}(t, \omega) = [A_1(t) - j\omega I]\underline{\alpha}(t, \omega) + \underline{b}_1(t) \tag{26}$$

where $\underline{\dot{\alpha}}$ denotes the differentiation with respect to $t$ for fixed $\omega$ and $I$ denotes the identity matrix. The choice of initial conditions for $\underline{\alpha}$ is considered in the example in the next section. From this it follows that $A_t(\omega)$ may be written

$$A_t(\omega) = \underline{c}^T \underline{\alpha}(t, \omega) \tag{27}$$

which is equivalent to the result in (23).

It is interesting to note that approximate solutions for $\underline{\alpha}$ follow from considerations of (26) (see ref. [6]). A crude approximation (but which is sometimes useful) is obtained by simply setting $\underline{\dot{\alpha}}$ to zero and solving for $\underline{\alpha}$ as

$$\underline{\alpha}(t, \omega) = -[A_1(t) - j\omega I]^{-1} \underline{b}_1(t) \tag{28}$$

This is commented upon also in the next section where it is referred to as the 'zero'th order approximation'.

## 5. EVOLUTIONARY SPECTRAL ANALYSIS OF VEHICLES ON ROUGH GROUND

To demonstrate an application of the above theory, consider the case of a vehicle running over rough ground (this is reported in full in [5]). Figure 1 depicts a highly simplified model.

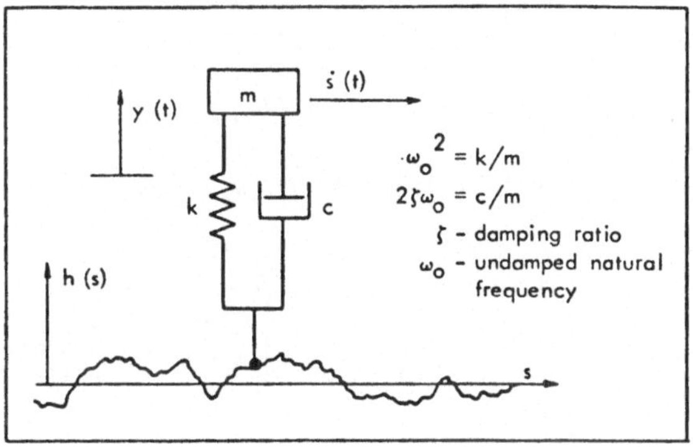


Fig. 1. Simplified vehicle model

The rough ground will be assumed spatially homogeneous and since the vehicle encounters the ground as a function of time, the temporal structure of the input excitation as 'perceived' by the vehicle depends on the vehicle's velocity. If the velocity is constant, the input is a sample function from a stationary process; if the vehicle velocity changes, the input is nonstationary. The vehicle response is stationary only when the velocity is constant and any starting transients have died out, so in general the response is nonstationary. The objective here is to obtain descriptions of the evolutionary spectral density of the vehicle response for an accelerating vehicle.

The vehicle system model chosen is a single degree of freedom, second order linear system. The equations of motion, assuming point contact with the ground is given by:

$$\ddot{y} + 2\zeta\omega_o\dot{y} + \omega_o^2 y = 2\zeta\omega_o\dot{\tilde{h}} + \omega_o^2\tilde{h} \tag{29}$$

where $y(t)$ is the absolute displacement of the mass and $\tilde{h}$ is the ground profile $h(s)$ regarded as a function of time. The other parameters are defined on the figure. (We note that models of greater complexity including nonlinearity and multi-wheels are described in ref. [5]).

In order to obtain the state space form it is necessary to model the surface roughness and we shall use the spatial autocovariance function

$$R_{hh}(\xi) = \sigma^2 e^{-\beta|\xi|} \tag{30}$$

where $\xi = s_1 - s_2$ is the spatial lag variable. The justification for such a model (and others) is given in [5]. The process $h(s)$ may therefore be modelled as the output of a white noise excited, space domain filter whose equation is

$$\frac{dh}{ds} = -\beta h + \sqrt{2\beta}\sigma\, w(s) \tag{31}$$

where

$$E[w(s_1)w(s_2)] = \delta(s_1 - s_2) \tag{32}$$

If we now combine (31) with (29) and use $\frac{d}{dt}(\ ) = \frac{d}{ds}(\ ).\frac{ds}{dt}$, we can write

$$\frac{d}{dt}\begin{bmatrix} y \\ \dot{y} \\ \tilde{h} \end{bmatrix} = \begin{bmatrix} 0 & 1 & 0 \\ -\omega_o^2 & -2\zeta\omega_o & \omega_o^2 - 2\zeta\omega_o\beta\dot{s}(t) \\ 0 & 0 & -\beta\dot{s}(t) \end{bmatrix}\begin{bmatrix} y \\ \dot{y} \\ \tilde{h} \end{bmatrix} + \begin{bmatrix} 0 \\ 2\zeta\omega_o\dot{s}(t) \\ \dot{s}(t) \end{bmatrix}\sqrt{2\beta}\sigma w[s(t)] \tag{33}$$

Now $w[s(t)]$ is replaced by its covariant equivalent form $\dfrac{w_1(t)}{\sqrt{\dot{s}(t)}}$ and $y$ and $\tilde{h}$ are conveniently normalized with respect to $\sigma$ to give:

$$\frac{d}{dt}\begin{bmatrix} x_1 \\ x_2 \\ x_3 \end{bmatrix} = \begin{bmatrix} 0 & 1 & 0 \\ -\omega_o^2 & -2\zeta\omega_o & \omega_o^2 - 2\zeta\omega_o\beta\dot{s}(t) \\ 0 & 0 & -\beta\dot{s}(t) \end{bmatrix}\begin{bmatrix} x_1 \\ x_2 \\ x_3 \end{bmatrix} + \begin{bmatrix} 0 \\ 2\zeta\omega_o \\ 1 \end{bmatrix}\sqrt{2\beta\dot{s}(t)}\, w_1(t) \tag{34}$$

where $x_1 = y_1/\sigma$, $x_2 = \dot{y}_1/\sigma$, $x_3 = h_1/\sigma$, $y_1$, $\dot{y}_1$, $h_1$ denote *covariance equivalent* processes, i.e., $y_1$ is covariance equivalent to $y$, etc. These are then renamed $x_1$, $x_2$, $x_3$ for convenience.

The evolutionary spectral density for $y_1(t)$, say, is obtained by evaluating the integral (23) or equivalently solving the differential equation (26), using the above to define $A_1(t)$, $\underline{b}_1(t)$, etc.

## Initial conditions for $\underline{\alpha}(t, \omega)$

Care must be taken in defining the initial condition for $\underline{\alpha}(t, \omega)$ if the vehicle is to be allowed to start from rest. In fact this may be conveniently considered as a limiting case of the vehicle being in its steady state with some initial velocity at $t = t_o$, then allowing the velocity profile to change. The case of the vehicle having an initial velocity, $v_o$, is straightforward since the initial condition on $\underline{\alpha}(t, \omega)$ is given by the steady-state, frequency response function vector, at $v_o$.

Considering the case of the vehicle standing on the rough ground and starting from rest, the initial velocity is zero, and so only conditions on $x_1$ and $x_3$ need be addressed. By considering the rest state as a limiting case of the vehicle being in a steady state with constant velocity ($v_o \to 0$) then initial conditions on $\underline{\alpha}(t, \omega)$ may be argued to be (see [5] for details)

$$\underline{\alpha}(t_o, \omega) = \begin{bmatrix} \delta(\omega) \\ 0 \\ \delta(\omega) \end{bmatrix} \sqrt{\frac{2}{\beta}} \tag{35}$$

The factor of $\sqrt{2/\beta}$ is retained to ensure correct spectral magnitudes. The delta function obviously cannot be accommodated numerically so for this case only solutions for $\omega \neq 0$ are available.

## Results and discussions

In this section only the case of the vehicle starting from rest is considered (the finite operating time and other problems being dealt with in ref. [5]).

Two cases are considered here, for identical vehicle and ground parameters but at two different values of constant acceleration. The parameters are: $\omega_o = 10$ rad/s, $\zeta = 0.2$, $\beta = 0.2$ rad/m and the two values of acceleration are $\ddot{s} = 5\ \text{m/s}^2$ and $\ddot{s} = 10\ \text{m/s}^2$. The velocity is allowed to vary over the range 0-100 m/s in both cases. It should be noted that the theory allows any velocity profile; the linear profile is chosen here simply for ease of interpretation.

Figure 2a shows the exact evolution of the spectrum of the absolute displacement of the mass for the case when $\ddot{s} = 10\ \text{m/s}^2$. Figure 2b is the zero'th order approximation to this spectrum and Figure 2c shows their difference (approximate minus true). The moving spectra for the less severe case ($\ddot{s} = 5\ \text{m/s}^2$) are omitted since they are very similar in appearance to Figures 2a and 2b. However, their difference is shown in Figure 2d to the same scale as Figure 2c.

During early time (low velocity) the bulk of the excitation energy is concentrated at low frequency and so there is little or no response in the vicinity of the resonance ($\omega_o$ = 10 rad/s). As time progresses (velocity increases) there is a spreading out, along the frequency axis, of the excitation energy, hence the peak response builds up. However, as the velocity increases, to unrealistically high values, the overall spectral amplitude gradually decreases. This is due to the constant variance nature of the excitation process so that, at any time, the area under the excitation spectrum must be constant; hence, as the energy spreads out, so the amplitude must decrease. In the limit as $\dot{s}(t)$ approaches infinity, the evolutionary excitation spectrum tends to zero.

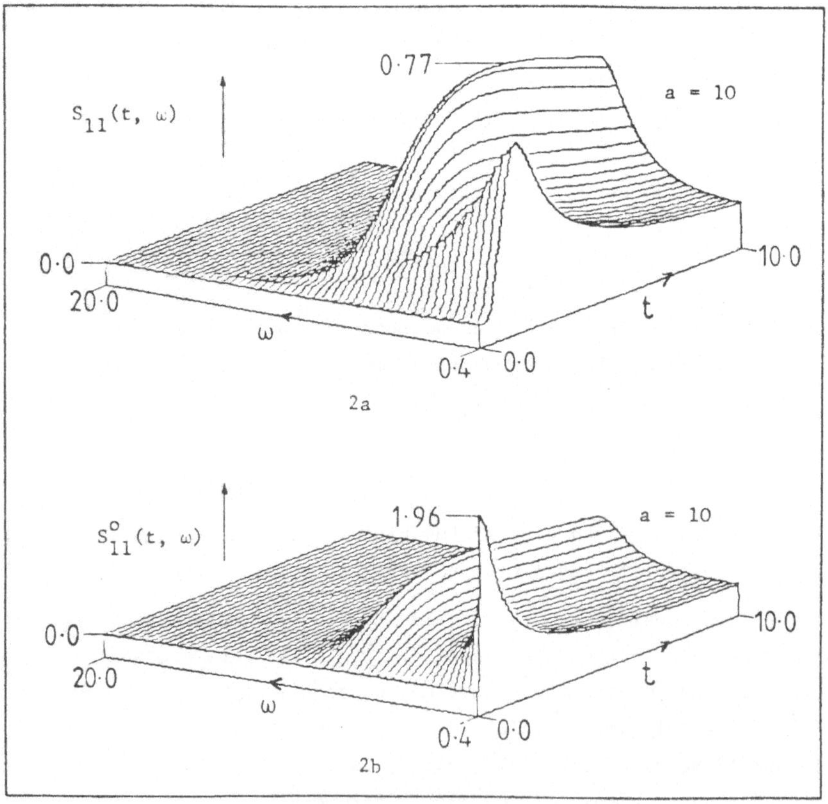


Fig. 2 (a and b) Exact evolutionary spectrum and zero'th order approximation for absolute vehicle motion - constant acceleration.

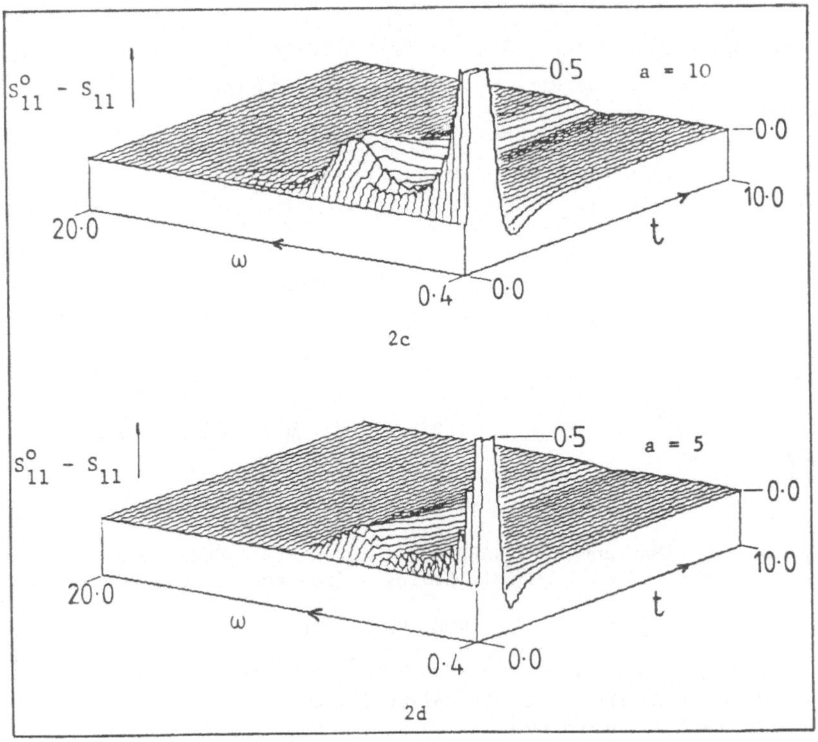


**Fig.** 2 (c and d) Error in zero'th order approximation for two values of acceleration.

Figures 2a and 2b are very similar in appearance, except at very low frequencies and velocity. Figure 2c, the difference between the approximate and true spectra, is perhaps clearer. (The prominent 'spike' has been truncated for convenience.) The zero'th order approximation is in considerable error in the region of ($t = t_o$, $\omega = 0$) and also in the vicinity of the peak response (for moderate velocities). This error, however, becomes less severe as the velocity increases, approaching zero asymptotically.

Figure 2d shows that the error for the less severe case is correspondingly less severe over the same velocity range, although close to the origin it again exhibits singular behaviour.

The fact that the zero'th order approximation is such a poor match at low frequencies and velocities can be shown to be predictable using an argument described in [5], based on some results due to Tsao [6].

## 6. OTHER APPLICATIONS OF COVARIANCE EQUIVALENT FORMS

The vehicle application described above is considered in greater depth in [5] and includes a discussion of the effect of nonlinearity on the dynamic system.

A different application is that relating to acoustic processes, i.e., the sound perceived by an observer due to a moving source. This was noted in [4] and described fully by Tsao in [6].

## 7. REFERENCES


1. M.B. PRIESTLEY 1965 J.R. Stat. Soc. B27. Evolutionary spectra and non-stationary processes.
2. M.B. PRIESTLEY 1966 J.R. Stat. Soc. B28. Design relations for non-stationary processes.
3. L. ZADEH 1950 Proc. IRE, 38. Frequency analysis of variable networks.
4. J.K. HAMMOND, Y.H. TSAO and R.F. HARRISON 1983 Proc. ICASSP, Boston. Evolutionary spectral density models having a frequency modulated structure.
5. R.F. HARRISON 1983 Ph.D. Thesis, ISVR, University of Southampton, England. The nonstationary response of vehicles on rough ground.
6. Y.H. TSAO 1983 Ph.D. Thesis, ISVR, University of Southampton, England. Frequency-time methods in acoustics.

# A GENERAL CLASS OF ESTIMATORS FOR THE WIGNER-VILLE SPECTRUM OF NON-STATIONARY PROCESSES

Patrick Flandrin [1)]
Wolfgang Martin [2)]

1) Laboratoire de Traitement du Signal (LA 346 CNRS)
ICPI, 25 rue du Plat, 69288 LYON Cedex 02 FRANCE.

2) Botanisches Institut der Universität,
Kirschallee 1, D-5300 BONN F.R.G.

ABSTRACT

The Wigner-Ville spectrum is known to be the unique generalized spectrum for the time-varying spectral analysis of harmonizable processes. This time-frequency representation of a process is based on the covariance function and, for quasi-stationary processes, estimators can be defined by means of local time-averaging. We propose here a general class of such estimators relying on an arbitrary weighting function and discuss their first and second order properties in an unifying way. When specifying the arbitrary function, conventional estimators such as short-time periodograms and pseudo-Wigner estimators are recovered and can be compared. This generalized framework emphasizes the versatility of smoothed pseudo-Wigner estimators, especially for uncoupled time and frequency behaviors : they overcome the uncertainty relations of short-time periodograms which only can improve the performances in one direction of the time-frequency plane at the expense of a loss in the other one.

## 1. INTRODUCTION.

The spectral analysis of non-stationary processes has recently regained interest through the rediscovery of the so-called Wigner-Ville distribution. Such a time-frequency representation was initially introduced for deterministic signals [5, 6] but it has been shown that most of its attracting properties directly carry over to harmonizable processes [1]. Furthermore, the resulting Wigner-Ville spectrum has been proved to be unique under natural conditions [2] and hence constitutes the convenient candidate for a time-varying spectrum. This new approach to the spectral analysis of non-stationary processes requires then new estimators : we propose here a general class of such estimators which allows to discuss in an unifyied way the properties of special conventional estimators such as short-time periodograms or pseudo-Wigner estimators.

## 2. ESTIMATION OF THE WIGNER-VILLE SPECTRUM.

### 2.1. The Wigner-Ville spectrum.

Let X(t) be a discrete-time, centered, analytic random process which is supposed to be harmonizable, i.e. that its covariance $K_x$ admits a two-dimensional spectral representation. As shown in [2], there is a unique solution for defining a time-varying spectrum of X which both :

(i) preserves the linear time-frequency dualism,

(ii) is compatible with linear filtering and modulations, and

(iii) gives the expected instantaneous frequency and group delay as local moments.

This solution, which only discards the non-negativity property of an ordinary spectrum, defines the Wigner-Ville spectrum :

$$W_x(t,\omega) := 2 \sum_{\tau=-\infty}^{\infty} K_x(t+\tau,t-\tau)\, e^{-i2\omega\tau} \tag{1}$$

The analyticity [3] of X is supposed both for obtaining (iii) and avoiding aliasing in the discrete equation (1). This allows also to recover the ordinary (unfolded) spectral density if the process should happen to be stationary.

The estimation of the Wigner -Ville spectrum requires then an estimation of the covariance and, to handle efficiently this problem, further assumptions on the underlying process are necessary. In order to replace ensemble averages by time averages without destroying the non-stationary features of the process, the best way is to restrict to the class of quasi-stationary processes. We just recall (and refer to [3] for further details) that we will consider a non-stationary process as a quasi-stationary one if it can be locally approximated by tangential stationary processes for which estimation procedures relying on time averages are possible. In this context, estimation of the local times of stationarity can be performed by an informal procedure based on an Akaike-type criterion [3] and the above mentioned assumptions enable us to propose now admissible estimators of (1).

### 2.2. The general class of estimators.

Since the process is now supposed to be quasi-stationary, its covariance can be locally estimated by means of weighted sums of products of realizations ot the process. If x denotes a realization of X, the most general estimator $\hat{K}_x$ of $K_x$ may be written as :

$$\hat{K}_x(t+\tau,t-\tau) := \sum_{\tau'=-\infty}^{\infty} \Phi(\tau',2\tau)x(t+\tau'+\tau)x^*(t+\tau'-\tau) \tag{2}$$

where $\Phi$ is an arbitrary data window, the shape and the duration of which will determine the nature of the time averaging performed on $x(t+\tau)x^*(t-\tau)$ in order to estimate $K_x(t+\tau,t-\tau)$.

Replacing in (1) $K_x$ by (2), we obtain the desired expression of a general estimator of the Wigner-Ville spectrum :

$$\hat{W}_x(t,\omega;\phi) := 2 \sum_{\tau=-\infty}^{\infty} \sum_{\tau'=-\infty}^{\infty} \Phi(\tau',2\tau)x(t+\tau'+\tau)x^*(t+\tau'-\tau)e^{-i2\omega\tau} \tag{3}$$

with :
$$\Phi(\tau',2\tau) := \frac{1}{2\pi}\int_{-\pi}^{\pi} \phi(\Omega,2\tau)e^{i\Omega\tau'}d\Omega \tag{4}$$

Eqs.(3) and (4) are the exact discrete-time version of the general time-frequency representation of x [4]. The properties of such general representations have been extensively discussed in the case of continuous-time deterministic signals with finite energy [5, 6] and some of them can be easily carried over to our problem. For instance, the estimator (3) will be real-valued if and only if :

$$\phi(\Omega,\tau) = \phi^*(-\Omega,-\tau)$$

However, and to adapt to processes with finite power, new normalizations of $\phi$ will be required, which differ from those used in the finite energy case. The arbitrary weighting function $\phi$ characterizes now the various estimators and peculiar choices of $\phi$ give special estimators of interest which will be considered in Sec.3.
In some cases, it will be simpler to deal with Fourier transforms of $\phi$ and we let :

$$\Pi(t,\omega) := 2 \sum_{\tau=-\infty}^{\infty} \Phi(t,2\tau)e^{-i2\omega\tau} \tag{5}$$

$$\Psi(\Omega,\omega) := 2 \sum_{\tau=-\infty}^{\infty} \phi(\Omega,2\tau)e^{-i2\omega\tau} = \sum_{t=-\infty}^{\infty} \Pi(t,\omega)e^{-i\Omega t} \tag{6}$$

## 2.3. First order properties.

According to (3), the expectation value of the general estimator of the Wigner-Ville spectrum expresses as :

$$E\{\hat{W}_x(t,\omega;\phi)\} = 2 \sum_{\tau'=-\infty}^{\infty} \sum_{\tau=-\infty}^{\infty} \Phi(\tau',2\tau)K_x(t+\tau'+\tau,t+\tau'-\tau)e^{-i2\omega\tau}$$

Using now the property of the Wigner-Ville spectrum that [5] :

$$K_x(t+\tau,t-\tau) = \frac{1}{4\pi}\int_{-\pi}^{\pi} W_x(t,\tfrac{\omega}{2})e^{i\omega\tau}\, d\omega \tag{7}$$

we finally get :

$$E\{\hat{W}_x(t,\omega;\phi)\} = \frac{1}{2\pi}\sum_{t'=-\infty}^{\infty}\int_{-\pi/2}^{\pi/2} W_x(t',\omega')\Pi(t'-t,\omega-\omega')\, d\omega' \tag{8}$$

In the general case, the estimator appears as being doubly biased, in both time and frequency, the bias depending directly on $\Pi$.
Furthermore, a good normalization of the weighting function imposes :

$$\frac{1}{2\pi}\sum_{t=-\infty}^{\infty}\int_{-\pi/2}^{\pi/2} \Pi(t,\omega)\, d\omega = 1 \tag{9}$$

i.e. the simple condition : $\phi(0,0) = 1$.

## 2.4. Second order properties.

Starting again from (3), the covariance of the estimator is given by :

$$C := \mathrm{Cov}\{\hat{W}_x(t_1,\omega_1;\phi),\hat{W}_x(t_2,\omega_2;\phi)\}$$

$$= 4 \sum_{\tau_1'} \sum_{\tau_2'} \sum_{\tau_1} \sum_{\tau_2} \Phi(\tau_1',2\tau_1)\phi^*(\tau_2',2\tau_2)e^{-i2(\omega_1\tau_1 - \omega_2\tau_2)} .$$

$$. \ \mathrm{Cov}\ \{x(t_1+\tau_1'+\tau_1)x^*(t_1+\tau_1'-\tau_1),x(t_2+\tau_2'+\tau_2)x^*(t_2+\tau_2'-\tau_2)\}$$

which reduces to :

$$C = 4 \sum_{\tau_1'} \sum_{\tau_2'} \sum_{\tau_1} \sum_{\tau_2} \Phi(\tau_1',2\tau_1)\phi^*(\tau_2',2\tau_2)e^{-i2(\omega_1\tau_1 - \omega_2\tau_2)} .$$

$$. \ K_x(t_1+\tau_1'+\tau_1,t_2+\tau_2'+\tau_2)\ K_x^*(t_1+\tau_1'-\tau_1,t_2+\tau_2'-\tau_2) \tag{10}$$

if we suppose that the process is gaussian and analytic.
The covariances which appear in (10) can be written :

$$\begin{aligned} K_x(t_1+\tau_1'+ \tau_1,t_2+\tau_2'+\tau_2) &=: K_x(t_a+\tau_a/2,t_a-\tau_a/2) \\ K_x(t_1+\tau_1'- \tau_1,t_2+\tau_2'-\tau_2) &=: K_x(t_b+\tau_b/2,t_b-\tau_b/2) \end{aligned} \tag{11}$$

and, if using the hypothesis of quasi-stationarity in the domain defined by $\Phi$, we can approximate in (10) the covariances (11) by :

$$K_x(t_a+\tau/2,t_a-\tau/2) \sim K_x(t_b+\tau/2,t_b-\tau/2) \sim K_x(t_0+\tau/2,t_0-\tau/2) \tag{12}$$

with : $t_0 := (t_1+t_2)/2$.
Such quasi-stationary covariances can now be expressed by means of the spectral density $f_{t_0}$ of the tangential stationary process in $t_0$ as :

$$K_x(t_0+\tau/2,t_0-\tau/2) = \frac{1}{2\pi}\int_{-\pi}^{\pi} f_{t_0}(\omega)e^{i\omega\tau}d\omega \tag{13}$$

and some calculations yield the final result, where the sign "$\sim$" stems from the approximation used in (12) :

$$\mathrm{Cov}\{\hat{W}_x(t_1,\omega_1;\phi),\hat{W}_x(t_2,\omega_2;\phi)\} \sim$$

$$\frac{1}{2\pi^2}\iint_{-\pi}^{\pi} \Psi(2\Omega,\omega_1-\omega')\Psi^*(2\Omega,\omega_2-\omega')f_{t_0}(\omega'-\Omega)f_{t_0}(\omega'+\Omega)e^{-i2\Omega(t_1-t_2)}d\Omega\ d\omega' \tag{14}$$

This general expression shows that, in the general case, the considered estimators give correlated estimations, in both time and frequency.
If we now restrict to the time dependence, we get the simplified expression :

$$\mathrm{Cov}\{\hat{W}_x(t_1,\omega_1;\phi),\hat{W}_x(t_2,\omega_2;\phi)\} \sim\ 2\ W_{f_{t_0}}(t_1-t_2,\omega;\phi_1) \tag{15}$$

with :

$$\phi_1(\Omega,\tau) := \sum_{\tau'=-\infty}^{\infty} \phi(\Omega,\tau')\phi^*(\Omega,\tau'-\tau) \tag{16}$$

and where :

$$W_{f_t}(t,\omega;\phi) = \frac{1}{4\pi^2} \iint_{-\pi}^{\pi} \Psi(2\Omega,\omega-\omega_1) f_t(\omega_1-\omega_2) f_t(\omega_1+\omega_2) e^{-i2\omega_2 t} d\omega_1\, d\omega_2 \tag{17}$$

is the general time-frequency representation of $f_t$ in the sense of (3), but expressed in the frequency domain.

Hence, it appears that the covariance in the time direction of the general estimator is described by a related general time-frequency representation of the spectral density of the stationary process tangential to X at the time midpoint of the considered interval.

Setting now $t_1=t_2=t$ in (15), we finally get for the variance of the general estimator the expression :

$$\mathrm{Var}\{\hat{W}_x(t,\omega;\phi)\} \sim 2\, W_{f_t}(0,\omega;\phi_1) \tag{18}$$

Since this quantity can be expressed by means of (17) with $\Psi_1=|\Psi|^2$, it clearly appears that the approximated variance is ensured to be non-negative.

Further simplifications can be achieved when supposing that the time of stationarity is much larger than the correlation time. This ensures $\Psi(\Omega,\omega)$ to be a peaked function in $\Omega$ and leads for (14) to the approximated form :

$$\mathrm{Cov}\{\hat{W}_x(t_1,\omega_1;\phi),\hat{W}_x(t_2,\omega_2;\phi)\} \sim$$
$$\frac{1}{2\pi} \sum_{t=-\infty}^{\infty} \int_{-\pi/2}^{\pi/2} f_{t_0}^2(\omega)\,\Pi(t-t_1,\omega_1-\omega)\,\Pi^*(t-t_2,\omega_2-\omega)\, d\omega \tag{19}$$

This equation, which expresses that the correlation vanishes only when the $\Pi$ are non overlapping, reduces for the variance to :

$$\mathrm{Var}\{\hat{W}_x(t,\omega;\phi)\} \sim \{\frac{1}{2\pi} \sum_{t=-\infty}^{\infty} \int_{-\pi}^{\pi} |\Pi(t,\omega')|^2\, d\omega'\}.f_t^2(\omega) \tag{20}$$

General first and second order properties have then been stated and special cases can now be recovered and discussed when specifying the arbitrary weighting function $\phi$.

## 3. CLASSICAL ESTIMATORS AS SPECIAL CASES.

### 3.1. Short-time periodograms.

Such an estimator, which is certainly the most popular one [7], can be expressed as :

$$J_{2N-1}^2(t,\omega) := \frac{1}{2N-1} \left| \sum_{t'=-\infty}^{\infty} x(t')h(t'-t)e^{-i\omega t'} \right|^2 \tag{21}$$

where $h(t)$ is a window with 2N-1 non-zero values.

According to (3) and (4), such a choice corresponds to :

$$\phi_{STP}(\Omega,2\tau) = \frac{1}{2N-1} \sum_{t=-\infty}^{\infty} h(t+\tau)h^*(t-\tau)e^{i\Omega t} =: A_h(\Omega,2\tau) \tag{22}$$

i.e.: the associated weighting function is an ambiguity function $A_h$.

First and second order properties of (21) as an estimator of the Wigner-Ville spectrum follow then directly from the properties of (22).

Replacing first $\phi$ by $\phi_{STP}$ in (8), we obtain :

$$E\{J^2_{2N-1}(t,\omega)\}=\frac{1}{2\pi}\frac{1}{2N-1}\sum_{t'=-\infty}^{\infty}\int_{-\pi/2}^{\pi/2} W_x(t',\omega')W_h(t-t',\omega-\omega')\,d\omega' \tag{23}$$

which expresses that the short-time periodogram is a doubly biased estimator of the Wigner-Ville spectrum, the time and frequency bias being described by the Wigner-Ville distribution $W_h$ of the window h. A correct normalization of this window is now just imposed by (9) together with (22), which gives :

$$\frac{1}{2N-1}\sum_{t=-N+1}^{N-1}|h(t)|^2 = 1 \tag{24}$$

In order to derive the second order properties from (14), we must first evaluate $\Psi_{STP}$ which leads to :

$$\Psi_{STP}(2\Omega,\omega) = \frac{1}{2N-1} H(\omega-\Omega)H^*(\omega+\Omega) \tag{25}$$

with :

$$H(\Omega) = \sum_{t=-N+1}^{N-1} h(t)e^{-i\Omega t} \tag{26}$$

Setting then (25) in (14), we get :

$$\mathrm{Cov}\{J^2_{2N-1}(t_1,\omega_1),J^2_{2N-1}(t_2,\omega_2)\} \sim \frac{1}{4\pi^2(2N-1)^2}\left|\int_{-\pi}^{\pi} f_t(\omega)H(\omega_1-\omega)H^*(\omega_2-\omega)e^{-i\omega(t_1-t_2)}d\omega\right|^2 \tag{27}$$

Under the reasonable assumption that the window is large enough to ensure H to be a peaked function, we obtain the desired result :

$$\mathrm{Cov}\{J^2_{2N-1}(t_1,\omega_1),J^2_{2N-1}(t_2,\omega_2)\} \sim \left|f_{(t_1+t_2)/2}\left(\frac{\omega_1+\omega_2}{2}\right)\right|^2.\left|A_h(\omega_2-\omega_1,t_1-t_2)\right|^2 \tag{28}$$

From this equation, it appears that the short-time periodogram gives estimates which are correlated in both time and frequency, this correlation being described by the ambiguity function $A_h$ of the chosen window.

Finally, if the window is normalized so as to verify (24), the variance of the short-time periodogram expresses as :

$$\mathrm{Var}\{J^2_{2N-1}(t,\omega)\} \sim |f_t(\omega)|^2 \tag{29}$$

All these results clearly show a strong interdependence of the short-time periodogram estimates in both time and frequency directions, which is by no way satisfactory. The same results could have been directly obtained from the definition (21) [3] but the general formulation of the problem gives now a hint how to elaborate new estimators in order to overcome the disadvantages of the short-time periodograms.

### 3.2. Pseudo-Wigner estimators.

Looking on one hand at the general expressions of first and second order moments (eq.

(8) and (14)), and on the other hand at the special form of (22), it appears that the bad-looking behavior of the short-time periodogram as estimator of the Wigner-Ville spectrum essentially comes out from the fact that the chosen weighting function is governed by "uncertainty relations". Increasing the performances in one direction of the time-frequency plane is only possible at the expense of a decrease in the other one. This suggests to replace the ambiguity function (22) by a new function which is now separable in the time and frequency variables in order to control independently the behavior of the corresponding estimator in the time and frequency directions. Such a choice may be expressed as :

$$\phi_{PW}(\Omega,2\tau) := |h_N(\tau)|^2 . \sum_{\tau'=-\infty}^{\infty} g_M(\tau')e^{-i\Omega\tau'} \tag{30}$$

where $h_N$ and $g_M$ are windows with respectively 2N-1 and 2M-1 non-zero values. Explicitation of (3) with (30) defines the associated estimator :

$$PW_{2N-1}^{2M-1}(t,\omega) := 2 \sum_{\tau=-N+1}^{N-1} |h_N(\tau)|^2 . \sum_{\tau'=-M+1}^{M-1} g_M(\tau')x(t+\tau'+\tau)x^*(t+\tau'-\tau)e^{-i2\omega\tau} \tag{31}$$

This corresponds to the recently proposed (smoothed) pseudo-Wigner estimators [3]. Most of their properties which have been up to now discussed on the basis of (31) can now be simply restated as special cases of the results of Sec.2.

First of all, the correct normalization imposes now :

$$|h_N(0)|^2 . \sum_{t=-M+1}^{M-1} g_M(t) = 1 \tag{32}$$

It follows from (4) and (30) that :

$$\Phi_{PW}(t,2\tau) = |h_N(\tau)|^2 . g_M(t) \tag{33}$$

which, when inserting in (8), gives :

$$E\{PW_{2N-1}^{2M-1}(t,\omega)\} = \frac{1}{2\pi} \sum_{\tau=-\infty}^{\infty} \int_{-\pi/2}^{\pi/2} g_M(\tau-t)W_h(0,\omega-\omega')W_x(\tau,\omega')\, d\omega' \tag{34}$$

As expected, the double bias in (34) is now split into two separate biases which can be controlled independently. Especially, the case M=1 (which corresponds to the unsmoothed pseudo-Wigner estimator [3]) yields :

$$E\{PW_{2N-1}^{2M-1}(t,\omega)\} = \frac{1}{2\pi} \int_{-\pi/2}^{\pi/2} W_x(t,\omega')W_h(0,\omega-\omega')\, d\omega' \tag{35}$$

and such an estimator is now only biased in the frequency direction.

Introducing :

$$\Psi_{PW}(\Omega,\omega) = H_N^{(2)}(\omega) . G_M(\Omega) \tag{36}$$

where :

$$H_N^{(2)}(\omega) = 2 \sum_{t=-N+1}^{N-1} |h_N(t)|^2 e^{-i2\omega t} \tag{37}$$

and :

$$G_M(\Omega) = \sum_{\tau=-M+1}^{M-1} g_M(\tau)e^{-i\Omega\tau} \tag{38}$$

the covariance of the pseudo-Wigner estimators can be written down by using (14) :

$$\mathrm{Cov}\{PW_{2N-1}^{2M-1}(t_1,\omega_1),PW_{2N-1}^{2M-1}(t_2,\omega_2)\} \sim$$
$$\frac{1}{2\pi^2}\int_{-\pi}^{\pi} H_N^{(2)}(\omega_1-\omega)H_N^{(2)*}(\omega_2-\omega)W_{f_{t_0}}(t_1-t_2,\omega;\phi_2)\,d\omega \tag{39}$$

with :
$$\phi_2(2\Omega,\tau) = |G_M(\Omega)|^2 \tag{40}$$

Assuming again that the window $h_N$ is large enough to ensure $H_N$ to be a peaked function, we obtain for (39) the simplified form :

$$\mathrm{Cov}\{PW_{2N-1}^{2M-1}(t_1,\omega_1),PW_{2N-1}^{2M-1}(t_2,\omega_2)\} \sim$$
$$2\,W_{f_{t_0}}(t_1-t_2,\frac{\omega_1+\omega_2}{2};\phi_2)\cdot\frac{1}{4\pi^2}\int_{-\pi}^{\pi} H_N^{(2)}(\omega_1-\omega)H_N^{(2)*}(\omega_2-\omega)\,d\omega \tag{41}$$

Comparing (41) with (28), it now appears that the integral at the right hand-side of (41) only concerns a correlation in the frequency direction. Since $H_N^{(2)}$ has been supposed to be highly concentrated, this integral will vanish for a proper spacing $\omega_1-\omega_2$ of the order of $\pi M/N$. This expresses that, for an appropriate spacing, pseudo-Wigner estimators give nearly uncorrelated estimates in the frequency direction.
If the window is normalized so as :

$$\sum_{t=-N+1}^{N-1} |h_N(t)|^4 = 1 \tag{42}$$

one finally gets :

$$\mathrm{Cov}\{PW_{2N-1}^{2M-1}(t_1,\omega_1),PW_{2N-1}^{2M-1}(t_2,\omega_2)\} \sim \begin{cases} 0 & ;\ |\omega_1-\omega_2|\geqq\pi M/N \\ 2\,W_{f_{t_0}}(t_1-t_2,\omega;\phi_2) & ;\ \omega_1=\omega_2=\omega \end{cases} \tag{43}$$

and the variance of the estimator is obtained with $t_1=t_2=t$ :

$$\mathrm{Var}\{PW_{2N-1}^{2M-1}(t,\omega)\} \sim 2\,W_{f_t}(0,\omega;\phi_2) \tag{44}$$

These expressions (43) and (44) have already been obtained by direct calculations under a slightly different form of weighted finite sums in the case of a rectangular window $g_M$ [8]. This can however be shown to be completely equivalent and the form (43) has the advantage of directly pointing out how the variance behaves for different M. In the case M=1, $\phi_2=1$ and we recover the known result [3] :

$$\mathrm{Var}\{PW_{2N-1}^{1}(t,\omega)\} \sim 2\,W_{f_t}(0,\omega) \tag{45}$$

whereas, if M is large enough, we have (according to (32)) :

$$\phi_2 \sim 1/(2M-1)$$

and :
$$\mathrm{Var}\{PW_{2N-1}^{2M-1}(t,\omega)\} \sim \frac{2}{2M-1}\,|f_t(\omega)|^2 \tag{46}$$

according to the marginal property of the Wigner-Ville distribution [5] :

$$\lim_{M\to\infty} \sum_{t'=-M+1}^{M-1} W_{f_t}(t',\omega) = |f_t(\omega)|^2$$

Smoothing over time naturally diminishes the variance but also increases the bias (cf. eq. 34)). A compromise is then necessary for the choice of a proper M : it has been shown elsewhere [8] that an efficient optimization procedure can be proposed on the basis of an Akaike-type criterion, in our context of quasi-stationary processes.

## 4. CONCLUSION.

Since the Wigner-Ville spectrum now appears as a convenient tool for the time-varying spectral analysis of harmonizable processes, estimators are to be defined. Under the assumption of quasi-stationarity, a general class of such estimators has been proposed, which basically takes into account estimations of the local covariances. This has provided a general framework to handle statistical properties of the estimators in an unifyied way : they only depend on an arbitrary weighting function. Specification of this weighting function has allowed to recover conventional estimators and to emphasize the usefulness of pseudo-Wigner estimators, according to their uncoupled properties with regard to time and frequency. Given this general formulation, it will be conversely possible to consider the design of special estimators with specified properties.

## 5. REFERENCES.


1. Martin W. : "Time-frequency analysis of random signals", Proc. ICASSP, 1329-1332, Paris, 1982.
2. Flandrin P., Martin W. : "Sur les conditions physiques assurant l'unicité de la représentation de Wigner-Ville comme représentation temps-fréquence", 9ème Colloque GRETSI, Nice, 1983.
3. Martin W., Flandrin P. : "Analysis of non-stationary processes : short-time periodograms versus a pseudo-Wigner estimator", in Schüssler H.(Ed.), EUSIPCO-83, North Holland, Amsterdam, 1983.
4. Escudié B., Gréa J. : "Sur une formulation générale dans l'analyse en temps et fréquence des signaux d'énergie finie", Comptes Rendus, A, 283, 1049-1051, 1976.
5. Claasen T.A.C.M., Mecklenbräuker W.F.G. : "The Wigner distribution - a tool time-frequency signal analysis", Philips J. Res., 35, 217-250, 276-300, 372-389, 1980.
6. Flandrin P., Escudié B. : "Time and frequency representation of finite energy signals : a physical property as a result of an hilbertian condition", Signal Proc., 2, 93-100, 1980.
7. Allen J.B., Rabiner L.R. : "A unified approach to short-time Fourier analysis and synthesis", Proc. IEEE, 65, 1558-1564, 1977.
8. Flandrin P., Martin W. : "Pseudo-Wigner estimators for the analysis of non-stationary processes", Proc. ASSP Spectrum Estimation Workshop II, 181-185, Tampa, 1983.

# BAYESIAN ESTIMATION OF A SPECTRUM OF A NONSTATIONARY AUTOREGRESSIVE PROCESS

Maciej Niedźwiecki
Technical University of Gdańsk
Institute of Computer Science
ul.Majakowskiego 11/12
80-952 Gdańsk , Poland

## SUMMARY

The new parametric spectrum estimator for the purpose of nonstationary autoregressive process analysis is presented.The proposed estimator is obtained by minimization of the Bayesian risk function corresponding to the normalized mean square spectral error measure.The obtained results concern the two most frequently used models of process parameters' variation : the Kalman filter model and the fadding memory (exponential forgetting) one.The efficient computational algorithms are indicated and the results of computer simulation are presented.

## 1. INTRODUCTION

The problem of the parametric estimation of a power spectrum of an autoregressive ( AR ) process has gained more and more attention in the recent years.The main reason is that a variety of processes we find in practice can be described well by the autoregressive model.Since the autoregressive process can be viewed as a result of passing of the white noise sequence through the linear all-pole filter its theoretical spectrum can be easily evaluated using the well known results of the linear systems theory.When the true filter coefficients occuring in the expression for the theoretical spectrum are replaced by the respective estimates the classical autoregressive spectrum estimator results.
The alternative approach based on the Bayesian reasoning is presented. The obtained estimator minimizes (in the set of all continuous square integrable spectra) the Bayesian risk function corresponding to the normalized mean square spectral error measure.Its explicit (approximate) form is given in the case of a Gaussian posterior densities.Since in most of the practical applications,including speech and EEG processing [8],[9],we deal with the nonstationary processes all the conside-

rations concern this case.The obtained results can be considered an extension of the two most frequently used methods of the nonstationary process identification : the Kalman filter approach and the approach based on the exponential weighting of the past data.

## 2. BAYESIAN SPECTRUM ESTIMATOR

Let $\{y_t\}$ denote the univariate autoregressive Gaussian process,i.e., the process that can be described by the following difference equation

$$y_t = \sum_{i=1}^{r} a_{i_t} y_{t-i} + n_t \quad , \quad E\left[n_t^2\right] = \rho_t \tag{1}$$

or equivalently

$$y_t = \alpha_t^T s_t + n_t$$

where $\alpha_t = \left[a_{1_t} , \dots , a_{r_t}\right]^T$ , $s_t = \left[y_{t-1} , \dots , y_{t-r}\right]^T$ .In (1) $\{n_t\}$ denotes the unobservable noise sequence,made up of zero mean independent Gaussian variables having a nonzero and possibly time-dependent variance $\rho_t$.As concerns the autoregressive coefficients $a_{i_t}$ in (1) we will assume that they are randomly varying with time ; the more detailed assumptions concerning the process nonstationarity will be made in the next section.

Let $\mathcal{Y}(N) = \{y_N, y_{N-1}, \dots, y_1, \mathcal{Y}(0)\}$ ,where $\mathcal{Y}(0) = \{y_0, \dots, y_{1-r}\}$ is the set of initial conditions,denote the observation history of the process at the instant N.We intend to obtain (in terms of the available data) the best estimate of the unknown spectral density function of the process $\{y_t\}$ at instant N.Since the process (1) can be treated as a result of passing of the white noise sequence through the linear all-pole filter of the time-dependent transfer function

$$A_t(z) = A(\alpha_t, z) = 1 - \sum_{i=1}^{r} a_{i_t} z^{-i}$$

its true spectrum at instant N can be defined as (see Grenier [1] for further justification)

$$S_N = \frac{\rho_N}{\left|A(\alpha_N , e^{i\omega})\right|^2} \tag{2}$$

where $\omega$ denotes the normalized angular frequency.Note that since the process is nonstationary it does not have a power spectrum in the usual sense.The same concerns the concept of the "time-varying transfer function.

Let $\hat{S}_N(\omega)$ denote any estimator of the spectral density function.We will use the following normalized mean square D-measure to determine quantitatively the distance between $S_N(\omega)$ and $\hat{S}_N(\omega)$:

$$D_N = \frac{1}{2\pi}\int_{-\pi}^{\pi}\left(\frac{S_N(\omega) - \hat{S}_N(\omega)}{S_N(\omega)}\right)^2 d\omega \tag{3}$$

We note that if the values of $\hat{S}_N(\omega)$ do not deviate significantly from $S_N(\omega)$ the D-measure can be treated as a good approximation of the widely accepted mean square log spectral measure L

$$L_N = \frac{1}{2\pi}\int_{-\pi}^{\pi}\left(\ln S_N(\omega) - \ln \hat{S}_N(\omega)\right)^2 d\omega \tag{4}$$

Actually,for $\hat{S}_N(\omega)\approx S_N(\omega)$ we have

$$\ln \hat{S}_N(\omega) - \ln S_N(\omega) = \ln\left(1 + \frac{\hat{S}_N(\omega) - S_N(\omega)}{S_N(\omega)}\right) \approx \frac{\hat{S}_N(\omega) - S_N(\omega)}{S_N(\omega)}$$

In the case where the reference spectrum is parametrized by the vector of random process coefficients $\Phi_N = [\alpha_N^T , \rho_N]^T$ ,i.e.

$$S_N(\omega) = S(\Phi_N , \omega)$$

the Bayesian spectrum estimator $S_N^*(\omega)$ corresponding to the assumed error measure can be defined as that minimizing the expected value of $D_N = D(\Phi_N)$ .Since the minimum should be attained for any data set $\mathcal{Y}(N)$ and any $\omega$ one arrives at the expression

$$S_N^*(\omega) = \arg\min_{\hat{S}_N(\omega)} D_N' \tag{5}$$

where $D_N'$ denotes the conditional risk function

$$D_N' = \int\left(\frac{S(\Phi_N , \omega) - \hat{S}_N(\omega)}{S(\Phi_N , \omega)}\right)^2 p(\Phi_N|\mathcal{Y}(N))\, d\Phi_N \tag{6}$$

Minimization of (6) with respect to $\hat{S}_N$ is relatively easy.By requiring

$$\left.\frac{\partial D_N'}{\partial \hat{S}_N}\right|_{S_N^*} = 2\int\frac{S_N^*(\omega) - S(\Phi_N , \omega)}{S^2(\Phi_N , \omega)}\, p(\Phi_N|\mathcal{Y}(N))\, d\Phi_N = 0$$

one obtains

$$S_N^*(\omega) = \frac{E\left[H(\Phi_N, \omega) \middle| \mathcal{Y}(N)\right]}{E\left[H^2(\Phi_N, \omega) \middle| \mathcal{Y}(N)\right]} \tag{7}$$

where

$$H(\Phi_N, \omega) = \frac{1}{S(\Phi_N, \omega)} = \frac{\left|A(\alpha_N, e^{i\omega})\right|^2}{\rho_N} \tag{8}$$

denotes the inverse spectrum and the expectation is taken with respect to the posterior probability of $\Phi_N$. Note that the risk (6) is minimized not in the set of autoregressive parametric spectra, but in the general set of continuous square integrable spectra, i.e. it <u>is not</u> assumed that $\hat{S}_N(\omega)$ has the form $S(\hat{\Phi}_N, \omega)$.

## 3. EVALUATION OF THE POSTERIOR PROBABILITY DENSITY OF $\Phi_N$

### 3.1 Explicit model of parameters' variation - the Kalman filter approach

Let us suppose that the variance of the process $\{n_t\}$ takes a constant known value $\rho^o$ and that the vector of autoregressive coefficients $\alpha_t$ obeys the following difference equation

$$\alpha_{t+1} = A\alpha_t + v_t \tag{9}$$

where A is the rxr transition matrix and $\{v_t\}$ denotes the sequence of independent Gaussian vectors having zero mean and covariance matrix $V^o$. It is assumed that the sequences $\{n_t\}$ and $\{v_t\}$ are uncorrelated.

If the assumptions made above are accepted the evaluation of the posterior density $p(\Phi_N | \mathcal{Y}(N))$ can be carried out by means of the Kalman filtering theory. It is sufficient to note that (9) and (1) can be viewed as the state equation and the observation (output) equation in the standard filtering problem formulation [3] (see [4] for some more comments on the applicability of the Kalman filtering equations to (1), (9)).

Hence, for the Gaussian prior density

(A1) $\quad p(\alpha_0) = N(\tilde{\alpha}_0, \tilde{P}_0)$

and under the assumption

(A2) $\quad p(\mathcal{Y}(0) | \alpha_0) = p(\mathcal{Y}(0))$

the posterior density of $\alpha_N$ is also Gaussian

$$p(\alpha_N | \mathcal{Y}(N)) = N(\tilde{\alpha}_N, \tilde{R}_N)$$

with the mean $\tilde{\alpha}_N$ and the covariance matrix $\tilde{R}_N$ given by the well known Kalman filter recursions.In fact,the assumption that the transition matrix A in (9) is known is not very realistic.Of course one may try to overcome this difficulty by using the adaptive filtering approach (joint parameter and state estimation) but the nonlinear filtering problem which arises in such case [2] is computationally prohibitive. The "preliminary" estimation of A using the long data record is another possibility but it also requires a lot of computation.Fortunately,it was verified experimentally that in many practical situations the "rough estimate" of A can be used

$$A = I \tag{10}$$

without substantially affecting the obtained results [4], [5].
Under (10) the Kalman filter equations are

$$\tilde{\alpha}_N = \tilde{\alpha}_{N-1} + \tilde{P}_N s_N \tilde{e}_N$$

$$\tilde{e}_N = y_N - \tilde{\alpha}_{N-1}^T s_N \tag{11}$$

$$\tilde{P}_N = \frac{V^o}{\rho^o} + \left[\tilde{P}_{N-1}^{-1} + s_N s_N^T\right]^{-1} = \frac{V^o}{\rho^o} + \tilde{P}_{N-1} - \frac{\tilde{P}_{N-1} s_N s_N^T \tilde{P}_{N-1}}{1 + s_N^T \tilde{P}_{N-1} s_N}$$

The conditional covariance matrix $\tilde{R}_N$ is equal to $\rho^o \tilde{P}_N$.

<u>Remark</u>

If the noise covariances $V^o$ and $\rho^o$ are unknown they can be estimated in a simple way provided that a long data record is available [4].

## 3.2 Implicit model of parameters' variation - the subjective probability approach

Let us suppose that $\{\phi_t\}$ is a stochastic process independent of $\{y_t\}$. Similarly as in the preceding subsection we will assume that the variation of the parameter vector can be described well by the first order Markov model,i.e.

$$p(\phi_{t+1} | \Phi(t), \mathcal{Y}(t)) = p(\phi_{t+1} | \phi_t) \tag{12}$$

where $\Phi(t) = \{\phi_t, \ldots, \phi_0\}$.In order to obtain an explicit expression for $p(\phi_t | \mathcal{Y}(t)) = p(\phi_t, \mathcal{Y}(t)) / p(\mathcal{Y}(t))$ we will look for a recursion on $p(\phi_t, \mathcal{Y}(t))$.Note that

$$p(\phi_{t+1}, \mathcal{Y}(t+1)) = p(y_{t+1} | \phi_{t+1}, \mathcal{Y}(t)) \, p(\phi_{t+1}, \mathcal{Y}(t)) \tag{13}$$

where,according to (12)

$$p(\phi_{t+1}, \mathcal{Y}(t)) = \int p(\phi_{t+1} | \Phi(t), \mathcal{Y}(t)) \left[\int p(\Phi(t), \mathcal{Y}(t)) \, d\phi_0 \ldots d\phi_{t-1}\right] d\phi_t =$$

$$= \int p(\Phi_t, \mathcal{Y}(t))\, p(\Phi_{t+1} | \Phi_t)\, d\Phi_t \qquad (14)$$

If the parameter vector did not change in time one would have $p(\Phi_{t+1} | \Phi_t) = \delta(\Phi_{t+1} - \Phi_t)$, where $\delta(\cdot)$ denotes the Dirac's delta function.Consequently

$$\int p(\Phi_t, \mathcal{Y}(t))\, p(\Phi_{t+1} | \Phi_t)\, d\Phi_t = p(\Phi_t = \Phi_{t+1}, \mathcal{Y}(t))$$

The term $p(\Phi_t = \Phi_{t+1}, \mathcal{Y}(t))$ should be interpreted as "the density $p(\Phi_t, \mathcal{Y}(t))$ with the vector $\Phi_t$ replaced by $\Phi_{t+1}$ ".Note that in a general case this <u>is not</u> the joint probability density of $\Phi_{t+1}$ and $\mathcal{Y}(t)$. If $\Phi_t$ is changing sufficiently slowly $p(\Phi_{t+1} | \Phi_t)$ is still "delta-like" function located around $\Phi_t$ ,causing the effect of reproducing of the probability density function transformed in (14) .However,since $p(\Phi_{t+1} | \Phi_t) \neq 0$ for $\Phi_{t+1} \neq \Phi_t$ at least in some neighbourhood of $\Phi_t$ , the additional effect of equalizing or flattening of the integrated density must be observed.For this reason if nothing is known about the variation of $\Phi_t$ except that the changes are "sufficiently slow" it seems reasonable to put [6]

$$\int p(\Phi_t, \mathcal{Y}(t))\, p(\Phi_{t+1} | \Phi_t)\, d\Phi_t \approx \mathcal{F}_t \left[ p(\Phi_t, \mathcal{Y}(t)) \right] \qquad (15)$$

where $\mathcal{F}_t$ denotes the flattening operator

$$\mathcal{F}_t \left[ p(\Phi_t, \mathcal{Y}(t)) \right] = \frac{\left( p(\Phi_t = \Phi_{t+1}, \mathcal{Y}(t)) \right)^{\theta}}{\int \left( p(\Phi_t = \Phi_{t+1}, \mathcal{Y}(t)) \right)^{\theta} d\Phi_t \, d\mathcal{Y}(t)} \qquad (16)$$

The parameter $\theta$ , $0 < \theta < 1$ , in (16) is usually called the forgetting constant;its value should be chosen in accordance with the "rate of nonstationarity" of the investigated process.
Combining (13) , (14) and (15) one arrives at

$$p(\Phi_{t+1}, \mathcal{Y}(t+1)) \sim p(y_{t+1} | \Phi_{t+1}, \mathcal{Y}(t))\, p(\Phi_t = \Phi_{t+1}, \mathcal{Y}(t)) \qquad (17)$$

where $\sim$ denotes proportionality.Applying this recursion succesively for $t=0,\dots,N-1$ and using the relationship $p(\Phi_N | \mathcal{Y}(N)) = p(\Phi_N, \mathcal{Y}(N)) / p(\mathcal{Y}(N))$ one obtains the explicit expression for the posterior density of $\Phi_N$

$$p(\Phi_N | \mathcal{Y}(N)) \approx c_N \left( p(\Phi_0 = \Phi_N, \mathcal{Y}(0)) \right)^{\theta^N} \prod_{t=1}^{N} \left( p(y_t | \Phi_t = \Phi_N, \mathcal{Y}(t-1)) \right)^{\theta^{N-t}}$$

where $c_N$ does not depend on $\Phi_N$.The following assumptions concerning prior distributions will be made

(B1) $p(\Phi_0)$ tends to the uniform distribution over the parameter space $\Omega = \{ \mathbb{R}^r, \mathbb{R}_+ \}$ (the noninformative prior distribution,see e.g. Peterka [11] )

(B2) $p(\mathcal{Y}(0)|\Phi_0) = p(\mathcal{Y}(0))$

Under the assumptions (B1) and (B2) the following approximate expression for the limiting posterior distribution of $\Phi_N$ can be derived

$$p(\Phi_N|\mathcal{Y}(N)) = D_N(2\pi\rho_N)^{-\frac{k_N}{2}} \exp\left\{-\frac{1}{2\rho_N}\sum_{t=1}^{N}\theta^{N-t}(y_t-\alpha_N^T s_t)^2\right\} \quad (18)$$

where

$$k_N = \sum_{t=1}^{N}\theta^{N-t} = \frac{1-\theta^N}{1-\theta} \approx \frac{1}{1-\theta} \quad \text{for large N}$$

denotes the effective number of observations.After certain amount of calculations one can rewrite (18) in the form $p(\rho_N|\mathcal{Y}(N))$ x x $p(\alpha_N|\rho_N, \mathcal{Y}(N))$ :

$$p(\rho_N|\mathcal{Y}(N)) = \frac{1}{\Gamma(l_N-1)}\left(\frac{E_N^*}{2}\right)^{l_N-1}(\rho_N)^{l_N}\exp\left\{-\frac{E_N^*}{2\rho_N}\right\} \quad (19)$$

$$p(\alpha_N|\rho_N, \mathcal{Y}(N)) = \frac{1}{|2\pi\rho_N P_N^*|^{1/2}}\exp\left\{-\frac{1}{2}\|\alpha_N-\alpha_N^*\|_{(\rho_N P_N^*)^{-1}}\right\} \quad (20)$$

where

$$l_N = \frac{k_N-r}{2} \quad (21)$$

$$E_N^* = \sum_{t=1}^{N}\theta^{N-t}(y_t - \alpha_N^{*T}s_t)^2 \quad (22)$$

$$P_N^* = \left[\sum_{t=1}^{N}\theta^{N-t}s_t s_t^T\right]^{-1} \quad (23)$$

, $\Gamma(\cdot)$ denotes Euler's gamma function and

$$\alpha_N^* = P_N^*\left(\sum_{t=1}^{N}\theta^{N-t}y_t s_t\right) \quad (24)$$

can be recognized as the exponentially weighted least squares estimate of $\alpha_N$ (= the exponentially weighted maximum likelihood estimate [7]). All the quantities involved in (21) - (24) can be computed recursively

$$\alpha_N^* = \alpha_{N-1}^* + P_N^* s_N e_N^*$$

$$e_N^* = y_N - \alpha_{N-1}^{*T} s_N$$

$$P_N^* = \left[\theta P_{N-1}^{*-1} + s_N s_N^T\right]^{-1} = \frac{1}{\theta}\left[P_{N-1}^* - \frac{P_{N-1}^* s_N s_N^T P_{N-1}^*}{\theta + s_N^T P_{N-1}^* s_N}\right] \tag{25}$$

$$E_N^* = \theta\left[E_{N-1} + \frac{e_N^{*2}}{\theta + s_N^T P_{N-1}^* s_N}\right]$$

$$k_N = \theta k_{N-1} + 1$$

Remark

The expression for the posterior density of $\phi_N$ derived by Peterka [6] is slightly different from the one given here. This was caused by the use of the "noninformative" prior distribution instead of the "conjugate" one.

## 4. EVALUATION OF THE SPECTRUM ESTIMATE $S_N^*(\omega)$

We remind that $S(\phi,\omega)$ defines the spectral density function of a stationary AR process characterized by the vector of coefficients $\phi$. For this reason when defining the instantaneous spectral density of a nonstationary AR process in the form

$$S_t(\omega) = S(\phi_t, \omega) \tag{26}$$

one makes the silent assumption that the instantaneous parameter vector $\phi_t$ in (26) represents the stable filter, i.e. that it belongs to the stability subspace $\Omega_s = \{\alpha \in \mathbb{R}^r$ : all zeros of $A(\alpha,z)$ lie inside the unit circle ; $\rho \in \mathbb{R}_+\}$ of the parameter space $\Omega = \{\mathbb{R}^r, \mathbb{R}_+\}$. Within the Bayesian framework this restriction can be easily taken into account by introducing the prior density $p_s(\phi_0)$ that vanishes for $\phi_0 \notin \Omega_s$. Let $p(\phi_N|\mathcal{Y}(N))$ denote the posterior density of $\phi_N$ given $\mathcal{Y}(N)$ corresponding to the prior distribution $p(\phi_0)$, and let $p_s(\phi_0) = c_0\eta(\phi_0)p(\phi_0)$ where

$$\eta(\phi_0) = \begin{cases} 1 \text{ for } \phi_0 \in \Omega_s \\ 0 \text{ for } \phi_0 \notin \Omega_s \end{cases} , \quad c_0 = 1 / \int_{\Omega_s} p(\phi_0)\, d\phi_0$$

denote the prior distribution "restricted" to the stability subspace $\Omega_s$. As easy to find the corresponding posterior distribution of $\phi_N$ is also restricted to $\Omega_s$ and takes the form

$$p_s(\phi_N|\mathcal{Y}(N)) = c_N\, \eta(\phi_N) p(\phi_N|\mathcal{Y}(N)) , \quad c_N = 1/\int_{\Omega_s} p(\phi_N|\mathcal{Y}(N))\, d\phi_N$$

Hence, (7) can be rewriten as

$$S_N^*(\omega) = \frac{\int_{\Omega_s} H(\Phi_N, \omega) p(\Phi_N | \mathcal{Y}(N)) d\Phi_N}{\int_{\Omega_s} H^2(\Phi_N, \omega) p(\Phi_N | \mathcal{Y}(N)) d\Phi_N} \tag{27}$$

4.1 Cautious spectrum estimator

From the technical point of view the integration carried out in (27) is difficult because of the rather complicated form (for $r > 2$) of the stability subspace $\Omega_s$. We note, however, that as far as

$$\int_{\Omega_s} p(\Phi_N | \mathcal{Y}(N)) d\Phi_N \gg \int_{\Omega - \Omega_s} p(\Phi_N | \mathcal{Y}(N)) d\Phi_N \tag{28}$$

$S_N^*(\omega)$ given by (27) can be approximated well by the quantity

$$S_N^{**}(\omega) = \frac{\int_{\Omega} H(\Phi_N, \omega) p(\Phi_N | \mathcal{Y}(N)) d\Phi_N}{\int_{\Omega} H^2(\Phi_N, \omega) p(\Phi_N | \mathcal{Y}(N)) d\Phi_N} \tag{29}$$

For the reasons that will be given later we will call $S_N^{**}(\omega)$ the cautious spectrum estimator.

Several notions will be made before we derive the explicit expressions for $S_N^{**}(\omega)$. We will rewrite $A(\alpha, e^{i\omega})$ in the form

$$A(\alpha, e^{i\omega}) = 1 - \alpha^T h(\omega)$$

where

$$h(\omega) = \left[e^{i\omega}, \ldots, e^{ri\omega}\right]^T$$

The following theorem, proved in Appendix 1, will be useful

Theorem 1

Let $p(\alpha) = N(\alpha_o, R_o)$. Then

$$\int \left|A(\alpha, e^{i\omega})\right|^2 p(\alpha) d\alpha = \left|A(\alpha_o, e^{i\omega})\right|^2 + h^T(\omega) R_o h^+(\omega) = \mathcal{Z}_1\left(A(\alpha_o, e^{i\omega}), R_o\right)$$

$$\int \left|A(\alpha, e^{i\omega})\right|^4 p(\alpha) d\alpha = \left(\left|A(\alpha_o, e^{i\omega})\right|^2 + h^T(\omega) R_o h^+(\omega)\right)^2 +$$

$$+ 4\left(\mathrm{Re}\left[A(\alpha_o, e^{i\omega}) h^{+T}(\omega)\right]\right) R_o \left(\mathrm{Re}\left[h^+(\omega) A(\alpha_o, e^{i\omega})\right]\right) +$$

$$+\left|h^T(\omega)R_o h(\omega)\right|^2 + \left(h^T(\omega)R_o h^+(\omega)\right)^2 = \mathscr{Z}_2\left(A(\alpha_o, e^{i\omega}), R_o\right)$$

where $h^+(\omega)$ denotes the complex conjugate of $h(\omega)$.

Finally,we will introduce the quantity

$$G(\phi, e^{i\omega}) = \frac{A(\alpha, e^{i\omega})}{\sqrt{\rho}}$$

which easily relates to the inverse spectrum

$$H(\phi,\omega) = G(\phi, e^{i\omega})G^+(\phi, e^{i\omega}) = \left|G(\phi, e^{i\omega})\right|^2$$

Case 1 - The Kalman filter approach

We remind that the posterior distribution of $\phi_N$ obtained within the Kalman filter approach was Gaussian with the mean $\tilde{\alpha}_N$ and the covariance matrix $\rho^o\tilde{P}_N$ .Hence,according to the Theorem 1 :

$$S_N^{**}(\omega) = \frac{\mathscr{Z}_1\left(G(\tilde{\phi}_N, e^{i\omega}), \tilde{P}_N\right)}{\mathscr{Z}_2\left(G(\tilde{\phi}_N, e^{i\omega}), \tilde{P}_N\right)} \tag{30}$$

where $\tilde{\phi}_N = [\tilde{\alpha}_N^T, \rho^o]^T$.

Case 2 - The subjective probability approach

According to (20) the conditional distribution of $\alpha_N$ given $\rho_N$ and $\mathcal{Y}(N)$ is Gaussian with the mean $\alpha_N^*$ and the covariance matrix $\rho_N P_N^*$ . Hence,according to the Theorem 1 :

$$E\left[H(\alpha_N, \rho_N, \omega) \,\middle|\, \rho_N, \mathcal{Y}(N)\right] = \mathscr{Z}_1\left(G(\alpha_N^*, \rho_N, e^{i\omega}), P_N^*\right) \tag{31}$$

$$E\left[H^2(\alpha_N, \rho_N, \omega) \,\middle|\, \rho_N, \mathcal{Y}(N)\right] = \mathscr{Z}_2\left(G(\alpha_N^*, \rho_N, e^{i\omega}), P_N^*\right) \tag{32}$$

Evaluating the expectation of (31) and (32) with respect to $\rho_N$ and taking into consideration the fact that ( see (19) )

$$E\left[\frac{1}{\rho_N}\,\middle|\,\mathcal{Y}(N)\right] = \frac{1}{\rho_N^*} \quad , \quad E\left[\frac{1}{\rho_N^2}\,\middle|\,\mathcal{Y}(N)\right] = \frac{l_N}{l_N-1}\,\frac{1}{(\rho_N^*)^2}$$

where $\rho_N^* = E_N^* / (k_N-r-2)$, one arrives at the expression

$$S_N^{**}(\omega) = \frac{\mathcal{Z}_1\left(G(\phi_N^*, e^{i\omega}), P_N^*\right)}{\mathcal{Z}_2\left(G(\phi_N^*, e^{i\omega}), P_N^*\right) + \frac{\left|G(\phi_N^*, e^{i\omega})\right|^4}{l_N - 1}} \qquad (33)$$

where $\phi_N^* = \left[\alpha_N^{*\,T}, \rho_N^*\right]^T$.

Remark 1

Some general properties of the estimators derived above are worth noticing. Suppose for the moment that the matrix $\widetilde{P}_N$ in (30) is zero, i.e. that the vector of parameter estimates $\widetilde{\alpha}_N$ is not "uncertain" (in the case of the subjective probability approach the same arguments are valid). Since

$$\mathcal{Z}_1\left(G(\widetilde{\Phi}_N, e^{i\omega}), 0\right) = H(\widetilde{\Phi}_N, \omega), \quad \mathcal{Z}_2\left(G(\widetilde{\Phi}_N, e^{i\omega}), 0\right) = H^2(\widetilde{\Phi}_N, \omega)$$

the spectrum estimator (30) would be the same as the "classical" autoregressive estimator $\hat{S}_N(\omega) = S(\widetilde{\Phi}_N, \omega)$. For any positive definite matrix $\widetilde{P}_N$

$$\mathcal{Z}_1\left(G(\widetilde{\Phi}_N, e^{i\omega}), \widetilde{P}_N\right) > \mathcal{Z}_1\left(G(\widetilde{\Phi}_N, e^{i\omega}), 0\right)$$

$$\mathcal{Z}_2\left(G(\widetilde{\Phi}_N, e^{i\omega}), \widetilde{P}_N\right) > \left(\mathcal{Z}_1\left(G(\widetilde{\Phi}_N, e^{i\omega}), \widetilde{P}_N\right)\right)^2$$

Hence, for any $\omega$ $S_N^{**}(\omega) < \hat{S}_N(\omega)$. The last inequality means that the spectral plot corresponding to $S_N^{**}(\omega)$ will be always placed <u>below</u> the one corresponding to the AR estimate $\hat{S}_N(\omega)$. This justifies the term "cautious spectrum estimator" introduced earlier. Note that the degree of cautiousness of $S_N^{**}(\omega)$ depends on the degree of uncertainty of the parameter estimate $\widetilde{\Phi}_N$, reflected by the "magnititude" of the covariance matrix $\widetilde{P}_N$.

Remark 2

Since the quantity $k_N P_N^*$ can be considered an estimate of the inverse covariance matrix of $\{y_t\}$, the term $h^T(\omega) P_N^* h(\omega)$ (appearing in $\mathcal{Z}_1$ and $\mathcal{Z}_2$ in (33)), when multiplied by $k_N$, can be recognized as the Capon or the so called "ML estimate" of $S_N^{-1}(\omega)$. Hence, according to Burg [12], it is proportional to the sum of inverse spectra corresponding to all lower order autoregressive models.

Remark 3

It is worth noticing that the terms $1/\mathcal{Z}_1\left(G(\widetilde{\Phi}_N, e^{i\omega}), \widetilde{P}_N\right)$ and $1/\mathcal{Z}_1\left(G(\phi_N^*, e^{i\omega}), P_N^*\right)$ can be interpreted as the Bayesian estimators associated with the measure

$$I_N = \frac{1}{2\pi}\int_{-\pi}^{\pi}\left(S_N^{-1}(\omega) - S_N^{-1}(\omega)\right)^2 d\omega$$

## 4.2 Relation between the cautious and the Bayesian spectrum estimator

As it was already said,as far as (28) holds the cautious spectrum estimate $S_N^{**}(\omega)$ can be considered a good approximation of the Bayesian estimate $S_N^{*}(\omega)$.The formal verification of (28)is difficult.Generally speaking,the condition (28) is fulfilled if the posterior density of $\Phi_N$ is a sufficiently "narrow" function (what takes place if $\Phi_N$ is changing sufficiently slowly) and/or if the distance of $\Phi_N$ from the boundary of the stability region is sufficiently large (i.e. if the zeros of the polynomial $A_N(z)$ lie far enough from the unit circle).One should realize that for the time-varying processes the condition (28) may be fulfilled at some time instants (in some time intervals) and may not be fulfilled at the others.

Note that the normalized mean square error measure penalizes more strongly the differences occuring in the low-magnititude part of the reference spectrum than the differences occuring in its high-magnititude part. For this reason the "cautiousness" of the estimator $S_N^{*}(\omega)$ results in part from the properties of the applied measure of fit.The mean square log spectral measure (4) does not suffer from the drawback mentioned above.Note however that the corresponding Bayesian spectrum estimator

$$S_N^{\#}(\omega) = \exp\left\{E \ln\left[S(\Phi_N, \omega) \middle| \mathcal{Y}(N)\right]\right\} \tag{34}$$

is <u>extremely sensitive</u> to the form of the prior distribution of $\Phi_0$ in the close neibourhood of the boundary of the stability region.Since $\ln S(\Phi_N, \omega)$ tends to infinity if $\Phi_N$ approaches the boundary of $\Omega_s$ this effect is easy to explain.

## 5. EFFICIENT COMPUTATION OF SPECTRA

Let

$$\begin{aligned}
z(\omega) &= \left(P^{1/2}\right)^T h(\omega) = z_1(\omega) + i\, z_2(\omega) \\
g(\omega) &= \frac{1 - \alpha^T h(\omega)}{\rho^{1/2}} = g_1(\omega) + i\, g_2(\omega) \\
v_{11}(\omega) &= z_1^T(\omega) z_1(\omega), \quad v_{22}(\omega) = z_2^T(\omega) z_2(\omega) \\
v_{12}(\omega) &= z_1^T(\omega) z_2(\omega)
\end{aligned} \tag{35}$$

where $P^{1/2}$ denotes the square root (the lower triangular matrix) of the positive definite matrix P .As easy to check

$$\mathcal{Z}_1\left(G(\Phi, e^{i\omega}), P\right) = g_1^2(\omega) + g_2^2(\omega) + v_{11}(\omega) + v_{22}(\omega) = s(\omega) \tag{36}$$

$$\mathcal{Z}_2\left(G(\Phi,\ e^{i\omega}),P\right) = s^2(\omega)+4v_{11}(\omega)g_1^2(\omega)+4v_{22}(\omega)g_2^2(\omega)+ \\ + 8v_{12}(\omega)g_1(\omega)g_2(\omega)+2v_{11}^2(\omega)+2v_{22}^2(\omega)+4v_{12}^2(\omega) \tag{37}$$

Both, the scalar $g(\omega)$ and the vector $z(\omega)$ can be efficiently computed using the FFT-based procedure. Actually, applying the fast Fourier transform to the sequence

$$\{0,\ a_1,\ \dots,\ a_r,\ \underbrace{0,\ \dots,\ 0}_{2^{K+1}-r-1\ \text{-times}}\}$$

one gets as the result the values of $\mathrm{Re}\left[\alpha^T h(\omega_k)\right]$ and $\mathrm{Im}\left[\alpha^T h(\omega_k)\right]$ computed at $2^K$ equidistant frequencies $\omega_k=(k-1)\pi/2^K$ ,$k=1,\dots,K$, from which the values of $g_1(\omega_k)$ and $g_2(\omega_k)$ can be easily obtained. Since the j-th coordinate of the vector $z(\omega)$ is $z^{(j)}(\omega) = p_j^T h(\omega)$ ,where $p_j$ denotes the j-th column of the matrix $P^{1/2}$ ,the values of $z(\omega_k) = \left[z^{(1)}(\omega_k),\ \dots,\ z^{(r)}(\omega_k)\right]^T$ can be computed in the same way. Note, that if the algorithms (11) and (25) are replaced by their square root filtering counterparts (i.e. by the algorithms updating the matrices $\tilde{P}^{1/2}$ and $\overset{*}{P}^{1/2}$ instead of $\tilde{P}$ and $P^*$, respectively [10]) the triangularization of the matrix P in (35) is not necessary.

## 6. EXAMPLE

The second order nonstationary AR process was generated in order to check experimentally the obtained results. At each instant $t=1,\dots,500$ the values of the AR coefficients were chosen so that the characteristic polynomial $A_t(z)$ had 2 complex roots given by $r\cdot e^{\pm i\varphi_t}$ , $\varphi_t=\pi t/1000$ (see Fig.1a). The approach based on the exponential forgetting of past data was applied. The true and the estimated spectra were computed at 128 frequencies by the FFT-based procedure and used to obtain the approximate values $\tilde{D}_t$ and $\tilde{L}_t$ of $D_t$ and $L_t$, respectively. The final results obtained for various estimates, expressed in terms of cumulative spectrum estimation errors (obtained by summing the values of $\tilde{D}_t$ and $\tilde{L}_t$ computed at instants $t=100,110,\dots,500$) are shown below. The presented results correspond to the two different values of r (0.8 and 0.95) and the forgetting constant $\theta$ equal to 0.975.

| | AR | $1/\mathcal{Z}_1$ | $\mathcal{Z}_1/\mathcal{Z}_2$ | AR | $1/\mathcal{Z}_1$ | $\mathcal{Z}_1/\mathcal{Z}_2$ |
|---|---|---|---|---|---|---|
| $\sum\tilde{D}_t$ | 10.572 | 7.937 | 5.339 | 157.740 | 86.332 | 48.773 |
| $\sum\tilde{L}_t$ | 7.341 | 6.933 | 6.774 | 27.657 | 24.795 | 22.206 |
| | r = 0.80 | | | r = 0.95 | | |

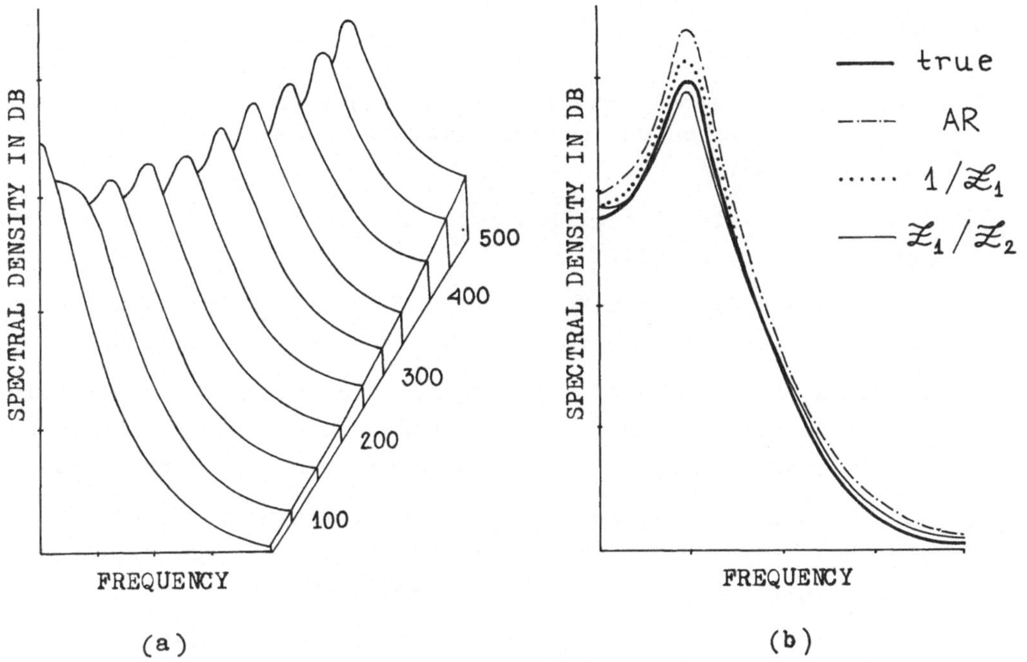


Fig.1 (a) The evolution of a spectrum of a generated nonstationary process (r=0.8) .
(b) Comparison of various spectrum estimates obtained at the instant t=250.

## APPENDIX

Let $\alpha=\alpha_o+\Delta\alpha$ , $1-\alpha_o^T h(\omega)=a$ and $-\Delta\alpha^T h(\omega)=b$.Hence, $1-\alpha^T h(\omega) = a+b$ where a is a constant and b is a zero mean complex Gaussian variable.We notice that

$$|A(\alpha,e^{i\omega})|^2 = |a+b|^2 = |a|^2 + |b|^2 + ab^+ + a^+b$$

$$|A(\alpha,e^{i\omega})|^4 = |a+b|^4 = |a|^4 + |b|^4 + a^2(b^+)^2 + (a^+)^2b^2 + $$
$$+ 4|a|^2|b|^2 + 2a|a|^2b + 2a^+|a|^2b + 2ab^+|b|^2 + 2a^+b|b|^2$$

Since the odd order moments of the zero mean Gaussian variables are zero

$$E\left[|a+b|^2\right] = |a|^2 + E\left[|b|^2\right] \tag{A}$$
$$E\left[|a+b|^4\right] = |a|^4 + E\left[|b|^4\right] + a^2E\left[(b^+)^2\right] + (a^+)^2E\left[b^2\right] + 4|a|^2E\left[|b|^2\right]$$

As easy to check : $E\left[|b|^2\right] = h^TR_oh^+$ , $E\left[b^2\right] = h^TR_oh$ , $E\left[(b^+)^2\right] = (h^+)^TR_oh^+$ , $E\left[|b|^4\right] = 2\left(h^TR_oh^+\right)^2 + |h^TR_oh|^2$ .Combining these relationships with (A) and grouping the terms properly one arrives at the expressions for $\mathcal{Z}_1$ and $\mathcal{Z}_2$ .

REFERENCES

[1] Y.Grenier,"Rational non-stationary spectra and their estimation", 1-st ASSP Workshop on Spectral Estimation,Hamilton,Ontario,Canada, Aug.17-18,1981.

[2] P.Eykhoff,"System Identification : Parameter and State Estimation" London : Wiley , 1974.

[3] A.P.Sage and J.L.Melsa,"Estimation Theory with Applications to Communications and Control" , New York : McGraw-Hill,1971.

[4] T.Bohlin,"Four cases of identification of changing systems",in: System Identification : Advances and Case Studies,R.K.Mehra and D.G.Lainiotis,eds.,pp.441-518,Academic Press,New York,1976.

[5] A.Isaksson,"On the time variable properties of EEG signals examined by means of a Kalman filter method",Royal Institute of Technology,Stockholm,Technical Report No.95,April 1975.

[6] V.Peterka,"Subjective probability approach to real-time identification",4th IFAC Symposium on Identification and System Parameter Estimation,Tbilisi,Sept.1976,Part 3,pp.83-96.

[7] M.Niedźwiecki,"On the localized estimators and generalized Akaike's criteria",to appear in IEEE Trans.Autom.Contr. see also 20th IEEE Conference on Decision and Control,San Diego, Dec.16-18,1981,pp.56-61 .

[8] J.D.Markel and A.H.Gray,Jr.,"Linear Prediction of Speech",New York : Springer-Verlag,1976.

[9] A.Isaksson,"Examples of EEG signals with time-varying spectra analysed by means of a Kalman filter method",Royal Institute of Technology,Stockholm,Technical Report No.96,April 1975.

[10] V.Strejc,"Least squares parameter estimation",Automatica,vol.16, pp.535-550,1980.

[11] V.Peterka,"Bayesian system identification",Automatica,vol.17, pp.41-53,1981.

[12] J.P.Burg,"The relationship between maximum entropy spectra and maximum likelihood spectra",Geophysics,vol.37,pp.375-376,April 1972.

Session 2

# STABILITY I

# STABILITÉ I

# INTERPRETATION OF THE ROBUST STABILITY CONDITIONS APPEARING IN ADAPTIVE CONTROL

R. Ortega*, I. Landau

GRECO (69) Systèmes Adaptatifs (C.N.R.S.)
et
Laboratoire d'Automatique de Grenoble
E.N.S.I.E.G.
B.P. 46, 38402 - Saint-Martin-d'Hères

## ABSTRACT

Some recent results on robust stability of direct discrete time adaptive controllers are analyzed through simple examples. The simultaneous existence of reduced order modelling and bounded output disturbances which is the rule and not the exception in practice, is shown to lead the estimator to a tradeoff between disturbance rejection and performance (in the sense of model following).

## 1. INTRODUCTION

The purpose of this communication is to interpret, through simple examples, some recent results on robust stability of discrete time adaptive controllers [1-4]. We consider the problem of preserving closed-loop stability of the usual pole-zero placement adaptive schemes in spite of reduced order modelling (ROM) and the existence of bounded output distrubances (BOD). We assume the process to be linear time invariant, stably invertible and with known delay. The latter assumption has proved to be fundamental to prove that adaptive controllers posses a certain margin of stability to other classes of model process mismatch besides parametric uncertainty [4].

The main technical device to study the robustness of the adaptive schemes is a conicity preperty [5] of the parameter adaptation algorithm (PAA) first established in [6] and later used in [9,4] to analyze the error model stability. It has been shown in [3] that normalization of the signals feeded into the PAA is compulsory to insure that the PAA conic sector has a radius which is independent of the level of excitation. This modification provides a "margin of stability" to the adaptive controller since it assures a non-vanishing radius to the conic sector. For the considereded type of systems the conditions to insure robust stability reduce to the existence of parameter values for the chosen regulator structure such that: the resulting closed-loop transfer function (denoted $H_2$) is "sufficiently close" to the desired reference model and the internal model of the BOD is incorporated in the regulator. The concept of vicinity of the transfer functions is formaly given in terms of a conicity condition and basically implies that its quotient is passive. Since the analysis of stability is carried on a $L_2$-framework, frequency domain interpretations are available. In particular the conicity condition defines a family of disks in the complex plane containing the Nyquist locus of $H_2$. An example is used to illustrate this idea.

The requirement of robust servo behaviour considerably reduces the possibility of the adaptive scheme to satisfy the model-following objective since the search is constrained to the parameter subspace containing the BOD internal model. The deleterious effect of the BOD on performance is also illustrated with an example.

* R. Ortega's work is sponsored by the National University of Mexico.

The robust stability conditions are given in the following Section and the examples that furnish insight into them are presented in Section 3.

## 2. MAIN STABILITY THEOREM

Let the process be described by

$$A(q^{-1})Y_t = q^{-d} B(q^{-1})U_t + \xi_t \tag{2.1}$$

where $A,B \in R[q^{-1}]$ have unknown degrees and coefficients, A is monic, $\xi_t: Z_+ \to R$ and d is known.

Our desired objective is

$$Y_t - \frac{1}{C_R} \omega_t \to 0 \tag{2.2}$$

where $\omega_t: Z_+ \to R$ is a reference sequence known d-steps ahead and $C_R \in R[q^{-1}]$, that is the reference model is $q^{-d}/C_R$. We propose the following adaptive controller

$$\omega_{t+d} = \hat{S}_t(q^{-1})U_t + \hat{R}_t(q^{-1})Y_t \triangleq \hat{\theta}_t^T \phi_t \tag{2.3 a}$$

$$\phi_t = [U_t, U_{t-1}, \ldots, U_{t-n_S} ; Y_t, Y_{t-1}, \ldots Y_{t-n_R}]^T \tag{2.3 b}$$

$$\hat{\theta}_t = \hat{\theta}_{t-d} + \lambda''_t F_t \phi_{t-d} e_t ; e_t \triangleq C_R Y_t - \omega_t \tag{2.3 c}$$

$$F_t^{-1} = \lambda'_t F_{t-d}^{-1} + \lambda''_t \phi_{t-d} \phi_{t-d}^T ; \lambda'_t \in (0,1] ; \lambda''_t = \rho_t^{-1} \tag{2.3 d}$$

$$\rho_t = \mu\rho_{t-1} + \phi_{t-d}^T \phi_{t-d} ; \mu \in (0,1) \tag{2.3 e}$$

Define an operator

$$H_1: e_t \to \psi_t \triangleq (\hat{\theta}_{t-d} - \theta_*)^T \phi_{t-d} , \theta_* \in R^n , n \triangleq n_S + n_R + 2 \tag{2.4}$$ [1]

combining equations (2.1)-(2.4) we get the error model (see [3] for further details)

$$e_t = - H_2 \psi_t + e_t^* \tag{2.5 a}$$

$$H_2 = C_R C_*^{-1} B \tag{2.5 b}$$

$$e_t^* \triangleq (H_2-1)\omega_t + C_R C^{-1} S_* \xi_t \tag{2.5 c}$$

$$C_* \triangleq AS_* + q^{-d} BR_* \tag{2.5 d}$$

where $S_*$, $R_* \in R[q^{-1}]$ of orders $n_S$ and $n_R$ respectively. Notice that $H_2$ is the transfer function of the process in closed loop with a fixed regulator ($\theta_*$) multiplier by $C_R$, e.q. $C_R Y_t^*/\omega_t$. The signal $e_t^*$ is the filtered tracking error (see 2.3c) of that linear scheme. Remark that in the full order case $H_2 = 1$ and $e_t^*$ contains only a filtered version of the BOD, which might eventually cancel if $S_*$ contains the internal model of the BOD.

---

[1] In the sequel the asterisk will be used to denote all operator and signals resulting from a fixed gain regulator.

Remark

The appearance of the operator $H_2$ in the error model can be explained noting that while in the ideal case we can insure the existence of $S_*$, $R_*$ verifying (2.5 d) with $C_* \equiv C_R B$, e.g.

$$\frac{Y_t^*}{U_t} = q^{-d}\frac{B}{A} = \frac{q^{-1}S_*}{C_R - q^{-d}R_*} \qquad \text{(non mismatched case)}$$

which lead to a linear in the parameters regressor form. In the ROM case $C_*$ may not equal $C_R B$ hence the reparametrization is bilinear in the parameters, that is

$$\frac{Y_t^*}{U_t} = q^{-d}\frac{B}{A} = \frac{q^{-d}BS_*}{C_* - q^{-d}BR_*} \qquad \text{(ROM case)}$$

and consequently

$$Y_t^* = C_*^{-1} B\, \theta_*^T \phi_{t-d}$$

The impossibility to get a linear-in the parameters error model hampers the application of the usual PAA and stability formulations (Lyapunov or Popov based). Notice that our adaptation error coincides, up to a filtering factor, with the tracking error. The sector stability formulation may be viewed as a natural extension of the feedback decomposition-based hyperstability approach used in [6].

It is reasonable to expect that the stability conditions for the adaptive scheme should reflect the need of being able to "approach" $H_2$ to 1, e.g. $Y_t^* \to 1/C_R \omega_t$ and cancel $e_t^*$. The result is summarized in the following theorem. The proof is given in [3].

Theorem. Consider the system (2.1) and the adaptive controller (2.3). If for the given $n_S$, $n_R$ and $\mu$ there exists $\theta_*$ such that

i) $|\bar{\sigma} H_2(\mu^{-1/2} e^{-j\theta}) - 1| < \sqrt{1-\bar{\sigma}}$ , $\forall\, \theta \in [0,\pi]$ (conicity condition) (2.6)

$1 > \bar{\sigma} > \{1 + \| \lambda_t' \, [\lambda^{max}(F_{t-d})]^{-1} \|_\infty \}$ (2.7)

ii) $e_t^* \in L_2$ (robust servo behaviour condition)

then $\psi_t$, $e_t \in L_2$ (hence $\to 0$) and $\phi_t \in L_\infty$ for all $\omega_t \in L_\infty$ .

The linear regulator verifying these conditions will be refered to as tuned regulator.

## 3. ROBUST STABILITY CONDITIONS INTERPRETATION

The two conditions derived above are now illustrated by simple examples.

Example 1 (ROM)

Consider a process

$$Y_t = q^{-2}\frac{b_0 + b_1 q^{-1}}{1 + a_1 q^{-1}} U_t$$

and let $n_S = 1$, $n_R = 0$ and $C_R = 1 + C_1 q^{-1}$. Notice that it difers from the ideal choice

$n_S = n_B+d-1$, $n_R = n_A-1$ which insures a unique solution to the Bezout identity. Such a situation may arise on systems with time delay not multiple of the sampling time, that gives rise to an unmodeled zero ranging on $[-\infty,0]$. A root locus analysis of the characteristic equation

$$1 + R^* \frac{q + b_1/b_o}{q^2(q+a_1)(q+S_1^*/S_o^*)} = 0 \qquad (2.8)$$

furnishes insight into the robustness problem. In the non-mismatched case ($n_S = 2$, $n_R = 0$), the regulator adds two zeros at the origin, a pole at the open-loop zero and an additional pole such that for the gain $R_*$ the poles are placed at the desired locations e.g. $q(q+C_1)$ (see Fig. 1).

Fig. 1

It is clear that when $n_S = 1$ only one zero (at the origin) and one pole are added and there is no possible ubication of this pole such that the resulting root-locus contains the desired characteristic polynomial if $C_1 \neq a_1$. In Fig.2 the root-locus with the open-loop zero cancelled is shown.

Fig. 2

The stability condition of robust servo behaviour implies the existence of S and R verifying

$$[(1 + C_1 q^{-1} - q^{-2} R)(b_o + b_1 q^{-1}) - (S_o^* + S_1^* q^{-1})(1 + a_1 q^{-1})]\ \omega_t = 0$$

See (2.5 a,c). If the reference sequence is a constant the tuned regulator is (if it exists!) inside the following set

$$\Theta_i \triangleq \{ \theta_* \in R^n \mid S_o^* + S_1^* = K(1 - R_*)\}$$

where K is the open-loop process steady-state gain.

Besides the requirement of $e_t^* \in L_2$ the tuned regulator must also verify the conicity condition ii) of Theorem 2.1. A first necessary condition is that the poles of the tuned system ($H_2$) lie within a disk of radius $\mu^{1/2}$, that is $\theta_*$ contained in

$$\Theta_\mu \triangleq \{ \theta_* \in R^n \mid C_*(q) \neq 0 \ , \forall\ q \in C \ , |q| \geq \mu^{1/2} \} \qquad (2.9)$$

The dotted circle of Fig.2 defines the allowable pole region for the considered example.

The conicity condition (2.6) restricts the tuned transfer function $H_2(\mu^{-1/2} e^{-j\theta})$ Nyquist locus to be inside the family of disks shown in Fig.3.

Fig. 3

Notice that the sector is contained in the passivity sector and always encircles the point 1. Henceforth if there exists a $\theta_*$ such that the transfer function $Y_t^*/\omega_t$ "approaches" the reference model the conicity condition will be verified. For our example if $a_1 = c_1$, no bandwidth improvement, an open-loop control with $S_* = B$ will exactly verify the objective, that is $H_2 \equiv 1$. As the desired closed loop pole approaches the origin it becomes increasingly difficult to insure a pole location for $H_2$ verifying (2.6).

We cannot overestimate the importance of noting that $H_2$ is a proper transfer function (see 2.5 b,d) hence its global phase shift (when taking $\theta \in [0,\pi]$) is zero for all stably invertible processes. This in its turn implies that by suitable filtering we can always provide the required phase shift to insure (2.6). [3,4].

It is clear from (2.5 c) that when no BOD are present $e_t^* \in L_2$ and "approaching" $H_2$ to 1 are complementary requirements. The situation is considerably worse in the former case as will be seen in the example below.

Example 2 (ROM and BOD)

A continuous time process consisting of a dominant pole and two parasitic complex poles has the pole-zero pattern shown in the Fig. 4 as the parasitics grow faster.

Fig. 4

Assume the process is subject to a constant output disturbance and that $n_S = 1$, $n_R = 0$. In order to cancel the BOD the search of the PAA is restricted to

$$\Theta_{ii} = \{ \theta_* \varepsilon R^n | S_o^* = - S_1^* \}$$

The root locus in this case is given in Fig.5. It is well known from linear control theory that adding an integrator reduces the overall relative stability. In the adaptive context we clearly see that the μ-stability condition (2.8), which implies a certain degree of relative stability, is extremely difficult to verify when the PAA incorporates the BOD internal model. The convergence of the estimates in this case, besides being extremely slow usually will exhibit a "dither " behaviour around the unitary circle.

Fig. 5

Explicit incorporation of the BOD internal model, as being treated in [7], preserves the stability analysis and may help to improve the robust stability. In Fig.6 the root locus is shown when an integrator is added, hence $n_R = 1$, $n_S = 0$ and the zero added by $R_*$ is assumed to cancel the open loop pole. Compare with Fig.5.

Fig. 6

REFERENCES

[1] Kosut, R.L., Friedlander, B. "Performance robustness properties of adaptive control systems". 21st IEEE CDC Orlando Fl. Dec. 8-10, 1982.

[2] Gawthrop, P.J. "On the stability and convergence of a self-tuning controller". Int. J. of Control, Vol. 31, N°5, pp.973-98.

[3] Ortega, R., Praly, L., Landau I.D. "Robustness of discrete adaptive controllers: quantification and improvement". IEEE Trans. on Aut. Cont. (submitted).

[4] Ortega, R. "Robustness enhancement of adaptive controllers by incorporation of process a priori knowledge". Syst. and Control Letters. (To appear).

[5] Safonov, M.G. "Stability robustness of multivariable feedback systems". MIT Press 1980.

[6] Landau, I.D. "Adaptive control- the model reference approach". New York, Dekker, 1979.

[7] Ortega, R., M'Saad, M. Canudas, C. "Practical requirements and theoretical results in robust adaptive control". 9th IFAC Congress. Budapest, Hun. 2-6 July, 1984.

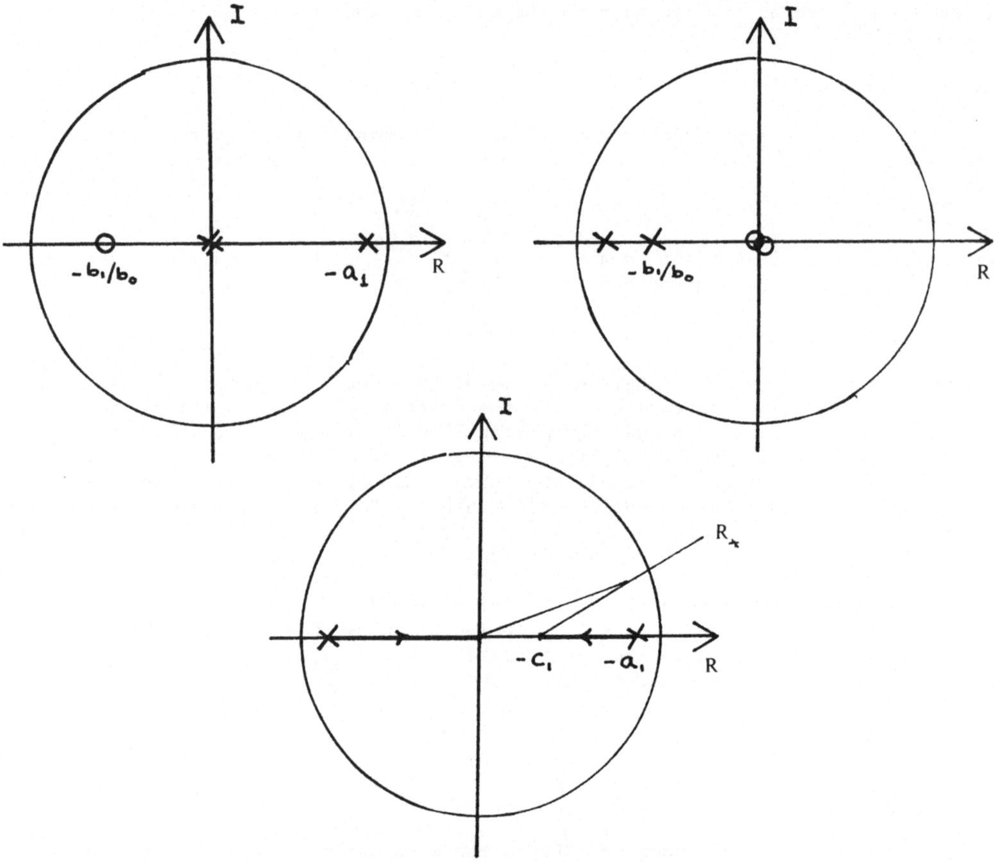


Fig. 1

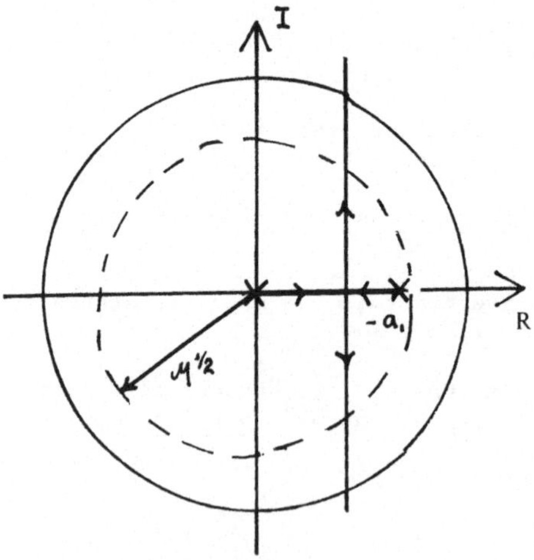


Fig. 2

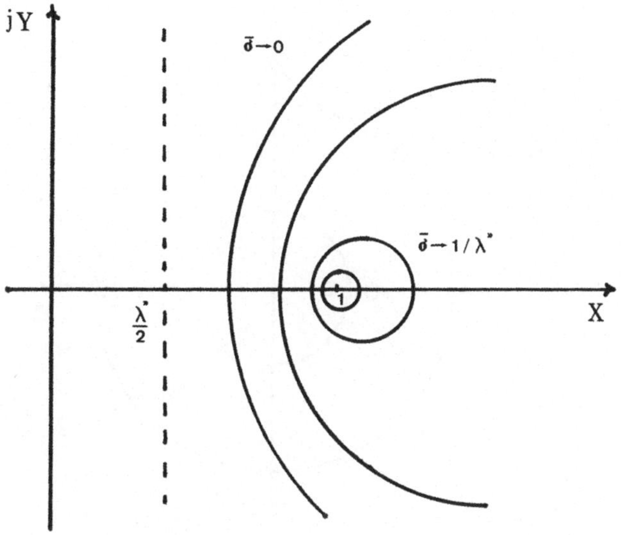
jY
$\bar{\sigma} \to 0$
$\bar{\sigma} \to 1/\lambda^*$
1
$\frac{\lambda^*}{2}$
X

Fig. 3

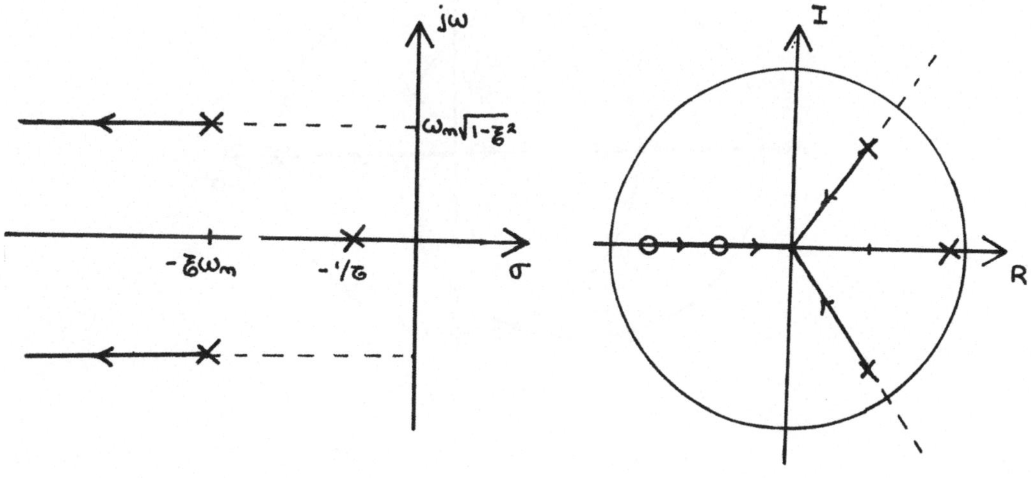
jω
$\omega_m\sqrt{1-\xi^2}$
$-\xi\omega_m$
$-1/\tau$
σ
I
R

Fig. 4

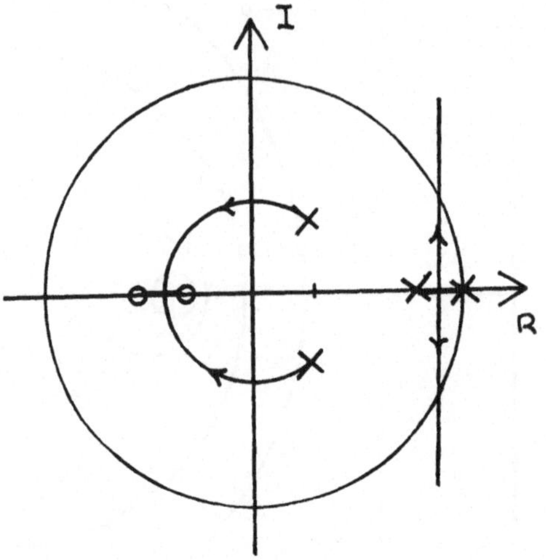


Fig. 5

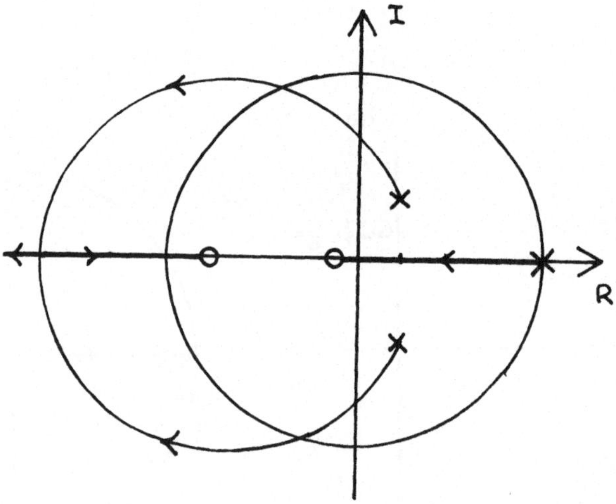


Fig. 6

# GLOBAL ADAPTIVE STABILIZATION IN THE ABSENCE OF INFORMATION ON THE SIGN OF THE HIGH FREQUENCY GAIN

by

J.C. WILLEMS
Mathematics Institute
University of Groningen
P.O. Box 800
9700 AV Groningen
The Netherlands

and

C.I. BYRNES*
Department of Applied Science
Harvard University
Cambridge MA 02138
USA

ABSTRACT

An adaptive control algorithm is presented which globally stabilizes any n-th order linear time-invariant system having (n-1) left half plane zeros. Knowledge of the sign of the high frequency gain is not required for convergence.

INTRODUCTION

The conditions under which one can show that an adaptive control algorithm stabilizes a system, or is self-tuning, demand certain a priori knowledge on the to be controlled plant. Typically for single input/single output continuous time linear time-invariant systems one requires knowledge of such things as its *order* (or an upper bound for it), the *number of zeros*, the *minimum phase* property requiring these zeros to be in the left half plane and, finally, the *sign* of the leading coefficient of the numerator of its transfer function (i.e., the sign of the *high frequency gain*, or, equivalently in the time domain, the sign of the *'instantaneous gain'*: one should know whether the response to a positive unit step is positive or negative for t sufficiently small).

Whatever be the practical consequences of having to impose such rather unphysical and unrobust conditions, it is important, in order to come to grips with the fundamental problems of adaptive control, to understand in how far these conditions are intrinsic and in how far they depend on the particular algorithms used. In other words, for theoretical reasons, it is important to understand in how far these conditions impose fundamental limitations.

Morse [1] has suggested that these conditions are to some extent intrinsic. In particular, he conjectured that the linear plant

---

*Research supported in part by the Netherlands Organisation for the Advancement of Pure Research (ZWO), and by the Air Force Office of Scientific Research under Grant No. AFOSR 810054.

$$\Sigma : \dot{y} = \alpha y + \beta u \tag{1}$$

*cannot* be globally adaptively stabilized by any feedback controller

$$\Sigma_f : \dot{z} = f(z,y) \;;\; u = h(z,y) \tag{2}$$

which is independent of the sign of $\beta$. More precisely, he conjectured that there donot exist q and smooth $(f,h): \mathbb{R}^{q+1} \to \mathbb{R}^{q+1}$ such that the closed loop system

$$\Sigma_{cl} : \begin{array}{ll} \dot{y} = \alpha y + \beta h(z,y) & \text{(3a)} \\ \dot{z} = f(z,y) & \text{(3b)} \end{array}$$

is globally adaptively stabilized, meaning that for all $\alpha \in \mathbb{R}$, $0 \neq \beta \in \mathbb{R}$, and initial conditions $(y(0),z(0))$, there holds: $\lim_{t\to\infty} y(t) = 0$ and $z(t)$ bounded on $[0,\infty)$.

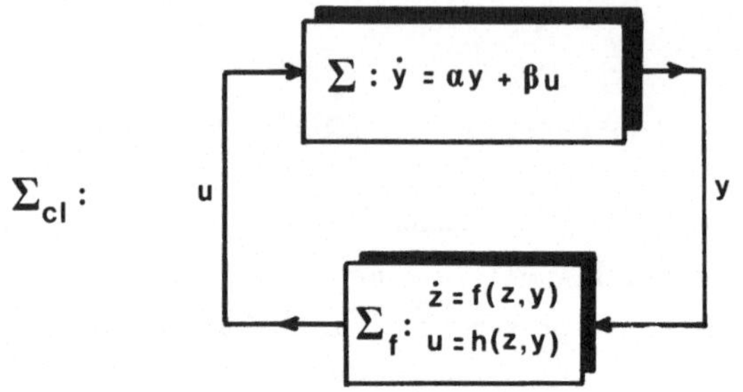


FIG. 1

In addition to the fact that verification of this conjecture would undoubtly have depended on the allowed smoothness of f and h and on the order (and certainly the finite-dimensionality) of z, the state of the feedback controller $\Sigma_f$, it seemed to us that this conjecture was unlikely to be true. This view was based on the observation that it is clearly possible to identify the sign of $\beta$ by an off-line experiment. Once the sign of $\beta$ is known it is easy to see how to proceed. Indeed, as is well-known, for $\beta > 0$

$$\Sigma_f : \dot{k} = y^2 \;;\; u = -ky \tag{4a}$$

will globally adaptively stabilize the plant, while for $\beta < 0$

$$\Sigma_f : \dot{k} = y^2 \;;\; u = ky \tag{4b}$$

achieves this. The philosophy of the control algorithm (4a) or (4b) is clear. If $\beta > 0$ then high negative gain feedback will stabilize the plant. The control law (4a) simply turns the feedback gain up until stability is achieved. The controller (4b) does the analogous thing for the case $\beta < 0$.

## A GLOBALLY ADAPTIVE CONTROLLER FOR A FIRST ORDER PLANT

In a recent article, Nussbaum [2] proves Morse's conjecture for first order controllers $\Sigma_f$ with *rational* f and h. However, more importantly, he constructed a globally adaptively stabilizing controller, thus disproving Morse's conjecture.

Nussbaum's controller is first order and uses the following (analytic) functions f and h:

$$\dot{z} = y(z^2 + 1) \ ; \ u = y(z^2 + 1)\tilde{h}(z)$$

with $\tilde{h}$: $\mathbb{R} \to \mathbb{R}$ even and differentiable and such that

$$\sup_{z \geq 0} \int_0^z \tilde{h}(\nu)d\nu = \infty \text{ and } \inf_{z \geq 0} \int_0^z \tilde{h}(\nu)d\nu = -\infty$$

($\tilde{h}(z) = (\cos\frac{1}{2}\pi z) \exp z^2$ is an example of such a function).

Since it is not easy to see from a system theoretic point of view why this controller achieves its purpose, we attempted to combine the features of the adaptive controller (4) with Nussbaum's result. This led us to consider

$$\Sigma_f: \dot{k} = y^2 \ ; \ u = s(k)ky \tag{5}$$

Choose now s such that it takes on *both* positive and negative values, to **make** sure to always hit a stabilizing gain. If in addition the sign of s(k) is kept constant for sufficiently long periods, one can also give the system time to stabilize out. Under suitable conditions, this can indeed be achieved:

Theorem 1: *Let* s: $\mathbb{R} \to \mathbb{R}$ *be bounded on compact sets and be such that*

$$\dot{y} = \alpha y + \beta s(k)ky \tag{6a}$$

$$\dot{k} = y^2 \tag{6b}$$

*has a unique (absolutely continuous) solution for all* $(y(0),k(0)) \in \mathbb{R}^2$. *Define* $S(k) := \int_0^z s(\sigma)\sigma d\sigma$. *Assume now that*

$$\sup_{k \geq 1} \frac{1}{k} S(k) = \infty \text{ and } \inf_{k \geq 1} \frac{1}{k} S(k) = -\infty \tag{7}$$

*Then, for all* $y(0) \in \mathbb{R}$, $k(0) \in \mathbb{R}$, $\alpha \in \mathbb{R}$, *and* $0 \neq \beta \in \mathbb{R}$, *there holds:*

(i) $\lim_{t \to \infty} y(t) = 0$

(ii) $\lim_{t \to \infty} k(t)$ *exists and is finite.*

Proof: First observe that (7) implies that for all a and b $\neq 0$,

$$\sup_{k \geq 1}(ak + bS(k)) = \infty \text{ and } \inf_{k \geq 1}(ak + bS(k)) = -\infty \tag{8}$$

Indeed, since S is continuous and satisfies (7), there exist $\{k'_n\}$ and $\{k''_n\}$, $n = 1,2,\ldots$, with $k'_n, k''_n \xrightarrow[n\to\infty]{} \infty$, such that $S(k'_n) = nk'_n$ and $S(k''_n) = -nk''_n$. This yields $ak'_n + bS(k'_n) = (a + bn)k'_n$ and $ak''_n + bS(k''_n) = (a - bn)k''_n$. Letting $n \to \infty$ gives, since $b \neq 0$, (8).

We now proceed to prove the theorem. Consider the evolution of $y^2$. Using (6) yields:

$$\frac{1}{2}\frac{dy^2}{dt} = (\alpha + \beta s(k)k)\frac{dk}{dt} = \frac{d}{dt}(\alpha k + \beta S(k))$$

Hence

$$\tfrac{1}{2}y^2(t) - \tfrac{1}{2}y^2(0) = \alpha k(t) + \beta S(k(t)) - \alpha k(0) - \beta S(k(0)) \tag{9}$$

Consequently, $\alpha k(t) + \beta S(k(t)) \geq \alpha k(0) + \beta S(k(0)) - \frac{1}{2}y^2(0)$ for $t \in \mathbb{R}^+ := [0,\infty)$. Hence $\alpha k(t) + \beta S(k(t))$ remains bounded from below for $t \in \mathbb{R}^+$. Since k is also continuous this implies, by (8), that there exists $K < \infty$ such that $k(t) \leq K$ for $t \in \mathbb{R}^+$.

Since, by (6b), k(t) is monotone nodecreasing this implies that $\lim_{t\to\infty} k(t)$ exists and is finite, proving (ii). Using (6b) again, we conclude that $y \in L_2(\mathbb{R}^+; \mathbb{R})$. Together with $k \in L_\infty(\mathbb{R}^+; \mathbb{R})$ this implies, by (6a), that $\dot{y} \in L_2(\mathbb{R}^+; \mathbb{R})$. Now $\{y, \dot{y} \in L_2(\mathbb{R}^+; \mathbb{R})\}$ implies $\{\lim_{t\to\infty} y(t) = 0\}$, proving (i).□

It is easy to construct (bounded analytic) examples of functions satisfying (7). Take for instance $s(k) = \sin\sqrt{|k|}$. An interesting example which preserves much of the intuition of the control law (4) is to choose

$$s(k) = \begin{cases} +1 & n^2 \leq |k| < (n+1)^2 \qquad n = 0,2,4,\ldots \\ -1 & n^2 \leq |k| < (n+1)^2 \qquad n = 1,3,5,\ldots \end{cases}$$

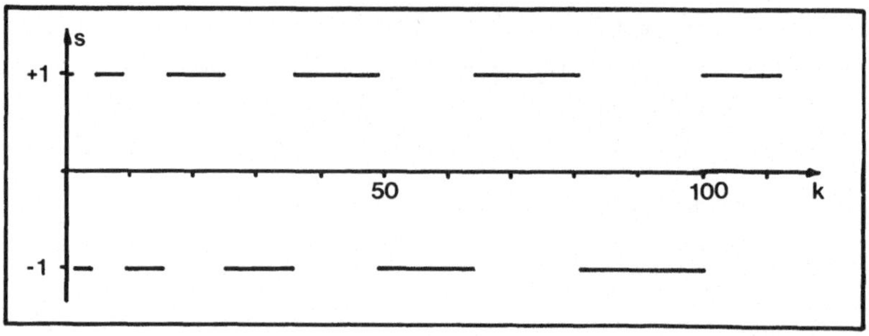


FIG. 2

It is easily verified that this function indeed satisfies condition (7). Figure 3 shows some typical simulation examples obtained with the above control law. The somewhat disturbing high gain features of this procedure is obvious from these graphs. However Theorem 1 is of theoretical interest only.

A GLOBALLY ADAPTIVE CONTROLLER FOR AN N-TH ORDER PLANT

It is possible to generalize the result obtained in Theorem 1 to a class of n-th order systems. Let

$$\Sigma: \dot{x} = Ax + Bu \qquad y = Cx \tag{10}$$

be a controllable observable single imput/single output n-th order system and assume that it has (n-1) zeros, all in the open left half plane.

Choosing a basis compatible with the sum decomposition of the state space $X$ into $X = X_1 + X_2$ with $X_1 = \ker C = X^*_{\ker C}$, the supremal controlled invariant subspace contained in ker C, and $X_2 = \operatorname{im} B$ yields

$$\begin{aligned} \dot{x}_1 &= A_{11}x_1 + A_{12}y \\ \dot{y} &= A_{21}x_1 + \alpha y + \beta u \end{aligned} \tag{11}$$

Since we assume that (10) has left half plane zeros, we know that $A_{11}$ has its eigenvalues in the open left half plane. Also, since (10) is controllable, $\beta \neq 0$. However, we donot assume that we know the sign of $\beta$. The decomposition (ii) follows also easily from the feedback decomposition of (10) (shown in Figure 4):

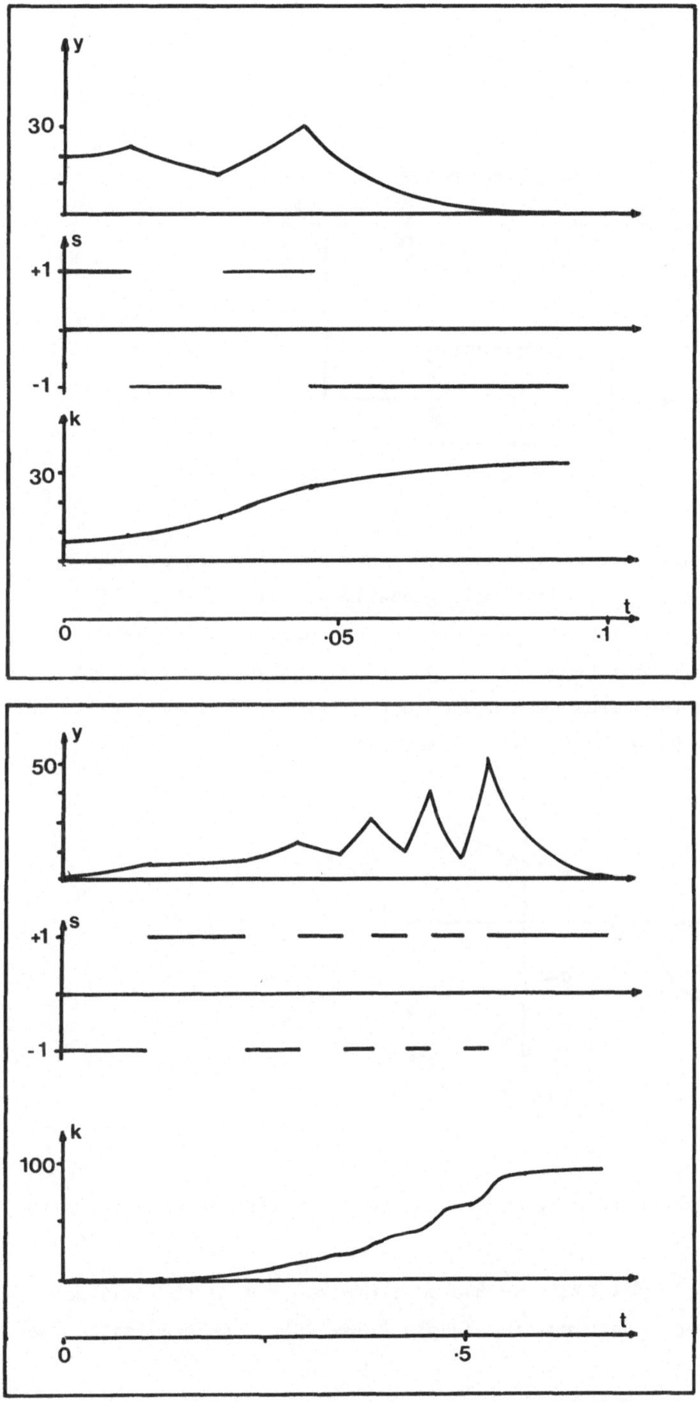


Simulation of (6) with s as in Figure 2

$\alpha = -.23$
$\beta = 2.8$
$y(0) = 24$
$k(0) = 4$

$\alpha = 4.9$
$\beta = -.5$
$y(0) = 2.7$
$k(0) = 2.2$

FIG. 3

$$G(s) = \frac{\beta}{s-\alpha}\left(1 + \frac{\beta}{s-\alpha}F(s)\right)^{-1}$$

with $G(s) = C(Is - A)^{-1}B$ and $F(s) = A_{21}(Is - A_{11})^{-1}A_{12}$

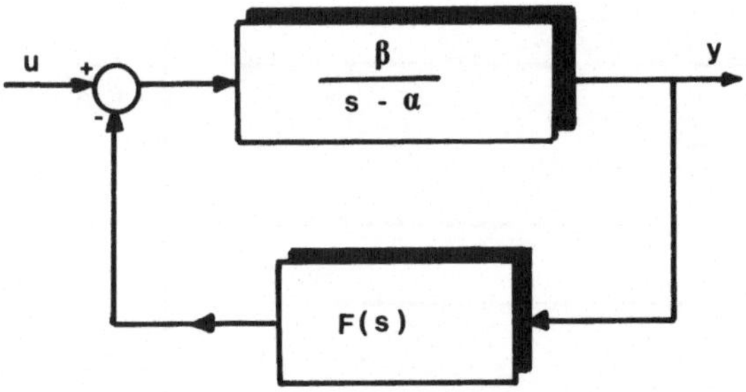


FIG. 4

If the sign of $\beta$ were known, it is again easy to see how to proceed. Indeed, when $\beta > 0$, the feedback law (4a) will adaptively globally stabilize, while if $\beta < 0$, (4b) will achieve this. The mechanism of these high gain controllers is obvious if one considers the root locus associated with the plant (10): high gain feedback drives (n-1) of the poles to the zeros (which are in the left half plane), while the last remaining pole will go to $-\infty$ along the real axis.

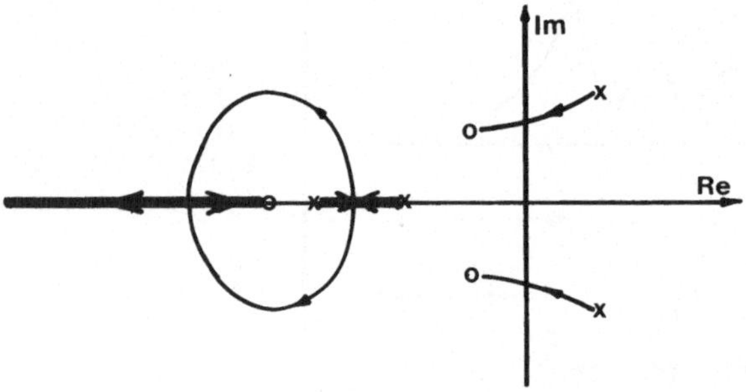


FIG. 5

The feedback law (5) also achieves global adaptive stabilization when the sign of $\beta$ is not known:

Theorem 2: *Assume that the plant* (10) *is n-th order, minimal, and with* (n-1) *zeros in the open left half plane. Consider now the feedback law* (5). *This results in the closed loop system*

$$\Sigma_{cl}: \quad \dot{x} = Ax + Bs(k)kCx \qquad (12a)$$
$$\dot{k} = (Cx)^2 \qquad (12b)$$

*Let* $s: \mathbb{R} \to \mathbb{R}$ *be bounded on compact sets and be such that* (12) *has a unique solution for all* $x(0) \in \mathbb{R}^n$ *and* $k(0) \in \mathbb{R}$. *Assume that* s *also satisfies* (7). *Then, for all*

$x(0) \in \mathbb{R}^n$ *and* $k(0) \in \mathbb{R}$, *there holds:*

(i) $\lim_{t\to\infty} x(t) = 0$

(ii) $\lim_{t\to\infty} k(t)$ *exists and is finite*

Proof: In terms of the decomposition (ii), (12) may be rewritten as

$$\dot{x}_1 = A_{11}x_1 + A_{12}y \tag{13a}$$
$$\dot{y} = (\alpha + \beta s(k)k)y + A_{21}x_1 \tag{13b}$$
$$\dot{k} = y^2 \tag{13c}$$

Using (13b) and (13c) (following the proof of Theorem 1) yields

$$\alpha k(t) + \beta S(k(t)) \geq \alpha k(0) + \beta S(k(0)) - \tfrac{1}{2}y^2(0) - \int_0^t y(\tau)A_{21}x_1(\tau)d\tau \tag{14}$$

for $t \in \mathbb{R}^+$. By the lemma of the Appendix this implies that for some $M_1$ and $M_2$ (depending only on $A_{11}, A_{12}$ and $A_{21}$) there holds:

$$\alpha k(t) + \beta S(k(t)) \geq \alpha k(0) + \beta S(k(0)) - \tfrac{1}{2}y^2(0) - M_1\|x_1(0)\|^2 - M_2\int_0^t y^2(\tau)d\tau$$

Hence, by (13c), $(\alpha + M_2)k(t) + \beta S(k(t)) \geq (\alpha + M_2)k(0) + \beta S(k(0)) - \tfrac{1}{2}y^2(0) - M_1\|x_1(0)\|^2$ for $t \in \mathbb{R}^+$. By (8) and the continuity of k, this implies that k is bounded from above on $\mathbb{R}^+$. Since k is also monotone nondecreasing by (13c), (ii) follows.

By (13c) this implies $y \in L_2(\mathbb{R}^+; \mathbb{R})$. This and the asymptotic stability of $A_{11}$ in (13a) yield $x_1 \in L_2(\mathbb{R}; \mathbb{R})$. Hence $x = (x_1, y) \in L_2(\mathbb{R}^+; \mathbb{R}^n)$. Together with $k \in L_\infty(\mathbb{R}; \mathbb{R})$ this implies, by (12a), that $\dot{x} \in L_2(\mathbb{R}^+; \mathbb{R}^n)$. Hence $\{x, \dot{x} \in L_2(\mathbb{R}^+; \mathbb{R}^n)\}$, which implies $\{\lim_{t\to\infty} x(t) = 0\}$. The theorem is now completely proven.□

A simulation using a third order example and the s of Figure 1 is shown in Figure 6. The high gain features are again obvious from these graphs.

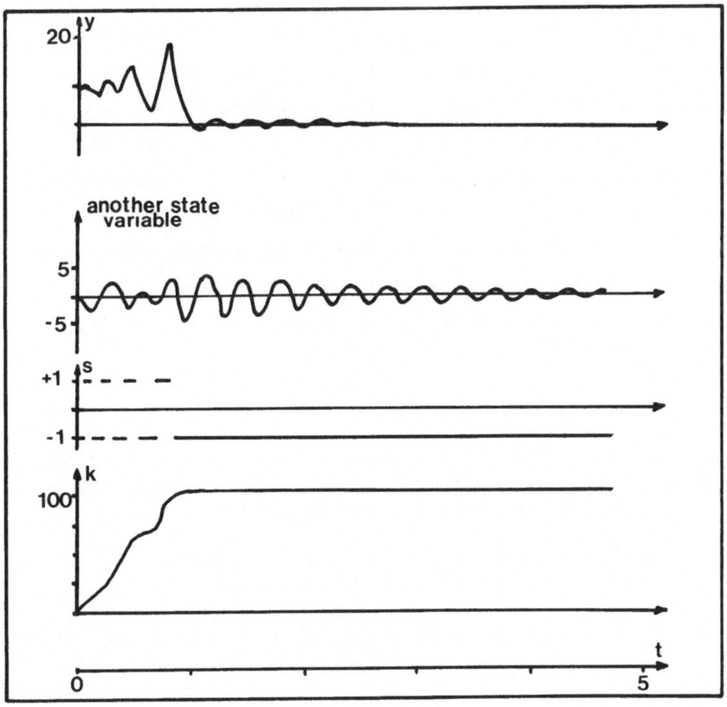


Simulation of the plant with transfer function

$$\frac{.15\,s^2 + .3\,s + 6.2}{s^3 + 1.5\,s^2 + .36\,s - 22}$$

with suitable initial conditions and control law (5), and with s as in Figure 2.

FIGURE 6

CONCLUDING COMMENTS

In this paper we have shown that it is possible to adaptively stabilize a plant with n poles and (n-1) left half plane zeros. The sign of the high frequency gain need not be know, however. These results provide evidence for the existence of a globally adaptively stabilizing or self tuning controller without minimum phase conditions. Similar results as those presented here have recently been obtained by Morse [3].

We will close with some general comments pertaining to the questions which motivated the conjecture discussed in this paper.

Our first remark concerns the high gain features of the feedback control law (4) or (6). It can be seen that to some extent this is a necessary evil. Indeed, eventhough the class of plants (1) or (11) is assumed to be minimal ($\beta \neq 0$), it contains in its closure ($\beta \to 0$) non minimal unstable ($\alpha > 0$) systems. From there it follows that no feedback control law (2) which defines a finite gain input/output stable system can be stabilizing for all these plants simultaneously. Consequently (quadratic growth or similar) high gain behavior of the globally adaptively stabilizing control laws is not so much a consequence of adaptation as it is a consequence of requiring stabilization for plants which in the limit may be uncontrollable or unobservable. All this also shows that Morse's conjecture could equally well have been called a conjecture in *robust* control as in *adaptive* control.

Our second remark is concerned with the 'passive' control policy which underlies the control law (6). This control law switches back and forth between the control laws (4). The parameter space is divided into two regions $\beta > 0$ and $\beta < 0$ and in each of the regions a certain control policy should be used. Now (6) proposes a certain switching control law in which switching is induced as long as instability is still being observed. In addition care is taken of the fact that sufficient time is allowed between switchings so that the system is given a chance to settle down. It is worth trying this idea out in other situations, in which the parameter space may be divided into a finite or countably infinite union of regions which each ask for a different feedback control regime. Using this idea it may in fact be possible to generalize the results of the present paper to the multivariable case.

REFERENCES


[1] A.S. Morse: "Recent Problems in Parameter Adaptive Control", pp. 733-740, Vol.3, *Outils et Modèles Mathématiques pour l'Automatique, l'Analyse de Systèmes et le Traitement du Signal*, Ed. I.D. Landau, Editions du CNRS, 1983.

[2] R.D. Nussbaum: "Some Remarks on a Conjecture in Parameter Adaptive Control", *Systems and Control Letters*, Vol. 3, No. 5, pp. 243-246, 1983.

[3] A.S. Morse: "An Adaptive Control for Globally Stabilizing Linear Systems with Unknown High-Frequency Gains", Department of Electrical Engineering, Yale Univ., Report No. 8402, January 1984.

APPENDIX

Consider the response of the single input/single output finite dimensional linear time-invariant system $\dot{w} = Fw + Gv$ ; $r = Hw$ ; $w \in \mathbb{R}^k$, with F a matrix with its eigenvalues in the open left half plane. There holds:

Lemma: There exist $M_1, M_2 < \infty$ such that for all $w(0) \in \mathbb{R}^k$, $T \in \mathbb{R}^+$, and $v \in L_2([0,T];\mathbb{R})$, there holds

$$|\int_0^T v(t)r(t)dt| \le M_1\|w(0)\|^2 + M_2\int_0^T |v(t)|^2 dt$$

Proof: $r(t) = He^{Ft}w(0) + \int_0^t He^{F(t-\tau)}Gv(\tau)d\tau$. In the obvious operator notation

$$r = L_1 w(0) + L_2 v$$

Since F has its eigenvalues in the open left half plane $L_1$ is a bounded linear operator from $\mathbb{R}^k$ into $L_2(\mathbb{R}^+;\mathbb{R})$, while $L_2$ is a bounded linear operator on $L_2(\mathbb{R}^+;\mathbb{R})$. Now

$$\langle r,v\rangle_{L_2(\mathbb{R}^+;\mathbb{R})} = \langle L_1 w(0),v\rangle_{L_2(\mathbb{R}^+;\mathbb{R})} + \langle L_2 v,v\rangle_{L_2(\mathbb{R}^+;\mathbb{R})}$$

Hence

$$|\langle r,v\rangle_{L_2(\mathbb{R}^+;\mathbb{R})}| \le \|L_1\|\|w(0)\|\|v\|_{L_2(\mathbb{R}^+;\mathbb{R})} + \|L_2\|\|v\|^2_{L_2(\mathbb{R}^+;\mathbb{R})}$$

$$\le \|L_1\|\|W(0)\|^2 + (\|L_1\| + \|L_2\|)\|v\|^2_{L_2(\mathbb{R}^+;\mathbb{R})}$$

Using this inequality for v's with support on [0,T] yields the lemma. □

# AN ADAPTIVE CONTROL FOR GLOBALLY STABILIZING LINEAR SYSTEMS WITH UNKNOWN HIGH-FREQUENCY GAINS†

A. S. Morse
Department of Electrical Engineering
Yale University
New Haven, Ct. 06520/USA

ABSTRACT

This paper presents an algorithm for adaptively controlling a single-input, single-output process admitting an n-dimensional, minimum phase linear model of relative degree 1 or 2, with unknown parameters. Apriori knowledge of the sign of the model's high frequency gain is not required, and a sufficiently rich probing signal is not needed for stabilization.

## Introduction

It is well-known that techniques now exist {e.g., [1],[2]} for adaptively stabilizing any process which can be modelled by a linear system with unknown, strictly proper transfer function of the form $g_p\alpha_p(s)/\beta_p(s)$, where $\alpha_p(s)$ and $\beta_p(s)$ are monic, coprime polynomials and $g_p$ is a nonzero constant, provided it can be assumed that

i. $\alpha_p(s)$ is a strictly stable polynomial

ii. a bound $n \geq$ degree $\beta_p(s)$ is known

iii. the relative degree $n^* =$ degree $\beta_p(s)$ - degree $\alpha_p(s)$ is known

iv. the sign of $g_p$ is known.

In view of the difficulties encountered in attempting to extend these techniques to larger classes of systems, it is natural to ask if the preceding assumptions are in some sense fundamental. In [3] it was speculated that iv is actually necessary for adaptive stabilization with a smooth controller. More specifically, it was conjectured that it is impossible to stabilize the one-dimensional system $\dot{y} = y + gu$, with g unknown, using a controller of the form

† The work of the author was supported in part, by the National Science Foundation, Grant No. ECS7916871; portions of this work were done while the author was visiting the Department of Electrical Engineering, University of Southern California, Los Angeles, California.

$$\left.\begin{aligned}\dot{x} &= f(x,y) \\ u &= v(x,y)\end{aligned}\right\} \quad (1)$$

where $x \in \mathbb{R}^m$ and f and v are continuous in x and y. The conjecture was shown to be true for m = 1 if f and v are constrained to be polynomials in x and y of degree $\lesssim 2$.

In a recent paper [4], Nussbaum addressed the conjecture of [3] by considering the problem of finding a smooth controller (1) for the general one-dimensional system

$$\dot{y} = ay + gu \quad (2)$$

where both g and a are unknown and $g \neq 0$. Nussbaum first showed that if m = 1 and if f and v are constrained to be rational in x and y, then for any a > 0 there does not exist a controller such as (1) which will stabilize (2) in the sense that $(x(t),y(t))$ is bounded on $[0,\infty)$ and $y(t) \to 0$ as $t \to \infty$.

Nussbaum's second and somewhat surprising result was to show that there do in fact exist a whole family of smooth one-dimensional controllers which will stabilize (2). One controller with this property, somewhat simpler in structure than the type proposed by Nussbaum, is given by $f = y^2$, $v = x^2\cos(x)y$. The asymptotic behaviour of the resulting closed-loop system

$$\dot{x} = y^2 \quad (3a)$$

$$\dot{y} = (a + gx^2\cos(x))y \quad (3b)$$

can be easily understood by first noting that, $dy^2/dx = 2(a + gx^2\cos(x))$; by integrating with respect to x, it follows that for any initial state $(x_0,y_0)$

$$y^2(t) = \pi(x(t)) + y_0^2 - \pi(x_0) \quad (4)$$

where $\pi(z) = 2az + 2g(z^2\sin(z) + 2(z\cos(z) - \sin(z)))$. Since (3a) implies that x(t) is monotone nondecreasing, x(t) must either approach a finite limit or grow without bound. However $\pi(z) + y_0^2 - \pi(x_0)$ has infinitely many zeros for $z \geq x_0$ so if x(t) were to grow without bound, it would eventually pass through a value $x_1$ for which in view of (4), $y^2 = 0$; at this time y = 0 and (3) would be in equilibrium state $(x_1,0)$ so $x(t)$ could grow no further - a contradiction. Clearly then, $x(t)$ is bounded between $x_0$ and $x_1$, and thus from (4), $y(t)$ is bounded as well. In addition, (3a) implies that $y(t)$ is square integrable on $[0,\infty)$ and (3b) implies that $\dot{y}(t)$ is bounded so $y(t)$ must approach zero as $t \to \infty$.

In spite of their mathematical simplicity, such "universal controllers" are capable of regulating <u>any</u> stabilizable, one-dimensional linear system; for this reason, the idea behind them due to Nussbaum, must be regarded as an important advance in system theory. Generalization to higher dimensional systems is the obvious next step. It is the purpose of this paper to present such a generalization, applicable to systems satisfying assumptions i-iii for $n^* \leq 2$.

## 1. Process Model Parameterization

The process to be controlled is assumed to satisfy i-iii above. It is known [1] that assumption ii implies that the relationship between process input u and output y can be modelled by an n-dimensional, observable, stabilizable system of the form

$$\left.\begin{aligned} \dot{x}_p &= (A_0 + h_p c_0)x_p + b_p g_p u \\ y &= c_0 x_p \end{aligned}\right\} \qquad (5)$$

where $(A_0,b_0,c_0)$ is any n-dimensional, canonical system, preselected so that $A_0$ is strictly stable, and $h_p$ and $b_p$ are vectors of unknown parameters. The control objective is to cause the tracking error

$$\left.\begin{aligned} e &= y - c_r x_r \\ \dot{x}_r &= A_r x_r + b_r r \end{aligned}\right\} \qquad (6)$$

between y and the output $c_r x_r$ of prespecified reference system (6) to approach zero as $t \to \infty$, while at the same time insuring that the controller's state together with $x_p$ remain bounded on $[0,\infty)$. Here $r(t)$ is any piecewise-continuous reference signal, bounded on $[0,\infty)$, and $(A_r,b_r,c_r)$ is any canonical realization of $1/\beta_r(s)$ where $\beta_r(s) = \prod_{i=1}^{n^*} (s+\lambda_{i-1})$ and $\lambda_i > 0$. Reference model (6) is the first of several component subsystems of the controller we propose to examine.

A second subsystem is described by the equations

$$\left.\begin{aligned} \dot{\theta}_u &= A_0\theta_u + b_0u \\ \dot{\theta}_y &= A_0\theta_y + b_0y \\ \theta &= [\theta_u', \theta_y', r]' \end{aligned}\right\} \quad (7)$$

where prime denotes transpose. The significance of <u>sensitivity function</u> $\theta$ {cf. [1]}, is that together with assumptions i and iii, it allows one to reparameterize (5)-(7) as

$$\left.\begin{aligned} e &= c_r x_e \\ \dot{x}_e &= A_r x_e + b_r g_p(u - k_p'\theta + \bar{\varepsilon}) \end{aligned}\right\} \quad (8a)$$

$$\left.\begin{aligned} \theta &= \bar{C}\bar{x} + \bar{d}r \\ \dot{\bar{x}} &= \bar{A}\bar{x} + \bar{b}(u - k_p'\theta + \bar{\varepsilon}) \end{aligned}\right\} \quad (8b)$$

where $\bar{x} = [\theta_u', \theta_y', x_p']'$, $(\bar{A}, \bar{b}, \bar{C}, \bar{d})$ is an unknown but strictly stable system, $k_p$ is an unknown vector of parameters, and $\bar{\varepsilon}$ is an unknown linear combination of decaying exponentials. Use will be made of this parameterization in studying closed-loop system behaviour.

## 2. Control Equations

The remainder of the controller consists of a <u>filtered sensitivity function</u> $\phi$,

$$\left.\begin{aligned} \phi &= \theta \text{ if } n^* = 1 \\ \dot{\phi} &= -\lambda_1\phi + \theta \text{ if } n^* = 2 \end{aligned}\right\} \quad (9)$$

an <u>augmented error</u> $\bar{e}$,

$$\bar{e} = e + \sigma \quad (10a)$$

$$\dot{\sigma} = -\lambda_0\sigma - (x^2 + \phi'\phi)\bar{e} \quad (10b)$$

$$x = \bar{e}^2/2 + z \quad (10c)$$

$$\dot{z} = (\lambda_0 + x^2 + \phi'\phi)\bar{e}^2 \quad (10d)$$

a <u>Nussbaum Gain</u> N,

$$N(x) = x\cos(x) \quad (11)$$

a parameter adjustment law

$$\dot{\hat{k}} = N\phi\bar{e} \tag{12}$$

and a feedback law

$$u = \hat{k}'\theta + v \tag{13a}$$

$$v = \begin{cases} 0 & \text{if } n^* = 1 \\ N\phi'\phi\bar{e} & \text{if } n^* = 2 \end{cases} \tag{13b}$$

The resulting closed-loop adaptive control system determined by these equations is shown in Figure 1.

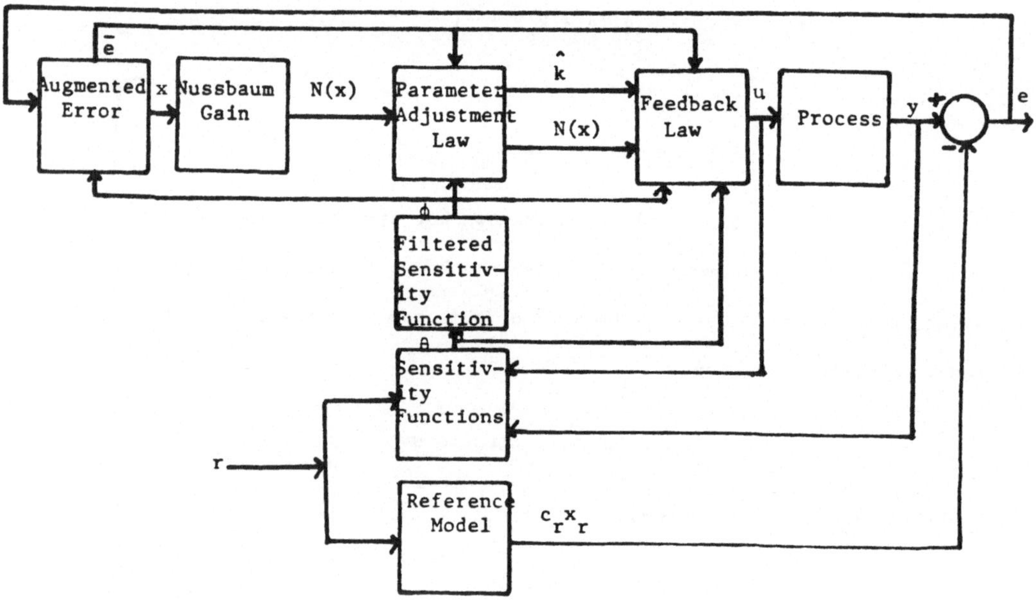


Figure 1: Closed-Loop System

Our main result is as follows:

Theorem 1: For each initial state and each piecewise-continuous input r, bounded on $[0,\infty)$, the state response $X = [x_p', \theta_u', \theta_y', x_r', \hat{k}', \sigma, z, \phi']'$ of the adaptive control system defined by (5)-(7) and (9)-(13), exists and is bounded on $[0,\infty)$ and the tracking error e approaches zero as $t \to \infty$.†

†If $n^* = 1$, $\phi$ is not really a state component and thus should be deleted from the definition of X.

The remainder of this paper is devoted to a proof of this theorem.

## 3. Stability Analysis

We begin by defining <u>parameter error</u> $k = \hat{k} - k_p$ so that from (12)

$$\dot{k} = N\phi\bar{e} \tag{14}$$

whereas from (8b) and (13)

$$\left.\begin{aligned} \theta &= \bar{C}\bar{x} + \bar{d}r \\ \dot{\bar{x}} &= \bar{A}\bar{x} + \bar{b}(k'\theta + v + \bar{\varepsilon}) \end{aligned}\right\} \tag{15}$$

In addition, using (8a),(10a) and (10b) it is easy to verify {cf. [1]} that

$$\dot{\bar{e}} = -(\lambda_0 + x^2 + \phi'\phi)\bar{e} + g_p(k'\phi + \varepsilon) \tag{16}$$

where $\varepsilon$ is an unknown linear combination of decaying exponentials.

Observe that (14)-(16) together with (9), (10b)-(10d), (11) and (13b) define a dynamical system of the form $\dot{X}_e = f(X_e,\bar{\varepsilon},\varepsilon,r)$ where f is a continuous, nonlinear function and $X_e = [k',\bar{e},\sigma,\bar{x}',z,\phi']'$ if $n^* = 2$ or $X_e = [k',\bar{e},\sigma,\bar{x}'z]'$ if $n^* = 1$. In the sequel it will be shown that in either case, $X_e$ exists and is bounded on $[0,\infty)$ and that both $\bar{e}$ and $\sigma$ approach zero as $t \to \infty$. The latter together with (10a) imply that e also goes to zero as $t \to \infty$. Since reference model state $x_r$ is clearly bounded on $[0,\infty)$, and since $\hat{k} = k + k_p$ and $\bar{x} = [\theta_u',\theta_y',x_p']'$, boundedness of $X_e$ implies boundedness of adaptive control system state $X = [x_p',\theta_u',\theta_y',x_r',\hat{k}',\sigma,z,\phi']'$. In other words, to prove Theorem 1, it is enough to prove that $X_e$ is bounded on $[0,\infty)$ and that $\bar{e}$ and $\sigma$ go to zero.

Continuity of f guarantees that for any initial state $X_e(0)$, there must be an interval $I = [0,t_1]$ of positive length on which a solution to $\dot{X}_e = f(X_e,\bar{\varepsilon},\varepsilon,r)$ exists. To establish existence and boundedness of $X_e$ on $[0,\infty)$, it is therefore enough to show that $\|X_e\|$ is bounded on I by a constant not depending on $t_1$.†

In the sequel, when we say a function $\zeta(t)$ is bounded on I, we mean that $\|\zeta(t)\|$ is bounded on I by a constant not depending on $t_1$. We sometimes write $\alpha(t) = \beta(t)$ $(\mu)$

---

†Here and elsewhere, for $q \in \mathbb{R}^n$, $\|q\|$ denotes the Euclidean norm $\sqrt{q'q}$.

if $\|\alpha(t) - \beta(t)\|$ is bounded on I, and $\|\alpha\| \epsilon\, L^2[I]$ if $\int_0^t \|\alpha(\tau)\|^2 d\tau$ is bounded on I.

<u>Proposition</u> 1: $\|k\|$, $\|x\|$, $\|z\|$ <u>and</u> $\|\bar{e}\|$ <u>are bounded on</u> I <u>and</u> $\|\bar{e}\|$ <u>and</u> $\|\phi\bar{e}\|$ <u>are in</u> $L^2[I]$.

<u>Proof</u>: Using (14), $d\|k\|^2/dt = 2N\bar{e}k'\phi$; hence by solving (16) for $k'\phi$ and substituting in the preceding, there results $d\|k\|^2/dt = 2N\bar{e}((\dot{\bar{e}} + (\lambda_0 + x^2 + \phi'\phi)\bar{e})/g_p - \varepsilon)$. Since (10c) and (10d) imply that $\dot{x} = \bar{e}(\dot{\bar{e}} + (\lambda_0 + x^2 + \phi'\phi)\bar{e})$, there follows $d\|k\|^2/dt = 2((N\dot{x})/g_p - N\varepsilon\bar{e})$. Thus by completing the square

$$\frac{d\|k\|^2}{dt} = \frac{2N\dot{x}}{g_p} + \frac{N^2\bar{e}^2}{|g_p|} + |g_p|\varepsilon^2 - \left(\frac{N}{\sqrt{|g_p|}}\bar{e} + \sqrt{|g_p|}\,\varepsilon\right)^2$$

$$\leq \frac{2N\dot{x}}{g_p} + \frac{N^2\bar{e}^2}{|g_p|} + |g_p|\varepsilon^2$$

From (11) and then (10d), $N^2\bar{e}^2 = x^2\cos^2(x)\bar{e}^2 \leq (\lambda_0 + x^2 + \phi'\phi)\bar{e}^2 = \dot{z}$, so

$$\frac{d\|k\|^2}{dt} \leq \frac{2N\dot{x}}{g_p} + \frac{\dot{z}}{|g_p|} + |g_p|\varepsilon^2$$

Integrating, and then noting from (10c) that $z \leq x$, there follows

$$\|k(t)\|^2 - \|k(0)\|^2 \leq \frac{2}{g_p}\int_0^t N\dot{x}dt + \frac{1}{|g_p|}(z(t) - z(0)) + |g_p|\int_0^t \varepsilon^2(\tau)d\tau$$

$$\leq \frac{2}{g_p}\int_{x(0)}^{x(t)} \lambda\cos(\lambda)d\lambda + \frac{1}{|g_p|}(x(t) - z(0)) + |g_p|\int_0^\infty \varepsilon^2(\tau)d\tau$$

Hence if we define the function $\pi(\cdot)$ so that

$$\pi(w) = \frac{2}{g_p}(w\sin(w) + \cos(w)) + \frac{w}{|g_p|} + C$$

where

$$C = \|k(0)\|^2 - \frac{2}{g_p}(x(0)\sin(x(0)) + \cos(x(0))) - \frac{z(0)}{|g_p|} + |g_p|\int_0^\infty \varepsilon^2(\tau)d\tau$$

then

$$\|k(t)\|^2 \leq \pi(x(t)), \quad t \in I \tag{17}$$

Examination of $\pi(w)$ shows that there must exist a closed, bounded interval $[x^-, x^+]$ containing $x(0)$, for which both $\pi(x^-)$ and $\pi(x^+)$ are negative. Since (17) implies that for any $t \in I$, $x(t)$ cannot pass through either $x^-$ or $x^+$, it must be true that $\|x(t)\| \leq \max\{\|x^-\|, \|x^+\|\}$. Therefore $\|x\|$ is bounded on I. In addition, since $\pi(w)$ is continuous, it follows from (17) that $\|k\|$ is bounded on I as well.

From (10c) and (10d), $x(t) - z(0) = \bar{e}^2/2 + \int_0^t (\lambda_0 + x^2(\tau) + \phi'(\tau)\phi(\tau))\bar{e}^2(\tau)d\tau$. Thus $\|\bar{e}\|$ must be bounded on I and $\|\bar{e}\|$ and $\|\phi\bar{e}\|$ must be in $L^2[I]$. The latter together with (10d), therefore show that $\|z\|$ is also bounded on I. ∇

The following proposition deals with the remaining components of $X_e$ for which boundedness on I is to be established.

Proposition 2: *$\|\bar{x}\|$, $\|\phi\|$ and $\|\sigma\|$ are bounded on I.*

To prove this proposition, we shall need

Lemma 1: *Let $R = \bar{c}e^{\bar{A}t}\bar{b}$. Then*

$$\phi = R*(k'\phi) \qquad (\mu) \tag{18}$$

*where * denotes convolution product.*

Proof: From (15),

$$\theta = R*(k'\theta + v) \qquad (\mu) \tag{19}$$

If $n^* = 1$, $v = 0$ and $\phi = \theta$ so (18) follows from (19).

Suppose $n^* = 2$, in which case (9) implies $\phi = e^{-\lambda_1 t}*\theta\ (\mu)$.

Thus using (19) there follows

$$\begin{aligned}\phi &= e^{-\lambda_1 t}*(R*(k'\theta + v)) \qquad (\mu)\\ &= R*(e^{-\lambda_1 t}*(k'\theta + v))\end{aligned} \tag{20}$$

But from (9),(13), and (14)

$$\begin{aligned}\frac{d(\phi'k)}{dt} &= -\lambda_1\phi'k + \theta'k + \phi'\dot{k}\\ &= -\lambda_1\phi'k + \theta'k + v\end{aligned}$$

so $\phi'k = e^{-\lambda_1 t}*(\theta'k + v)$ $(\mu)$. Substitution in (20) yields (18) which is the desired result. ∇

Proof of Proposition 2: Solving (16) for $k'\phi$, there results

$$k'\phi = \frac{1}{g_p}(\dot{\bar{e}} + (x^2 + \lambda_0)\bar{e}) - \varepsilon + \phi'\phi\frac{\bar{e}}{g_p}$$

Since $x^2$, $\bar{e}$ and $\varepsilon$ are bounded on I, and since R is the weighting pattern of a strictly proper system, $R*(\frac{1}{g_p}(\dot{\bar{e}} + (x^2 + \lambda_0)\bar{e}) - \varepsilon)$ is bounded on I. Thus $R*k'\phi = R*(\phi'\phi\frac{\bar{e}}{g_p})$ $(\mu)$. Hence from Lemma 1,

$$\phi = R*(\phi'\phi\frac{\bar{e}}{g_p}) \qquad (\mu)$$

Lemma 5 of [2] asserts that there must exist constants $c_1$ and $c_2$, not depending on $t_1$, such that

$$\|\phi(t)\|^2 \leq c_1 + c_2\int_0^t (\phi'\phi)^2\bar{e}^2 d\tau \quad , \qquad t \in I$$

$$= c_1 + c_2\int_0^t \|\phi\bar{e}\|^2\|\phi\|^2 d\tau \ , \qquad t \in I$$

Hence, by the Bellman-Gronwall Lemma

$$\|\phi(t)\|^2 \leq c_1 e^{c_2\int_0^t \|\phi\bar{e}\|^2 d\tau} \qquad t \in I$$

Since Proposition 1 states that $\|\phi\bar{e}\| \in L^2[I]$, it follows that $\|\phi\|$ is bounded on I.

If $n^* = 1$, $\phi = \theta$ so $\|\theta\|$ is bounded on I. Since $\|k\|$ is also bounded on I and in this case $v = 0$, it follows from (15) that $\|\bar{x}\|$ is bound on I.

If $n^* = 2$, $\theta = \dot{\phi} + \lambda_1\phi$, so $\theta'k = \dot{\phi}'k + \lambda_1\phi'k = d(\phi'k)/dt + \lambda_1\phi'k - \phi'\dot{k}$. Since $\phi'k$ is bounded on I and R is the weighting pattern of a strictly proper system, $R*(d(\phi'k)/dt + \lambda_1\phi'k)$ is bounded on I. Hence $R*\theta'k = -R*(\phi'\dot{k})$ $(\mu)$; but $v = \phi'\dot{k}$ so $R*(\theta'k + v)$ is bounded on I. Therefore from (15), $\|\bar{x}\|$ is bounded on I for $n^* = 2$.

Boundedness of $\sigma$ on I now follows from (10b), together with the boundedness of $\bar{e}$, x and $\phi$. $\nabla$

Thus far it has been established, by means of Propositions 1 and 2 that state response $X_e$ is bounded on I. As noted previously, we can conclude that $X_e$ is bounded on $[0,\infty)$. It follows from Proposition 1 that $\bar{e}$ is in $L^2[0,\infty]$. In addition, from (16) it can be seen that $\dot{\bar{e}}$ is bound on $[0,\infty)$. Therefore $\bar{e}$ must go to zero as $t \to \infty$. Furthermore, (10b) shows that $\sigma$ is the output of a strictly stable system with input $(x^2 + \phi'\phi)\bar{e}$, which in turn is in $L^2[0,\infty]$;

clearly $\sigma$ must go to zero as well.

The preceding thus constitutes a proof of Theorem 1.

Concluding Remarks

The essential difference between the stabilizing adaptive controller of [1] and the controller considered here is that the former utilizes a parameter adjustment law of the form $\dot{k} = -\,\mathrm{sign}(g_p)\phi\bar{e}$, whereas the latter uses (12). It is primarily because of this difference {i.e., the use of N(x) instead of $-\,\mathrm{sign}\,(g_p)$} that the present controller can stabilize independent of the sign of $g_p$. Other possible choices for N(x) {e.g., $N(x) = x^2 \cos(x)$} also result in stabilizing controllers. What seems to be important about such functions is that they oscillate in sign and grow in magnitude with increasing $|x|$.

In independent research, Willems and Byrnes [5] have recently proposed an alternative controller, capable of stabilizing systems of relative degree one, without the need of assumption iv. Their controller also uses a Nussbaum gain, but in a way which appears to be quite different from the way in which it is used here. It would be useful to obtain a clearer understanding of the similarities and differences between the two structures.

The present controllers are applicable only to systems of relative degrees one and two. It is however possible to achieve stability for arbitrary $n^* \geq 1$ using a modified version of the controller of [6]. Results along these lines will appear elsewhere.

Finally, we remark that since system robustness has not been addressed, the results obtained here should be viewed only as a first step toward the development of a practical adaptive controller for regulating systems without knowledge of sign $(g_p)$. Since the present controller contains a state variable, namely z, which cannot be driven to zero, even with a probing signal, robustness is likely to be a particularly interesting issue.

References

[1] A. S. Morse, "Global Stability of Parameter-Adaptive Control Systems," IEEE Trans. Auto. Control, AC-25(3), June 1980, pp. 433-439.

[2] K. S. Narendra, Y. H. Lin, L. S. Valavani, "Stable Adaptive Controller Design, Part II: Proof of Stability," ibid, pp. 440-448.

[3] A. S. Morse, "Recent Problems in Parameter Adaptive Control," Proc. CNRS Colloquium on Development and Utilization of Mathematical Models in Automatic Control," Belle-Isle, France, Sept. 1982.

[4] R. D. Nussbaum, "Some Remarks on a Conjecture in Parameter Adaptive Control," Systems and Control Letters, to appear.

[5] J. C. Willems and C. I. Byrnes, "Global Adaptive Stabilization in the Absence of Information on the Sign of the Instantaneous Gain," Sixth Int. Conf. on the Analysis and Optimization of Systems, Nice, June 1984, to appear.

[6] A. Feuer and A. S. Morse, "Adaptive Control of a Single-Input, Single-Output Linear System," IEEE Trans. Auto. Control, AC-23, Aug. 1978, pp. 557-569.

Acknowledgment

The author would like to thank David Mudgett for useful discussion contributing to this work.

# The Stabilization of Single Input Uncertain Linear Systems Via Linear Control†

by

Ian R. Petersen
Formerly at the Department of Electrical Engineering, University of Rochester, Rochester, NY, 14627, USA.
Now at the Department of Systems Engineering, Australian National University, Canberra, ACT, 2601, Australia.
Telephone (062)492461.

B. Ross Barmish
Department of Electrical Engineering, University of Rochester, Rochester, NY, 14627, USA. Telephone (716)2755930.

ABSTRACT: This paper investigates the problem of stabilizing a single input uncertain linear system using linear state feedback control. The uncertain system is described by a linear state equation which contains uncertain parameters which are unknown but bounded. A quadratic Lyapunov function is used to establish the stability of the closed loop system.

A number of papers have appeared in recent years in which the desired stabilization of the system is achieved using nonlinear state feedback control. The main result of this paper demonstrates that for a class of single input uncertain linear systems, linear state feedback control can equally well be used to achieve this stabilization. That is, we describe a class of single input uncertain linear systems which have the following property: If the system can be stabilized using nonlinear state feedback control, then it can also be stabilized using linear state feedback control.

## I. INTRODUCTION

In the design of a stabilizing feedback controller, one is often faced with the problem that no accurate model of the plant is available. One approach to this problem is to view the plant as an uncertain system containing unknown but bounded parameters; e.g., see [1]-[11] and [13]. Consequently, a quadratic Lyapunov function is exploited to establish the stability of the resulting closed loop uncertain system; e.g. see [1]-[11]. The use of a quadratic Lyapunov function motivates the formal definition of "quadratic stabilizability" given in Section II; see also [8].

†This work was supported by the National Science Foundation under Grant Number ECS-8108804.

In [1]–[3], [8] and [10], the stabilization of an uncertain linear system is achieved through the use of nonlinear state feedback control. However, from the point of view of reliability and ease of implementation, it is often desirable to use linear state feedback control. This motivates consideration of the following question: If an uncertain linear system is stabilizable via nonlinear control, does it follow that the system is also stabilizable via linear control? It has recently been shown that if we restrict attention to quadratic Lyapunov functions, then the answer to this question is negative. In fact, there exists a counterexample which demonstrates the following fact: Quadratic stabilizability via nonlinear control does not imply quadratic stabilizability via linear control; see [9]. Given this fact, we are motivated to look at additional assumptions which can be made so that this implication will hold. It has been shown in [5]-[7] that if we consider uncertain linear systems for which the input matrix contains no uncertain parameters, then quadratic stabilizability implies quadratic stabilizability via linear control.

The results of this paper apply to a class of uncertain linear systems which is in some sense complementary to the class of systems considered in [5]-[7]. In this paper, we consider a single input uncertain linear system in which the system matrix $A(\cdot)$ contains no uncertain parameters. The input connection vector $b(\cdot)$, however, is uncertain and a euclidean ball is used to bound the unknown parameters. The main result of this paper is the fact that for systems of this type, quadratic stabilizability implies quadratic stabilizability via linear control.

## II. SYSTEM, ASSUMPTIONS AND DEFINITIONS

We consider a class of single input uncertain linear systems described by a state equation of the form

$$\dot{x}(t) = Ax(t) + (b_0 + \sum_{i=1}^{\ell} b_i s_i(t))u(t);$$

$$s(t) \in S \triangleq \{s \in R^{\ell}: \| s \| \leq \bar{s}\} = B(\bar{s}) \qquad (\Sigma)$$

where $x(t) \in R^n$ is the state; $u(t) \in R$ is the control; $b_i \in R^n$ for $i = 0,1,\ldots,\ell$; $s(t) \in R^{\ell}$ is the vector of input connection uncertainty parameters and $s_i(t)$ denotes the $i^{th}$ component of the vector $s(t)$ and $\bar{s} > 0$ is a prescribed uncertainty bound. In the expression defining the set† $S$, $\| \cdot \|$ refers to the standard euclidean norm on $R^{\ell}$. It is assumed that the uncertain vector function $s(\cdot): R \to S$ is Lebesgue measurable. The two definitions to follow are also given in [11] but we include them here for the sake of completeness.

†A discussion concerning the special form of the set $S$ can be found in the remark following the proof of Theorem 3.1.

Definition 1: The system ($\Sigma$) is said to be quadratically stabilizable if there exists a continuous feedback control $p(\cdot)$: $R^n \to R$ with $p(0) = 0$, an nxn positive-definite symmetric matrix P and a constant $\alpha > 0$ such that the following condition is satisfied: Given any admissible uncertainty $s(\cdot)$, the Lyapunov derivative corresponding to the closed loop system (with control $u(t) = p(x(t))$ and the Lyapunov function $V(x) = x'Px$ satisfies the inequality

$$L(x,t) \triangleq x'[A'P + PA]x + 2x'P(b_0 + \sum_{i=1}^{\ell} b_i s_i(t))p(x)$$

$$\leq -\alpha||x||^2 \tag{2.1}$$

for all pairs $(x,t) \varepsilon R^n x R$.

Remark: If inequality (2.1) is satisfied, then it can be proven that given any admissible uncertainty $s(\cdot)$, the corresponding closed loop system has $x = 0$ as an asymptotically stable equilibrium point; e.g., see [1]-[3].

Definition 2: The system ($\Sigma$) is said to be quadratically stabilizable via linear control if the system ($\Sigma$) is quadratically stabilizable and furthermore, there exists a linear control function $p(\cdot)$ which satisfies Definition 1. That is, $p(\cdot)$ can be written in the form $p(x) = kx$ where k is a constant 1xn vector.

In order to present a condition which is necessary for quadratic stabilizability, we require a preliminary definition.

Definition 3; The function $\lambda(\cdot)$: $R^n x R^{nxn} \to R$ and the set $N \subset R^n$ are defined as follows:

$$\lambda(\eta,S) \triangleq \eta'[AS + SA']\eta;$$

$$\tilde{N} \triangleq \{\eta \varepsilon R^n: (b_0 + \sum_{i=1}^{\ell} b_i s_i)'\eta = 0 \text{ for some } s \varepsilon \mathcal{B}(\bar{s})\}.$$

Lemma 2.1: The following condition is necessary for the system ($\Sigma$) to be quadratically stabilizable: There exists a positive-definite symmetric matrix S such that

$$\lambda(\eta,S) < 0 \tag{2.2}$$

for all non-zero vectors $\eta \varepsilon \tilde{N}$:

Proof: Suppose that inequality (2.1) is satisfied with positive-definite symmetric matrix P and let $\eta \varepsilon R^n$ be any non-zero vector in $\tilde{N}$. Letting $S = P^{-1}$ and $x = S\eta$, it follows that

$$(b_0 + \sum_{i=1}^{\ell} b_i s_i)'Px = 0$$

for some $s\varepsilon S$. Hence, inequality (2.1) implies

$$x'[A'P + PA]x \leq -\alpha||x||^2.$$

That is,

$$\eta'[AS + SA']\eta \leq -\alpha||S\eta||^2 < 0.$$

However, the non-zero vector $\eta\varepsilon\tilde{N}$ was arbitrary. Consequently,

$$\lambda(\eta,S) < 0$$

for all non-zero vectors $\eta\varepsilon\tilde{N}$. □

Remark: It has been shown in [8] that the condition presented in the lemma above is both necessary and sufficient for quadratic stabilizability. However, in this paper, we only require the necessity result.

## III. THE EXISTENCE OF A LINEAR STABILIZING CONTROL

The following theorem establishes the equivalence between quadratic stabilizability of $(\Sigma)$ and quadratic stabilizability of $(\Sigma)$ via linear control.

Theorem 3.1: The system $(\Sigma)$ is quadratically stabilizable if and only if it is quadratically stabilizable via linear control.

Before proceeding to prove this theorem, we present some additional notation and preliminary lemmas.

Definition 4: The functions $\Delta(\cdot): R^n \to R$ and $\delta(\cdot): RxR^{\ell} \to R$ are defined as follows:

$$\Delta(\eta) \triangleq |b_0'\eta| - \bar{s}(\sum_{i=1}^{\ell} (b_i'\eta)^2)^{\frac{1}{2}};$$

$$\delta(\alpha,a) \triangleq |\alpha| - ||a||.$$

Furthermore, given any nxn matrix S, the function $\theta_S(\cdot): R^n \to R$ is defined by

$$\theta_S(\eta) \triangleq \frac{\lambda(\eta,S)}{2|b_0'\eta|\Delta(\eta)} .$$

In the following lemmas, we establish some useful properties of the functions $\Delta(\cdot)$, $\delta(\cdot)$ and $\theta_S(\cdot)$.

Lemma 3.1: $\tilde{N} = \{\eta\varepsilon R^n: \Delta(\cdot) \leq 0\}$.

Proof: Let $\eta\varepsilon R^n$ be a vector such that $\Delta(\eta) \leq 0$. Taking $a \triangleq [b_1'\eta \; b_2'\eta \; \ldots \; b_\ell'\eta]'$, it follows that

$$|b_0'\eta| \leq \bar{s}||a||. \tag{3.1}$$

We consider two cases.

Case 1: $a=0$. In this case, $b_0'\eta = 0$. It follows immediately that $\eta\varepsilon\tilde{N}$.

Case 2: $a\neq 0$. In this case, let

$$s \triangleq -\frac{ab_0'\eta}{||a||^2}$$

Inequality (3.1) implies that $s\varepsilon\mathcal{B}(\bar{s}) = \mathcal{S}$. Furthermore,

$$b_0'\eta = -s'a = \sum_{i=1}^{\ell} b_i'\eta s_i$$

That is,

$$(b_0 + \sum_{i=1}^{\ell} b_i s_i)'\eta = 0$$

which implies that $\eta\varepsilon\tilde{N}$.

Conversely, suppose $\eta\varepsilon\tilde{N}$ is given. Letting $a \triangleq [b_1'\eta \; b_2'\eta \; \ldots \; b_\ell'\eta]'$, the definition of $\tilde{N}$ implies that there exists a vector $s\varepsilon\mathcal{B}(\bar{s})$ such that $b_0'\eta = s'a$. Therefore,

$$|b_0'\eta| \leq ||s||\cdot||a|| \leq \bar{s}(\sum_{i=1}^{\ell} (b_i'\eta)^2)^{\frac{1}{2}}.$$

That is, $\Delta(\eta) \leq 0$. □

Lemma 3.2: Given any positive-definite symmetric nxn matrix S, the function $\theta_S(\cdot)$ is continuous on the set $\{\eta\varepsilon R^n: b_0'\eta\neq 0;\ \Delta(\eta) \neq 0\}$.

Proof: We observe that the functions $\Delta(\cdot)$ and $\lambda(\cdot,S)$ are both continuous. Hence, the function $\theta_S(\cdot)$ is continuous at all points $\bar{\eta}$ such that $\Delta(\bar{\eta}) \neq 0$ and $b_0'\eta \neq 0$. □

The somewhat technical proof of the lemma below can be found in [9].

Lemma 3.3 (See [9] for proof): Suppose that the pairs $(\alpha,a)\epsilon R\times R^{\ell}$ and $(\beta,b)\epsilon R\times R^{\ell}$ are such that $\delta(\alpha,a) < 0$, $\delta(\beta,b) > 0$ and $\delta(\alpha + \beta,a + b) = 0$. Then

$$\delta(\alpha|\beta|\delta(\beta,b) + \beta|\alpha|\delta(\alpha,a),a|\beta|\delta(\beta,b) + b|\alpha|\delta(\alpha,a)) \leq 0.$$

Lemma 3.4: Suppose that the vectors $\eta_1$, $\eta_2 \epsilon R^n$ are such that $\Delta(\eta_1) < 0$, $\Delta(\eta_2) > 0$ and $\Delta(\eta_1 + \eta_2) = 0$. Then

$$\Delta(|b_0'\eta_2|\Delta(\eta_2)\eta_1 + b_0'\eta_1|\Delta(\eta_1)\eta_2) \leq 0.$$

Proof: Given $\eta_1$ and $\eta_2$ as above, let $\alpha \triangleq b_0'\eta_1$, $\beta \triangleq b_0'\eta_2$, $a \triangleq [\bar{s}b_1'\eta_1\ \bar{s}b_2'\eta_1 \ldots \bar{s}b_{\ell}'\eta_1]'$ and $b \triangleq [\bar{s}b_1'\eta_2\ \bar{s}b_2'\eta_2 \ldots \bar{s}b_{\ell}'\eta_2]'$. Hence, $\delta(\alpha,a) = \Delta(\eta_1) < 0$, $\delta(\beta,b) = \Delta(\eta_2) > 0$ and $\delta(\alpha + \beta,a + b) = \Delta(\eta_1 + \eta_2) = 0$. Using Lemma 3.3, we conclude that

$$\Delta(|b_0'\eta_2|\Delta(\eta_2)\eta_1 + |b_0'\eta_1|\Delta(\eta_1)\eta_2) \leq 0. \quad \square$$

Lemma 3.5: Suppose that there exists a positive-definite symmetric matrix S such that

$$\lambda(\eta,S) < 0. \tag{3.2}$$

for all non-zero vectors $\eta\epsilon\tilde{N}$. Furthermore, assume that the set $X \triangleq \{\eta\epsilon R^n: b_0'\eta \neq 0, \Delta(\eta) < 0, ||\eta|| = 1\}$ is non-empty. Then there exists a vector $\hat{\eta}\epsilon X$ such that

$$\theta_S(\hat{\eta}) = \inf\{\theta_S(\eta): \eta\epsilon X\}.$$

Proof: Let $\eta^*$ be any point contained in X. It follows from Lemma 3.1 that $\eta^*\epsilon\tilde{N}$. Hence, we conclude that $\lambda(\eta^*,S) < 0$. Therefore, $\theta_S(\eta^*) > 0$. Let the non-empty set $X_0$ be defined by

$$X_0 \triangleq \{\eta\epsilon R^n: b_0'\eta \neq 0, \Delta(\eta) < 0, \theta_S(\eta) \leq \theta_S(\eta^*), ||\eta|| = 1\}.$$

Claim: The set $X_0$ is compact.

To establish this claim, we must show that the set $X_0$ is closed and bounded. The boundedness of this set is immediate. To establish closedness, let $\langle\eta_k\rangle_{k=1}^{\infty}$ be a sequence of points in $X_0$ such that $\eta_k \to \bar{\eta}$. We must show that $\bar{\eta}\epsilon X_0$. To this end, let

$$\zeta \triangleq -\max\{\lambda(\eta,S): \Delta(\eta) \leq 0, \|\eta\| = 1\}.$$

It follows from Lemma 3.1 and inequality (3.2) that $\zeta > 0$. Furthermore, for each integer k, we have

$$\lambda(\eta_k,S) > 2\theta_S(\eta^*)|b_0'\eta_k|\Delta(\eta_k).$$

Hence, for each k,

$$|b_0'\eta_k|\Delta(\eta_k) \leq \frac{-\zeta}{2\theta_S(\eta^*)}$$

If we let $k \to \infty$ in the above inequality, we conclude that

$$|b_0'\bar{\eta}|\Delta(\bar{\eta}) < 0.$$

Consequently, $b_0'\bar{\eta} \neq 0$ and $\Delta(\bar{\eta}) < 0$. Furthermore, using Lemma 3.2 and the fact that $\theta_S(\eta_k) \leq \theta_S(\eta^*)$ for all k, it follows that $\theta_S(\bar{\eta}) \leq \theta_S(\eta^*)$. Also, $\|\eta_k\| = 1$ for all k and hence $\|\bar{\eta}\| = 1$. Therefore, $\bar{\eta}\varepsilon X_0$ which completes the proof of the claim.

To complete the proof of the lemma, we observe that

$$\inf\{\theta_S(\eta): \eta\varepsilon X\} = \inf\{\theta_S(\eta): \eta\varepsilon X_0\}.$$

However, the function $\theta_S(\cdot)$ is continuous on the set $X_0$ and this set is compact. Therefore, there exists a point $\hat{\eta}\varepsilon X_0 \subset X$ such that

$$\theta_S(\hat{\eta}) = \inf\{\theta_S(\eta): \eta\varepsilon X_0\};$$

e.g., see page 89 of [12]. This completes the proof of the lemma. □

Lemma 3.6: *Suppose that there exists a positive-definite symmetric matrix S such that*

$$\lambda(\eta,S) < 0 \tag{3.3}$$

*for all non-zero vectors* $\eta\varepsilon\tilde{N}$. *Furthermore, assume that there exists a vector* $\eta^*\varepsilon R^n$ *such that* $\|\eta^*\| = 1$ and $\lambda(\eta^*,S) \geq 0$. *Then there exists a vector* $\hat{\eta}\varepsilon R^n$ *such that*

$$\theta_S(\hat{\eta}) = \sup\{\theta_S(\eta): \eta\varepsilon\tilde{N}^c \text{ and } \|\eta\| = 1\} \geq 0.$$

*where* $\tilde{N}^c$ *denotes the complement of* $\tilde{N}$.

Proof: Let the non-empty set $V$ be defined by

$$V \triangleq \{\eta \varepsilon R^n: \lambda(\eta,S) \geq 0 \text{ and } ||\eta|| = 1\}.$$

We observe that $V$ is compact since $\lambda(\cdot,S)$ is continuous. Moreover, using inequality (3.3), it follows that $V \subset \tilde{N}^c$. Using the definition of $\theta_S(\cdot)$ and Lemma 3.1, it is clear that $\theta_S(\eta) < 0$ for all $\eta \varepsilon N^c \cap V^c$ such that $||\eta|| = 1$. Therefore,

$$\sup\{\theta_S(\eta): \eta \varepsilon \tilde{N}^c, ||\eta|| = 1\} = \sup\{\theta_S(\eta): \eta \varepsilon V\} \geq 0.$$

Now, using the fact that a continuous function on a compact set attains its supremum (see, for example, page 89 of [12]), there must be a vector $\hat{\eta} \varepsilon V$ such that

$$\theta_S(\hat{\eta}) = \sup\{\theta_S(\eta): \eta \varepsilon V\}.$$

This completes the proof of the lemma. □

Proof of Theorem 3.1: The fact that quadratic stabilizability via linear control implies quadratic stabilizability is a trivial consequence of the definitions. It is the converse of this statement which constitutes the main result of this paper. Indeed, suppose that the system $(\Sigma)$ is quadratically stabilizable. Then using Lemma 2.1, it follows that there exists an nxn positive-definite symmetric matrix S such that

$$\lambda(\eta,S) < 0 \tag{3.4}$$

for all non-zero vectors $\eta \varepsilon \tilde{N}$.

We will show that the system $(\Sigma)$ is quadratically stabilizable via linear control of the form

$$p(x) = -\gamma b_0' Px; \quad \gamma \geq 0. \tag{3.5}$$

The quadratic Lyapunov function which will be used is $V(x) = x'Px$. The construction of the constant $\gamma \geq 0$ is accomplished by considering two cases.

Case 1: $\lambda(\eta,S) < 0$ for all $\eta \varepsilon R^n$ such that $||\eta|| = 1$. In this trivial case, the system is quadratically stabilizable via the linear control $p(x) \equiv 0$. Hence, we set $\gamma = 0$.

Case 2: There exists a point $\eta^* \varepsilon R^n$ such that $||\eta^*|| = 1$ and $\lambda(\eta^*,S) \geq 0$. In order to construct $\gamma$ for this case, we first establish a claim.

Claim:

$$\sup\{\theta_S(\eta): \eta\varepsilon\tilde{N}^c, \|\eta\| = 1\}$$

$$< \inf\{\theta_S(\eta): b_0'\eta \neq 0, \Delta(\eta) < 0, \|\eta\| = 1\}.$$

We first observe that if the set $\{\eta\varepsilon R^n: b_0'\eta \neq 0, \Delta(\eta) < 0, \|\eta\| = 1\}$ is empty, then the claim holds automatically. Otherwise, we use Lemmas 3.5 and 3.6 to conclude that this inequality is equivalent to the inequality

$$\max\{\theta_S(\eta): \eta\varepsilon\tilde{N}^c, \|\eta\| = 1\}$$

$$< \min\{\theta_S(\eta): b_0'\eta \neq 0, \Delta(\eta) < 0, \|\eta\| = 1\}. \tag{3.6}$$

We establish the claim by contradiction. Suppose that (3.6) does not hold. Then there exists points $\tilde{\eta}_1, \tilde{\eta}_2 \ \varepsilon R^n$ such that

$$\tilde{\eta}_1\varepsilon\{\eta\varepsilon R^n: b_0'\eta \neq 0, \Delta(\eta) < 0, \|\eta\| = 1\},$$

$$\tilde{\eta}_2\varepsilon\{\eta\varepsilon R^n: \eta\varepsilon\tilde{N}^c, \|\eta\| = 1\},$$

$$\theta_S(\tilde{\eta}_1) = \min\{\theta_S(\eta): b_0'\eta \neq 0, \Delta(\eta) < 0, \|\eta\| = 1\}$$

and

$$\theta_S(\tilde{\eta}_1) \leq \theta_S(\tilde{\eta}_2). \tag{3.7}$$

Using Lemma 3.1, it is apparent that $\Delta(\tilde{\eta}_2) > 0$. We now define a continuous function $h(\cdot)$: $R \to R$ as follows:

$$h(\lambda) \triangleq \Delta(\lambda\tilde{\eta}_1 + (1-\lambda)\tilde{\eta}_2).$$

Since $h(0) = \Delta(\tilde{\eta}_2) > 0$ and $h(1) = \Delta(\tilde{\eta}_1) < 0$, it follows from the Intermediate Value Theorem (see, for example, page 93 of [12]), that there exists a point $\lambda^*\varepsilon(0,1)$ such that

$$h(\lambda^*) = \Delta(\lambda^*\tilde{\eta}_1 + (1-\lambda^*)\tilde{\eta}_2) = 0.$$

Letting $\eta_1 = \lambda^*\tilde{\eta}_1$ and $\eta_2 = (1-\lambda^*)\tilde{\eta}_2$, we have

$$\Delta(\eta_1 + \eta_2) = 0. \tag{3.8}$$

Furthermore, from the definition of $\Delta(\cdot)$,

$$\Delta(\eta_1) = \lambda^*\Delta(\tilde{\eta}_1) < 0 \tag{3.9}$$

and

$$\Delta(\eta_2) = (1-\lambda^*)\Delta(\tilde{\eta}_2) > 0. \tag{3.10}$$

Given (3.8)-(3.10) in conjunction with Lemma 3.4, it follows that

$$\Delta(|b_0'\eta_2|\Delta(\eta_2)\eta_1 + |b_0'\eta_1|\Delta(\eta_1)\eta_2) \leq 0.$$

In view of Lemma 3.1, we conclude that the vector

$$\hat{\eta} \triangleq |b_0'\eta_2|\Delta(\eta_2)\eta_1 + |b_0'\eta_1|\Delta(\eta_1)\eta_2$$

belongs to the set $\tilde{N}$.

In order to complete the proof of the claim, we now exploit $\hat{\eta}\varepsilon\tilde{N}$ to obtain a contradiction to inequality (3.7). Since

$$\theta_S(\lambda\eta) = \theta_S(\eta)$$

for all $\lambda > 0$ and all $\eta\varepsilon R^n$, inequality (3.7) implies that

$$\theta_S(\eta_1) \leq \theta_S(\eta_2).$$

Using the definition of $\theta_S(\cdot)$ and inequalities (3.9) and (3.10), we obtain

$$|b_0'\eta_1|\Delta(\eta_1)\lambda(\eta_2,S) \leq |b_0'\eta_2|\Delta(\eta_2)\lambda(\eta_1,S). \tag{3.11}$$

For convenience, we re-write this inequality as

$$\xi_2\eta_1'M\eta_1 + \xi_1\eta_2'M\eta_2 \geq 0 \tag{3.12}$$

where $\xi_1$ and $\xi_2$ are the strictly positive constants

$$\xi_1 \triangleq -|b_0'\eta_1|\Delta(\eta_1),$$

$$\xi_2 \triangleq |b_0'\eta_2|\Delta(\eta_2)$$

and

$M \stackrel{\Delta}{=} AS + SA'$.

Considering equation (3.8) and Lemma 3.1, we note $(\eta_1 + \eta_2)\varepsilon N$. Hence, inequality (3.4) requires that

$$\lambda(\eta_1 + \eta_2, S) < 0.$$

Furthermore, using inequality (3.12),

$$0 > \lambda(\eta_1 + \eta_2, S)$$

$$= (\eta_1 + \eta_2)'M(\eta_1 + \eta_2)$$

$$\geq \eta_1'M\eta_1 + 2\eta_1'M\eta_2 + \eta_2'M\eta_2$$

$$- \frac{\xi_1 + \xi_2}{\xi_1\xi_2}(\xi_2\eta_1'M\eta_1 + \xi_1\eta_2'M\eta_2)$$

$$= \frac{-\xi_2}{\xi_1}\eta_1'M\eta_1 + 2\eta_1'M\eta_2 - \frac{\xi_1}{\xi_2}\eta_2'M\eta_2$$

$$= -\frac{1}{\xi_1\xi_2}(\eta_1\xi_2 - \eta_2\xi_1)'M(\eta_1\xi_2 - \eta_2\xi_1)$$

$$= -\frac{1}{\xi_1\xi_2}\lambda(\eta_1\xi_2 - \eta_2\xi_1, S)$$

$$= -\frac{1}{\xi_1\xi_2}\lambda(\hat{\eta}, S).$$

That is,

$$\lambda(\hat{\eta}, S) > 0. \tag{3.13}$$

This contradicts inequality (3.4). The claim is now established.

To complete the proof of the theorem, we now prove that any real constant $\gamma$ satisfying

$$\sup\{\theta_S(\eta): \eta\varepsilon N^c, \|\eta\| = 1\}$$

$$< \gamma < \inf\{\theta_S(\eta): b_0'\eta \neq 0, \Delta(\eta) < 0\} \tag{3.14}$$

will suffice for the controller in (3.5). Indeed, given any admissible uncertainty $s(\cdot)$ and any pair $(x,t)\varepsilon R^n x[0,\infty)$, the Lyapunov derivative is bounded as follows:

$$L(x,t) = x'[A'P + PA]x - 2\gamma x'P(b_0 + \sum_{i=1}^{\ell} b_i s_i(t))b_0'Px$$

$$= x'[A'P + PA]x - 2\gamma(b_0'Px)^2$$

$$- 2\gamma(b_0'Px) \sum_{i=1}^{\ell} (b_i'Px)s_i(t)$$

$$\leq x'[A'P + PA]x - 2\gamma(b_0'Px)^2$$

$$+ 2\gamma\bar{s}|b_0'Px| (\sum_{i=1}^{\ell} (b_i'Px)^2)^{\frac{1}{2}}$$

$$\triangleq L_{max}(x).$$

This function $L_{max}(\cdot)$ is referred to as the maximal Lyapunov derivative. The next step of the proof will be to show that $L_{max}(x) < 0$ for all non-zero $x\varepsilon R^n$. Given the positive homogeneity[†] of $L_{max}(\cdot)$ and the non-singularity of S, it suffices to prove that

$$L_{max}(S\eta) < 0$$

for all $\eta\varepsilon R^n$ such that $||\eta|| = 1$. Indeed, for any $\eta\varepsilon R^n$,

$$L_{max}(S\eta) = \eta'[AS + SA']\eta$$

$$- 2\gamma|b_0'\eta|[|b_0'\eta| - \bar{s}(\sum_{i=1}^{\ell} (b_i'\eta)^2)^{\frac{1}{2}}]$$

$$= \lambda(\eta,S) - 2\gamma|b_0'\eta|\Delta(\eta).$$

We now consider four cases.

Case A: $\Delta(\eta) = 0$; $||\eta|| = 1$. In this case, $\eta\varepsilon N$ and using inequality (3.4),

$$L_{max}(S\eta) = \lambda(\eta,S) < 0.$$

Case B: $|b_0'\eta| = 0$; $||\eta|| = 1$. In this case,

$$\Delta(\eta) = -(\sum_{i=1}^{\ell} (b_i'\eta)^2)^{\frac{1}{2}} \leq 0.$$

Therefore, by inequality (3.4), $\eta\varepsilon\tilde{N}$ and $\lambda(\eta,S) < 0$. Consequently,

---

[†] $L_{max}(\lambda x) = \lambda^2 L_{max}(x)$ for all $x\varepsilon R^n$ and all $\lambda \geq 0$.

$L_{max}(S\eta) < 0.$

Case C: $\eta \varepsilon \tilde{N}^c$; $\|\eta\| = 1$. In this case, it follows from the construction of $\gamma$ that

$$\gamma > \theta_S(\eta) = \frac{\lambda(\eta,S)}{2|b_0'\eta|\Delta(\eta)}$$

Furthermore, $\Delta(\eta) > 0$ by Lemma 3.1. Therefore,

$$2\gamma|b_0'\eta|\Delta(\eta) > \lambda(\eta,S)$$

which implies that

$$L_{max}(S\eta) < 0.$$

Case D: $\Delta(\eta) < 0$; $|b_0'\eta| \neq 0$; $\|\eta\| = 1$. In this case, it follows from the construction of $\gamma$ that

$$\gamma < \theta_S(\eta) = \frac{\lambda(\eta,S)}{2|b_0'\eta|\Delta(\eta)}.$$

Therefore

$$2\gamma|b_0'\eta|\Delta(\eta) > \lambda(\eta,S)$$

which implies that

$$L_{max}(S\eta) < 0.$$

We have now established that $L_{max}(S\eta) < 0$ for all vectors $\eta \varepsilon R^n$ such that $\|\eta\| = 1$. Hence, $L_{max}(x) < 0$ for all non-zero vectors $x \varepsilon R^n$. Our next objective is to bound $L_{max}(\cdot)$ by a class K function of the form $c_0(\|x\|) = \alpha\|x\|^2$. Noting that the function $L_{max}(\cdot)$ is continuous, we define a constant $\alpha > 0$ by

$$\alpha \triangleq -\max\{L_{max}(x): \|x\| = 1\}.$$

Therefore†, given any admissible uncertainty $s(\cdot)$,

$$L(x,t) \leq L_{max}(x) \leq -\alpha\|x\|^2$$

for all pairs $(x,t) \varepsilon R^n x[0,\infty)$. Hence, the system $(\Sigma)$ is quadratically stabilizable via linear control. □

---

†We use the positive homogeneity property of $L_{max}(\cdot)$ described in the previous footnote.

Remarks: From the proof above it can be seen that if $V(x) = x'Px$ is used to establish the quadratic stabilizability of $(\Sigma)$, then this same Lyapunov function can be used when proving quadratic stabilizability via linear control. If, however, we relax the assumption that $\mathcal{S}$ is a euclidean ball then this property does not hold. Indeed, consider the linear uncertain system described by the state equations

$$\dot{x}_1(t) = -5x_1(t) + s_1(t)u(t);$$

$$\dot{x}_2(t) = x_2(t) + (4 + s_2(t))u(t);$$

$$\begin{bmatrix} s_1(t) \\ s_2(t) \end{bmatrix} \in \mathcal{S} \triangleq \operatorname{conv}\left\{ \begin{bmatrix} -16 \\ 4 \end{bmatrix}, \begin{bmatrix} 16 \\ 4 \end{bmatrix}, \begin{bmatrix} -2 \\ -3 \end{bmatrix}, \begin{bmatrix} 2 \\ -3 \end{bmatrix} \right\}.$$

It is straightforward to establish the quadratic stabilizability of this system using the Lyapunov function $V(x) = x_1^2 + x_2^2$. However, this Lyapunov function cannot be used to establish quadratic stabilizability via linear control.

It is of interest to note that in constructing the linear control, the gain constant $\gamma$ must satisfy both an upper and lower bound; see inequality (3.14). This indicates that when stabilizing systems containing uncertainty in the input matrix, a straightforward "high gain approach" may not succeed. Instead, the gain must in some sense be "tuned" to the amount of uncertainty in the system.

## REFERENCES


[1] G. Leitmann, "Guaranteed Asymptotic Stability for Some Linear Systems with Bounded Uncertainty," ASME Journal of Dynamical Systems, Measurement and Control, Vol. 101, no. 3, 1979.

[2] G. Leitmann, "On the Efficacy of Nonlinear Control in Uncertain Linear Systems," ASME Journal of Dynamical Systems, Measurement and Control, Vol. 103, no. 2, 1981.

[3] S. Gutman and Z. Palmor, "Properties of Min-Max Controllers in Uncertain Dynamical Systems," SIAM Journal on Control and Optimization, Vol. 20, no. 6, 1982.

[4] J.S. Thorp and B.R. Barmish, "On Guaranteed Stability of Uncertain Linear Systems via Linear Control," Journal of Optimization Theory and Applications, Vol. 35, no. 4, 1981.

[5] B.R. Barmish, I.R. Petersen and A. Feuer, "Linear Ultimate Boundedness Control of Uncertain Dynamical Systems," Automatica, Vol. 19, no. 5, 1983.

[6] A.M. Meilakhs, "Design of Stable Control Systems Subject to Parametric Perturbation," Automatika i Telemekhanika, no. 10, 1978.

[7] C.V. Hollot and B.R. Barmish, "Optimal Quadratic Stabilizability of Uncertain Linear Systems," Proceedings of the 18th Allerton Conference on Communication, Control and Computing, University of Illinois, Monticello, 1980.

[8] B.R. Barmish, "Necessary and Sufficient Conditions for Quadratic Stabilizability of an Uncertain Linear System," to appear in the Journal of Optimization Theory and Applications.

[9] I.R. Petersen, "Investigation of Control Structure in the Stabilization of Uncertain Dynamical Systems," PhD dissertation, Department of Electrical Engineering, University of Rochester, Rochester, New York, 1983.

[10] B.R. Barmish, "Fundamental Issues in Guidance and Control of Uncertain Systems," Proceedings of the American Control Conference, Arlington, Virginia, 1982.

[11] B.R. Barmish, "Stabilization of Uncertain Systems via Linear Control," Proceedings of the IEEE Asilomar Conference on Computers, Circuits and Systems, Monterey, 1981; see also IEEE Transactions on Automatic Control, Vol. AC-28, no. 8, 1983.

[12] W. Rudin, Principles of Mathematical Analysis, third edition, McGraw Hill, New York, 1976.

[13] J.L. Willems and J.C. Willems, "Robust Stabilization of Uncertain Systems," SIAM Journal on Control and Optimization, Vol. 21, no. 3, 1983.

Session 3

# UTILITY SYSTEMS

# RÉSEAUX DE SERVICE

# OPTIMISATION ET ACHEMINEMENT DYNAMIQUE DANS LES RESEAUX TELEPHONIQUES

J. Bernussou, F. Le Gall, J.M. Garcia
Laboratoire d'Automatique et d'Analyse des Systèmes du C.N.R.S.
7, avenue du Colonel Roche
31077 Toulouse Cédex

## I. Introduction

C'est un lieu commun que de dire que les progrès en matière technologique et notamment sur les équipements digitaux permettent la maîtrise de systèmes de complexité et dimension croissante. Pour le problème qui nous intéresse ici, du routage dynamique dans les réseaux à commutation de circuits l'introduction de moyens numériques tant en transmission qu'en commutation se généralise. C'est par exemple la mise en place d'autocommutateurs à programme enregistré qui remplacera les vieux équipements électro-mécaniques des centraux de commutation. C'est également la possibilité future de l'utilisation d'un réseau de transmission de données, dit réseau à canal sémaphore, qui pourrait permettre un échange d'information entre les différents centraux de commutation dans le réseau. Pouvant disposer d'organes numériques de traitement de l'information au niveau de chaque central de commutation et éventuellement d'un réseau de communication entre ces organes il est réaliste de se poser le problème de la détermination d'une politique de routage dynamique pour les appels dans les réseaux téléphoniques.
Par routage dynamique, il est entendu le principe de base de l'automatique : la commande temps réel capable de s'adapter et réagir aux aléas ou perturbations dont peut souffrir le système. Dans le cas présent (les réseaux téléphoniques) il s'agit essentiellement des variations au niveau de la demande et des perturbations structurelles du réseau (pannes de centraux, ruptures d'artères, etc...)

Dans les réseaux de communication on peut distinguer deux grands types de réseaux : les réseaux dits à commutation de paquets et ceux à commutation de circuits. Pour les premiers, il s'agit de véhiculer des messages d'un noeud origine à un noeud destination, l'acheminement se fait selon la technique dite du "store and forward". En chaque noeud les

appels sont stockés sur des files d'attente , attendent là jusqu'à leur envoi à un noeud successeur choisi en fonction de la destination finale. Pour arriver à destination, tout message subit un certain retard.
Pour les réseaux téléphoniques, réseaux à commutation de circuits, l'acheminement d'un appel consistera à trouver entre le noeud origine et destination un chemin tel que aucun de ses arcs constitutifs ne soit bloqué et à utiliser un circuit sur ce chemin durant toute la durée de conversation. Si un tel chemin libre n'est pas trouvé, l'appel est rejeté (réseau à pertes).

A côté de ces différences, si l'on s'intéresse à un niveau microscopique de description on voit que ces deux types de réseaux admettent une représentation de type systèmes discrets. Il s'agit en effet dans les deux cas d'acheminer ou traiter des évènements discrets caractérisés par un instant d'apparition et une durée de service. On peut, par exemple, sous quelques hypothèses classiques concernant les lois de distribution de ces phénomènes aléatoires (apparition, durée de service) - (en régime stationnaire) dériver des modèles sous forme chaînes de Markov. Une telle description est utilisable à des fins d'analyse hors ligne pour des réseaux de taille réduite et servir dans une optique d'évaluation de performances, détermination de politiques optimales stationnaires et au test sur des cellules simples de règles de routage fixées a priori de manière heuristique [1, 2, 3]. Cette approche, à laquelle a été donnée la terminologie de "processus de décision Markoviens" se révèle tout à fait inadaptée pour le cas de la commande dynamique de réseaux réels (difficultés dues à la dimension de modèle auxquelles viennent s'ajouter celles liées aux contraintes de décentralisation pour la commande).

Plus réalistes, sont les approches qui se basent sur l'utilisation de méthodes en programmation mathématique. Elles nécessitent toutefois de travailler sous l'hypothèse de quasi-stationnarité ou de variations quasi-statiques des trafics à acheminer. Vu la nature aléatoire de ces trafics, il s'agit alors de faire la synthèse d'une politique de commande capable de s'adapter aux variations du trafic en moyenne et non pas sur une base appel par appel, utilisateur par utilisateur. Ces hypothèses sont en fait suffisamment réalistes pour le cas des grands réseaux où, vu la multiplicité des utilisateurs, la détermination d'un comportement moyen représentatif est possible. Dans le cadre des réseaux à commutation de paquets ces idées générales ont été à la base des travaux de

Gallager et Bertsekas [4, 5], (voir aussi [6, 7]) qui ont proposé des structures de routage distribuées par résolution, au moyen d'algorithmes distribués, de problèmes d'optimisation de type multiflots, définis sur des intervalles de temps pendant lesquels il est possible de supposer que le trafic à écouler reste constant. Nous reprenons la même démarche pour les réseaux à commutation de circuits, en l'adaptant à ce problème particulier et en incluant l'utilisation en ligne de mesures sur le réseau afin d'aboutir à une politique de routage distribuée et à caractère "boucle fermée" plus affirmée que dans les travaux antérieurs sur la commutation de paquets.

Dans le paragraphe suivant nous présentons le problème du routage dans les réseaux à commutation de circuits : ses hypothèses, ses contraintes et ses caractéristiques structurelles. Vient ensuite un paragraphe qui rappelle certains résultats obtenus en modélisation : (modèle de la moyenne) ce qui permet de définir le problème d'optimisation associé à celui de l'acheminement des appels. Le paragraphe quatre propose quelques algorithmes pour la résolution du problème d'optimisation.

Il est enfin montré comment, à partir des conditions nécessaires d'optimalité du problème, arriver à une structure de commande distribuée intégrant calculs et mesures temps réel. Enfin dans un dernier paragraphe de conclusion, nous proposons, sans les développer, quelques idées pour la synthèse de structures distribuées de commande de type "multilayer" intégrant les actions quasi-statiques précédemment développées à des actions à caractère nettement plus temps réel.

## II. Le problème de l'acheminement des appels dans un réseau téléphonique

La plupart des réseaux téléphoniques présentent une structure hiérarchisée dans laquelle on peut discerner des réseaux locaux (peu maillés) qui sont chargés d'écouler le trafic local "courte distance" et, au dessus; un réseau interurbain ou interrégional qui a pour but de relier les réseaux locaux et d'acheminer le trafic longue distance. Il s'agit d'un réseau relativement maillé. C'est sur celui-ci que le problème de commande se pose tout d'abord.

Si l'on considère le réseau français, il est constitué de quatre niveaux hiérarchiques, les deux premiers constituant le réseau interurbain (fig. 1)

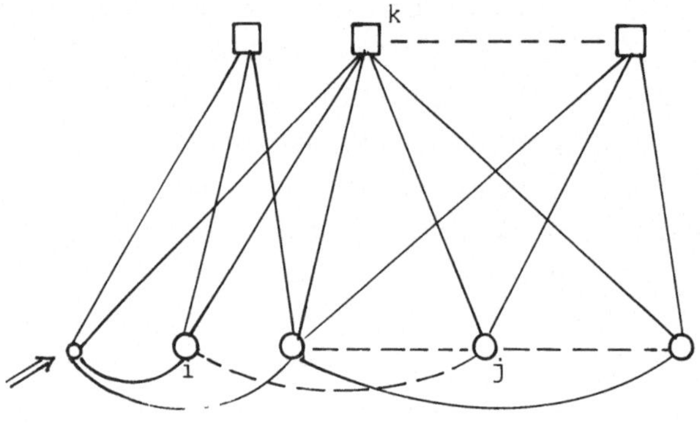


Figure 1 : le réseau interurbain

Il n'y a pas d'arrivée de trafic au niveau des CTP, tout le trafic "entre" et "sort" au niveau des CTS. Pour les réseaux à commutation de circuits, une règle qui s'impose à l'évidence pour l'acheminement des appels, consiste à interdire à tout appel un chemin trop long dans le réseau : la longueur d'un chemin étant définie comme le nombre de faisceaux (à arcs) qui constituent ce chemin. C'est cette règle qui est à l'origine de la structure du réseau ci-dessus évoqué et qui impose des contraintes pour l'acheminement des appels en restreignant le nombre de transits possibles dans le réseau.

La politique pour l'acheminement des appels à l'heure actuelle est une politique dite "à deux choix". Pour l'établissement d'une communication entre deux CTS (i.j.) une route "premier choix" est d'abord testée (c'est le faisceau direct entre i et j si celui-ci existe); si elle est bloquée, une autre route est alors choisie passant par un centre de transit CTP (k). Il ne semble pas qu'il soit avantageux de multiplier le nombre de choix, c'est la raison pour laquelle la contrainte "politique à deux choix" est retenue même pour le cas d'une commande dynamique. L'hypothèse de faisceaux unidirectionnels, ce qui est le cas actuel, est retenue. Ces remarques et hypothèses font que l'on peut adopter une représentation éclatée et simplifiée du réseau interurbain en ne prenant que les faisceaux sur lesquels est appliquée une "vraie" commande (omettant par exemple les liaisons directes inter CTP) et en considérant des CTS origine et des CTS destination qui en réalité sont physiquement les mêmes.

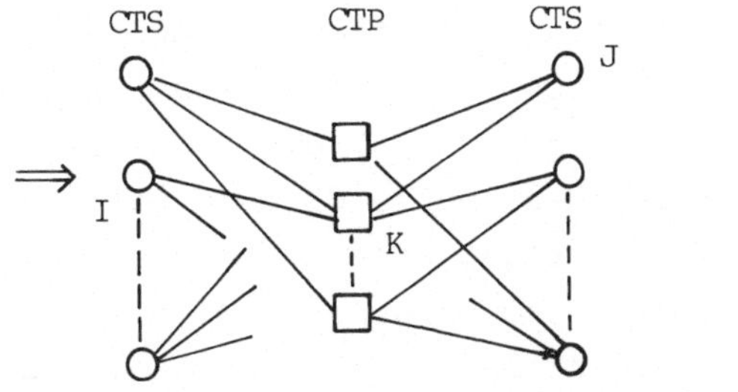


Figure 2 : représentation schématique du réseau interurbain.

Dans ce qui suit les indices I,K,J seront réservés respectivement aux centres origine, transit, destination.
Acheminer un appel de I vers J consiste donc à choisir un chemin non saturé, c'est-à-dire deux faisceaux en série IK et KJ disposant chacun d'au moins un circuit libre.

La politique de gestion dynamique proposée et étudiée ici, est basée, en premier choix, sur une politique dite de partage de charge, qui consiste à diviser les flux d'appels sur les diverses routes possibles pour leur routage. Pour le flot (I,J) le choix des routes (IKJ) est effectué proportionnellement à un paramètre $\alpha_{ikj}$, dit paramètre de partage de charge. Tout le flot proposé au chemin (IKJ) ne va pas être acheminé par ce dernier à cause du phénomène de saturation des faisceaux, donc une portion de ce flot sera "débordé" et tentera un autre choix qui pourra également se faire proportionnellement à un paramètre.

$$\beta_{IKK'J}\ , \quad K' \neq K$$

C'est par rapport à ce type général de loi paramétrique de routage dynamique que nous avons développé le modèle, donné au paragraphe suivant.

## III. Modèle de la moyenne : agrégation

La détermination d'une politique de routage dynamique est faite dans l'hypothèse de variations lentes des caractéristiques du trafic téléphonique qui permet une approche dite de programmation mathématique. Elle consiste à la résolution de problèmes d'optimisation pour la recherche de paramètres optimaux. Pour celà, il est bien sûr nécessaire de modéliser l'ensemble réseau - commande - trafic téléphonique d'une manière suffisamment précise mais simple à la fois, pour permettre une utilisation pseudo-temps réel.

La plupart des modélisations existantes sur le trafic téléphonique étaient faites à des fins de planification de réseau et adoptèrent en général une procédure "faisceau par faisceau" sans s'attacher à une modélisation globale du réseau. Certainement une des plus anciennes et encore largement utilisée est celle dite du faisceau équivalent proposée par Wilkinson [8]. Il s'agit d'une modélisation à deux moments où, en régime stationnaire le processus aléatoire qu'est le trafic téléphonique est caractérisé au second ordre : moyenne, variance.

Nous avons développé un modèle global du réseau : un modèle du premier ordre.

Nous rappelons ici les idées fondamentales qui ont conduit à l'élaboration d'un tel modèle. Toutes précisions peuvent être trouvées dans [9, 10, 11]. Tout d'abord, nous dirons qu'il s'agit d'un modèle de type fluide : la variation de l'état d'occupation d'un faisceau est égal au flux entrant auquel on soustrait le flux sortant.

Si on considère un faisceau isolé de capacité N auquel est offert un trafic poissonien (hypothèse classique pour le trafic téléphonique) avec $\lambda$ comme taux d'arrivée moyen des appels et T durée moyenne de conversation, on peut écrire :

$$\overset{\circ}{X}(t) = -\frac{X(t)}{T} + \lambda\,(1 - P_N(t)) \qquad (1)$$

où

- X est l'état moyen d'occupation du faisceau (nombre d'appels en moyenne servis)
- $\lambda\,[1-P_N(t)]$ est le trafic "entrant" : taux d'arrivée x probabilité de trouver un circuit libre sur le faisceau.
- $X/_T$ est le trafic sortant; taux d'arrêt (fin) des appels
- $P_N(t)$ dans (1) est la probabilité de blocage du faisceau de capacité N; $P_N(t) = \Pr\left[X(t) = N\right]$.

En régime stationnaire ($\overset{\circ}{X} = 0$) cette probabilité de blocage dépend du trafic offert - $A = \lambda T$ et de la capacité N

$$P_N(t) \cong P_N(\infty) = E\left[A, N\right]$$

et $E\,(A,N) = \dfrac{A^N/N!}{\sum_{i=1}^{N} A^i/i!}$ est appelée formule d'Erlang.

En régime quasi-statique il est possible d'approximer $P_N(t)$ par utilisation de la formule d'Erlang valide dans le cas stationnaire [12] .
On a alors :

$$\overset{\circ}{X}(t) = - \frac{X(t)}{T} + \lambda (1 - \tilde{P}_N(t))$$

où

$$\tilde{P}_N(t) = E\ (Y(t),N) \text{ avec } E\ (Y(t),N) \text{ donnée par} \tag{2}$$
$$X = Y(t)\ \left[1 - E\ (Y(t),N)\right]$$

La généralisation de ce modèle a été faite à un réseau tel que celui représenté figure (2), et l'on a

$$\overset{\circ}{X}_{IK} = - \frac{X_{IK}}{t} + \sum_J \Big\{ \alpha_{IKJ}\ \lambda_{IJ}\ (1 - P_{IK})\ (1 - P_{KJ})$$
$$+ \sum_{K' \neq K} \alpha_{IK'J}\ \beta_{IK'KJ}\ \lambda_{IJ}\ (P_{IK'} - P_{IK';IK})\ (1 - P_{KJ}) \Big\} \tag{3}$$

et

$$\overset{\circ}{X}_{KJ} = - \frac{X_{KJ}}{T} + \sum_I \Big\{ \alpha_{IKJ}\ \lambda_{IJ}\ (1 - P_{IK})\ (1 - P_{KJ})$$
$$+ \sum_{K' \neq K} \alpha_{IK'J}\ \beta_{IK'KJ}\ \lambda_{IJ}\ (P_{IK'} - P_{IK',IK})\ (1 - P_{KJ}) \Big\}$$

où

. $P_{IK}$, $P_{KJ}$ sont les probabilités de blocage des faisceaux IK et KJ respectivement. Elles sont estimées comme précédemment et

$$P_{LM} = E\ (Y_{LM}\ ,\ N_{LM}) \text{ avec } X_{LM} = Y_{LM} \left[1 - E\ (Y_{LM},\ N_{LM})\right] \tag{4}$$

. $P_{IK',JK}$ est la probabilité conjointe (Probabilité que IK et IK' soient bloqués). Ce type de probabilité conjointe est nécessaire à la modélisation du trafic qui passe en 2ème choix. Elle est déterminée en utilisant la formule d'Erlang sur la cellule simple constituée par les faisceaux IK' et IK isolément.

$$P_{IK',IK} \cong E\ (Y_{IKK'},\ N_{iK} + N_{IK'}) \text{ avec}$$
$$X_{IK} + X_{IK'} = Y_{IKK'} \left[1 - E\ (Y_{IKK'},\ N_{IK} + N_{IK'})\right] \tag{5}$$

L'établissement du modèle (3) outre les approximations précédentes, suppose l'indépendance des probabilités de blocage sur les faisceaux en série IK et KJ. Ceci n'est bien sûr pas vrai mais constitue une approximation supportable dans le cas de grands réseaux où un grand nombre de flots (I.J) différents viennent se mélanger sur chacune des lignes de transmission.

Le modèle (3), (4), (5) est un modèle non linéaire, nécessitant la résolution d'équationsalgébriques implicites. L'ensemble des équations du modèle en régime stationnaire ($\overset{\circ}{X}_{LM} = 0$) et l'approximation des probabilités de blocage par l'utilisation de trafics offerts fictifs se ramène à un système équivalent (non linéaire) sur ces trafics fictifs. Ce modèle s'écrit sous une une forme générale :

$$Y = \mathcal{F}(Y, N, A)$$

Y ayant autant de composantes qu'il y a de probabilités de blocage à calculer. Pour résoudre ce système les propriétés de contraction de l'opérateur $\mathcal{F}$ permettent de définir des algorithmes efficaces d'approximations successives (recherche du point fixe) tel que :

$$Y^{i+1} = \mathcal{F}(Y^i, N, A), \quad Y^\circ \geqslant 0$$

La rapidité de convergence[ 9] obtenue, est, bien sûr un atout important pour l'utilité d'un tel modèle. Sa qualité du point de vue précision a été testée sur de nombreuses simulations évènement par évènement de type Monte Carlo; et, pour de petits réseaux, les résultats purent être confrontés à ceux, exacts, donnés par résolution du modèle markovien associé. Il apparaît une très bonne précision de ce modèle vis à vis du trafic écoulé et donc, vis à vis des pertes encourues sur chacun des flots (I.J). Cette qualité résulte pour une bonne part de l'introduction de trafics offerts fictifs $Y_{IK}$, $Y_{IKK'}$ pour l'évaluation des diverses probabilités de blocage. Il a pu être montré par exemple [10, 11] sur des cellules élémentaires que ces trafics offerts fictifs tenaient implicitement compte de la nature des trafics écoulés (poissonnien, sous ou sur-variants...) ce qui est d'une très grande importance dans l'évaluation des pertes.

Enfin, nous tenons là un modèle global qui lie les divers trafics sur l'ensemble du réseau. Nous sommes donc en mesure de définir le problème d'optimisation suivant : (minimisation des pertes ou maximisation du trafic écoulé)

$$C = \max_{[\alpha],[\beta]} \sum_{IK} X_{IK} \qquad \left(\text{ou } \sum_{KJ} X_{KJ}\right)$$

sous

$$\left.\begin{array}{l} - (3) \quad (\overset{\circ}{X}_{IK}, \overset{\circ}{X}_{KJ} = 0\text{; hypothèse comportement quasi-statique}) \\ - (4) \\ - (5) \\ - \sum_K \alpha_{IKJ} = 1 \quad \forall\ I.J \\ - \alpha_{IKJ} \geqslant 0 \end{array}\right. \qquad (6)$$

- $\sum_{K'} \beta_{IKK'J} = 1 \quad \forall I,K,J$ (6)

- $\beta_{IKK'J} \geqslant 0$

## IV. Algorithmes et commande

Dans un premier temps nous nous intéressons purement au problème programmation mathématique - optimisation que nous avons résolu au moyen d'un algorithme classique de type direction admissible et ce, de manière centralisée. Ceci nous a permis de tirer quelques conclusions partielles sur la nature du problème étudié et sur son point d'opération optimal, conclusions qui guident pour une réalisation distribuée de l'algorithme d'optimisation pour finalement aboutir à une commande de type boucle fermée distribuée.

### IV.1. Un algorithme de directions admissibles - version centralisée

Regardant le problème d'optimisation (3) et compte tenu de la structure simple des contraintes linéaires portant sur les paramètres $\alpha$, $\beta$, un algorithme de type Frank-Wolfe a été utilisé pour la résolution centralisée [12] . On peut penser voyant l'expression complexe définissant le modèle du système et donc le critère C qu'il serait plus intéressant de chercher au niveau de la définition de la direction admissible de descente un algorithme plus évolué que celui de Frank-Wolfe. En fait, le modèle (3 - 4 - 5) présente une complexité plus apparente que réelle. Les fonctions non-linéaires de type Erlang sont relativement douces de telle sorte que des techniques simples de type relaxation permettent une solution rapide (en nombre d'itérations) du système (3 - 4 - 5) de même que celui nécessaire pour la détermination des vecteurs gradients. $\frac{dC}{d\alpha}$, $\frac{dC}{d\beta}$. Le sous-problème de la méthode de Frank-Wolfe est un ensemble de sous-problèmes découplés qui sont en fait très faciles à résoudre. Il n'est pas possible de rentrer ici dans le détail des calculs qui peuvent, par ailleurs, être trouvés dans [12, 9].

Parmi les conclusions qui peuvent être retirées d'un certain nombre de résultats numériques on peut noter :

i - Le critère présente un minimum relativement plat par rapport aux paramètres $\alpha$

ii - Le critère présente un minimum extrêmement plat par rapport aux paramètres $\beta$ lorsque les paramètres $\alpha$ se trouvent au voisinage de leurs valeurs optimales.

Ces deux conclusions qui sont d'ailleurs plus une confirmation qu'une découverte du fait que dans des problèmes de transport les politiques de routage du type partage de charge conduisent à une bonne robustesse des performances du système permettent de prendre les décisions :

i - de s'affranchir de l'optimisation des paramètres $\beta$ , qui peuvent être fixés, après optimisation par rapport aux paramètres $\alpha$ , d'une manière heuristique en fonction de ceux-ci (ex : proportionnellement).

ii - de ne pas rechercher une précision très grande de convergence sur ces paramètres $\alpha$ , ou même d'utiliser pour leur détermination des modèles approchés [13].

## IV.2. Version distribuée d'un algorithme de gradient projeté

Nous nous restreignons à l'optimisation des paramètres de partage de charge premier choix. Le problème d'optimisation devient :

$$\max_{[\alpha]} \quad C = \sum_I \sum_J \sum_K \alpha_{IKJ} \cdot \lambda_{IJ} (1 - P_{IK}) (1 - P_{KJ}) \tag{7}$$

sous

(4)

(5)

$$\Sigma \alpha_{IKJ} = 1, \qquad \alpha_{IKJ} \geqslant 0$$

Nous nous proposons la réalisation d'un algorithme distribué basé sur une méthode primale (gradient projeté, par exemple) :

$$\forall_{IKJ}; \quad \alpha^{n+1}_{IKJ} = \alpha^{n}_{IKJ} + a_{ij} \left[\frac{dC}{d\alpha}\right] \tag{8}$$

où $\left[\frac{dC}{d\alpha}\right]$ représente la projection du vecteur gradient $\frac{dC}{d\alpha}$ sur l'ensemble défini par les contraintes

$$\Sigma \alpha_{IKJ} = 1 \qquad \alpha_{IKJ} \geqslant 0 \tag{9}$$

Il est aisé de voir qu'une telle projection se décompose par flux (I,J) et que la réactualisation des paramètres $\alpha_{IKJ}$ par (7) fait intervenir des termes tels que $\frac{dC}{d\alpha_{IKJ}}$ et $\frac{dC}{d\alpha_{IK'J}}$ .

Pour une réalisation distribuée de l'algorithme il nous faut voir quels types d'information sont nécessaires pour le calcul des termes $\frac{dC}{d\alpha_{IKJ}}$ localement au niveau du centre origine I.

$$\frac{dC}{d\alpha_{IKJ}} = \lambda_{IJ} (1 - P_{IK}) (1 - P_{KJ}) - \sum_{J'} \alpha_{IKJ'} \lambda_{IJ'} (1 - P_{KJ'}) \frac{\partial P_{IK}}{\partial \alpha_{IKJ}} = \sum_{I'} \alpha_{I'KJ} \lambda_{I'J} (1 - P_{I'K}) \frac{\partial P_{KJ}}{\partial \alpha_{IKJ}} \qquad (10)$$

En fait, il n'apparait pas de décomposition bien évidente et la solution distribuée consiste en un transfert d'information total en chaque centre origine pour, localement, résoudre le problème global. Ceci n'est, bien sûr, pas satisfaisant pour des raisons de quantité d'informations à véhiculer, de lourdeur des calculs. De plus certaines grandeurs telles que $\lambda_{IJ}$ sont difficilement exploitables en ligne. Cet échec relatif à l'obtention d'une structure de calcul distribuée efficace pour la résolution du problème d'optimisation (6) peut s'expliquer par le fait que le critère pertes est fonction de la probabilité de blocage d'un chemin qui fait intervenir le produit des probabilités de blocage sur chacun des faisceaux. Cette forme multiplicative du critère empêche une décomposition par faisceaux comme c'est le cas par exemple pour les travaux sur les circuits à commutation de paquets [5, 6]. En effet, dans ce type de réseau si l'on prend comme critère le temps d'attente moyen des messages, il apparait comme une forme additive des différents retards encourus à chaque file d'attente.

## IV.3. Routage dynamique distribué

Une expression équivalente à (9) peut être fournie qui, si elle ne change rien quant aux conclusions relatives à un algorithme de calcul distribué, permet d'entrevoir une réalisation de type boucle fermée où les calculs seraient remplacés par des mesures périodiques sur le réseau. En effet (10) peut s'écrire :

$$\frac{dC}{d\alpha_{IKJ}} = \lambda_{IJ} (1 - P_{IK}) (1 - P_{KJ}) + X_{IK} (1 - P_{IK})^{-1} \frac{\partial P_{IK}}{\partial \alpha_{IKJ}} + X_{KJ} (1 - P_{KJ})^{-1} \frac{\partial P_{KJ}}{\partial \alpha_{IKJ}} \qquad (11)$$

Cette expression montre que la détermination de la composante $\frac{dJ}{d\alpha_{IKJ}}$ du gradient ne fait intervenir que des grandeurs relatives à la seule route IKJ. On peut alors songer à remplacer le calcul de ces diverses grandeurs et expressions par une mesure correspondante sur le chemin. Le calcul des dérivées partielles des probabilités de blocage

$\frac{\partial P_{IK}}{\partial \alpha_{IKJ}}$ et $\frac{\partial P_{KJ}}{\partial \alpha_{IKJ}}$ présente cependant quelques difficultés. Plusieurs options et approximations sont envisageables pour la détermination de ce terme. L'une d'elles consiste à en faire une évaluation itérative, par exemple selon

$$\frac{\Delta P_{KJ}}{\Delta \alpha_{IKJ}} = \frac{P^{n+1}_{KJ} - P^{n}_{KJ}}{\alpha^{n+1}_{IKJ} - \alpha^{n}_{IKJ}}$$

Une autre consiste à utiliser une expression établie dans le cas d'un faisceau isolé et montrant que la dérivée de la probabilité de blocage par rapport au trafic offert est, autour du point de fonctionnement, une fonction proportionnelle à la capacité résiduelle du faisceau (par capacité résiduelle on entend le nombre de circuits libres sur le faisceau).

D'autre part, dans les conditions où la commande dynamique se révèle intéressante et où le réseau est relativement chargé, le premier terme dans l'expression de $\frac{dC}{d\alpha_{IKJ}}$ (11) devient prépondérant devant les seconds ($\frac{\partial P}{\partial \alpha} \cong 0$)

$$\frac{dC}{\partial \alpha} \cong \lambda_{IJ} (1 - P_{IK}) (1 - P_{KJ}) \tag{12}$$

Cette nouvelle approximation se trouve renforcée par le fait que, nous l'avons dit précédemment, une politique de partage de charge assure une bonne robustesse des performances obtenues, vis à vis de toutes les imprécisions sur le modèle, la connaissance du trafic. De plus, cette approximation présente l'énorme avantage d'aboutir à une réalisation distribuée simple et de type boucle fermée. En effet, on peut aisément voir que la loi d'adaptation des paramètres proposée en (8) peut s'affranchir de la connaissance du type $\lambda_{IJ}$ qui est assez délicat à estimer en ligne. Avec l'approximation (12) $\lambda_{IJ}$ devient un terme multiplicatif pour tous les termes correctifs des $\alpha_{IKJ}$, $\forall K$ $\left[ a_{ij} \left[ \frac{dC}{d\alpha} \right] \right]$ et donc peuvent être inclus dans le pas $a_{IJ}$ de réactualisation. On obtient alors, un schéma de réactualisation du type

$$\alpha^{n+1}_{IKJ} = \alpha^{n}_{IKJ} + a_{ij} \, \mathcal{P} \left[ \varepsilon_{I1J}, \varepsilon_{I2J}, \dots \varepsilon_{IKJ} \dots \right]$$

$$\varepsilon_{IKJ} = (1 - P_{IK}) (1 - P_{KJ})$$

où $\mathcal{P}$ résulte de la projection sur les contraintes (9)

Dans cette loi de commande l'adaptation des paramètres de partage de charge se fait en fonction des probabilités de succès sur les routes, probabilité qui peut être "mesurée" à partir des probabilités de bloca-

ge sur les différents faisceaux du réseau. Rappelons que c'est sous l'hypothèse de quasi stationnarité qu'un tel schéma est applicable car il est alors possible d'approximer les moyennes "statistiques" par des moyennes "temporelles" calculées et estimées sur des intervalles de temps suffisamment longs. (environ une dizaine de minutes se trouve être une période convenable pour de bons résultats).
Il serait beaucoup trop long (espace) de présenter ici des résultats chiffrés d'expérimentations numériques (voir pour cela les références données). Toutefois nous pouvons affirmer que la loi de commande proposée dans ce dernier paragraphe fournit des performances très comparables à celles obtenues pour des lois singulièrement plus complexes (par exemple celle centralisée du paragraphe IV.1).
Enfin, il est à noter que ce type de commande peut être mis en parallèle avec le routage dynamique à base d'automates à apprentissage [14] .
Son mode de fonctionnement est de type périodique, moyen terme et non plus, appel par appel comme pour les automates qui réactualisent les paramètres de la loi de commande à chaque arrivée d'appel,(l'intervention "appel par appel" s'avérant au vu des simulations, "redondante).

## V. Remarques et conclusion

L'étude présentée dans cette communication a été développée pour une politique de partage de charge qui consiste, en premier choix, à effectuer une division du trafic téléphonique sur les diverses routes disponibles à l'acheminement des appels.
Une autre politique est envisageable, c'est celle dite de débordement. Elle consiste à utiliser une route et une seule pour le premier choix et "déborder" éventuellement sur une route, unique également, en deuxième choix. Avec une telle politique, l'introduction du caractère dynamique au niveau de la commande consisterait donc à effectuer, en temps réel, un choix et un classement de 2 routes pour chaque couple origine destination en fonction de l'état du réseau. La détermination de ces deux routes se prête très mal à une approche "programmation mathématique" contrairement à la politique de partage de charge. En effet, le problème d'optimisation associé est de caractère combinatoire complexe, de grande dimension et qui ne se prête guère à une solution décentralisée ou distribuée même approchée. Seules des heuristiques plus ou moins élaborées sont envisageables. Elle peuvent d'ailleurs conduire à des commandes relativement peu stables même en régime stationnaire (comportement de type cycle limite). Cet inconvénient est réduit en imposant un choix fixe pour la route de premier choix, mais alors le gain en performances se trouve réduit. Un autre avantage plaidant en faveur d'une politique

du type partage est fourni par la bonne robustesse que présente ce type de commande vis à vis de perturbations structurelles dans le réseau (ruptures de lignes de transmission, pannes de centraux...). En effet, cette robustesse s'explique simplement par le fait que le partage de charge augmente le nombre de routes possibles pour l'acheminement.
La commande de type partage de charge est particulièrement adaptée aux actions de premier choix, qui peuvent être considérées comme celles qui vont déterminer le comportement à moyen terme du trafic écoulé dans le réseau. L'action de deuxième choix (dont le taux d'existence est, du fait de l'optimisation faite en premier choix, faible) peut apparaître plus comme une action temps réel. Dans ce sens, on peut envisager une commande de structure distribuée agissant sur deux horizons de temps différents (multilayer - décomposition fonctionnelle).

- un niveau de moyen terme (paragraphe IV.3) pour ce premier choix ($\alpha$)
- un niveau de court terme pour le deuxième choix ($\beta$), basée par exemple sur des mesures "temps réel" de type capacités résiduelles et rapatriées aux centraux de commutation à une cadence de quelques dizaines de seconde (canal semaphore).

L'étude proposée ici a été faite en s'appuyant sur un critère unique qui est la minimisation des pertes globales sur l'ensemble du réseau. Comme dans beaucoup de cas pratiques, l'évaluation des performances de la commande obtenue ne peut se faire par rapport à ce seul critère. D'autres critères doivent rentrer en ligne de compte : économique, simplicité, etc... Il en est un autre qui dans le cas des réseaux téléphoniques assurant un service public est particulièrement important : l'uniformité de la qualité de service pour tous les flux IJ. On peut montrer que pour un réseau "bien" dimensionné la commande proposée en IV.3 assure effectivement que les degrés de qualité de service pour chacun des couples IJ sont très voisins (peu de dispersion [15]).

References

1 A. EPHREMIDES, P. VARAIYA, J.C. WALRAND. A simple dynamic routing problem, IEEE Trans. A.C., 25, n° 4, Aug. 80.
2 Z. ROSBERG, P. VARAIYA, J.C. WALRAND. Optimal control of service in tandem queues. IEEE Trans. AC, 27, n° 3, 600-610, June 82.
3 J.M. GARCIA, B. GOPINATH, P. VARAIYA. Routing trafic in telephone networks, Conf. Dec. Control. San Diego, Dec. 81.
4 R. GALLAGER. A minimum delay routing algorithm using distributed computation. IEEE Trans. on Communications 85, n° 1, 73-85, 1977.
5 D.P. BERTSEKAS. Distributed dynamic programming. IEEE Trans. Aut. Contr. 27, n° 3, 610-616, 1982.
6 T.E. STERN. A class of decentralized routing algorithm using relaxation. IEEE Trans. on Com., 25, n° 10, 1092-1102, 1977.
7 G. AUTHIE, J. BERNUSSOU, D. ELBAZ. Distributed asynchronous iterative control algorithms. Optimal routing application. IFAC Symposium CIDCS, Dec. 82, Paris.

8 R.I. WILKINSON. Theories for toll traffic engineering in the USA, Bell System Technical Journal. 421-434, March 56.

9 F. LE GALL. Contribution à la modélisation et la commande de réseaux téléphoniques, Thèse de Docteur-Ingénieur, UPS n° 796, Toulouse, 1982.

10 F. LE GALL, J. BERNUSSOU. An analytical formulation for grade of service determination in telephone networks. IEEE Trans. on Com. March 1983.

11 F. LE GALL, J.M. GARCIA, J. BERNUSSOU. A one moment model for telephone traffic. Application to bloking estimation and resource allocation. 2nd Ial Workshop on the Performance of Computer Communication Systems. IBM INRIA, 21-23 Mars 84, Zurich.

12 J.M. GARCIA, Problèmes liés à la modélisation du trafic et à l'acheminement des appels dans un réseau téléphonique. Thèse Docteur-Ingénieur, UPS, Toulouse, 1980.

13 J. BERNUSSOU, J.M. GARCIA, I.S. BONATTI, F. LE GALL. Modelling and control of large scale telecommunications networks. 8th Ial World Congress, IFAC Kyoto, 1981.

14 P.M. SRIKANTAKUMAR, K.S. NARENDRA. A learning model for routing in telephone network. SIAM Journal of Control, vol. 20, n° 1, Janv. 82

15 I.S. BONATTI. Gestion de réseaux de service : application au réseau téléphonique interurbain. UPS Toulouse n° 767, 1981.

# UN FEEDBACK GLOBAL POUR LA PLANIFICATION DU PARC DE PRODUCTION ELECTRIQUE FRANCAIS.

P. LEDERER, Ph TORRION, JP BOUTTES
Electricité de France - Etudes Economiques Générales
2, Rue Louis Murat - 75384 PARIS CEDEX 08 - FRANCE

Résumé : On présente la nouvelle "Maquette" du système français de production consommation d'électricité.

Cette nouvelle représentation de la gestion annuelle, à partir d'une statégie d'exploitation à pas de temps journalier, où chaque type de moyen de régulation existant du côté de l'offre (réserves hydrauliques et de pompage), comme du côté de la demande (nouvelle options tarifaires gérées près du temps réel), est géré explicitement, permettra de progresser dans la problèmatique en développement de régulation globale du système offre-demande. Elle donne en effet des résultats nécessaires pour définir les conditions d'un développement harmonieux des différents moyens de régulation.

## LE POIDS CROISSANT DES ALEAS DANS LE SYSTEME ELECTRIQUE

Les principales tendances qui caractérisent l'évolution à long terme de la demande électrique nationale sont les suivantes (voir tableau 1) :

- amélioration du facteur de charge journalier grâce à la politique tarifaire mise en oeuvre à EDF (il est dès aujourd'hui supérieur à 0,9) ;

- sensibilité accrue de la courbe de charge à la température : ainsi, en hiver, la puissance minimale appelée un jour froid sera supérieure à la puissance maximale d'un jour clément ;

- accroissement de la modulation saisonnière en raison de la part croissante des usages thermiques dans la demande globale.

Il en résulte que la période de pointe de la demande sera constituée de la quasi-totalité des heures des quelques jours les plus froids de l'hiver, jours qui ne seront connus, un hiver donné, qu'a posteriori.

## LA DIFFERENCIATION ACCRUE DES COUTS DE PRODUCTION

Pour satisfaire cette demande, on disposera alors d'un parc de production majoritairement thermique (stagnation du productible hydraulique en absolu), caractérisé par :

- une proportion de plus en plus importante de nucléaire (en 2000, 75 % en puissance, 90 % en énergie)
- des coûts de combustible fortement différenciés du nucléaire -1- à la turbine à gaz -25- en 2000)
- un aléa de disponibilité des équipements non négligeable bien que très inférieur à celui de la demande (3,5 GW contre 10 GW)

Tableau 1 : Caractéristiques de l'évolution de la demande

| | 1982 | | 2000 | |
|---|---|---|---|---|
| | Jour d'hiver le plus chargé | Jour ouvrable d'été le moins chargé | Jour d'hiver en espérance le plus chargé | Jour ouvrable d'été en espérance le moins chargé |
| Puissance Moyenne du jour GW | 38 | 22 | 86 | 41 |
| Ecart-type GW | (3,5) | (1) | 10 | 1,5 |
| Modulation journalière GW | 10 | 7 | 13 | 10 |

## SOULIGNENT LE BESOIN D'UNE REGULATION DU SYSTEME OFFRE-DEMANDE

Ainsi, en 2000, hors action sur la demande et hors le jeu des réserves qui existent dans le parc de production (hydraulique gravitaire et de pompage), la structure du système production-consommation sera telle que le coût marginal de production pourra prendre, un jour quelconque d'hiver, n'importe quelle valeur entre le coût proportionnel du nucléaire et le coût de défaillance, du seul fait de la réalisation des aléas. Qualitativement ce phénomène sera déjà significatif en 1990 bien qu'avec une probabilité moindre.

C'est pour transmettre cette information aux consommateurs susceptibles d'en profiter que l'Etablissement s'oriente vers le développement d'options tarifaires qui comprennent des périodes définies par l'exploitant; à travers la gestion de stocks de "demande effaçable", elles se traduisent par la possibilité de moduler la demande en fonction de la réalisation des aléas.

Du côté de la production, les moyens de régulation sont constitués par les réserves (hydraulique gravitaire et de pompage). Ces moyens ne représentent certes, en terme d'énergie, qu'un volant de manoeuvre limité, mais qui permet de délivrer sur le réseau, instantanément, des puissances qui sont de l'ordre de grandeur des fluctuations dues aux aléas qui affectent le système électrique.

Satisfaire la demande au moindre coût pour la collectivité dépendra ainsi, d'abord, de notre capacité à gérer au mieux, dans un tel contexte, ces différents moyens, et à trouver leur développement respectif optimal à partir d'une juste valorisation de leurs services rendus.

## LES MOYENS DE CETTE REGULATION ETANT TRES VARIES

- On peut caractériser chacun de ces moyens par leur type (hydraulique gravitaire, pompage, options tarifaires type EJP) et leur constante de temps (i.e. le temps de vidange de la réserve pleine à la puissance maximale). Il est légitime, pour clarifier le problème, d'agréger les réserves de même type et de même constante de temps. On ramène alors la dimension du système français d'environ 60 à 4 ou 5. Le tableau 2 donne la typologie ainsi obtenue :

TABLEAU 2 : Moyens de régulations

| Type | Constante de temps ( heures) |
|---|---|
| Lac saisonnier | 1 100 |
| Pompage inter-hebdomadaire | 190 |
| Pompage hebdomadaire | 35 |
| EJP (effacement jour de pointe) | 400 |
| HCM (heures creuses modulables) | 4 000 - 5 000 (?) |

## BIEN GERER LE SYSTEME C'EST LEUR APPLIQUER CHAQUE JOUR UNE STRATEGIE GLOBALE.

On voit en effet que la diversité des types et des constantes de temps qui caractérisent les différents moyens de régulation implique l'existence d'interactions complexes entre leurs modes de gestion respectifs.

Gérer au mieux le système production-consommation, c'est donc appliquer une statégie d'exploitation :

. globale (par rapport à l'ensemble de ces moyens, qu'ils soient du côté de l'offre comme de la demande) :

- La loi de commande doit être fonction des réalisations des aléas couplants de demande et disponibilité. Ces aléas affectant l'ensemble du système et de façon importante, il faut mettre en oeuvre à chaque instant une gestion d'ensemble du système.

. journalière - Le pas de temps élémentaire de prise de décision doit en effet être de l'ordre de la journée si l'on veut être en mesure de représenter et d'utiliser au mieux l'information disponible sur les aléas - comme la température - qui ne peuvent être prévus correctement plusieurs jours à l'avance.

## LES METHODES CLASSIQUES NE SONT PAS ADAPTEES A CE PROBLEME

L'optimisation des grands systèmes stochastiques est en effet un problème qui, s'il a fait l'objet de nombreuses études, n'a pas encore été totalement résolu.

En principe, seule la programmation dynamique permettrait de résoudre complètement de tels problèmes. Malheureusement elle ne peut être mise en oeuvre pour plus de trois variables d'état en raison de la croissance exponentielle du temps de calcul avec le nombre de réserves.

En l'absence de solution parfaitement satisfaisante, plusieurs compromis ont été proposés par le passé. Or, les compromis "classiques" ne sont pas adaptés à notre problème :

i) Les techniques déterministes ne conviennent évidemment pas puisqu'elles éludent totalement l'importance des aléas dans le système électrique.

ii) Si l'on agrège tous les moyens de régulation en une réserve unique (Meslier, 1978), on perd toute l'information relative aux contraintes locales. Cette pratique peut être justifiée dans le cas d'un système composé de réserves semblables i. e. des réserves ayant à peu près la même constante de temps -temps de vidange de la réserve (pleine) à la puissance maximale- et des régimes d'apport similaires. Cependant ceci ne correspond pas à la situation française. (voir Tableau 2).

L'agrégation en une réserve unique permet d'obtenir facilement une stratégie globale, i. e. un feedback par rapport aux aléas couplants comme la demande, mais n'est évidemment pas admissible pour les raisons évoquées ci-dessus.

iii) Les techniques de relaxation peuvent être utilisées pour déterminer la gestion optimale par feedback local d'un système de réserves en parallèle, en tenant compte de l'aléa hydraulique (Pronovost et Boulva (1978), Delebecque et Quadrat (1978) ou Colleter, Lederer, Ortmans (1981)). En effet la relaxation gère succéssivement chaque réserve, le comportement des autres moyens étant connu seulement en moyenne ou en loi et on obtient une stratégie admissible. Cependant, dans le cas présent, où les phénomènes aléatoires les plus importants -température et indisponibilité fortuite du thermique- couplent la gestion des différentes réserves, la seule façon d'utiliser la relaxation consisterait à résoudre le problème à nouveau au début de chaque période (à chaque pas de temps), mettant ainsi en oeuvre une solution en "boucle ouverte adaptée". Une telle approche peut servir en exploitation, mais non en vue d'une planification des investissements où le raisonnement en boucle fermée s'impose.

iv) On peut avoir dans certains cas une idée a priori de la forme d'un "bon" feedback. Cette information a priori permet de transformer le problème initial en un problème plus simple. Le feedback est paramétrisé et ces paramètres sont optimisés en boucles ouverte (Colleter et Lederer, 1981 ; Quadrat, 1982). Malheureusement nous avons à gérer des réserves de constantes de temps très différenciées, et l'information a priori dont nous disposons pour le moment est très pauvre.

CE QUI SE TRADUIT PAR UN DECALAGE CROISSANT ENTRE LES MODELES MIS AU POINT DANS LE PASSE ET LA REALITE DU SYSTEME

Les modèles anciennement disponibles à E.D.F. pour représenter en tout ou partie la gestion annuelle du système électrique, appliquent en effet l'une ou l'autre des approches évoquées ci-dessus.

On peut distinguer deux grandes familles de modèles E.D.F. :

- ceux qui gèrent un grand nombre de réserves avec un feedback local admissible ; grâce à des techniques de relaxation ou de paramétrisation de la commande :

- ceux qui fournissent un feedback global mais en agrégeant l'ensemble des réserves en une seule.

Par ailleurs tous ces modèles travaillent sur monotone hebdomadaire (le pas de temps élémentaire est donc la semaine) avec corrélation parfaite des aléas à l'intérieur de la semaine.

A l'évidence la représentation de la demande par des monotones hebdomadaires implique des inconvénients sérieux :

- D'une part, on veut obtenir une commande qui réponde bien à l'aléa de demande, or celui-ci est comme nous l'avons vu difficilement prévisible à plus de deux ou trois jours.

- D'autre part, le pas de temps naturel associé à la gestion EJP et du pompage est évidemment de l'ordre de la journée. En effet, le signal EJP incite un client à s'effacer pendant 18 heures d'une journée, et le pompage permet éventuellement un transfert des heures creuses aux heures pleines d'une journée ou un transfert d'un jour chaud à un jour froid.

## ON PROPOSE DONC UNE AUTRE APPROCHE

Pour obtenir une stratégie de gestion journalière, globale et admissible d'un système comprenant 4 ou 5 moyens de régulation (voir tableau 2), on a développé, sur la base d'une méthode exposée par Turgeon (1980), une approche nouvelle.

L'idée générale est la suivante : dans une première phase (optimisation) on calcule des valeurs du kWh en stock pour chacune des réserves à l'aide de modèles gérant en même temps chaque réserve et son complémentaire dans le système (i. e. l'ensemble des autres réserves agrégées en une seule). Dans une deuxième phase (simulation) on simule le comportement de l'ensemble du système en utilisant ces valeurs du kWh ; celles-ci permettent -avec les coûts de combustible et de défaillance- d'obtenir un ordre d'appel aux différents moyens de régulation et équipements disponibles ; la gestion ainsi calculée est globale bien que sous-optimale. De plus la mise au point d'une méthode de calcul très rapide de l'optimisation de transition (qui permet de passer d'un instant à un autre) a permis de passer à un pas de temps élémentaire de l'ordre de la journée *.

La description précise de la méthode utilisée figure dans la suite.

## LA MAQUETTE

### A - Modélisation du système

Comme il est d'usage dans les systèmes électriques fortement interconnectés, on se ramène à un modèle en un point, où l'ensemble des moyens de production doit satisfaire la demande totale d'électricité. Cette demande totale prend en compte les pertes sur le réseau.

(*) Voir note M. 745, "Calcul analytique de l'espérance du coût de gestion d'un parc électrique avec hydraulique gravitaire, pompage et thermique" (Lederer, Torrion ; rapport interne EDF).

. Courbes de charge

Les prévisions à long terme de la courbe de charge dans le système français se font à l'aide d'une décomposition par secteurs.

La prise en compte des usages thermiques détermine la consommation à température normale, et le gradient de température, et permet ainsi la simulation de l'aléa de demande dû aux fluctuations de température.

On utilise dans le modèle des courbes de charge journalières agrégées en trois niveaux chaque jour.

. Fonction coût

L'approximation habituelle des coûts thermiques par une courbe linéaire par morceaux est justifiée dans les études de long terme. Le coût de défaillance utilisé par EDF est une fonction de l'énergie manquante croissante avec la profondeur de la défaillance. Cette fonction est représentée dans le modèle sous la forme de centrales fictives permettant ainsi une gestion optimale des coupures.

. Le système hydraulique

On réduit la dimension du système en agrégeant l'ensemble des moyens de régulation de même type et de même constante de temps :

- selon une approche classique utilisée à EDF (Meslier, 1978), les usines hydrauliques conventionnelles sont représentées par trois réserves hydrauliques fictives : une réserve saisonnière, une réserve hebdomadaire et un fil de l'eau ;

- les ouvrages de pompage sont représentés par trois usines différentes de cycle saisonnier, hebdomadaire et journalier.

. L'option EJP (Effacement Jours de Pointe)

La réforme tarifaire d'EDF comprend l'option "Effacement Jours de Pointe" pour tous les consommateurs. 22 jours peuvent ainsi être sélectionnés près du temps réel par le dispatching national, avec pour 18 heures de ces journées un prix du kWh qui reflète les coûts marginaux de ce heures, beaucoup plus élevés que durant le reste de l'année.

Un kilowattheure souscrit en option EJP serait ainsi, idéalement, équivalent pour le producteur à une réserve de 400 kWh avec une puissance installée d'1 kW (pratiquement, le ratio est moins bon puisque l'efficacité du signal d'effacement à la pointe est inférieure à 100 %).

B - Le problème

Dans le système considéré on dispose de N réserves indépendantes ainsi que de paliers thermiques. Ces moyens de production doivent permettre de satisfaire la demande d'électricité en minimisant l'espérance mathématique du coût annuel de gestion et de défaillance sur une année.

Ces réserves sont définies par :

- leur stock maximal en énergie : $X_i^{max}$ ; $i = 1, N$ ;

- leur puissance maximale en turbine : $U_i^{max}$ ; i = 1, N ;
- leur puissance maximale en pompe (éventuellement nulle) :
. $U_i^{min}$ ; i = 1, N ;
- leurs apports naturels (aléatoires) : $a_i^t$ ; i = 1, N pendant la période t.

On note :

$x_i^t$ : l'énergie dans la réserve i au début de la période t ; $0 \leq x_i^t \leq X_i^{max}$ ;

$U_i^t$ : la production (qui peut être négative dans le cas d'un pompage) de la réserve i pendant la période t ; $U_i^{Min} \leq U_i^t \leq U_i^{Max}$ ;

$d_i^t$ : l'énergie déversée par la réserve, pendant la période t ;

$D^t$ : la demande d'énergie élétrique pendant la période t ;

.$C^t$ : le coût de gestion et de défaillance (qui dépend de la disponibilité thermique).

Le problème de départ comporte ainsi N variables d'état. Nous recherchons donc une gestion en boucle fermée telle que :

$$.(P) \qquad \underset{U}{\mathrm{Min}} \quad \sum_{t=0}^{T} E\, C^t \,(D^t - \sum_{i=1}^{N} U_i^t)$$

Si $U_i^t < 0$, la réserve i pompe ; l'équation d'état complète est en fait

$$x_i^{t+1} = x_i^t + a_i^t + U_i^{t-}/r - U_i^{t+} - d_i^t$$ avec r coefficient de rendement du pompage.

Sous les contraintes :

$$(C) \quad \begin{cases} x_i^{t+1} = x_i^t + a_i^t - U_i^t - d_i^t & \\ 0 \leq x_i^{t+1} \leq X_i^{max} & i = 1, \ldots, N \\ U_i^{min} \leq U_i^t \leq U_i^{max} & t = 0, \ldots, Tf \\ d_i^t \geq 0 & \end{cases}$$

Notons qu'étant donné l'importance des aléas de température et d'indisponibilité, la loi de commande doit être fonction de la demande d'électricité et de la disponibilité thermique.

Afin d'obtenir une telle commande, il faut résoudre le problème (P) par la programmation dynamique.

Malheureusement, N est ici plus grand que trois, la méthode est donc impraticable.

On veut avoir, d'une part une gestion admissible qui respecte les contraintes propres à chaque réserve, d'autre part une gestion en stratégie globale où l'on tient compte des autres réserves.

Ainsi, pour le calcul d'indicateurs économiques dans la phase optimisation, la loi de commande U doit être fonction au moins de :

- l'énergie dans la réserve i ;
- l'énergie totale dans l'ensemble des réserves.

Une fois obtenus ces indicateurs prenant en compte les aspects essentiels du système français, on peut construire des simulations qui respectent les contraintes propres à chaque réserve. Ces simulations sont effectuées sur des chroniques de demande, de disponibilité thermique et d'apports ; chaque simulation fournit ainsi l'enchaînement d'une gestion sur une année avec un pas de temps de l'ordre de la journée.

Pour ce faire, on a développé une idée de Turgeon (1980) qui propose un moyen d'obtenir une telle loi de commande d'un grand système comprenant plusieurs réserves.

## C - Optimisation : agrégations partielles

Cette idée consiste à passer du système initial de N réserves à N systèmes indépendants de deux réserves, puis à résoudre chacun de ceux-ci par la programmation dynamique stochastique.

Chaque problème (Pi) est construit à partir du problème (P) en agrégeant toutes les réserves sauf une, à savoir la réserve i.

Notons :

$$\bar{x}_i^t = \sum_{j\neq i} x_j^t, \quad \bar{X}_i^{max} = \sum_{j\neq i} X_j^{max} \text{ etc.}$$

Nous pouvons alors écrire le problème (Pi) :

$$(Pi) \qquad \min_{U_i^t,\, \bar{U}_i^t} \sum_{t=0}^{T} E\left[C^t\,(D^t - U_i^t - \bar{U}_i^t)\right]$$

avec les contraintes :

$$x_i^{t+1} = x_i^t + a_i^t - U_i^t - d_i^t$$

$$\bar{x}_i^{t+1} = \bar{x}_i^t + \bar{a}_i^t - \bar{U}_i^t - \bar{d}_i^t$$

$0 \leq x_i^t \leq x_i^{max}, \qquad 0 \leq \bar{x}_i^t \leq \bar{x}_i^{max}$

$U_i^{min} \leq U_i^t \leq U_i^{max}, \quad \bar{U}_i^{min} \leq \bar{U}_i^t \leq \bar{U}_i^{max}$

$d_i^t \geq 0 \; ; \; \bar{d}_i^t \geq 0$

$t = 0, \ldots, T.$

le vecteur d'état dans le problème (Pi) est de dimension deux. On peut résoudre totalement ce problème en utilisant l'équation de Bellman :

$$V_i^t\,(x_i^t, \bar{x}_i^t) = E \min_{U_i^t,\, \bar{U}_i^t} \left[ C^t\,(D^t - U_i^t - \bar{U}_i^t) + V_i^{t+1}\,(x_i^{t+1}, \bar{x}_i^{t+1}) \right]$$

où $U_i^t$, $\bar{U}_i^t$ doivent respecter les contraintes $(C_i)$ sur la période t.

Puisque, pour tout i, l'ensemble des contraintes (C) du problème (P) est plus grand que $(C_i)$, ensemble des contraintes du problème $(P_i)$, il est évident que

$$(2) \quad V_i^t\,(x_i^t, \bar{x}_i^t) \leq V^t\,(x_1^t, x_2^t, \ldots, x_N^t) \qquad \begin{array}{l} i = 1, \ldots, N \\ t = 0, \ldots, Tf \end{array}$$

Il faut remarquer que ces indications (les valeurs de Bellman $V_i$ associées à chaque modèle i) sont issues d'une gestion (de la réserve i et de son complémentaire) en stratégie globale face aux aléas couplants.

D - Simulation : le choix d'un feedback global :

La question est alors la suivante comment obtenir la meilleure boucle fermée avec pour seule information les N fonctions $V_i$ ?

On sait, en particulier, qu'en l'absence de contrainte sur la réserve i, le coût marginal est égal à la valeur de l'eau de cette réserve.

Ceci montre que l'ensemble <u>ordonné</u> des valeurs de l'eau des diverses réserves et des coûts de combustible constitue un ensemble mixte hydraulique-thermique d'indicateurs économiques qui permet de gérer les réserves. Dans la mesure où le système français est composé de réserves de type et de constante de temps très divers, il est évident que sa gestion optimale sera caractérisée à chaque instant t par un large spectre des valeurs de l'eau dans les différents stocks.

Comme nous l'avons mentionné ci-dessus, le calcul des vraies valeurs de l'eau $-\frac{\partial V}{\partial x_i}$, i = 1,..., N est impossible puisque l'on ignore la fonction V.

Considérons maintenant la dérivée $-\frac{\partial V_i}{\partial x_i}$ i = 1,..., N.

C'est clairement la vraie valeur de l'eau dans la réserve i, mais ceci quand les autres sont agrégées en une seule réserve. Ainsi elle prend en compte :

- les contraintes locales de la réserve i qui traduisent son type et sa constante de temps ;

- l'énergie totale dans l'ensemble des réserves, ce qui est, comme nous l'avons déjà remarqué, une bonne caractérisation de l'état du système.

C'est pourquoi nous avons choisi une gestion définie par :

$$(\text{PS}) \qquad \min_{U_1^t, U_2^t, \ldots, U_N^t} \; C^t\left(D^t - \sum_{i=1}^{N} U_i^t\right) - \sum_{i=1}^{N} \frac{\partial V_i^{t+1}}{\partial x_i} . U_i^t$$

. <u>Un commentaire sur l'article de Turgeon</u>

Turgeon suggère de choisir la gestion définie par :

$$(3) \quad \min_{U_1^t, \ldots, U_N^t} \; C^t\left(D^t - \sum_{i=1}^{N} U_i^t\right) + \frac{1}{N} \sum_{i=1}^{N} V_i^{t+1}(x_i^{t+1}, \bar{x}_i^{t+1})$$

en respectant les contraintes (C).

En d'autres termes, Turgeon suppose que la moyenne des N fonctions $V_i$ fournit une bonne approximation de la fonction inconnue V, i.e.

$$V^t(x_1^t, x_2^t, \ldots, x_N^t) \simeq \frac{1}{N} \sum_{i=1}^{N} V_i(x_i^t, \bar{x}_i^t) = V_m(x_1^t, \ldots, x_N^t)$$

En fait, il ne semble pas capable de donner un fondement général à cette affirmation. Celle-ci nous semble au contraire comporter quelques implications désagréables.

i) Considérant l'inégalité (2), $V_m$ ne sera à l'évidence jamais la meilleure estimation de V ;

ii) Partant de la définition de V comme fonction Bellman, on peut en déduire la valeur de l'eau associée pour la réserve i :

$$m_i = \frac{1}{N}\left[\frac{\partial V_i}{\partial x_i} + \sum_{j \neq i} \frac{\partial V_j}{\partial x_j}\right]$$

Ainsi $m_i$ est une moyenne entre :

. la vraie valeur de l'eau dans la réserve i, mais quand les autres sont agrégées en une seule réserve :

$$\frac{\partial V_i}{\partial x_i}(x_i, \bar{x}_i)$$

(N-1) valeurs de l'eau des autres réserves (complémentaire de la réserve j, j≠i) qui incluent la réserve i :

$$\frac{\partial V_j}{\partial \bar{x}_j}(x_j, \bar{x}_j)$$

Plus N est grand, plus les caractéristiques des réserves complémentaires xj sont proches ;

on en déduit que les dérivées sont **alors** d'autant plus proches et leurs poids dans m d'autant plus élevé. Ainsi, plus N est grand, plus les $m_i$ pour i = 1,2 .. N sont proches.

Un tel choix peut conduire à une stratégie voisine de l'optimum quand les caractéristiques des différentes réserves sont similaires (puisque dans ce cas les vraies valeurs de l'eau sont effectivement proches). Cependant comme nous l'avons déjà vu, ce n'est pas du tout le cas du système Français.

iii) Enfin, on peut signaler une autre conséquence générale gênante d'un tel choix : l'instabilité des $m_i$ (i=1 .. N) quand on ajoute une $(N=1)^{ième}$ réserve marginale, ce qui n'est évidemment par une propriété des vraies valeurs de l'eau des réserves i, i=1 .. N.

ANNEXE 1

## QUALIFICATION DE LA METHODE NUMERIQUE

Dans cette partie, on présente quelques résultats numériques sur la base des anciennes prévisions de consommation de l année 2000 et du parc de production optimal associé.

Plusieurs méthodes numériques sont comparées.

### A - Le système étudié

- Demande : 560 TWh
- Demande de pointe moyenne : 92 GW

PALIERS THERMIQUES

| TYPE DE CENTRALE PUISSANCE | DISPONIBLE EN MOYENNE (GW) | COUT DE COMBUSTIBLE (cF/kWh) |
|---|---|---|
| Nucléaire | 72,7 | 5,15 |
| Charbon récent | 7,5 | 22,0 |
| Charbon ancien | 4,5 | 22,2 |
| Fuel récent | 3,9 | 50,0 |
| Fuel ancien | 1,4 | 56,5 |
| Turbines à gaz | 2,2 | 111,0 |

MOYENS DE REGULATION

| TYPE DE RESERVES | Stock (GWh) | Puissance en pompe | Puissance en turbine |
|---|---|---|---|
| Hydraulique gravitaire saisonnier | 7 322 | - | 6,58 |
| Pompage saisonnier | 361 | 1,83 | 2,24 |
| Pompage hebdomadaire | 60 | 2,5 | 2,5 |

Pour estimer les coûts annuels de gestion, on a simulé 400 années. Comme on s'y attendait, les résultats se classent comme suit :

Coût de l'agrégation totale $<$ Coût de l'agrégation partielle i

$<$ Coût optimum (inconnu) $<$ Coût du système réel (maquette)

On fournit ainsi un minorant du vrai coût optimal. La sous-optimalité du feedback global proposé ici et de la gestion admissible associée peut être ainsi qualifiée par ce minorant.

| | AGREGATION TOTALE | AGREGATION PARTIELLE* | SYSTEME REEL |
|---|---|---|---|
| Coût de gestion (combustible + défaillance) $10^6$ . F | 31,530 | 32,410 | 32,670 |
| Coût de défaillance $10^6$ . F | 1,150 | 1,520 | 1,740 |
| Energie manquante GWh | 285 | 350 | 400 |
| Production totale de l'hydraulique GWh | 19,130 | 18,430 | 17,880 |
| Energie totale pompée GWh | 8,430 | 7,430 | 6,650 |

De là on peut déduire les différences relatives suivantes entre
- l'agrégation totale par rapport au système réel : - 3,5 %
- l'agrégation partielle par rapport au système réel : - 0,8 %

Ce dernier chiffre indique que le coût de gestion obtenu à partir de la méthode développée dans cette note est au plus à 0,8 % de l'optimum.

(*) Nous avons choisi ici le modèle de dimension deux qui conduit à la valeur $V°(X°, \bar{X}°)$ la plus forte afin d'obtenir après simulation le meilleur minorant.

## BIBLIOGRAPHIE


COLLETER-LEDERER-ORTMANS, Long Term Nuclear Scheduling in the French Power System, IFIP, September 1981.

COLLETER-LEDERER, Optimal Operation Feedbacks for the French Hydro-Power System, CORS-TIMS, ORSA, Nat.

DELEBECQUE-QUADRAT, Contribution of Stochastic Control Singular Pertubation Averaging and Team Theories to an Example of Large Scale System : Management of Hydropower Production, IEEE AC, April 1978.

ERNOULT-MESLIER, Analysis and Forecast of Electrical Energy Demand, R.G.E., April 1982.

MASSE, P., Les Réserves et la Régulation de l'Avenir, Hermann, Paris, 1944.

MESLIER, F., Simulation of the Optimal Control of a Power System : Latest Developments in the Greta Model, IPC, PSCC 6, 1978.

PRONOVOST, R., and J. BOULVA Long Range Operation Planning of a Hydro-Thermal System Modelling an Optimization. Meeting of the Canadian Electrical Association, Toronto, Ontario? March 1978.

QUADRAT J.P., On Optimal Stochastic Control Problem of Large Systems, Internal INRIA Report, 1982.

TURGEON, A., Optimal Operation of Multireservoir Power Systems with Stochastic Inflows, Water Resources Research, April 1980.

# OPTIMAL OPERATION OF THERMAL SYSTEMS WITH START-UP COSTS

J.C. Geromel
FEC/UNICAMP , C.P. 6166 13100 Campinas - SP - Brasil

L.F.B. Baptistella
CPqD/TELEBRÁS , C.P. 1579 13100 Campinas - SP - Brasil

*ABSTRACT* : An extension of the generalized Benders decomposition approach to solve the operation problem of electric power systems with thermal generation is proposed. The resulting minimal operation/start-up cost problem is partitioned into an economic dispatch problem and a pure integer non-linear programming problem. Some results regarding a numerical example are provided. Also, a multicriteria problem is defined in order to take into account a stochastic model for the electric demand.

## 1. Introduction

An electric power generation system has two means of production : the hydraulic generation and the non-hydraulic generation (including thermal production - fuel, gas, nuclear - and secondary energy import market). The optimal operation of this system depends on a compromise between immediate use of the water in the reservoirs and its conservation for future use, satisfying the load demand with minimum production cost (Arvanitidis N.V. and Rosing J., 1970), (Geromel J.C. and Luna H.P.L., 1981). The economic operating schedule of a hydrothermal system is consequently a function of the hydroplant management, due to the low cost of the hydraulic generation compared to the thermal generation cost.

However in a short horizon planning and with large capacity hydroplants, the optimal operation of the hydraulic system can be determined "a priori" and of course in this case it is necessary to determine the optimal generation of the thermal system in order to meet the production deficit with minimum cost.

Although physically coupled, these two systems have different cycles: an annual cycle of the hydraulic system (water cycle) and a shorter cycle (a week) of the thermal system (electric demand cycle). This suggests a division of the global economic planning problem into a short term problem and a long term one: the optimal operation of the later defines the amount of energy weekly necessary to be produced from the thermal system. The economic operation of the thermal units, generating this fixed amount of energy, has to take into account both the *operation cost*, which depends mainly on the burned fuel for the electric energy production, and the *start-up cost* which depends on the units down time (when a unit is restarted, fuel has to be burned to raise the temperature of the boiler up to its operating point) (Turgeon A., 1978). The thermal problem can be classified as a mixed-integer mathematical programming problem. In the literature it is solved by some methodologies : in (Turgeon A., 1978) a branch and bound method is proposed, in (Baptistella L.F.B. and Geromel J.C., 1980) the generalized Benders decomposition method, to solve simultaneously the hydrothermal system with short capacity hydroplants, is discussed, and more recently (Bertsekas D.P. and co-workers, 1983) proposed a dual method which appears to be powerfull when a large number of thermal unities is considered. This method, however can not be used, for instance, when the number of thermal units is relatively small since no feasible solution is available due to the *duality gap* (the problem is not *convex*).

In this paper we propose a method which is more adapted to large capacity hydroplants (the Brazilian case) and takes into account the system characteristics to simplify the numerical solution of the problem.

The paper is composed as follows : in section 2 the systems modelling is performed, in section 3 the decomposition method is obtained and a computational procedure is given in section 4. In section 5 a real example is solved in order to show the method convergence characteristics. Finally in section 6 an extension of the presented method is proposed in order to handle the model with stochastic load demand. It will be shown that the stochastic programming problem can be solved by means of a multicriteria approach.

## 2. System Modelling

Let the thermal system be composed of N generating units each one characterized by

$$\begin{cases} C_i(\xi) \text{ , production cost of } \xi \text{ MW in \$} \\ \underline{s}_i \leqslant \xi \leqslant \bar{s}_i \text{ , generation capacities in MW} \end{cases} \tag{1}$$

Let $s_i^t$ be the energy produced by unit i during the interval $[t, t+1)$. We need to satisfy.

$$\sum_{\substack{\text{units in} \\ \text{operation}}} s_i^t \geqslant L^t \tag{2}$$

where $L^t$ represents the energy production deficit, i.e., the difference between the hydraulic production and the electric demand. Considering a time horizon of T periods (a week hourly discretized for example) and defining the following logical variables

$$y_i = \begin{cases} 0 \text{ if the ith thermal unit is "off" during } \forall t \in [0, T-1] \\ 1 \text{ if the ith thermal unit is "on" during } \forall t \in [0, T-1] \end{cases} \tag{3}$$

We can formulate the following optimization problem (Turgeon A., 1978), (Baptistella L.F.B. and Geromel J.C., 1980):

$$\begin{aligned} &\text{Min} \sum_{t=0}^{T-1} \sum_{i=1}^{N} C_i(s_i^t) y_i + \sum_{i=1}^{N} P_i(y_i) \\ &\sum_{i=1}^{N} s_i^t \, y_i \geqslant L^t \qquad t = 0,1,\ldots(T-1) \\ &y \in Y \quad , \quad s_i^t \in S_i \end{aligned} \tag{4}$$

where $y' = [y_1 \ldots y_N] \in Y$ is the practical operating constraint set associated to the start-up and shunt-down constraints (Turgeon A., 1978), $S_i$ is the set of generation capacities defined by $S_i = \{\xi / \underline{s}_i \leqslant \xi \leqslant \bar{s}_i\}$ and $P_i(\cdot)$ represents the start-up cost of unit i which can be calculated from the cost $P_{oi}$ (in $) (to raise the temperature of the boiler of unit i to its operating point) and $y_i^-$, the state of unit i in the previous period. We can define the following table

| $y_i^-$ | $y_i$ | $P(y_i)$ |
|---|---|---|
| 0 | 0 | 0 |
| 0 | 1 | $P_{oi}$ |
| 1 | 0 | 0 |
| 1 | 1 | 0 |

(5)

which is equivalent to

$$P_i(y_i) = P_{oi} \cdot (1 - y_i^-) \cdot y_i \overset{\Delta}{=} \pi_i\, y_i \tag{6}$$

Finally, we can rewrite (4) as:

$$\begin{aligned} &\min_{\substack{y \in Y \\ s_i^t \in S_i}} \sum_{t=0}^{T-1} \sum_{i=1}^{N} C_i(s_i^t) y_i + \sum_{i=1}^{N} \pi_i\, y_i \\ &\qquad \sum_{i=1}^{N} s_i^t\, y_i \geqslant L^t \qquad t = 0,1,\ldots(T-1) \end{aligned} \tag{7}$$

This is a mixed-integer mathematical programming problem with $2^N$ possible combinations to the integer variables and N.T continuous variables. To solve (7) with the methodology that will be proposed in the sequel, it is necessary to assure convexity, of functions $C_i(\cdot)$, i=1...N which is normally assumed (Geromel J.C. and Luna H.P.L., 1981). In the following section (7) will be decomposed using Benders techniques considering by hypothesis that the production deficit $L^t$ is deterministic for all t of the planning horizon.

## 3. Decomposition

To solve problem (7) we can verify that the fixation of the logical variables $y_i$, i=1...N, produces an optimal dispatch problem as a sub-problem. Thus, it is apparent that some advantages are obtained if the projection onto the y-space, as proposed by (Geoffrion A.M., 1972), is accomplished. This projection gives

$$\min_{y \in Y \cap V} \sum_{i=1}^{N} \pi_i\, y_i + v(y) \tag{8}$$

where

$$\begin{aligned} v(y) = &\min_{s_i^t \in S_i} \sum_{t=0}^{T-1} \sum_{i=1}^{N} C_i(s_i^t) \cdot y_i \\ &\qquad \sum_{i=1}^{N} s_i^t\, y_i \geqslant L^t \qquad t = 0,1,\ldots(T-1) \end{aligned} \tag{9}$$

and $V \subset R^N$ is a set which assures that for every $y \in Y$, a feasible solution of problem (9) exists. Fortunately, it is easy to see that V is given here by

$$V = \{y \in R^N \ / \ \sum_{i=1}^{N} \bar{s}_i \, y_i \geq L^t \quad , \quad t = 0,1,\dots(T-1)\} \tag{10}$$

On the other hand, it is well known that the breaking-down probability of a thermal unit during this planning horizon is significant. If this fact is taken into account, a generation reserve is necessary and in place of (10) we must consider $\tilde{V} \subset V$ where

$$\tilde{V} = \{y \in R^N \ / \ \sum_{i=1}^{N} \bar{s}_i \, y_i - \max_i(\bar{s}_i \, y_i) \geq L^t \quad , \quad t = 0,1,\dots(T-1)\} \tag{11}$$

Let $\lambda^t$ be the price of the energy generated during the time interval $[t, t+1)$ (in \$/MW). Since (9) is convex, we have:

$$v(y) = \underset{\lambda^t \geq 0}{\text{Max}} \ \underset{s_i^t \in S_i}{\text{Min}} \ \sum_{t=0}^{T-1} \sum_{i=1}^{N} (C_i(s_i^t) - \lambda^t \, s_i^t) y_i + \sum_{t=0}^{T-1} \lambda^t \, L^t \tag{12}$$

Supposing that the Max-Min problem above was solved with $y_i^*$, i=1...N fixed, allowing the optimal price $\lambda_*^t$, t=0,1,...(T-1) we have:

$$v(y) \geq \sum_{t=0}^{T-1} \sum_{i=1}^{N} \min_{s_i^t \in S_i} \ (C_i(s_i^t) - \lambda_*^t \, s_i^t) y_i + \lambda_*^t \, L^t$$

$$\geq \sum_{i=1}^{N} a_i^* \, y_i + b^* \tag{13}$$

where $a_i^*$, i=1...N and $b^*$ are given by

$$a_i^* \triangleq \sum_{t=0}^{T-1} \min_{s_i^t \in S_i} \ (C_i(s_i^t) - \lambda_*^t \, s_i^t)$$

$$b^* \triangleq \sum_{t=0}^{T-1} \lambda_*^t \, L^t \tag{14}$$

We can note that (13-14) define an external linearization of the function v(y) (Geoffrion A.M., 1972). If we have p external linearizations, the following problem approximates (8):

$$\underset{\psi, y \in Y \cap V}{\text{Min}} \ \sum_{i=1}^{N} \pi_i \, y_i + \psi \tag{15}$$

$$\psi \geq \sum_{i=1}^{N} a_i^j \, y_i + b^j \qquad j = 1,2,\dots p$$

or, eliminating the continuous variable

$$\underset{y \in Y \cap V}{\text{Min}} \left( \sum_{i=1}^{N} \pi_i \, y_i + \underset{1 \leq j \leq p}{\text{Max}} \left\{ \sum_{i=1}^{N} a_i^j \, y_i + b^j \right\} \right) \tag{16}$$

If necessary, other external linearizations can be generated, which are added to the previous ones and as a result, the problem (16) approximates (8) in a better way. Problem (16) can be solved by implicit enumeration techniques (Dillon T.S., and co-workers, 1978), allowing $y_i^*$, i=1...N. To define problem (16) we have to calculate the coefficients $a_i$ and b, or in other words, solve the max-min problem (12). With the additional assumption that the functions $C_i(\cdot)$ are strictly convex, we can apply the classical dual method, where $\lambda_*^t$ is the solution of

$$\underset{\lambda \geq 0}{\text{Max}} \quad \phi(\lambda) \tag{17}$$

where $\lambda' = \left[\lambda^0, \ldots, \lambda^{T-1}\right] \in R^T$ and $\phi(\cdot) : R^T \to R$ is the dual function defined by

$$\phi(\lambda) = \sum_{t=0}^{T-1} \sum_{i=1}^{N} \min_{s_i^t \in S_i} \left(C_i(s_i^t) - \lambda^t \, s_i^t\right) y_i^* + \lambda^t \, L^t \tag{18}$$

observe that (18) is the optimal dispatch problem which can be easily solved for each fixed $\lambda^t$, t=0,1,...(T-1), allowing

$$\hat{s}_i^t = \max\{\underline{s}_i \, , \; \min\{\bar{s}_i, \, \tilde{s}_i^t\}\} \tag{19}$$

where $\tilde{s}_i^t$ is the solution of the algebric equation (see Appendix)

$$\left. \frac{dC_i(\xi)}{d\xi} \right|_{\xi = \tilde{s}_i^t} = \lambda^t \tag{20}$$

On the other hand, we must emphasize that the dual function is differentiable, due to the previous assumptions, and its gradient is given by

$$\frac{\partial \phi}{\partial \lambda^t} = L^t - \sum_{i=1}^{N} \hat{s}_i^t \, y_i^* \qquad t = 0,1,\ldots(T-1) \tag{21}$$

which can be used to solve (17), allowing the optimal energy prices $\lambda_*^t$, t=0, 1,...(T-1).

## 4. Computational Procedure

Bearing in mind the results of the previous sections, we can state the following procedure:

*Step 1* : Let $y_i^*$, i=1,...N be a known and feasible policy ($y^* \in Y \cap V$). Construct the set $I = \{i \mid y_i^* = 1\}$ and solve the dispatch problem (22), using the dual method (17-21)

$$\min_{s_i^t \in S_i} \sum_{t=0}^{T-1} \sum_{i \in I} C_i(s_i^t)$$
$$\sum_{i \in I} s_i^t \geq L^t \qquad t = 0,1,\ldots(T-1) \tag{22}$$

Let $\hat{s}_i^t$, $i \in I$ and $\lambda_*^t$, $t \in [0, T-1]$ be the optimal primal and dual solutions.

*Step 2* : Calculate the coefficients $a_i^*$ and $b^*$ given by (14) and the value $v(y^*) = \sum_{i \in I} a_i^* + b^* = BS$; make $p = 1$, $a_i^p = a_i^*$, i=1...N, $b^p = b^*$. BS is the upper bound of the optimal value of (7).

*Step 3* : Calculate the optimal solution $y_i^*$, i=1...N of problem (16) and

$$BI = \sum_{i=1}^{N} \pi_i\, y_i^* + \max_{1 \leq j \leq p} \left\{ \sum_{i=1}^{N} a_i^j\, y_i^* + b^j \right\} \tag{23}$$

BI is the lower bound of the optimal value of (7). If $BS - BI \leq \varepsilon$, where $\varepsilon > 0$ is arbitrarily small, stop. Otherwise redefine the set I and proceed to step 4.

*Step 4* : Solve the problem (22) and calculate the coefficients defined in (14) with the new set I. If $\overline{BS} - BI \leq \varepsilon$, where $\overline{BS} = \sum_{i \in I} a_i^* + b^*$, stop. Otherwise make $p = p+1$, $a_i^p = a_i^*$, i=1...N, $b^p = b^*$ and if $BS > \overline{BS}$ make $BS = \overline{BS}$. Return to step 3.

*Remarks* :

a) This procedure has a finite convergence due to the finite number of logical variables and due to the convexity of problem (7) for any temporary fixation of $y \in Y$ (Geoffrion A.M., 1972).

b) The numerical solution of problem (22) with the dual method is very simple, allowing a decomposition of that problem in N.T subproblems, each one with a unique decision variable.

c) If the number of feasible combinations of the logical variables in problem (16) is reduced, it can be solved by simple substitution, avoiding more sophisticated techniques like the implicit enumeration.

## 5. Application to an Energy Generation System

An example concerning an Electric Energy Generation System was solved using the presented approach and is presented here. We consider a network with ten generating thermal units, with an hourly operating policy for a period of a day, previously studied in (Turgeon A., 1978). The costs and others data are given in table 1.

TABLE 1 Characteristics of Thermal Units

| $i$ | $\underline{s}_i$ | $\bar{s}_i$ | $\alpha_i$ | $\beta_i$ | $\gamma_i$ | $P_{oi}$ |
|---|---|---|---|---|---|---|
| 1 | 0 | 120 | 1.40 | 0.0038 | 32 | 56 |
| 2 | 0 | 150 | 1.54 | 0.0021 | 29 | 68 |
| 3 | 0 | 150 | 1.33 | 0.0013 | 100 | 149 |
| 4 | 0 | 200 | 1.21 | 0.0015 | 82 | 136 |
| 5 | 0 | 60 | 1.40 | 0.0051 | 15 | 51 |
| 6 | 0 | 80 | 1.50 | 0.0040 | 25 | 60 |
| 7 | 0 | 100 | 1.35 | 0.0039 | 40 | 68 |
| 8 | 0 | 280 | 1.35 | 0.0026 | 72 | 105 |
| 9 | 0 | 520 | 1.39 | 0.0013 | 105 | 160 |
| 10 | 0 | 320 | 1.26 | 0.0029 | 49 | 112 |

The coefficients $\alpha_i$, $\beta_i$ and $\gamma_i$ are associated to the quadratic function

$$c_i(\xi) = \gamma_i + \alpha_i\,\xi + \beta_i\,\xi^2 \qquad i = 1,2,\ldots,10 \tag{24}$$

which represents the running cost of unit i. $P_{oi}$ is the start-up cost of unit i. It is assumed that no thermal units are operating initially implying that $y_i^- = 0$, $i=1\ldots N$ and for initialization purpose it was considered $y_i = 1$, $i=1\ldots N$.

The computational procedure terminates after three master iterations (Fig. 1) and the optimal trajectories are given in table 2. Although no special attention has been given to perform the iterations, the optimal operating policy was found in less than 20 sec. of CPU time on a IBM 370/168 computer.

## 6. A Stochastic Model

Up to now, the deficit $L^t$ was considered as deterministic, however it depends on the demand level which actually is a random variable (Galiana F.D. and co-workers, 1974). We suppose that $L^t$ is a gaussian variable with mean $\ell^t$ and standard deviation $\sigma^t$. The energy production constraint of problem (7) can be changed to

$$\text{prob}\left\{\sum_{i=1}^{N} s_i^t\, y_i \geq L^t\right\} \geq \mu^t \qquad t = 0,1,\ldots,(T-1) \tag{25}$$

where $\mu^t \in [0,\ 1]$. Calling $F_t(\cdot)$ the distribution function of $L^t$, the constraint (25) is equivalent to

$$\sum_{i=1}^{N} s_i^t\, y_i \geq F_t^{-1}(\mu^t) \qquad t = 0,1,\ldots(T-1) \tag{26}$$

Now, our problem is still to minimize the operation plus start-up costs but also to maximize the probability that the constraint in (4) is satisfied. With (26) and defining $z^t$ s.t. $F_t(z^t) = \mu^t$, our problem turns to be

$$\text{Min} \sum_{t=0}^{T-1} \sum_{i=1}^{N} C_i(s_i^t) y_i + \sum_{i=1}^{N} P_i(y_i)$$

$$\text{Max} \sum_{t=0}^{T-1} F_t(z^t) \qquad (27)$$

$$\sum_{i=1}^{N} s_i^t \, y_i \geqslant z^t \qquad t = 0,1,\ldots,(T-1)$$

$$y \in Y \quad , \quad s_i^t \in S_i \quad , \quad z^t \geqslant \ell^t$$

Note that the constraint $z^t \geqslant \ell^t$ has been introduced in order to garantee that at the optimum $\mu^t \geqslant 50\%$. (27) is a bi-criteria optimization problem and can be solved interactively (Baptistella L.F.B. and Olero A., 1980), by presenting to the decision-maker the optimal solution of the following monocriterium problem

$$\text{Min} \sum_{t=0}^{T-1} \sum_{i=1}^{N} C_i(s_i^t) y_i + \sum_{i=1}^{N} P_i(y_i) - \Theta \sum_{t=0}^{T-1} F_t(z^t) \qquad (28)$$

where (28) is subjected to the same constraints as (27) and $\Theta \geqslant 0$ defines the trade-off between the two criteria. We must emphasize that this problem can be solved by the proposed technique where (8) remains the same and v(y) admits the same external linearization (13) with $a_i^*$, i=1...N given by (14) and $b^*$ given by

$$b^* \triangleq - \sum_{t=0}^{T-1} \max_{z^t \geqslant \ell^t} (\Theta F_t(z^t) - \lambda_*^t z^t) \qquad (29)$$

It is interesting to note that (29) can help the decision-maker to set-up the trade-off $\Theta$ since supposing that $z^t > \ell^t (\mu^t > 50\%)$ its optimal solution satisfies

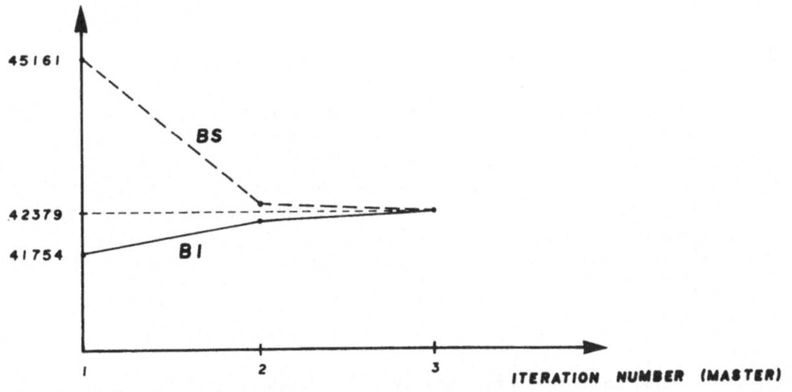


Fig. 1 - Convergence of the Procedure (BS-Upper and BI-Lower Bound)

TABLE 2 Optimal Energy Price and Thermal Production

| $t$ | $L^t$ | $s_1^t$ | $s_2^t$ | $s_3^t$ | $s_4^t$ | $s_5^t$ | $s_6^t$ | $s_7^t$ | $s_8^t$ | $s_9^t$ | $s_{10}^t$ | $\lambda_*^t$ |
|---|---|---|---|---|---|---|---|---|---|---|---|---|
| 1 | 1021 | 86 | 123 | - | 200 | 60 | 70 | 91 | - | 255 | 137 | 2.0 |
| 2 | 960 | 80 | 111 | - | 200 | 59 | 63 | 84 | - | 235 | 128 | 2.0 |
| 3 | 909 | 74 | 101 | - | 200 | 55 | 58 | 79 | - | 220 | 121 | 2.0 |
| 4 | 899 | 73 | 100 | - | 200 | 55 | 57 | 78 | - | 217 | 120 | 2.0 |
| 5 | 890 | 72 | 98 | - | 200 | 54 | 56 | 77 | - | 214 | 118 | 2.0 |
| 6 | 920 | 76 | 103 | - | 200 | 56 | 59 | 80 | - | 223 | 122 | 2.0 |
| 7 | 960 | 80 | 111 | - | 200 | 59 | 63 | 84 | - | 235 | 128 | 2.0 |
| 8 | 920 | 76 | 103 | - | 200 | 56 | 59 | 80 | - | 223 | 122 | 2.0 |
| 9 | 890 | 72 | 98 | - | 200 | 54 | 56 | 77 | - | 214 | 118 | 2.0 |
| 10 | 869 | 70 | 94 | - | 200 | 52 | 54 | 75 | - | 208 | 116 | 1.9 |
| 11 | 838 | 67 | 88 | - | 200 | 50 | 51 | 72 | - | 198 | 111 | 1.9 |
| 1 | 827 | 66 | 86 | - | 200 | 49 | 50 | 71 | - | 195 | 110 | 1.9 |
| 13 | 808 | 64 | 83 | - | 200 | 48 | 48 | 69 | - | 189 | 107 | 1.9 |
| 14 | 797 | 63 | 80 | - | 200 | 47 | 47 | 68 | - | 186 | 106 | 1.9 |
| 15 | 787 | 62 | 79 | - | 200 | 46 | 46 | 67 | - | 183 | 105 | 1.9 |
| 16 | 766 | 60 | 75 | - | 200 | 45 | 44 | 65 | - | 177 | 102 | 1.9 |
| 17 | 746 | 58 | 71 | - | 200 | 43 | 42 | 63 | - | 171 | 99 | 1.8 |
| 18 | 726 | 56 | 67 | - | 200 | 41 | 40 | 61 | - | 164 | 96 | 1.8 |
| 19 | 695 | 53 | 62 | - | 196 | 39 | 38 | 58 | - | 156 | 93 | 1.8 |
| 20 | 685 | 52 | 61 | - | 194 | 39 | 37 | 57 | - | 154 | 92 | 1.8 |
| 21 | 674 | 51 | 59 | - | 192 | 38 | 36 | 56 | - | 151 | 90 | 1.8 |
| 22 | 715 | 54 | 65 | - | 200 | 41 | 39 | 60 | - | 161 | 95 | 1.8 |
| 23 | 757 | 59 | 73 | - | 200 | 44 | 43 | 64 | - | 174 | 100 | 1.8 |
| 24 | 1021 | 86 | 123 | - | 200 | 60 | 70 | 91 | - | 255 | 137 | 2.0 |

$$0 \leq \lambda_*^t = \Theta \frac{dF_t}{dz^t} \leq \frac{1}{\sqrt{2\pi}\,\sigma^t} \cdot \Theta \qquad t = 0,1,\ldots,(T-1) \tag{30}$$

Meaning that the optimal energy price cannot be greater than the trade-off divided by a known constant. Of course this property can guide the decision-maker to define his preference point (non-inferior).

## 7. Conclusions

In this paper, a decomposition approach for determining the optimal operating schedule for thermal systems has been developed. The presented algorithm takes advantage of the structure of the problem, and significant features of the method include the upper and lower bound estimates of the optimal value of the objective function in any iteration. An examination of Fig. 1 reveals that the decomposition approach converges rapidly, allowing good and feasible solutions even at the beginning of the iterative procedure. With a few modifications, the same algorithm can handle stochastic models for the demand behavior.

*Acknowledgement* - This research was developed in part with the financial support of the Conselho Nacional de Desenvolvimento Científico e Tecnológico (CNPq-Brazil) under grant no. 301373/80 and Fundação de Amparo a Pesquisa do Estado de S.P Fapesp-Brasil.

## 8. References


Arvanitidis, N.V. and Rosing, J. (1970). Optimal Operation of Multireservoir Systems Using a Composite Representation - IEEE Transactions on PAS, Vol. PAS 89, Nº 2.

Baptistella, L.F.B. and Geromel, J.C. (1980). Decomposition Approach to the Problem of Unit Commitment Schedule for Hydrothermal Systems. Proceedings IEE, November.

Baptistella, L.F.B. and Ollero, A. (1980). Fuzzy Methodologies for Interactive Multicriteria Optimization. IEEE Transactions on Systems, Man and Cybernetics, Vol. SMC-10, Nº 7.

Bertsekas, D.P., Lauer, G.S., Sandell, Jr. N.R. and Posbergh, T.A. (1983). Optimal Short-Term Scheduling of Large-Scale Power Systems. IEEE Transactions on A.C. Vol. AC-28, Nº 1.

Dillon, T.S., Edwin, K.W., Kochs, H.D. and Taud, R.J. (1978). Integer Programming Approach to the Problem of Optimal Unit Commitment with Probabilistic Reserve Determination. IEEE Transactions on PAS, Vol. PAS 97, Nº 6.

Galiana, F.D., Handschin, E. and Fiechler, A.R. (1974). Identification of Stochastic Electric Load Models from Physical Data. IEEE Trans. on Automatic Control, Vol. AC-19, Nº 6.

Geoffrion, A.M. (1972). Generalized Benders Decomposition, Journal of Optimization Theory and Applications, Vol. 10, Nº 4.

Geromel, J.C. and Luna, H.P.L. (1981). Projection and Duality Techniques in Economic Equilibrium Models, IEEE Systems, Man and Cybernetics, Vol. SMC-11, Nº 5.

Turgeon, A. (1978). Optimal Scheduling of Thermal Generating Units. IEEE Transactions on A.C., Vol. AC-23, Nº 6.


## 9. Appendix

Clearly, the solution of (20) always exists only if

$$\frac{dC_i(\xi)}{d\xi} : (-\infty, +\infty) \to [0, +\infty) \qquad i = 1 \ldots N \tag{A1}$$

This is the case of the quadratic function (24) which represents the production costs for a practical system with good approximation.

In the general case, the optimal solution of (18) is given by

$$\hat{s}_i^t = \bar{s}_i \quad \text{if} \quad \left.\frac{dC_i(\xi)}{d\xi}\right|_{\xi=\bar{s}_i} \leqslant \lambda^t \tag{A2}$$

$$\hat{s}_i^t = \underline{s}_i \quad \text{if} \quad \left.\frac{dC_i(\xi)}{d\xi}\right|_{\xi=\underline{s}_i} \geqslant \lambda^t \tag{A3}$$

or if (A2) and (A3) are not satisfied then $\hat{s}_i^t$ must solve the equation

$$\left.\frac{dC_i(\xi)}{d\xi}\right|_{\xi=\hat{s}_i^t} = \lambda^t \tag{A4}$$

which in this case admits surely an unique solution.

# IDENTIFICATION OF COMBUSTION LOSSES AND AIR FLOW CONTROL IN POWER PLANTS BURNING INHOMOGENEOUS FUELS

Kari Lehtomäki Reijo Ramu
Tampere Univ. of Technology
Control Engineering Laboratory
P.O. Box 527
SF-33101 TAMPERE 10
Finland

Abstract. Combustion air flow control has a strong effect on the efficiency of a power plant. When using homogeneous fuels such as oil and natural gas the control can be based on an assumption that certain volumetric or mass flow of the fuel needs always a certain air flow for complete combustion. But with inhomogeneous fuels and in multi-fuel boilers this assumption is not relevant. For example composition, density and moisture of peat vary considerably and the combustion air flow cannot be fixed on the basis of fuel flow only. A common strategy is to use measurements of steam flow or pressure and of flue gas oxygen content as feedback information for air flow control. In this paper we will discuss new possibilities to solve the problem. Using a sensitive CO-measurement we can identify CO-loss curve and when we know both CO- and heat losses as functions of $O_2$-content we can determine optimum combustion air flow on-line. The losses and the optimum depend on the boiler load, the fuel proportions, the burner conditions, etc.

## 1. INTRODUCTION

The use of peat as a fuel of power plants and heating centres has increased rapidly in Finland during the last few years. A typical large peat power plant burns pulverized peat and has a mill-drying system shown in Fig. 1. Peat is fed from mill silos to drying and milling. Hot flue gases (700...1000°C) from top of the furnace are used for drying and transferring peat to combined oil/peat burners. The peat dries from 40... 65% initial moisture to about 5...25% moisture. Normally three milling lines are used each feeding peat to 3...4 burners. Slowly burning material like sticks etc. fall to an after-burning grate. Since peat is an inhomogenous fuel, oil is used as a supporting fuel (less than 10% of the total fuel power) to stabilize combustion. Peat is much cheaper than oil and, therefore reduction in the use of oil in normal operating situations is economically desirable. The water/steam part of the plant is conventional and consists e.g. of economizers, a natural circuit drum boiler, a three stage superheater, heat exchangers of district heating network, a turbine and a generator.

The main difficulties in the control of peat fueled power plants are caused by the large stochastic variation in the properties of the input peat flow. The effective heat value of peat can vary between 1200... 4600 MJ/$m^3$ due to the varying moisture, density and composition of peat. Peat feeders are usually volume feeders and thus the actual fuel power can vary widely although peat feed to the furnace is kept constant. The quality variations of peat appear in the operation of the plant as variations in the steam pressure and temperature, which the normal control loops cannot eliminate.

In power plants, which burn mainly one fuel, combustion air flow can be controlled using flue gas oxygen content as the only feedback information, because in this case the optimal set point of the oxygen content depends approximately only on boiler load. But if the plant uses several fuels the oxygen content set point is difficult to determine, because it depends strongly also on fuel proportions etc. In this case it is necessary to use some other measurement as an additional feedback. For this purpose carbon monoxide content in the flue gases can be used. New CO-monitors, which measure absorption of infrared light across the stack, provide sufficient sensitivity for this application.

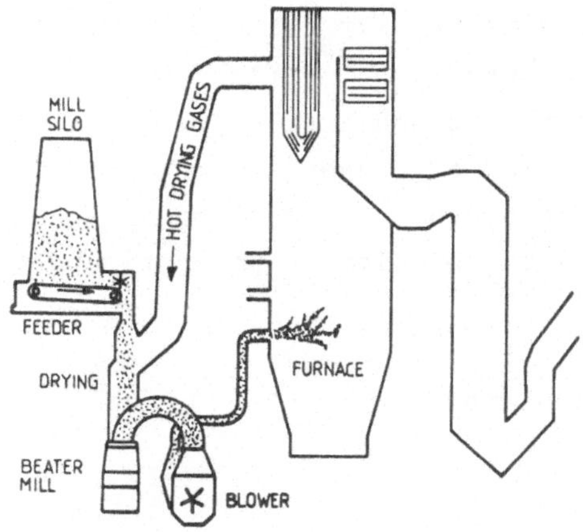


Fig. 1. The fuel drying system at a typical peat power plant.

## 2. COMBUSTION AIR FLOW CONTROL USING CO-FEEDBACK

New CO-analyzers give quite reliable information about how perfect combustion in the furnace is. A typical relation between CO- and $O_2$-contents in the flue gases is shown in Fig. 2. An empirical set point for CO-content is 150...300 ppm (parts per million) for most fuels. If CO-content increases losses due to unburned matter increase rapidly and if it decreases losses due to extra combustion air flow decrease the efficiency of the plant. Fig. 3 shows a simple way to use CO-measurement in control. It can be used only with homogeneous fuels like natural gas and oil.

Figures 4 and 5 show two strategies to use both $O_2$- and CO-measurements as feedback information. Set point for $O_2$-content is a function of the load of the boiler (i.e. steam flow) and therefore the selection of f(x) is important for good control. The first strategy uses CO-content only for fine tuning when $O_2$-content is in a desired range. In the second strategy both CO- and $O_2$-feedbacks are used simultaneously and CO-control changes the set point of $O_2$. If f(x) is chosen such that the set point of $O_2$-content is far too high there exists no significant CO-contents and the CO-controller is always out of its range.

## 3. OPTIMIZING THE COMBUSTION AIR FLOW

The flue gas losses can be divided into losses as unburned matter (mainly CO) and the heat losses of the flue gas. The CO-losses decrease rapidly when the air flow is increased but the heat losses in turn increase nearly linearly. The loss curve is usually drawn as a function of the air/fuel ratio but since the oxygen content in the flue gas is nearly linearly dependent on the air/fuel ratio at small $O_2$-content values the loss curve can also be drawn as a function of oxygen content (Fig. 6). The optimum oxygen content set point can now be obtained from the minimum value of the loss function. Thus the problem is to continuosly identify a changing loss function.

Heat losses can be calculated from equation

$$L_T = C_f(T_o - T_a)F_s , \qquad (1)$$

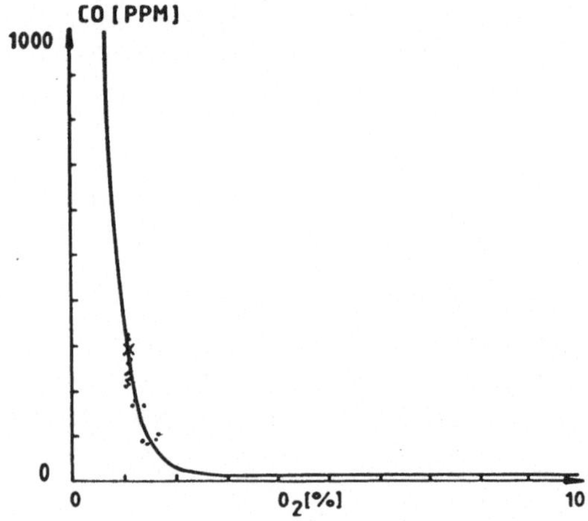


Fig. 2. A typical relation between CO- and $O_2$-contents in flue gases.

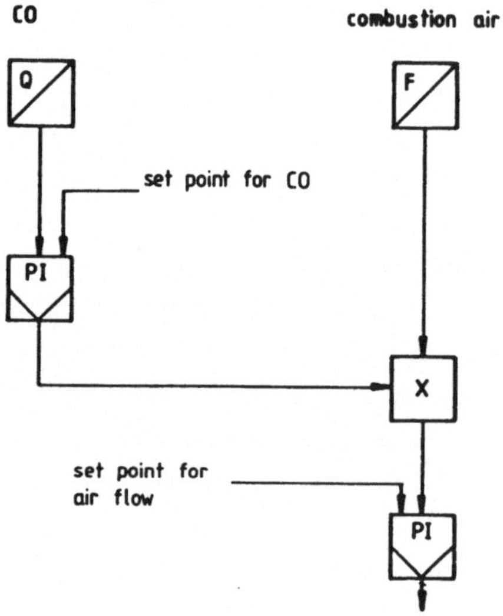


Fig. 3. Burning air flow control with correction from CO-controller

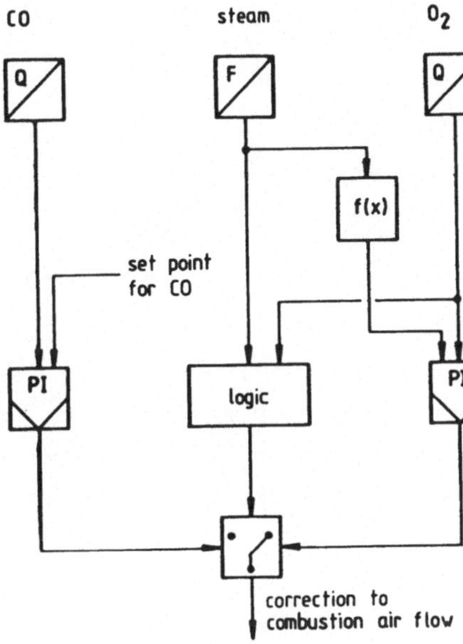


Fig. 4. Alternating $CO/O_2$-correction control.

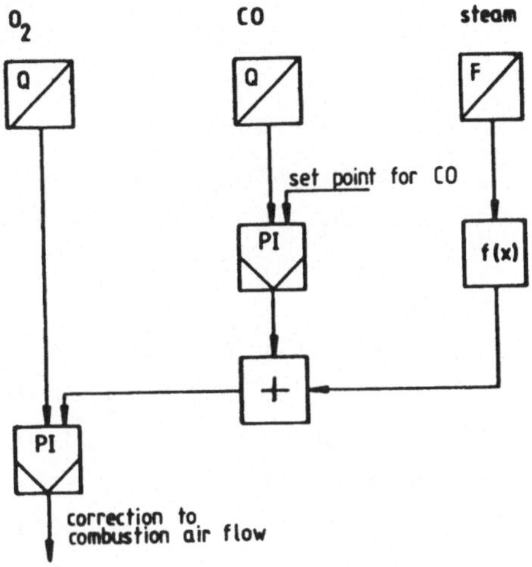


Fig. 5. $CO/O_2$-correction control

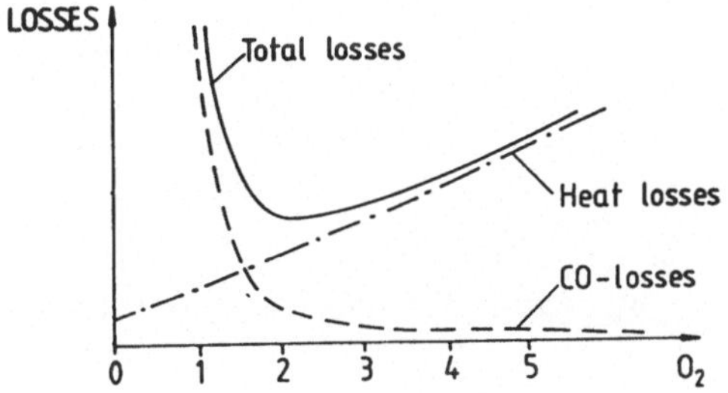


Fig. 6. The flue gas losses

where

$L_T$ = the heat losses, MW,
$C_f$ = the heat capacity of flue gases, $MJ/^0Cm^3$,
$T_0$ = the end temperature of flue gases, $^0C$,
$T_a$ = the outside air temperature, $^0C$,
$F_s$ = the flue gas flow, $m^3/s$.

Normally the flue gas flow is not measured and it must then be calculated from other quantities. If the power plant uses two different fuels (peat and oil) the flue gas flow can be calculated from four measurements (the oxygen content, the air flow, the support oil flow and the moisture of peat) using following equations(which take into account the compositions of fuels and combustion reactions)

$$m_s = \frac{20{,}95F_i - \left(100c_5 + \dfrac{20{,}95c_4X_{02}}{20{,}95-X_{02}}\right)m_0}{\dfrac{20{,}95(c_1M + c_2)X_{02}}{20{,}95 - X_{02}} + 100c_3(100-M)}, \qquad (2)$$

$$F_s = \frac{(c_1M + c_2)m_s + c_4m_0}{1 - \dfrac{X_{02}}{20{,}95}}, \qquad (3)$$

where

$m_s$ = the peat flow calculated as a fictive "standard peat" (composition constant), kg/s,

$m_0$ = the oil flow, kg/s,

$F_i$ = the combustion air flow, $m^3/s$,

$X_{02}$ = the $0_2$-content, vol-%,

M = the moisture of peat, weight-%

$c_1,\ldots,c_5$ = parameters depending on the compositions of fuels.

With most fuels the most important cause for losses as unburned matter is unreacted carbon monoxide in the flue gases. The losses can be calculated from equation

$$L_c = hX_{CO}\, F_s \cdot 10^{-6} \,, \tag{4}$$

where

$L_c$ = the CO-losses, MW,

$h$ = the heat value of CO ≈ 12,6 $MJ/m^3$,

$X_{CO}$ = the CO-content, ppm.

The losses can be given proportional to the power released in the furnace

$$P_c = ((100 - M)c_6 - c_7 M)m_s + c_8 m_0 \,, \tag{5}$$

where

$P_c$ = the fuel power released in the furnace, MW,

$c_6 \ldots c_8$ = constants.

The heat losses are

$$L_{T\%} = \frac{L_T}{P_c} \cdot 100\% \tag{6}$$

where

$L_{T\%}$ = the percentual heat losses, %

and the CO-losses

$$L_{c\%} = \frac{L_c}{P_c} \cdot 100\% \tag{7}$$

where

$L_{c\%}$ = the percentual CO-losses, %.

To obtain process data for testing the optimization scheme that will be presented in the sequel we made measurements at Kuopio Haapaniemi II peat power plant (60 MW electricity and 120 MW for district heating). Oxygen content in the flue gases was measured with a Kent in-situ $ZrO_2$ oxygen analyzer and CO- and $CO_2$-contents with an EDC across-stack IR analyzer. In Fig. 7 measurements at 43% load and in Fig. 8 at 70% load are shown. It can clearly be seen that decreasing $O_2$-content to about 1% causes quite abrupt jumps in CO-content. Figures 9 and 10 show calculated losses at the same experiments.

The problem in minimizing combustion losses is that the loss curves vary

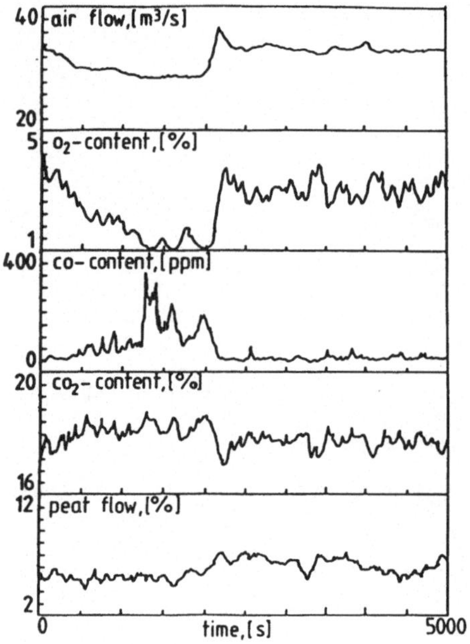


Fig. 7. Process measurements at 43% load level.

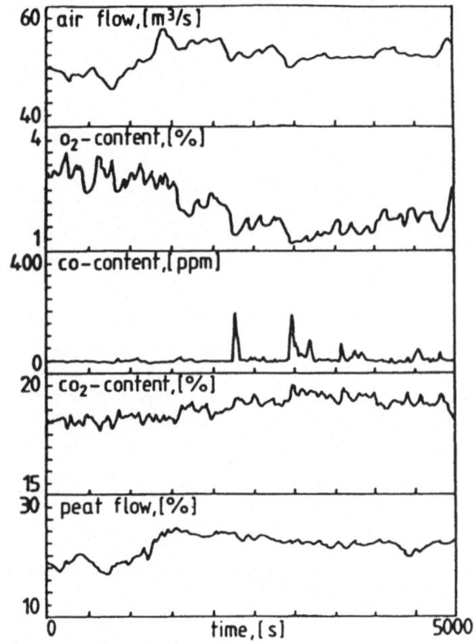


Fig. 8. Process measurements at 70% load level.

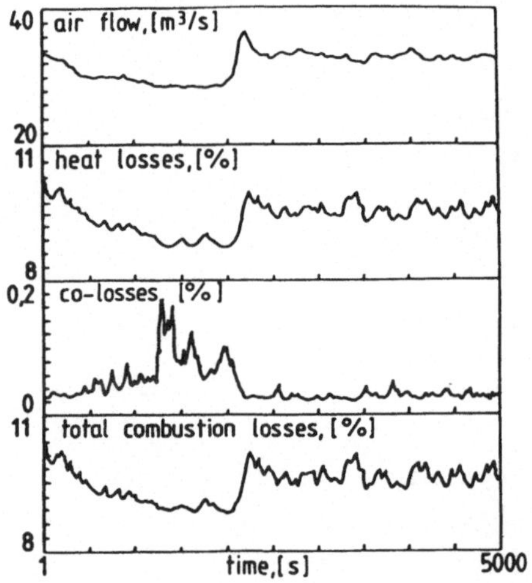


Fig. 9. Heat, CO- and total losses at 43% load level.

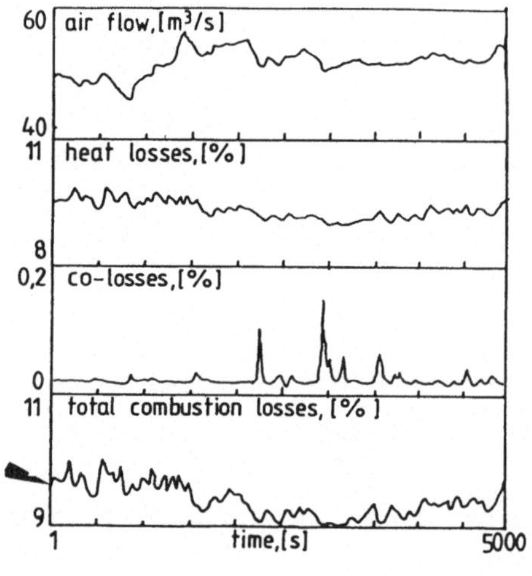


Fig. 10. Heat, CO- and total losses at 70% load level.

due to several indefinite and unmeasurable reasons and, therefore, the optimum set point for the $O_2$-content has to be calculated continuously. The loss curves can be identified when we have fast and reliable measurements of $O_2$- and CO-contents in the flue gases. This can be done by some recursive identification method.

For the identification of the loss curves we have used the recursive least squares method (RLS). Since the parameters are time varying the algorithm has to forget old data. The forgetting can be realized using an exponential or rectangular window. In the simulation studies it was shown that the normal exponential window forgetting scheme works quite badly in fast boiler load changes. On the contrary the algorithm with changing forgetting factor of Fortesque et al [1] gave clearly better results. The RLS-algorithm can be realized in different ways. The standard form is advantageous from the calculation burden view but the algorithm can produce numerical problems. Numerically more stable algorithms are the square root algorithm and the UD-algorithm. The algorithm presented here is the Fortesque algorithm in standard form. The algorithm is used separately for the CO-loss function and the heat loss function. The algorithm is

$$\begin{cases} \hat{y}_k = \underline{x}_k^T \hat{\underline{\theta}}_k \\ \underline{K}_{k+1} = \underline{P}_k \underline{x}_k / (1 + \underline{x}_k^T \underline{P}_k \underline{x}_k) \\ \hat{\underline{\theta}}_{k+1} = \hat{\underline{\theta}}_k + \underline{K}_{k+1} (y_k - \hat{y}_k) \\ \lambda = 1 - (1 - \underline{x}_k^T \underline{K}_{k+1})(y_k - \hat{y}_k)/\Sigma_o \\ \text{if } \lambda_k < \lambda_{min} \Rightarrow \lambda = \lambda_{min} \\ \underline{P}_{k+1} = (\underline{I} - K_{k+1} \underline{x}_k^T) \underline{P}_k / \lambda_k \end{cases} \tag{8}$$

where

$y_k$ = the measurement at time k,

$\hat{y}_k$ = estimate for the measurement at time k,

$\underline{x}_k$ = state vector at time k,

$\hat{\underline{\theta}}_k$ = estimate for the parameter vector at time k,

$\underline{K}_k$ = Kalman gain,

$\underline{P}_k$ = a matrix,

$\lambda_k$ = the forgetting factor,

$\lambda_{min}$, $\Sigma_o$ = tuning parameters.

For identification of CO-losses we assume that the dependence between $O_2$- and CO-contents is an exponential function

$$X_{CO} = a_1 e^{b_1 X_{O2}}, \tag{9}$$

where

$a_1$, $b_1$ = unknown parameters.

Equation (9) is linearized by taking logarithm from it,

$$\ln(X_{CO}) = a_2 X_{O2} + b_2, \tag{10}$$

where

$$\begin{cases} a_2 = b_1 \\ b_2 = \ln a_1. \end{cases} \tag{11}$$

The parameters of equation (10) can be identified by writing

$$\underline{X} = \begin{bmatrix} X_{02} \\ 1 \end{bmatrix},$$

$$y = \ell n(X_{CO}),$$

and

$$\underline{\hat{\theta}} = \begin{bmatrix} \hat{a}_2 \\ \hat{b}_2 \end{bmatrix}.$$

From estimates $\hat{a}_2$ and $\hat{b}_2$ we obtain $\hat{a}_1$ and $\hat{b}_1$ using equations (11).
Now

$$\hat{X}_{CO} = \hat{a}_1 e^{\hat{b}_1 X_{02}}, \tag{12}$$

where

$\hat{X}_{CO}$ = estimate for the CO-content, ppm.

The estimate $\hat{X}_{CO}$ is the identified curve, which describes the dependence between $O_2$- and CO-contents. Fig. 11 shows an identification run using the earlier process measurements at 43% load level. The picture shows $\hat{X}_{CO}$ at time 3500 s and last measurements with dots. Fig. 12 shows a corresponding run at 70% load level.

Estimate for the CO-losses s

$$\hat{L}_C = h\hat{X}_{CO}\, F_s \cdot 10^{-6} \tag{13}$$

where

$\hat{L}_C$ = estimate for CO-losses, MW.

Fig. 13 shows estimate for the percentual CO-losses at 43% load level.

At small $O_2$-contents the heat losses can be approximated to be a linear function of $O_2$-content

$$L_{T\%} = a_3 X_{02} + b_3, \tag{14}$$

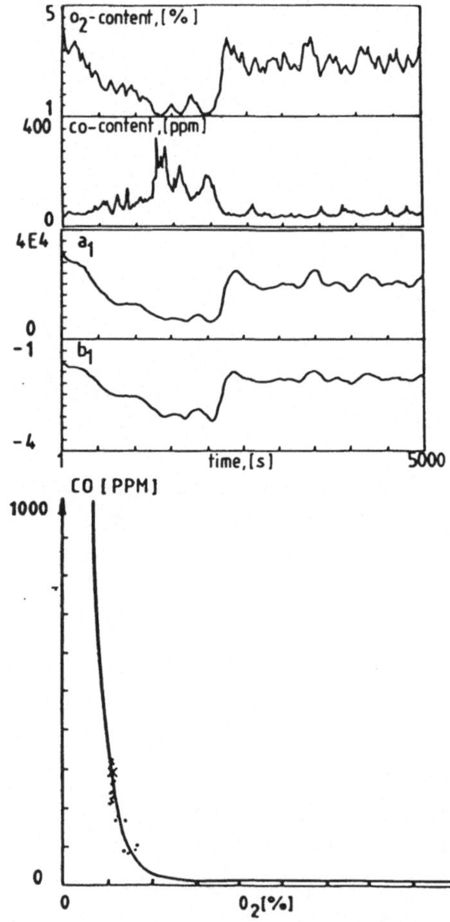


Fig. 11. $CO/O_2$-curve identification run at 43% load level

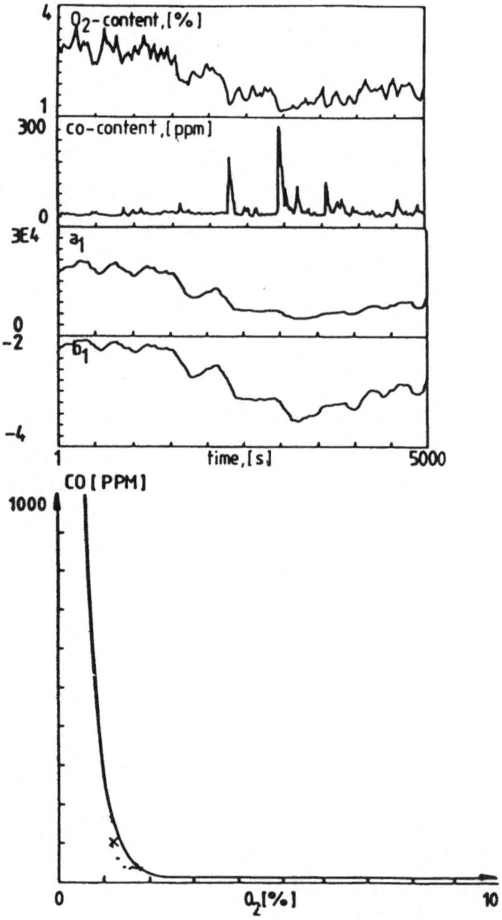


Fig. 12. $CO/O_2$-curve identification run at 70% load level.

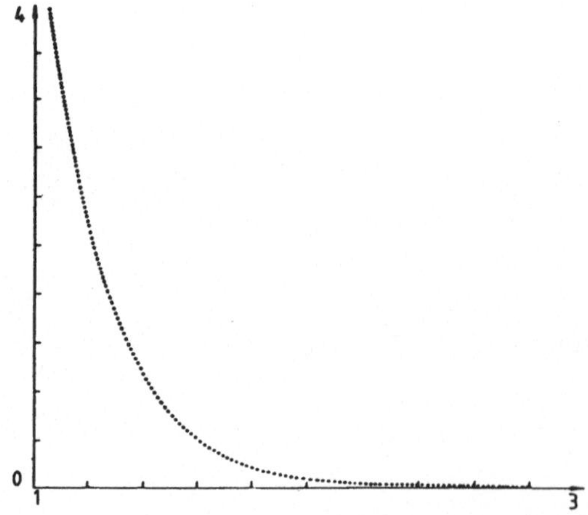


Fig. 13. Estimated CO-loss curve

where

$a_3, b_3$ = unknown parameters.

Now we write

$$\underline{X} = \begin{bmatrix} X_{02} \\ 1 \end{bmatrix} ,$$

$$y = L_{T\%} ,$$

and

$$\hat{\underline{\theta}} = \begin{bmatrix} \hat{a}_3 \\ \hat{b}_3 \end{bmatrix} .$$

Using the identification method (8) we now obtain

$$\hat{L}_{T\%} = \hat{a}_3 X_{02} + \hat{b}_3 , \tag{15}$$

where

$\hat{L}_{T\%}$ = estimate for the percentual heat losses, %.

Figures 14 and 15 show identification runs for heat loss estimation at 43% and 70% load levels.

The total combustion losses are the sum of CO- and heat losses

$$L_{\%} = L_{T\%} + \frac{L_c}{P_c} \cdot 100\% , \tag{16}$$

Where

$\hat{L}_{\%}$ = estimate for the total percentual combustion losses, %.

Figures 16 and 17 show the loss curves at one time instant based on the identification runs at load levels 43% and 70%, respectively.

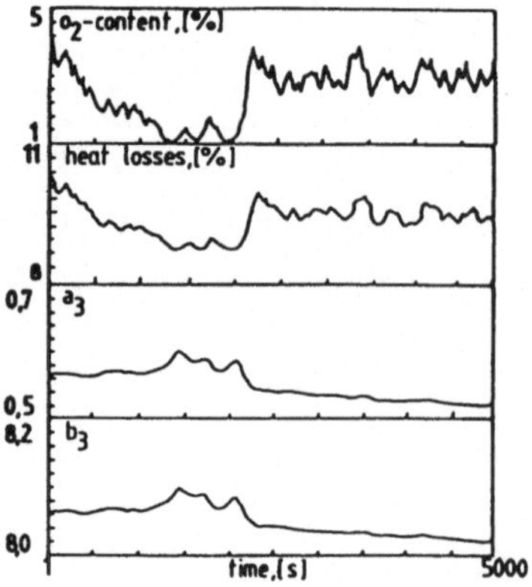


Fig. 14. Heat loss curve identification run at 43% load level

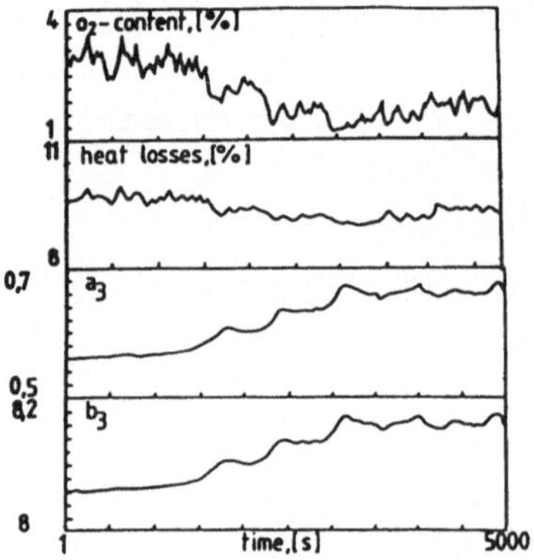


Fig. 15. Heat loss curve identification run at 70% load level

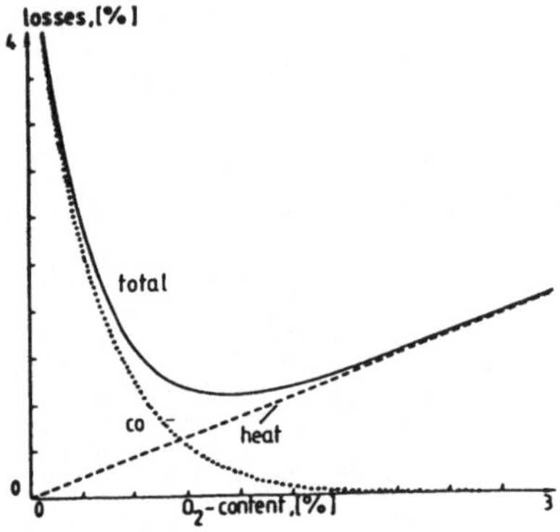


Fig . 16. Estimated losses at 43% load level

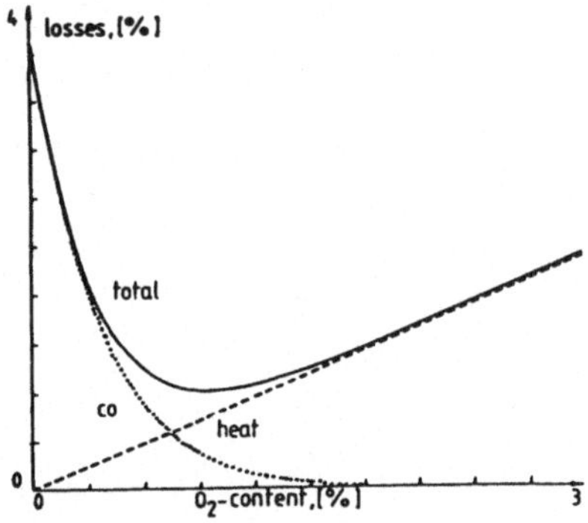


Fig. 17 . Estimated losses at 70% load level

Optimal control of the combustion air flow is based on continuos identification of total combustion losses. Result of the identification is the loss curve and the $0_2$-content which corresponds to the minimum losses. When circumstances change the optimum set point changes. Fig. 18 shows one possible solution for the optimal controller. The two calculation boxes include identification and search for the optimum $0_2$-content set point. The $0_2$-controller is a normal PI-type controller.

The combustion air flow optimization strategy has been programmed in a minicomputer system and will be tested at some peat power plant.

## 4. CONCLUSIONS

The biggest benefits of combustion air flow optimization strategy presented above are savings in fuel costs due to smaller combustion losses. Smaller extra combustion air flow means also smaller pressure and temperature variations in the furnace, which reduces stresses to furnace constructions. Also danger of operating at too small air flow and therefore incomplete combustion is prevented when the CO-content in the flue gases is measured and is taken into account in the control scheme.

The optimization principle presented in this paper together with the fuel flow control system that compensates the effects of fuel quality changes [2, 3] is a powerful tool at power plants burning inhomogeneous fuels like coal, peat, wood chips, etc. and in plants burning several fuels [4]. These both control strategies have been developed as a part of a larger research project together with Finnish industry.

## 5. REFERENCES


[1] Fortesque, T.R., Keshenbaum, L.S. & Udstie, B.E.: Implementation of Self-tuning Regulators with Variable Forgetting Factors. Automatica (17), 6. 1981.

[2] Lehtomäki, K., Kortela U. & Luukkanen J.: New Estimation and Control Methods for Fuel Power in Peat Power Plants. VIII IFAC World Congress, Kyoto, Japan, 1981.

[3] Lehtomäki, K., Wahlström, F., Luukkanen, J. & Kortela, U.: New Combustion Control Methods in Power Plants Burning Inhomogeneous Fuels, Fift Power Plant Dynamics, Control and Testing Symposium, March 21-23, 1983, Knoxville, Tennessee, USA.

[4] Kortela, U., Salmelin, B. & Wahlström, F.: A hierarchical Control Strategy for Multi-Fuel Multi-Boiler Systems, Fift Power Plant Dynamics, Control and Testing Symposium, March 21-23, 1983, Knoxville, Tennessee, USA.


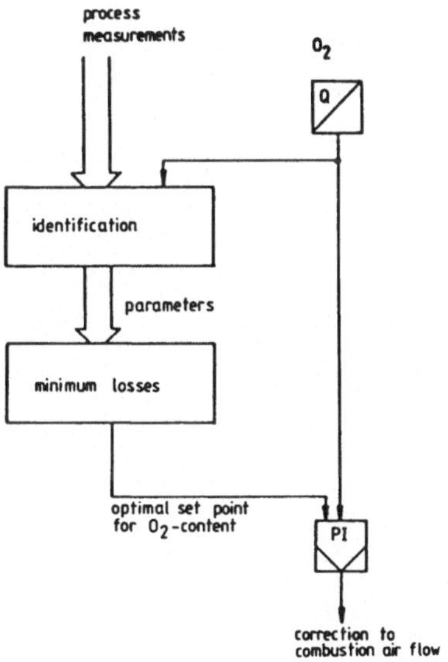


Fig. 18. The strategy for combustion air flow optimization.

Session 4

# DETECTION OF CHANGES IN SYSTEMS

# DÉTECTION DE CHANGEMENTS DANS LES SYSTÈMES

# DETECTION OF ABRUPT CHANGES IN SIGNALS AND DYNAMICAL SYSTEMS : SOME STATISTICAL ASPECTS.

A. BENVENISTE
IRISA/INRIA

M. BASSEVILLE
IRISA/CNRS

Campus Universitaire de Beaulieu
Avenue du Général Leclerc
35042 - RENNES Cédex
FRANCE.

## ABSTRACT

The aim of this paper is to present to the signal processing community some points of this detection problem, with a particular emphasis on the statistical aspects, leaving out the system theoretic aspects, which are of great importance in the control context, or, more generally, in the case of multichannel signal processing. A briel overview is presented of some of the issues developped in the CNRS - conference : "Détection de ruptures dans les Modèles Dynamiques de Signaux et Systèmes" held in Paris, on March 21-22 - (CNRS - 1984).

## 0 - INTRODUCTION

Detecting abrupt changes in signals or dynamical systems can be useful for several reasons.
The purpose can be the *segmentation* of a digital signal into homogeneous parts, as a tool for the modelling of these signals. This is typically the case for EEG, ECG, speech, or geophysical signals, but also for the line-by-line detection of edges in pictures.
The purpose can also be the detection of an *event* occurring in a dynamical system. This is the case for the detection of failures in control systems (a subject we will not develop here).
Finally, detecting a change can be simply a *tool* for improving the tracking properties of an adaptive algorithm when a fast change occurs on the identified system.

In this article, we shall only consider *one-line* methods, which are naturally used in connection with adaptive algorithms.

## 1 - A BASIC EXAMPLE : DETECTION OF CHANGES IN THE MEAN OF A SCALAR DIGITAL SIGNAL.

Let $(x_t)$ be a sequence of independent identically distributed scalar random variables with zero-mean probalility density f.
Assume we observe the signal

$$y_t = x_t + \theta(t) \tag{1}$$

where $\theta(t)$ in an unknown deterministic piecewise constant function. We are interested in detecting the jumps of $\theta(t)$ by observing $y_t$, i.e. we want to detect the changes in the mean of the digital signal $(y_t)$. The problem of interest in an one-line processing is the detection of a single jump in the mean of $(y_t)$ ; this problem can be formulated as follows.

*Test hypothesis $H_o$ :*

$$H_0 : y_1, \ldots, y_t \text{ are of mean } \theta_0 \tag{2}$$

*against hypothesis $H_1$*

$$H_1 : \text{for some } \tau<t,\ y_1, \ldots, y_\tau \text{ are of mean } \theta_0 \text{ and } y_{\tau+1}, \ldots, y_t \text{ are of mean } \theta_1. \tag{3}$$

Two situations of interest are to be considered :

*Ideal case* : $\theta_o$ and $\theta_1$ are known, but $\tau$, if it exists, is unknown, and has to be estimated.

*Pratical case* : $\theta_o$ is known, but $\tau$ and $\theta_1$ are unknown. This is indeed the pratical situation since $\theta_o$ can be recursively identified before the jump time $\tau$, for instance using an adaptive algorithm. The main techniques which will be presented here are related to Maximum Likelihood approaches.

## 1.1 - The Ideal Case : The Page-Hinkley Cumulative Sum (Cusum) test ([15],[22]).

Denote by f the distribution of $x_t$, and $\mathbb{P}_\theta$ the law of the process $(x_t+\theta)_{t\geq 1}$. The log-likelihood ratio of $H_1$ with respect to $H_o$ for a given value $\tau$ of the jump time is

$$\log \frac{\mathbb{P}_{\theta_o}(y_1,\ldots,y_\tau)\,\mathbb{P}_{\theta_1}(y_{\tau+1},\ldots,y_t)}{\mathbb{P}_{\theta_o}(y_1,\ldots,y_t)} = \sum_{s=\tau+1}^{t} (\log f(y_s-\theta_1) - \log f(y_s-\theta_o)) = S_\tau^t(\theta_o,\theta_1) \tag{4}$$

The maximum likelihood estimate $\hat{\tau}$ of the jump time knowing the sample $y_1, \ldots, y_t$ is

$$\hat{\tau}_t = \operatorname{Arg\,max}_{\tau} S_\tau^{\,t}(\theta_o,\theta_1) = \operatorname{Arg\,max}_{\tau} \{S_o^{\,t}(\theta_o,\theta_1) - \min_{s\leq\tau} S_o^{\,s}(\theta_o,\theta_1)\} \tag{5}$$

and a change is detected when

$$S_{\hat{\tau}_t}^{\,t}(\theta_o,\theta_1) > h \tag{6}$$

for some threshold $h>0$ to be chosen. The behavior of the test is depicted in various ways in the Figure 1.

When f is the Gaussian distribution $\mathcal{N}(0,\sigma^2)$, (4) reduces to the very simple Cusum test

$$S_\tau^{\,t}(\theta_o,\theta_1) = \frac{\theta_1-\theta_o}{\sigma^2}\sum_{\tau+1}^{t}\left(y_s - \frac{\theta_o+\theta_1}{2}\right) = \frac{\theta_1-\theta_o}{\sigma^2}\sum_{\tau+1}^{t}\left(y_s - \theta_o - \frac{\theta_1-\theta_o}{2}\right). \tag{7}$$

## 1.2 - The practical case : two basic approaches.

Recall that, in the practical case, $\theta_o$ can be considered as known, whereas $\theta_1$ is unknown at the change time. Among many others, two interesting approaches are the following.

*Extended Page-Hinkley CuSum test* ([2]).
Choose $\delta\theta>0$, which will represent twice the "minimum magnitude of jump" one is interested to detect, and use (5) and (6) for both following cusums :

$$S_\tau^{\,t}(\theta_o,\theta_o+\delta\theta)\;,\quad S_\tau^{\,t}(\theta_o,\theta_o-\delta\theta). \tag{8}$$

This Cusum test is quite simple and efficient. Moreover, the use of the form (7) of the Cusum, even in the non Gaussian case, is still a good test. See [3].

*The Generalized Likelihood Ratio (GLR) test* ([25],[26],[C8],[C3]).
The test is :

$$\max_{\tau,\theta} \; S_\tau^t(\theta_0,\theta) > h. \tag{9}$$

The double maximization in (9) significantly increases the computational cost of this procedure ; two ways of performing this procedure may be following :

- Perform first the maximization over $\tau$, for $\theta$ fixed using (5) ; then if the user constrains $\theta$ to take its values on a *finite* set H, the corresponding CuSums can be monitored in parallel, and a change is detected at the first time where one of these cuSums exceeds the threshold h.

- Perform first maximization over $\theta$, and then the maximization over $\tau$ ; this is efficient when the maximization over $\theta$ yields a closed formula, as it is the situation in the gaussian case, but the remaining maximization over $\tau$ is no more performed through the monitoring of a CuSum neither by a recursive scheme.

## 1.3 - Discussion (practical case)

*Comparison of the two approaches*
The advantage of the GLR over the Hinkley test in the practical situation is due to the following facts :

- The GLR test is basically the Maximum Likelihood approach of the testing problem which is known to have optimality properties ([16]).

- Moreover, in a sequential framework, estimating $\theta_1$ *after* the change as a part of the test is of significant help for an updating of the identification scheme after a detection occured ; the value $\theta_1$ can be chosen as a first estimate of the new value $\theta_0$ of the parameter before the next change to be detected.

On the other hand, even in the Gaussian case, the CuSum approach is far cheaper, and the choice among one of these techniques depends upon the tradeoff efficiency/computational cost in a given application.

*Comparison of the former approaches with other existing ones*
Other common techniques in signal processing for detecting a change in the mean of a signal are the following :
First a stable low pass filtering of the signal $y_t - \theta_0$ (which is zero-mean before the change) can be monitored, and compared with a threshold [14].
Nonparametric techniques can be also used, involving either rank statistics ([6]) or improved filtered derivatives ([2], [17], [18]). All those techniques are generally less efficient than the Hinkley test, especially when the signal between the changes is only approximately stationary, but only the filtered derivative techniques are simpler ; filtered derivatives techniques can be successfully used in the case of high signal to noise ratio.

For all these reasons, we shall concentrate in the sequel on the methods related to the likelihood approaches.

## 2 - A GENERAL DETECTION PROBLEM

Let $(y_t)_{t \geq 0}$ be a signal whose conditional distribution

$$\mathbb{P}\,(y_t \in dy \,/\, y_{t-1},\ldots, y_0) = f_\theta\,(y \,/\, y_{y-1},\ldots, y_0)\,dy \tag{10}$$

given the past, depends upon a vector parameter $\theta \in \mathbb{R}^k$ ; the problem is to detect a first change on $\theta$, and possibly to estimate relevant parameters like the change time, the magnitude of the jump. Again, we shall denote by $\theta_0$ the value of $\theta$ before the (first) change, and $\theta_1$ the value of $\theta$ after this change.

2.1 - The ideal case : $\theta_0$ and $\theta_1$ are known

Using Bayes rule,(4) can be generalized in the present case by setting

$$S_\tau^t(\theta_0,\theta_1) = \sum_{s=\tau+1}^{t} \left[ \log f_{\theta_1}(y_s/y_{s-1},\ldots,y_0) - \log f_{\theta_0}(y_s/y_{s-1},\ldots,y_0) \right] \qquad (11)$$

and (5), (6) define again the test.

2.2 - The practical case : $\theta_0$ is known, whereas $\theta_1$ is unknown.

We shall here indicate the methods which are related to the likelihood approach.

2.2.i - The likelihood approach ([C5], [C7], [25], [26], [C8], [16]).

The likelihood test, also known as GLR - Test, is given by

$$\max_{\tau,\theta} S_\tau^t(\theta_0,\theta) > h \quad . \qquad (12)$$

GLR tests have been extensively studied from a theoretical viewpoint. In [C7], an invariance principle is derived for the likelihood process

$$\{S_\tau^t(\theta_0,\theta)\}_{(\theta,\tau)\in \mathbb{R}^k \times [0,t]} ,$$

where $\theta_0$ and $t$ are fixed ; singularities occur when $\tau$ is close to t, which can be removed by convenient normalizations of $\theta - \theta_0$ and h when $\tau$ is close to t.

But the main problem in the on-line use of GLR tests lies in the high computational cost due to the double maximization in (12). A fundamental contribution has been given in this direction by Willsky and Jones ([26]) for the case of gaussian processes in State space form :

$$\begin{aligned} X_{t+1} &= F X_t + V_{t+1} + \delta(t-\tau_*) \,.\, \bar{X} \\ y_{t+1} &= H X_t + W_{t+1} + \delta(t-\tau_*) \,.\, \bar{y} \end{aligned} \qquad (13)$$

where H,F and the covariance matrices of $V_t$ and $W_t$ are known and fixed, $\delta(.)$ is the Dirac function, $\tau_*$ is the unknown jump time, and $(\bar{X},\bar{y})$ is the unknown jump. The key point it that *additive jumps on the state or observation are the only allowed jumps for a closed form of the maximization over* $\theta$ *in* (12) *to be available* (here, $\theta^T=(\bar{X}^T,\bar{y}^T)$ so that $\theta_0 = 0$). The basic ideas underlying Willsky and Jones approach are the following. Consider the Kalman filter associated to (13), under the hypothesis that there is no jump :

$$\begin{aligned} \hat{X}_{t+1} &= F \hat{X}_t + K_t e_{t+1} \\ e_{t+1} &= y_{t+1} - H \hat{X}_t \end{aligned} \qquad (14)$$

where $K_t$ is the Kalman gain, $e_t$ the innovation with covariance matrix $R_t$ (we dropped the Riccati equation here). Then we have

$$\begin{aligned} &\text{for} \quad t \le \tau_*,\ (e_t) \text{ is } \mathcal{N}(0,R_t) \text{ and white} \\ &\text{for} \quad t > \tau_*,\ (e_t - G(t,t_*).\,\theta) \text{ is } \mathcal{N}(0,R_t) \text{ and white,} \end{aligned} \qquad (15)$$

where the matrix $G(t,\tau)$ for $\tau < t$ has a closed forme definition, and *can be recursively updated for increasing* t *and* $\tau$ *fixed* ([25], [26]). Hence, referring to (12), and using (15), we get

$$-2\, S_{\tau}^{t}(0,\theta) = \sum_{s=\tau+1}^{t} e_s^{T} R_s^{-1} e_s - \sum_{s=\tau+1}^{t} (e_s - G(t,\tau)\theta)^{T} R_s^{-1}(e_s - G(t,\tau).\theta) \qquad (16)$$

thus allowing a closed form formula for $\hat{\theta}(t,\tau) = \arg\max_{\theta} S_{\tau}^{t}(0,\theta)$, with $\tau < t$ fixed. Then $\max_{\tau<t} S_{\tau}^{t}(0,\hat{\theta}(t,\tau))$ has to be performed, and it is sufficient in practice to search for the maximum inside a sliding window [t-T, t] , with T larger than the observability index of the pair (H,F). Furthermore, a complete reinitialization procedure is given for an updating of the Kalman filter after a jump has been detected.

This algorithm, together with a slight modification, has been very successfully used on geophysical signals in [4], where it appeared to be very robust when the assumed state space model is far from realistic. As a matter of fact, it is also possible to derive such formulas for detecting a jump $\theta$ in a prescribed subspace, thus allowing multiple hypothesis testing with a diagnosis of which failure occured, see [25]. Finally, when $\theta$ is constrained to take its values in a finite set, the resulting test is nothing but the multiple CuSum test described in the section (1.2), such a test was used in [10].

Finally, a new problem appears in the context of the present use of the GLR test, namely the prior problem of designing the state space model and the geometry of the possible failures when the original question is to monitor a given system. We shall not describe here the related *analytic redundancy* methods which are of greater interest in a control context ([8],[C9]).

2.2. ii - Multimodel approaches ([25],[C8]).

They can be used when a list of possible failed modes $\theta_1,..,\theta_N$ is available. In this case the CuSum tests based upon the statistics $S_{\tau}^{t}(\theta_o, \theta_i)$ for $i= 1,..., N$ can be monitored in parallel. This approach is often used in Bayesian framework, by including some a priori model for describing the possible changes from $\theta_i$ to $\theta_j$. Such methods are feasible only when N is not too large.

2.2. iii - CuSum tests derived from the Le Cam - Roussas's asymptotic expansion of the log likelihood ratio.([9], [23], [21], [C6]).

We now informally introduce this expansion for the reader. Denoting by $G_{\theta}(y/y_{t-1},\ldots,y_o)$ and $-F_{\theta}(y/y_{t-1},\ldots,y_o)$ respectively the first and second derivatives of $\log f_{\theta}(y/y_{t-1}, \ldots, y_o)$ with respect to $\theta$ , we have the expansion :

$$\text{Log } f_{\theta+\tilde{\theta}}(y/Y_{s-1}) - \log f_{\theta}(y/Y_{s-1})$$

$$= G_{\theta}^{T}(y/Y_{s-1}) \, . \, \tilde{\theta} - \tilde{\theta}^{T} F_{\theta}(y/Y_{s-1}) \, . \, \tilde{\theta} + o(||\tilde{\theta}||^2) \qquad (17)$$

where $Y_s$ denotes the observations $y_s \ldots y_o$.
Applying (17) with $\tilde{\theta} = t^{-1/2}\delta\theta$ and summing over s from 1 to t, we get the expansion

$$S_{o}^{T}(\theta,\theta + t^{-1/2}\delta\theta)$$

$$\cong t^{-1/2} \left( \sum_1^t G_\theta(y_s|Y_{s-1}) \right)^T . \delta\theta - \delta\theta^T \left(t^{-1} \sum_1^t F_\theta(y_s|Y_{s-1}) \right). \delta\theta$$

$$= t^{-1/2} \Delta_0^t (\theta)^T . \delta\theta - \delta\theta^T (t^{-1} F_0^t (\theta)).\delta\theta \quad . \tag{18}$$

Now, let $C$ be the ellipsoïd defined by

$$\delta\theta \in C \text{ iff } \delta\theta^T F(\theta_o)\delta\theta = \delta \tag{19}$$

where $F(\theta)$ is the Fis her information matrix, and consider the test whose rejection domain is given by

$$\max_{\delta\theta \in C,\ \tau < t} S_\tau^t (\theta_o, \theta_o + t^{-1/2}\delta\theta) > h \quad ; \tag{20}$$

we shall show that the maximization over $\delta\theta$ in (20) has a closed form solution due to (18) and (19).

First we have :

$$\max_{\delta\theta \in C} (\Delta_0^t (\theta_o))^T . \delta\theta = t^{-1/2}\delta^{1/2}(\Delta_0^t(\theta_o)^T F^{-1}(\theta_o)\Delta_0^t(\theta_o))^{1/2}, \tag{21}$$

and second, before the jump (i.e. under law $\mathbb{P}_{\theta_0}$), the law of large numbers gives

$$t^{-1} F_0^t (\theta_o) \sim F(\theta_o) \tag{22}$$

so that the second term of the righthandside of (18) is approximately a constant when $\delta\theta \in C$ in view of (19) and (22). Hence

$$\max_{\delta\theta \in C,\ \tau < t} S_0^t(\theta_o, \theta_o + t^{-1/2}\delta\theta) = t^{-1/2}\delta^{1/2}(\Delta_0^t(\theta_o)^T F^{-1}(\theta_o)\Delta_0^t(\theta_o))^{1/2} - \delta.$$

This shows that the test defined in (20) depends only upon the statistic :

$$X_\tau^t (\theta) = \Delta_\tau^t (\theta)^T . F^{-1}(\theta) . \Delta_\tau^t (\theta) \tag{23}$$

which is a central chi-square random variable before the jump occurs, and a noncentral one after the jump ; a practical test is then easely designed, based upon this statistic. For the case of Gaussian AR(p) processes, we get ((9),(21),(C6)) :

$$X_\tau^t(\theta) = \frac{1}{2}\left( \sum_{s=\tau+1}^t \left(\frac{e_s^2}{\sigma^2} - 1\right)\right)^2 + (V_\tau^t)^T R^{-1}(\theta) V_\tau^t$$

$$V_\tau^t = \frac{1}{\sigma^2} \sum_{s=\tau+1}^t Y_s e_s , \quad e_s = y_s - \sum_1^p a_i y_{s-i}$$

$$Y_s^T = (y_{s-1}, \ldots, y_{s-p}) , \quad R(\theta) = \sigma^{-2} \mathbb{E}_\theta (Y_s Y_s^T)$$

$$\theta^T = (a_1, \ldots, a_p ; \sigma). \tag{24}$$

But this approach can also be used for specific problems using nonstandard parametrizations. This general approach was used in various applications by NIKIFOROV (seismic processing, monitoring of industrial plants). It is a good tradeoff between efficiency and computational cost.

2.2. iv - Approximate GLR tests and related methods ([C4],[5])

Approximate GLR tests typically consist in replacing (12) by

$$\max_{\theta} \; S_{t-T}^{t} (\theta_o,\theta) > h \tag{25}$$

where T is a *fixed* lag. This leads in natural way to the so-called *two-model design methodology* ([7],[1],[5]) we shall explain now. Fit on the signal a (generally) long term reference model for estimating $\theta_o$, and a short term sliding model for estimating a candidate for $\theta_1$ according to the following possibilites depicted in Fig. 2, and monitor a convenient distance between the models $f_{\theta_1}$ and $f_{\theta_o}$. The first case in Fig. 2, together with the use of the loglikelihood as a distance, corresponds exactly to (27), and is mentioned in [5]. For the special case of the detection of changes in the characteristics of an AR signal, many other distances have been used. Apart from the loglikelihood mentioned before, the most interesting distances are the *cepstral distance* [13], and especially the *Kullback J-divergence* used in [5] we shall now briefly present. Let us modify the CuSum (11) as follows :

$$S_o^t (\theta_o,\theta_1) = \sum_{s=1}^{t} \Delta S_s(\theta_o,\theta_1) ,$$

$$\Delta S_s(\theta_o,\theta_1) = \log \frac{f_{\theta_1}(y_s|y_{s-1}, \ldots)}{f_{\theta_o}(y_s|y_{s-1}, \ldots)} - \int f_{\theta_o}(y|y_{s-1},\ldots) \operatorname{Log} \frac{f_{\theta_1}(y|y_{s-1},\ldots)}{f_{\theta_o}(y|y_{s-1},\ldots)} dy \tag{26}$$

by removing from (11) the conditional drift before the jump ; the conditional drift after the jump is nothing but the Kullback J- Divergence of the conditional distributions $f_{\theta_1}$ and $f_{\theta_o}$. In the Gaussian AR case (26) results in the simple CuSum

$$2.\Delta S_t(\theta_o,\theta_1) = 2 \frac{e_t^o e_t^1}{\sigma_1^2} - \left(1 + \frac{\sigma_o^2}{\sigma_1^2}\right) . \frac{(e_t^o)^2}{\sigma_o^2} - \left(\frac{\sigma_o^2}{\sigma_1^2} - 1\right) , \tag{27}$$

where $e_t^i$ (i = 0,1) is the prediction error corresponding to the model $\theta_i$, and $\sigma_i^2$ its variance. The models $\theta_o$ and $\theta_1$ are updated on-line according to the first scheme of the Fig. 2, and the behavior of (26,27) is shown in the figure 3 below.

The detection of the non-zero drift of $S_o^t$ can be improved using a simple Page-Hinkley test (7) ; the resulting test performs well for speech signal segmentation.

Let us also emphasize on a different approach by Appel and Brandt (see the next paper in this book), where joint distributions of the sample are monitored, rather than conditional ones, which results in a different setting of the detection problem in the case of a dependent process $(y_t)$.

## 3 - DISCUSSION

For general on-line testing problems, two competitive approaches are of interest when the complete form of the GLR test is too complex : the CuSum tests of section 2.2.iii and the two-models approaches related to the approximate GLR test described in section

2.2.iv . Both are of CuSum type from a computational viewpoint.

Because the estimation of $\theta_1-\theta_0$ is ignored, the former approach is computationally simpler and fairly general (only one model is needed to be adjusted) ; although explicit estimates of $\hat{\tau}$ are given, they can be largely corrupted by the lack of estimation $\theta_1-\theta_0$ in critical cases ; finally, no information is available for the updating of the identification procedure after a change has been detected.

On the other hand, when it is computationally tractable, the latter approach can offer improvements on both points.

Finally, we should acknowledge that simple nonparametric methods we have note described here are of help when the detection problem is very easy.

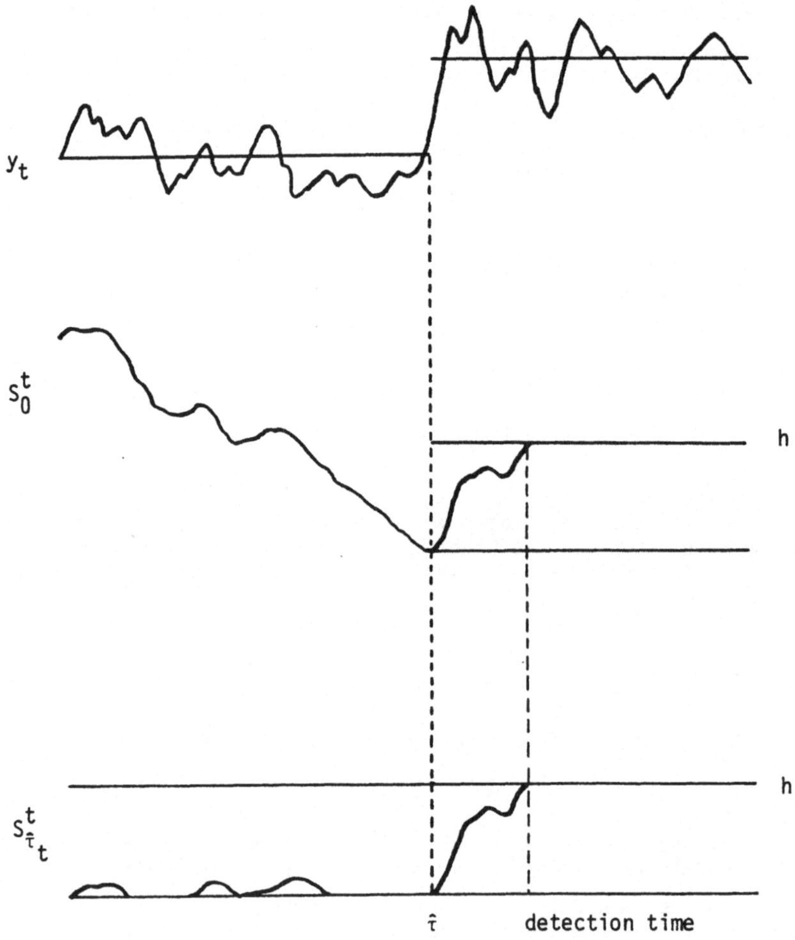


Fig.1

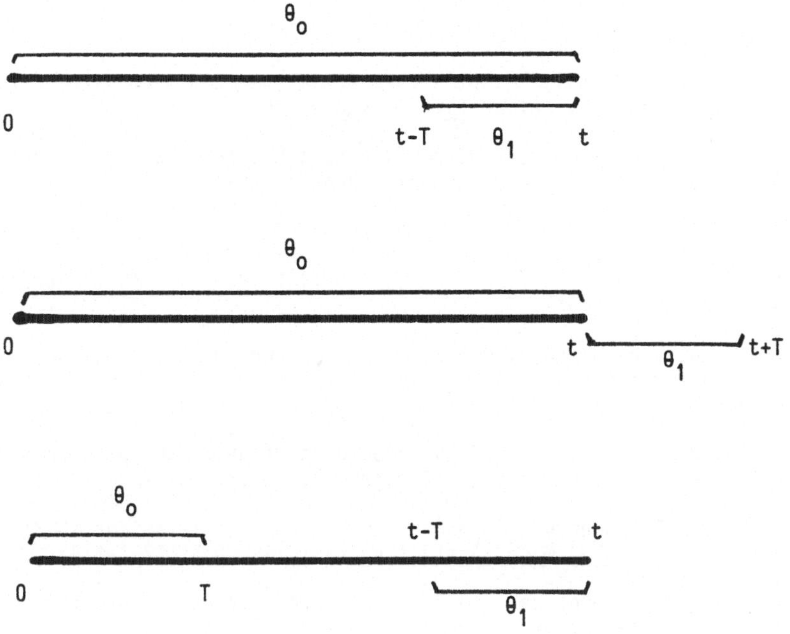


Fig.2

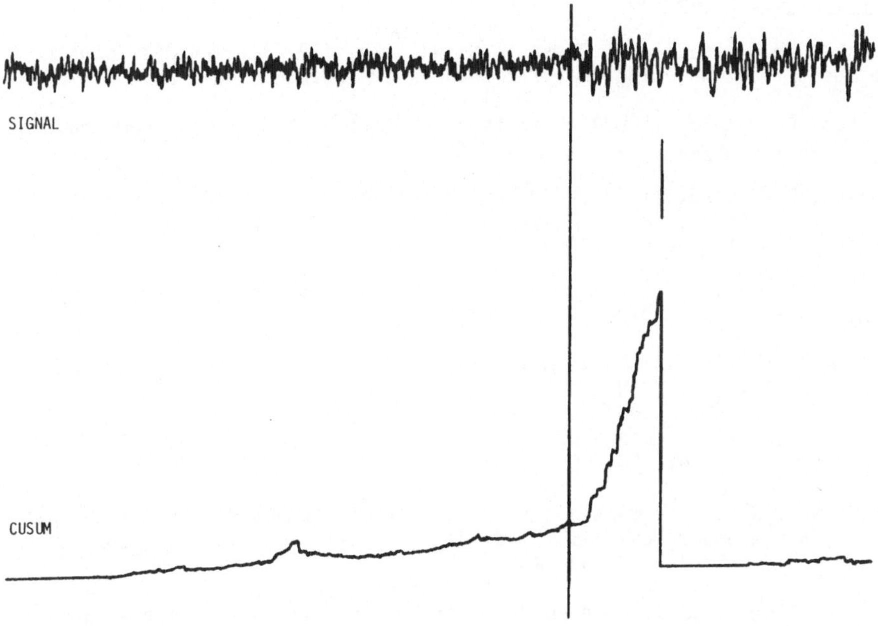


Fig.3

# REFERENCES


(1) U. Appel, A. Brandt, (1983). "Adaptive sequential segmentation of piecewise stationary time series". , Information Sciences, Vol. 28, April.

(2) M. Basseville, (1981) "Edge detection using sequential methods for change in level, part II : sequential detection of change in mean", IEEE - ASSP - 29 N°1 32-50

(3) M. Basseville, B. Espiau, J. Gasnier, (1981)., "Edge detection using sequential methods for change in level, part I : a sequential edge detection algorithm"., IEEE - ASSP - 29 n°1, 24-31.

(4) M. Basseville, A. Benveniste, (1983-a)., "Design and comparative study of some sequential jump detection algorithms for digital signals"., IEEE - ASSP - 31, N°3, June 1983.

(5) M. Basseville, A Benveniste, (1983-b)., "Sequential detection of abrupt changes in spectral characteristics of digital signals"., IEEE - IT - 24, Sept. 1983

(6) G.K Bhattacharya, R.A Johnson, (1968).n "Non parametric tests for shift at an unknown time point"., Ann. Math. Statistics, Vol. 39, n°5, 1731-1743.

(7) G. Bodenstein, H.M. Praetorius, (1977)., "Feature extraction form the encephalogram by adaptive segmentation"., Proc. IEEE, Vol. 65, 642-652.

(8) E.Y Chow, A.S Willsky, (1984)., "Analytical redundancy and the design of robust failure detection systems"., to appear, IEEE - AC, 1984.

(9) R.B. Davies, (1973)., "Asymptotic inference in stationary gaussian time series"., Adv. Appl. Proba. 5, 469-497.

(10) J.C. Deckert, M.N Desai, J.J. Deyst, A.S. Willsky, (1977)., "F8 DFWB sensor failure identification using analytic redundancy"., IEEE - AC - 22,N°5, 725-803 .

(11) J. Deshayes, D. Picard, (1982)., "Tests de rupture de régression, comparaison asymptotique"., Teoryia Ver. Prim. 95-108.

(12) J. Deshayes, D. Picard, (1983)., "Principe d'invariance sur les processus de vraisemblance"., Thèse d'état, Université d'Orsay, France, to appear 1984 in annales de l'institut Henri Poincaré.

(13) A.H. Gray, J.D. Markel, (1976)., "Distances measures for speech processing"., IEEE - ASSP - 24, N°5, 380-391.

(14) W.G.S. Hines, (1976)., "A simple monitor of a system with sudden parameter changes"., IEEE -IT - 22, N°2, 210-216.

(15) D.V. Hinkley, (1971)., "Inference about the change-point from cumulative sum-tests"., Biometrika, vol. 58, 509-523.

(16) I.A. Ibragimov, R.Z. Khas'minskii, (1972). "Asymptotic Behavior of Statistical Estimators in the Smooth case- I. Study of the Likelihood Ratio". Theory of Proba. and Appl. Vol 17 n°3. 445-462.

(17) B. Kedem, E. Slud, (1981)., "On goodness of fit of time series models, an application of high order crossing"., Biometrika, Vol. 68, N°2, 551-556.

(18) B. Kedem, E. Slud, (1982)., "Time series discrimination by higher order crossings"., Annals of Statistics, Vol.10, N°3, 786-794.

(19) I.V. Nikiforov, (1979).,"Cumulative sums for detection of changes in random process characteristics"., Autom. Remote control, Vol. 40, N°2, 192-202

(20) I.V. Nikiforov, (1980)., "Modification and analysis of the cumulative sum procedure"., Automatika i Telemekanikha, Vol. 41, N°9, 74-80.

(21) I.V. Nikiforov, (1983)., Sequential detection of abrupt changes in time series properties ; Nauka, Mascow.

(22) E.S. Page, (1954)., "Continuous inspection schemes"., Biometrika, Vol. 41, 100-114.

(23) G.G. Roussas, (1972)., Contiguity of probability measures, some applications in statistics., Cambridge University press.

(24) J. Segen, A.C. Sanderson, (1980)., "Detecting changes in time series"., IEEE-IT 26, N°2, 249-255.

(25) A.S. Willsky, (1976)., "A survey of design methods for failure detection in dynamic systems"., Automatica, Vol. 12, 601-611.

(26) A.S. Willsky, H.L. Jones, (1976)., "a generalized likelihood ratio approach to the detection and estimation of jumps in linear systems"., IEEE - AC - 21 N°1, 108-112.

CNRS - Conference : "Détection de Ruptures dans les Modèles Dynamiques de Signaux et Systèmes". Paris March 21-22, 1984.

(C1) R.André, M. Basseville, A. Benveniste : "un Exemple de Segmentation Temps-Réel du Signal de Parole".

(C2) M. Basseville : "Détection Séquentielle de Sauts de Moyenne".

(C3) M. Basseville : "Exemples d'Utililation de l'Algorithme GLR".

(C4) M. Basseville : "Quelques Algorithmes de Détection de Changements de Caractéristiques Spectrales Utilisés en Traitement du Signal".

(C5) J. Deshayes, D. Picard :"Méthodes Globales de test et d'Estimation de Ruptures : Points de Vue Asymptotiques".

(C6) I.V. Nikiforov : "Sequential Detection of changes in Times Series Properties Based on a Modified Cumulative Sum Algorithm".

(C7) D. Picard, J. Deshayes : "Comment utiliser les Statistiques de Vraisemblance dans un Problème de Rupture".

(C9) A.S. Willsky, E.Y. Chow, X.C. Lou, G.C. Verghese : "Redundancy Relations and Robust Failure Detection".

(C10) A.S. Willsky, P.C Doerschuk, R.R. Tenney ; "Estimation - Based Approaches to Rhythm Analysis in Electrocardiograms".

(C8) A.S. Willsky : "Detection of Abrupt changes in Dynamic Systems ".

# PERFORMANCE COMPARISON OF TWO SEGMENTATION ALGORITHMS USING GROWING REFERENCE WINDOWS

U. Appel and A. v.Brandt
Bundeswehr University / FB-ET
D-8014 Neubiberg

Abstract

Two procedures designed for the detection of parameter jumps in autoregressive gaussian distributed processes - the generalized likelihood ratio (GLR) algorithm and the cumulated sum (CUSUM) algorithm - are compared regarding their performance. Both algorithms share as a common feature a growing reference window and a sliding fixed length test window, but use different detection statistics. Some rough features of the algorithms are deducted using means instead of the stochastic signal itself. More detailed results are then obtained from extensive simulations performed with different types of parameter jumps in the test signals. As a general result, it is shown that the CUSUM procedure may perform slightly better with respect to the detection of spurious jumps, if direction and distance of the jump is known in advance. On the other hand, the GLR algorithm leads to much better results in the detection and particularly the positioning of jumps succeeding each other in a short time interval ("short segments"). Moreover, the GLR algorithm is more robust considering the application of a segmentation procedure under realistic assumptions.

## 1. Introduction

Among several models for nonstationary signals, the quasistationary autoregressive (AR) process has found widespread use in such different applications as speech signal, biomedical signal, seismic and surveillance signal processing. In this model, the AR-parameters are assumed to be held fixed within certain time intervals or "segments", and change abruptly to a new set of parameters when reaching a boundary of such a segment. In many applications, it will be necessary and sufficient to detect such parameter jumps (event detection), possibly with a minimum time delay; in other applications it might be necessary for modelling and coding purposes also to localize such segment boundaries with minimum bias and variance.

According to the growing demand in different applications, a considerable number of segmentation algorithms has been developed within the last few years / 1-9/. In most of these publications, it has been shown that the particular procedure outlined there works fairly well in the special application for which it was intended. However, there is a need for performance comparisons among such algorithms, which could lead to better insights into advantages and shortcomings of any particular algorithm from an application viewpoint. Due to the statistical nature of the problem, a comparison based on extensive simulations seems to be favourable, as an exact analytical assessment of the performance of segmentation algorithms is practically impossible.

One such comparison between adaptive, sequentially working algorithms all being able to process time series in real-time has been performed in /10/. In this paper, it has been shown that the generalized likelihood ratio (GLR) algorithm, which has been proposed originally in /11/ and was presented in an effective implementable version first in /1/, does lead to very good results compared to the two other procedures under test. It has been assumed that this good result is partly due to the fact that this algorithm uses a growing reference window and a sliding test window to define segments from which signal parameters are estimated, and therefore makes optimal use of the data available. In contrary to this, the two other algorithms use fixed reference windows, which in general will lead to worse statistical estimates.

Independantly, another algorithm using a growing reference window, too, but a cumulated sum (CUSUM) detection measure instead, had been developed /9/. GLR and CUSUM algorithm, therefore, are very similar in the parameter estimation process and only differ in the detection process. Hence, it will be interesting to study the differences in performance and in the practical implementation of these two procedures.

## 2. Two segmentation algorithms using a growing reference window

### 2.1 The generalized likelihood ratio (GLR) algorithm

This algorithm uses a generalized likelihood ratio test to decide at each new sample upon the detection of a new segment boundary /1,2/. To this purpose, for each time index t a growing window $[1:s-1]$ - starting from the last boundary detected - is defined as well as a sliding test window $[s:t]$ of constant length $L = t-s+1$ (fig. 1). Within both windows as well as within a "pooled" window $[1:t]$ formed by cocatenation of both, three sets of AR parameters are calculated using a harmonic mean covariance lattice algorithm: Sample covariance vectors $\underline{C} = [c_x(0),\ldots c_x(p)]$ with

$$c_x(i) = \sum_{k=u}^{t} x_k \cdot x_{k-i} \qquad i = 0,\ldots p$$

are computed recursively for each window, and the residual energies E(a:b) - where a and b are beginning and end of the respective data

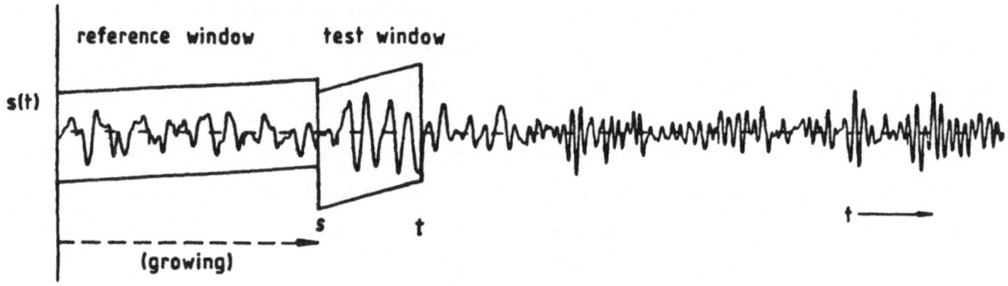


Fig. 1: Data windows for the GLR algorithm

window - are calculated using the algorithm depicted in tab. 1. From these energies, maximum logarithmic likelihood quantities

$$H(a:b) = (b-a+1)\cdot\ln\frac{E(a:b)}{b-a+1}$$

are calculated to derive a distance measure

$$d(t) = H(1:t) - H(s:t) - H(1:s-1)$$

This distance measure is - for gaussian distributed signals - a generalized-log-likelihood-ratio-test statistics, which makes optimum use of the data contained in these windows (under the practical restriction of a fixed length test window).

In order to optimally position a boundary detected e.g. at time index $t_D$, it is assumed that the position lies within the interval $[t_D-L+1:t_D]$. In the following the quantity

$$\Delta d(t)=H(1:r-1)+H(r:t)-H(1:t-L)-H(t-L+1:t)$$

is calculated at first for the initial estimate r of the boundary position, $r = t_D-L+1$, with $t = t_D+1$. This quantity is then contiguously calculated for $t = t_D+1, \dots t_D+L-1$, and the estimate r is replaced by a new value $r = t-L+1$

**Input variables:**

$C_x(i),\ x_{t-i},\ x_{p-i},\quad i=0,\dots,p$

where $C_x(i) = \sum_{k=p+1}^{t} x_k x_{k-i}$

For $i = 0,\dots,p$:

- $S_o(i) = 2\,C_x(i)$
- $G_o(i) = C_x(i)$
- $H_o(i) = C_x(i)$
- $f_o(t-i)=b_o(t-i)=x_{t-i}$
- $f_o(p-i)=b_o(p-i)=x_{p-i}$

$H'_o(0) = G_o(1)$

For $i = 0,\dots,p-1$:

- $S'_o(i)=S_o(i)-b_o(t-i)b_o(t)+b_o(p-i)b_o(p)$
- $H'_o(i+1)=H_o(i)-f_o(t-i)b_o(t)+f_o(p-i)b_o(p)$

For $m = 1,\dots,p$:

- $K_m = -\,2\,G_{m-1}(1)/S'_{m-1}(0)$
- $S_m(0)= (1+K_m^2)S'_{m-1}(0)+4K_mG_{m-1}(1)$
- If $(m=p)$ goto nextm.
- $b_m(t)=K_mf_{m-1}(t)+b_{m-1}(t-1)$
- $b_m(p)=K_mf_{m-1}(p)+b_{m-1}(p-1)$
- $S'_m(0)=S_m(0)-b_m^2(t)+b_m^2(p)$
- $G_m(1)=K_mS'_{m-1}(1)+G_{m-1}(2)+K_m^2H_{m-1}(1)$
- $H'_m(0)=G_m(1)$
- If $(m = p-1)$ goto nextm.
- For $i = 1,\dots,p-m-1$:
  - $f_m(t-i+1)=f_{m-1}(t-i+1)+K_mb_{m-1}(t-i)$
  - $f_m(p-i+1)=f_{m-1}(p-i+1)+K_mb_{m-1}(p-i)$
  - $b_m(t-i) = K_mf_{m-1}(t-i)+b_{m-1}(t-i-1)$
  - $b_m(p-i) = K_mf_{m-1}(p-i)+b_{m-1}(p-i-1)$
  - $S_m(i)=(1+K_m^2)S'_{m-1}(i)+ +2K_m\left[G_{m-1}(i+1)+H'_{m-1}(i)\right]$
  - $S'_m(i)=S_m(i)-b_m(t-i)b_m(t)+b_m(p-i)b_m(p)$
  - $H_m(i-1)=K_mS'_{m-1}(i-1)+H'_{m-1}(i-1)+ +K_m^2G_{m-1}(i)$
  - $H'_m(i)=H_m(i-1)-f_m(t-i+1)b_m(t)+ +f_m(p-i+1)b_m(p)$
  - $G_m(i+1)=K_mS'_{m-1}(i+1)+G_{m-1}(i+2)+ +K_m^2H'_{m-1}(i+1)$

nextm.

Tab. 1: Covariance lattice algorithm used for AR parameter estimation

whenever $\Delta d(t)$ assumes a positive value. At $t = t_D+L-1$, then, the current value of r is the optimized boundary position. All calculations necessary for $\Delta d(t)$ are obtained as the algorithm proceeds in time index using growing and sliding data windows as for the detection of a boundary; so practically no additional computations are necessary.

Application of the algorithm is simplified by the fact that only the test window length and the decision threshold have to be adjusted depending on the signal statistics; no other parameters are necessary.

## 2.2 The cumulated sum (CUSUM) algorithm

The basic idea in this segmentation procedure is to derive, for each new signal sample, a quantity with expectation value zero if no segment boundary is present, and with a strictly positive expectation value else /9/. By accumulating these local statistics, then, this sum will remain close to zero up to the time index of a new segment boundary, and then grow contiguously.

To this purpuse, a residual energy $\sigma_0^2$ is calculated within a growing data window $[1:t]$ using BURG's algorithm, and a residual energy $\sigma_1^2$ within a sliding window $[s:t]$ of constant length $L = t-s+1$ using the autocorrelation method (fig. 2). With these energies and the respective prediction error quantities (innovations) at time index t, $e_t^0$ and $e_t^1$, a distance

$$T_t = \frac{1}{2}\left[2\,\frac{e_t^0 \cdot e_t^1}{\sigma_1^2} - \left(1 + \frac{\sigma_0^2}{\sigma_1^2}\right)\cdot\frac{(e_t^0)^2}{\sigma_0^2} - \left(\frac{\sigma_0^2}{\sigma_1^2} - 1\right)\right]$$

is calculated. The accumulation of $T_t$ - with an a priori subtraction of a negative drift $\delta$ - then gives the decision variable d(t):

$$d(t) = \sum_{i=1}^{t} (T_t - \delta)$$

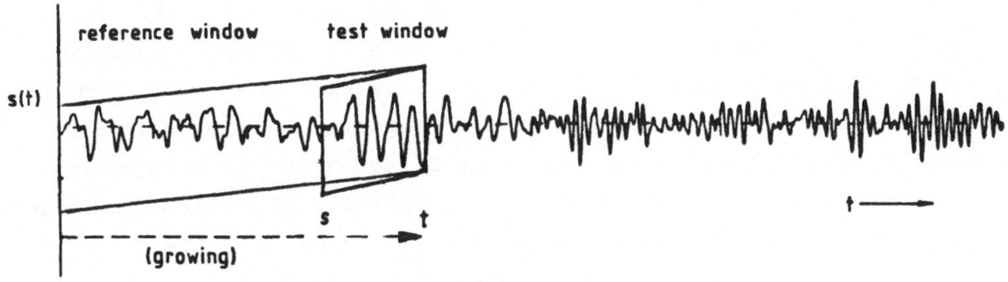


Fig. 2: Data windows for the CUSUM algorithm

A new boundary is detected whenever this variable exceeds a certain threshold relative to its last minimum. The position of such a detected boundary is assumed to be identical to the location of this minimum; thus no further calculations are necessary for positioning.

Ideally, the value of the drift $\delta$ should be half the minimum magnitude of $T_t$ in mean after a boundary for optimal positioning. However, as this value in general will not be known in advance in practical applications, a suboptimum value must be defined a priori; in /9/ it has been reported that a value $\delta = 0.1$ is a reasonable choice. Also, the values of the test window length L and the decision threshold have to be adjusted depending on the signal statistics.

## 3. Comparison results

Though the GLR algorithm uses three data windows and the CUSUM only two, the signal sets required in both procedures are identical. Moreover, the methods used to derive signal parameters are quite similar; at least if the data windows are sufficiently long ($\geq$100 samples), differences are not severe. Therefore, the main difference of both methods lies in the detection process in each procedure. In the following, a brief study of the decision statistics used for the detection of a parameter jump shall be given.

In fig. 3 and fig. 4, the decision variable d(t) for the CUSUM and the GLR algorithm is displayed as a function of the time index t, using expectation values. In these pictures, it is assumed that a segment boundary is located at t = 0 and that a new boundary at position $t = t_B$ is to be detected.

Fig. 3a shows d(t) for the CUSUM algorithm, as derived in /9/ under the (unrealistic) assumption that the two parameter sets belonging to the two segments are perfectly known in advance. In this case d(t) is falling continuously for $t < t_B$ due to the added negative drift $\delta$, while for $t > t_B$ d(t) is continuously growing with a slope being dependant upon the drift $\delta$ and the parameter distance between the two adjacend segments. Therefore, even minor parameter jumps should be detected with constant threshold after sufficient long delay. However, this ideal picture is no longer true under the more realistic situation that the parameters of the two segments have to be estimated from the growing reference window and the sliding test window, respectively. Even if this estimation would be ideal, d(t) will take on values as shown in fig. 3b: As the test window passes the jump point $t_B$, its new para-

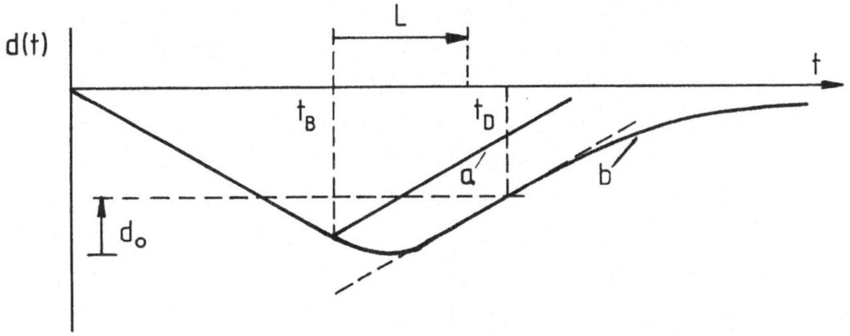


Fig. 3: Decision distance d(t) for the CUSUM algorithm:
a) New parameter set is known in advance
b) New parameter set is to be estimated

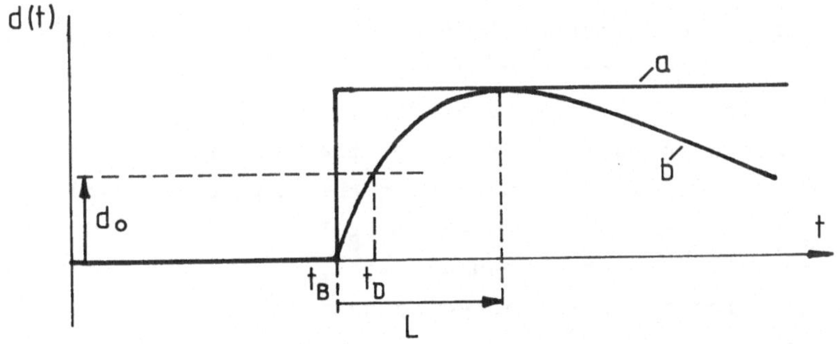


Fig. 4: Decision distance d(t) for the GLR algorithm:
a) New parameter set is known in advance
b) New parameter set is to be estimated

meter set is only slowly developing with growing t, while on the other hand the estimation of the parameter set of the reference window is increasingly falsified by the fact that the boundary lies within this window. It is at $t \approx t_B+L$ (L being the length of the test window) that both sets are distinguished with best possible distance. For $t > t_B+L$, the two parameter sets become more and more similar, leading finally to a negative slope of d(t) for $t \gg t_B$ again, as shown in fig. 3b. For a fixed decision threshold $d_o$, therefore, the detectability of a parameter jump depends on its distance, the value $\delta$ of the drift and the length L of the test segment.

A somewhat different situation will be found for the decision distance d(t) of the GLR algorithm, as shown in fig. 4. Again, if the two parameter sets were known in advance, d(t) would ideally jump from zero to a value depending on the distance of these two sets, as shown in fig. 4a. Under the realistic assumption that the parameters have to be estimated using the reference and the test window, however, this value is reached only at $t = t_B+L$. For $t_B < t < t_B+L$, d(t) is evolving slowly to this value, as the test window is filled more and more with data from the new segment. For $t > t_B+L$, on the other hand, the boundary lies in the reference window, and therefore the reference window parameter values approach those of the test window for $t \gg t_B$; the distance d(t), hence, decreases slowly back to zero again. According to this function, the basic detectability of a jump in this algorithm depends solely on the distance of the two segments and an appropriately choosen threshold.

The functions of fig. 3 and 4 certainly are not sufficient to predict the performance of both decision measures comnletely, as they only display the mean and neglect the variance of d(t). However, these curves at least give some hints on basic features of the decision properties. One such feature of the GLR algorithm is that the maximum of d(t) is reached for $t = t_B+L$. Any new boundary at $t > t_B+L$ therefore will not influence the detectability of the boundary at $t_B$; therefore, short segments (of approximate length L) are detected as faithfully as long segments with this algorithm, assuming a constant test window length L. This is not true in the CUSUM test as the maximum of d(t) will be achieved only for $t \gg t_B+L$, depending on the value of $\delta$ and the parameter distance. Therefore, one can expect that closely spaced boundaries influence each other in a sense that short segments are detected worse than long ones.

On the other hand, it is possible to detect even small parameter jumps with a high decision threshold (low false alarm rate) in the CUSUM algorithm due to the long cumulation of $T_t$. This is particularly true if the segments are very long. However, as can be seen from fig. 3, the minimum of d(t) - which defines the estimated boundary position - will have a substantial bias (delay) compared to the true position in this case. In the GLR algorithm, on the other hand, the threshold has to be lowered to obtain a high detection rate for segments with small parameter distance which in turn may lead to a higher false alarm rate.

In the following chapter, simulation results are presented comparing both algorithms; these will demonstrate up to which extend the qualitative results of this chapter can be confirmed.

## 4. Simulation results

In order to study the performance of the two segmentation algorithms empirically, a great number of simulations has been performed. Signals used in these simulations in general have been generated artificially using a suitable quasistationary signal source, as this allows an objective measure of performance; in comparison to this, with real world data generally only a (subjective) visual inspection of the segmented signals would be possible to judge on the correctness of the segmentation result.

For these simulations, a common data base has been established using a so-called "composite source": A noise source generating gaussian distributed, independent (white noise) signal samples is used as input to a pair of spectrum forming linear autoregressive filters. By generating a test time series from either one of these filters and abruptly switching from one to another whenever a new segment is supposed to begin, an ideal quasistationary AR process will be generated. In general, the filter gains are adjusted such that the output signal power of both filters is constant. Therefore, the signal power remains constant across segment boundaries (except in those simulations where a signal power jump is explicitely modelled), leading to segments which only differ in their spectral shape. In order to reduce the infinite number of possible parameter variations of a signal to a realistic amount of some typical test signals, all simulations have been performed using generating filters of order 2 (i.e. only one single pole pair). Filter parameters have been varied such that jumps of the center frequency and the bandwidth of the power spectrum as well as of the total power itself could be modelled.

With this setup, in all test time series a total number of 1000 samples has been processed in each run. In a first series of experiments, four boundaries have been generated in each time series at sample no. 200, 400, 700 and 800 leading to different segment lengths of 100, 200 and 300 samples in each series. Parameter states have been choosen such that - for both algorithms - the segmentation in close to perfect if boundaries would be well separated (long segments with length $>$4 times the test window length L = 70 samples). Therefore, this series of experiments is mainly a test of the ability of the algorithms to detect short segments ($<$300 samples).

In a second series of experiments, the detection of one single boundary located at sample no. 500 is analyzed. For long segments (length $>$4 times the test window length L = 100) as used in this test, boundaries are generally easier to detect; however, the cepstral distance of the two parameter states has been choosen to be much closer in this test. Therefore, these experiments give hints on the ability of each algorithm to detect spuriously appearing boundaries of small parameter jumps (e.g., failure detection).

For each test type and for each kind of parameter sets, a total number of 100 resp. 200 runs has been performed in order to obtain stable statistical results. For all runs of each of these experiments, cumulated histograms of detection location $t_D$ (the sample number where a boundary has been detected) and position location $t_B$ (the sample number where the boundary has been positioned in each individual time series) have been calculated (fig. 5-7 and 9-11). For an ideal segmentation algorithm, therefore, the position histogram should show a staircase-like curve irrespective of the individual detection locations, with a step of height 1 at each segment boundary (= 100% detection probability) and a horizontal slope in between a segment (= 0% false alarm rate). Moreover, if fast detection of a boundary is called for, the time delay between the detection and the position histogram curves should be as small as possible.

All tests have been performed using an AR model order of p=4 in the segmentation algorithms (this order in general should be at least as high as the highest order of the signal segments). As mentioned before, the test window length for both algorithms has been choosen to L = 100 for the long segment test and to L = 70 for the short segment test.

Fig. 5 shows the result of the first experiment: Four jumps of the pole location (= center frequency of the power spectrum) from $20^{o}$ to $40^{o}$ and vice versa. As can be seen, the GLR algorithm performs very well in this experiment with excellent positioning of detected boundaries even if there is a substantial detection delay. On the other hand, the positioning of detected boundaries is not very good for the CUSUM test if the segment is short (boundary at position 800), as could be expected from the considerations in chapter 3. Moreover, the result of any segmentation for this algorithm seems to be strongly dependant on the direction of

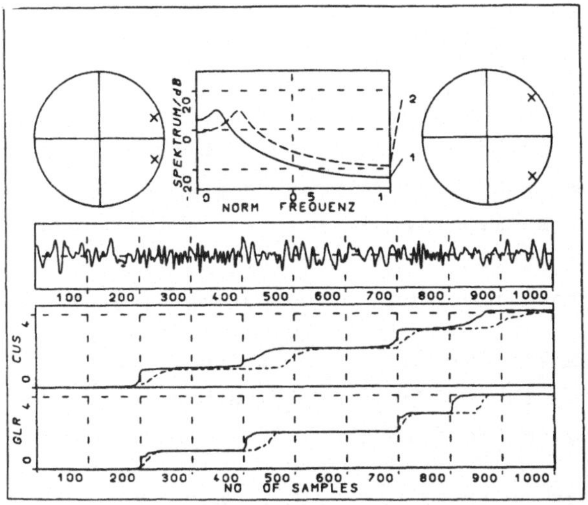


Fig. 5: Cumulated histogram of detection and position locations for pole frequency jumps $20^{\circ} \leftrightarrow 40^{\circ}$; pole radius $r = 0.9$

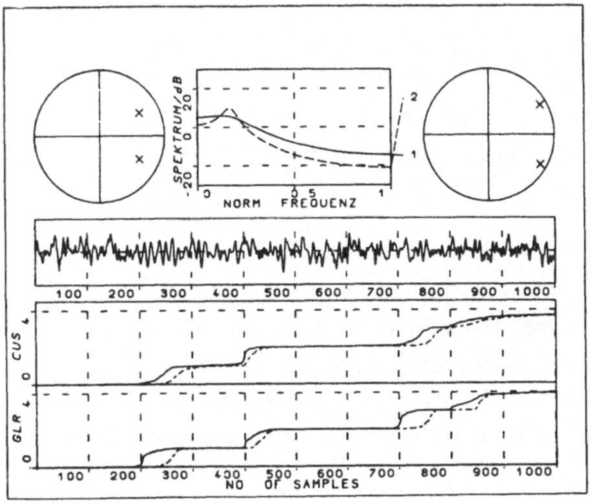


Fig. 6: Cumulated histogram of detection and position locations for pole radius jumps $0.7 \leftrightarrow 0.9$; pole frequency $\varphi = 30^{\circ}$

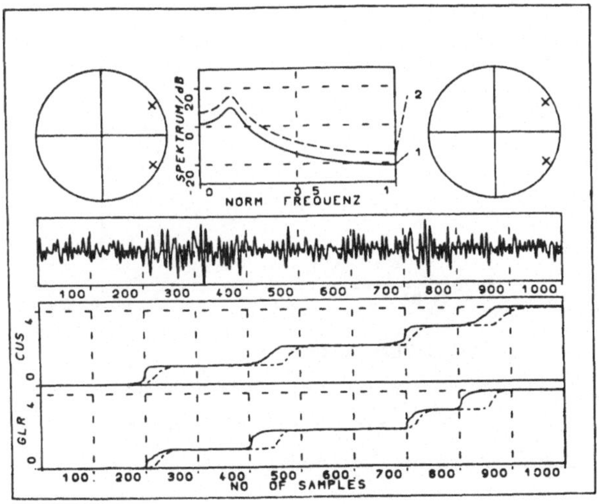


Fig. 7: Cumulated histogram of detection and position locations for signal energy jumps $\pm$ 6dB; pole radius $r = 0.9$, pole frequency $\varphi = 30^{\circ}$

the jump; this finding will be verified in the other experiments, too.

Fig. 6 shows a similar experiment with a jump in the pole radius (≙ bandwidth of the power spectrum) from 0.7 to 0.9 and vice versa. In this case, positioning of boundaries is not very good in both algorithms if the segment is only 100 samples long (boundary at sample no. 800). However, the GLR algorithm performs slightly better even in this case, and the quality of the boundary positioning for the CUSUM test again shows to be highly dependant on the direction of the parameter jump.

Fig. 7 finally shows results of a similar experiment using an energy jump by +6dB or vice versa. Again, the CUSUM algorithm shows a bad positioning of boundaries depending on the direction of the jump (-6dB).

As a general result of this series of experiments (all experiments have been performed with reversed time axis, too) the assumption made in chapter 3 that detection and particularly positioning of boundaries of short segments is difficult for the CUSUM algorithm, has been verified at least for the segments of length 100. Moreover, good positioning in the CUSUM algorithm turns out to be very much dependant on the direction of a parameter jump.

Another general result of this study (which cannot be extracted from these pictures) is that - though both algorithms are remarkably insensitive to variations of the decision threshold - the GLR algorithm in most experiments worked optimal for a wider range of threshold values than the CUSUM algorithm. To demonstrate this, in fig. 8 the detection probabilities $p_d$ and the false alarm probabilities $p_f$ are given for the

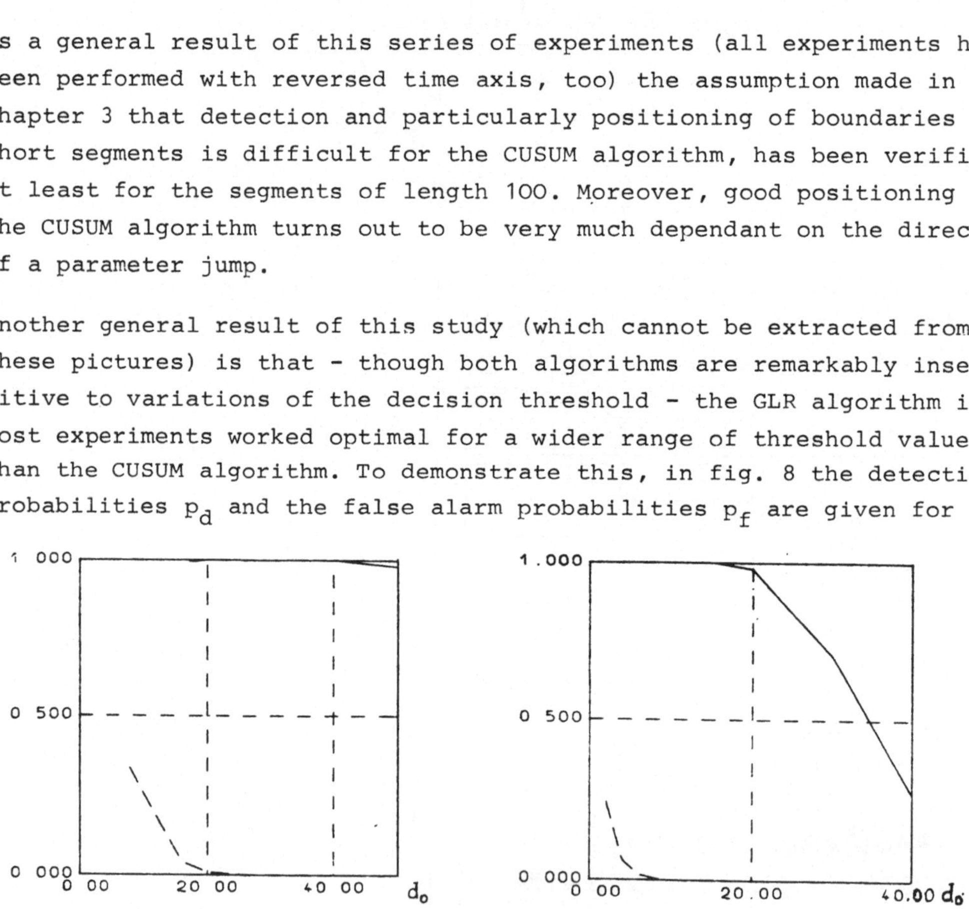


a) GLR algorithm b) CUSUM algorithm

Fig. 8: Detection probabilities $p_d$ (solid curve) and false alarm probabilities $p_f$ (dotted curve) as a function of the decision threshold $d_o$ for a pole frequency jump

pole frequency jump experiment and for thresholds between 5.0 and 40. ( a reasonable range for both algorithms); the detection probability in this case is calculated from all detected (not localized) boundaries within a 100 samples interval beginning with the true boundary location. As expected, for very high thresholds the detection rate is less than 1 (solid curve), while for very low thresholds the false alarm rate is high (dotted curve). The range where both probabilities are optimal is much wider in the GLR algorithm. Moreover, in other experiments with reduced parameter variation it turned out that the shift of this optimal range down to lower thresholds is higher in the CUSUM algorithm; with other words, the GLR algorithm will detect jumps of different distances more faithfully using only one fixed threshold, as it is the case in many practical applications.

In a final set of experiments, a pole frequency jump from $35^{\circ}$ to $40^{\circ}$ has been analyzed using the "long segment" test (only one single jump). As can be seen from fig. 9, for an optimum choice of thresholds both algorithms perform approximately equally good (the CUSUM "drift" $\delta$ has not been changed compared to the previous experiments). If $\delta$ is optimized, too, the detection of boundaries with the CUSUM algorithm is slightly better than with the GLR algorithm (fig. 10), however at the price of a slight increase in the false alarm rate and a poor positioning of detected boundaries.

In general, it turned out in this series of experiments that with a proper choice of $\delta$ and threshold, the CUSUM algorithm has a slightly better detection rate for spurious appearing boundaries. Another advantage is that - with these both parameters - it is possible to adjust the false alarm rate and the detection rate with somewhat less interaction than in the GLR algorithm. However, the choice of these parameters is very sensitive, and it is not possible to find optimal settings for a broader range of signal parameter jumps.

It is interesting to note in this context that the CUSUM algorithm fails completely if the direction of the jump is reversed ($40^{\circ}$ to $35^{\circ}$, see fig. 11). There is practically no correct detection of the boundary, but instead a high false alarm rate. Performance of the GLR algorithm, on the other hand, is approximately the same as before. This unsymmetry of the CUSUM algorithm seems to be a serious drawback for practical applications where the direction and type of a jump is not known in advance.

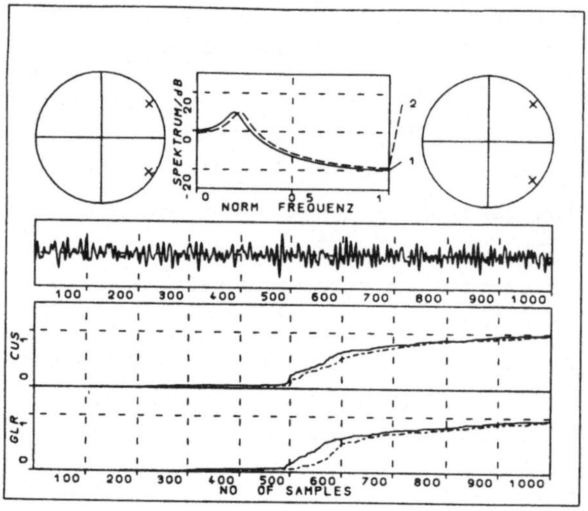


<u>Fig. 9:</u> Cumulated histogram of detection and position locations for a single pole frequency jump $35^{\circ} \leftrightarrow 40^{\circ}$; pole radius $r = 0.9$, CUSUM drift $\delta = 0.1$ (not optimized)

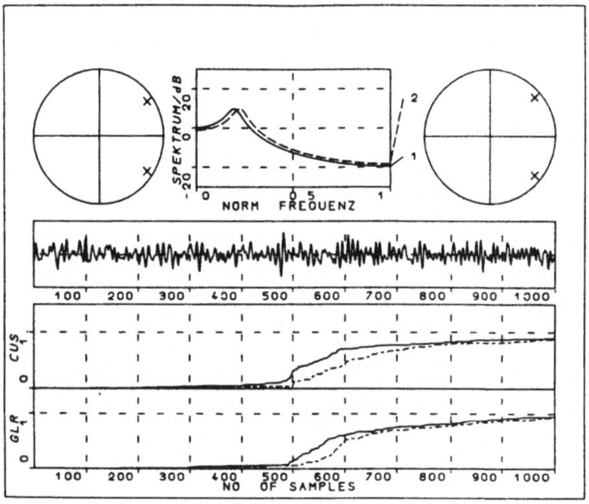


<u>Fig.10:</u> Cumulated histogram of detection and position locations for a single pole frequency jump $35^{\circ} \leftrightarrow 40^{\circ}$; pole radius $r = 0.9$, CUSUM drift $\delta = 0.07$ (optimized)

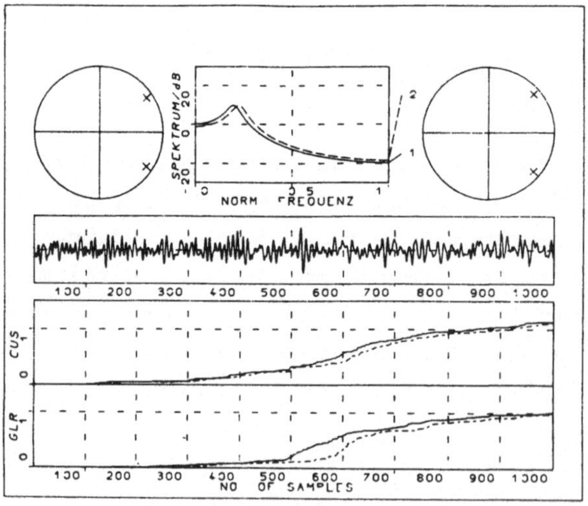


<u>Fig. 11:</u> Cumulated histogram of detection and position locations for a single pole frequency jump $40^{\circ} \leftrightarrow 35^{\circ}$; pole radius $r = 0.9$, CUSUM drift $\delta = 0.07$ (optimized)

## 5. Conclusion

Two recently published segmentation algorithms using both a growing reference window and a sliding test window - the generalized likelihood ratio (GLR) algorithm and the cumulated sum (CUSUM) algorithm - are compared against each other regarding their performance. Based on theoretical considerations and on extensive simulations, it can be stated that the GLR algorithm is favourable when short segments (length less than 4 times the test window length) are to be detected, and when good positioning of detected boundaries is called for. Moreover, the GLR algorithm is remarkably robust, allowing to detect a variety of different parameter jump types using one single threshold.

On the other hand, if type and distance of an expect ed parameter jump is known in advance and boundaries occur seldom ("long" segments), the CUSUM algorithm may lead to a higher detection probability if the two algorithm parameters (the drift $\delta$ and the threshold) are adjusted appropriately. Performance is very sensitive to the correct choice of these values.

As a consequence, if one is interested in combining good features of both algorithms, it seems reasonable to work with a GLR distance for a certain number of samples after the detection of a boundary. If no new boundary is detected in this interval, it is reasonable to switch over to the CUSUM "distance". This can simply be done as both algorithms use similar AR parameter estimates in the reference and the test window, allowing a smooth transition from one statistics to the other.

## References:


/1/ A. v.Brandt, "On-line Segmentation of Time Series Using a Pilot Segment", Proc. Int. Conf. on Digital Signal Processing, Florence, 1981, pp. 1111-1118

/2/ U. Appel, A. v.Brandt, "Adaptive Sequential Segmentation of Piecewise Stationary Time Series", Information Sciences 29, 1983, pp. 27-56

/3/ G. Bodenstein and H.M. Praetorius, "Feature extraction from the electroencephalogram by adaptive segmentation", Proc. IEEE, vol.65, no. 5, May 1977, pp. 642-652

/4/ B.H. Jansen, A. Hasman and R. Lenten, "Piecewise analysis of EEGs using AR-modeling and clustering", Computers and biomed. Res. 14, 1981, pp. 168-178

/5/ D. Michael and J. Houchin, "Automatic EEG analysis: A segmentation procedure based on the autocorrelation function", Electroenceph. clin. Neurophysiol., vol. 46, 1979, pp. 232-235

/6/ J.S. Barlow, O.D. Creutzfeld, D. Michael, J. Houchin and H. Epelbaum, "Automatic adaptive segmentation of clinical EEGs", Electroenceph. clin. Neurophysiol., 1981, vol. 51, pp. 512-525

/7/ J. Rissanen, "Modelling by shortest data description", Automatica, vol. 14, 1978, pp. 465-471

/8/ A.C. Sanderson, J. Segen and E. Richey, "Hierarchical modeling of EEG signals", IEEE Trans. Pattern Analysis, Machine Int., vol. PAMI-2, no. 5, Sept. 1980, pp. 405-415

/9/ M. Basseville, A. Benveniste, "Sequential Detection of Abrupt Changes in Spectral Characteristics of Digital Signals", IEEE Trans. Inf. Theory, Sept. 1983; (see also: Research Report no. 129, INRIA, Centre de Rennes, April 1982)

/10/ U. Appel, A. v.Brandt, "A Comparative Study of Three Sequential Time Series Segmentation Algorithms", To appear in Signal Processing 1984

/11/ A.S. Willsky and H.L. Jones, "A Generalized Likelihood Ratio Approach to the Detection and Estimation of Jumps in Linear Systems", IEEE Trans. on Automatic Control, vol. AC-21, 1976, pp. 108-112.

# ADAPTIVE FORGETTING IN RECURSIVE IDENTIFICATION THROUGH MULTIPLE MODELS

P. Andersson
Division of Automatic Control
Department of Electrical Engineering
Linköping University
S-581 83 Linköping, Sweden

## Abstract

A new recursive identification method, Adaptive Forgetting through Multiple Models - AFMM, is presented and evaluated using computer simulations. AFMM is especially suited for identification of systems with jumping parameters or parameters that change in an irregular fashion. It can be viewed as a particular way of implementing adaptive gains or adaptive forgetting factors for recursive identification. The new method essentialy consists of multiple Recursive Least Squares (RLS) algorithms running in parallel, each with a corresponding weighting factor. The simulations indicate that AFMM is able to track rapidly changing parameters well, and that the method is robust in several respects.

## 1. INTRODUCTION

Tracking of time-varying phenomena is an important problem in many application areas.

A basic underlying technique for these problems is recursive identification. See, e.g. Ljung and Söderström (1983) for a comprehensive discussion of this subject. A typical feature of a recursive identification algorithm is that the current parameter estimate is updated using a transformation of the last observation, multiplied by a certain gain factor (t):

$$\hat{\theta}(t) = \hat{\theta}(t-1) + \gamma(t) \cdot f\left(\hat{\theta}(t-1), y(t)\right) \qquad (1.1)$$

---

*) Now with Luxor AB, Motala, Sweden

The gain will determine how alert the algorithm will be to track changing properties of the system: A large gain means that it will quickly adapt to a new situation, but also that it will be sensitive to random errors in the signal y(t). The best choice of gain or forgetting is thus a trade-off between alertness and noise-sensitivity, and will depend on how much the system changes compared to the noise level in the measurements.

If these properties remain constant a good trade-off in the gain selection can be made a priori, either on formal grounds or by some initial tuning of the gain. However, very often both the measurement noise level and the rate of variation in the system may change. The system may, e.g., be subject to rare, but abrupt changes, or may change its dynamics considerably over a short period and then remain constant for a long while. In such cases also the choice of gain or forgetting must be adapted to thé changing environment.

In this paper we suggest one such approach to adaptive forgetting. We shall then concentrate on a particular model description for dynamical systems, viz the difference equation model

$$y(t)+a_1y(t-1)+\ldots+a_ny(t-n)=b_1u(t-1)+\ldots+b_mu(t-m)+e(t), \quad (1.2)$$

where $\{u(t)\}$ and $\{y(t)\}$ are the input and output, respectively of the system and $\{e(t)\}$ is a disturbance sequence of yet unspecified character.

A classical way of discounting old measurements is the wellknown Recursive Least Squares (RLS) algorithm with a forgetting factor $0<\lambda<1$, often chosen in the range 0.95-0.99 or a constant matrix, $R_1$, added to the covariance matrix.

A common problem of both the above modifications of the RLS algorithm is that in order to make the algorithm reasonably responsive to changes, the values of $\lambda$ (or $R_1$) have to be chosen rather small (large) and this will have a negative effect on the performance when no changes occur, since the estimates are based on few data. This means that a trade-off has to be made between the responsiveness when changes occurs and the performance when the parameters are constant.

Several suggestions of how to solve this problem have been made, see e.g. Kesten (1958), Fortescue et al (1981), Wellstead and Sanoff

(1981), Benveniste and Ruget (1982), Dumont (1982), Hägglund (1982, 1983), and Lozano (1983). In most of them some sort of algorithm to detect changes in the parameters is used and this information is then used to adjust $\lambda$ or $R_1$ in an intelligent way. The main problem with such a scheme is that in order to get a fast response, $\lambda$ (for example) has to be significantly decreased in one or a few samples which will essentially restart the algorithm. Hence, if the detection algorithm gives a false alarm (no change occured), the good information condensed in the parameter estimates will be lost. In this paper a new identification algorithm, Adaptive Forgetting through Multiple Models (AFMM), is proposed and investigated. The AFMM method is aiming at handling the problems mentioned above in a better way.

The paper is organized as follows:

In Section 2 the algorithm is presented and in Section 3 several simulations are performed in order to investigate the performance of the new method compared to others.

In Andersson (1983) more detailed information of AFMM is given and further simulations are presented.

## 2. A NEW METHOD - AFMM

A possible description of a time-discrete system with jumping parameters is the state space model:

$$\begin{cases} \theta(t+1) = \theta(t) + w(t) \\ y(t) = \varphi^T(t)\theta(t) + e(t) \end{cases} \tag{2.1}$$

where $\theta(t)$ is a vector containing the true parameters describing the system at time t, $\varphi(t)$ is a vector containing old inputs and outputs, and $e(t)$ and $w(t)$ are disturbances. The noise, $e(t)$, is supposed to be gaussian with zero mean and variance $R_2$. If the disturbance, $w(t)$, in (2.1) is gaussian with variance $R_1$, the best estimate of $\theta$ is given by a special case of the Kalman filter (cf. Ljung and Söderström (1983)).

When modelling changing parameters, one possibility would be to use a gaussian disturbance with varying variance. If the variation of the variance is known, with $R_1$ replaced by a time-varying matrix $R_1(t)$, the Kalman filter still gives the best estimate of $\theta$ (i.e. smallest possible error covariance matrix). For parameters that are constant most of the time, w(t) can be chosen as

$$w(t) = \begin{cases} v(t) & \text{w.p.} \quad q \\ 0 & \text{w.p.} \quad 1-q \end{cases} \tag{2.2}$$

where v(t) is gaussian with zero mean and covariance matrix $R_1$. Thus

$$R_1(t) = \begin{cases} R_1 & \text{w.p.} \quad q \\ 0 & \text{w.p.} \quad 1-q \end{cases} \tag{2.3}$$

As w(t) no longer is gaussian, the Kalman filter does not provide the optimal solution. Instead the problem (2.1)&(2.2) has become a non-linear filtering problem. There is an extensive literature on non-linear filtering, with many suggested approximate procedures, see e.g. Jazwinski (1970). We shall here pursue one approach based on finite-gaussian sum approximation.

Suppose that the posterior distribution of $\theta(t)$ given $y^{t-1}$ (i.e. all old y's up to and including y(t-1)) can be approximated with a sum of m gaussian density functions. This approach has previously been used in filtering theory, for example by Sorenson and Alspach (1971, 1972), J. T-H. Lo (1972) and Anderson and Moore (1979). The density function for $\theta(t)$ then becomes:

$$p(\theta(t)|y^{t-1}) = \frac{1}{(2\pi)^{n/2}} \cdot$$

$$\sum_{i=1}^{m} \alpha_i(t) \exp\{-\frac{1}{2}(\theta(t)-\bar{\theta}_i(t))^T P_i^{-1}(t)(\theta(t)-\bar{\theta}_i(t))\} \cdot |P_i(t)|^{-\frac{1}{2}} \tag{2.4}$$

where

$$\sum_{i=1}^{m} \alpha_i(t) = 1,$$

$\bar{\theta}_i(t)$ and $P_i(t)$ are the mean vectors and covariance matrices, respectively, of the different gaussian distributions at time t. They are functions of $y^{t-1}$ and so are the $\alpha_i$. Also, n is the dimension of $\theta$. Using Bayes' rule

$$p(x|z,y) = \frac{p(y|z,x)p(x|z)}{p(y|z)}$$

we can compute

$$p(\theta(t)|y^t) = p(\theta(t)|y(t),y^{t-1}).$$

$$p(y(t)|\theta(t),y^{t-1}) =$$

$$= \frac{1}{\sqrt{2\pi R_2}} \cdot e^{-\frac{1}{2}(y(t)-\theta^T(t)\varphi(t))^2 \cdot \frac{1}{R_2}} \tag{2.5}$$

Hence,

$$p(\theta(t)|y^t) = C_1 \cdot p(y(t)|\theta(t),y^{t-1}) \cdot p(\theta(t)|y^{t-1})$$

$$= C_1 \cdot \frac{1}{\sqrt{2\pi} \cdot (2\pi)^{n/2} \cdot \sqrt{R_2}} \cdot \sum_{i=1}^{m} \alpha_i(t) \cdot \exp\{-\frac{1}{2}[(y(t)-\theta^T(t)\varphi(t))^2 \cdot \frac{1}{R_2} +$$

$$+ (\theta(t)-\bar{\theta}_i(t))^T P_i^{-1}(t)(\theta(t)-\bar{\theta}_i(t))]\} \cdot |P_i(t)|^{-\frac{1}{2}} =$$

$$= C_2 \cdot \sum_{i=1}^{m} \alpha_i'(t) \cdot \exp\{-\frac{1}{2}(\theta(t)-\bar{\theta}_i'(t))^T P_i'(t)^{-1}(\theta(t)-\bar{\theta}_i'(t))\} \cdot$$

$$\cdot |P_i'(t)|^{-1/2} \tag{2.6}$$

where

$$P_i'(t) = P_i(t) - \frac{P_i(t)\varphi(t)\varphi^T(t)P_i(t)}{R_2+\varphi^T(t)P_i(t)\varphi(t)} \tag{2.7}$$

$$\bar{\theta}_i'(t) = \bar{\theta}_i(t) + \frac{1}{R_2} P_i'(t)\varphi(t)\varepsilon_i(t) \tag{2.8}$$

$$\alpha_i'(t) = C_n \cdot \frac{\alpha_i(t)}{\sqrt{R_2+\varphi^T(t)P_i(t)\varphi(t)}} \cdot e^{-\frac{1}{2} \cdot \frac{\varepsilon_i^2(t)}{R_2+\varphi^T(t)P_i(t)\varphi(t)}} \tag{2.9}$$

$$\varepsilon_i(t) = y(t) - \varphi^T(t)\bar{\Theta}_i(t) \tag{2.10}$$

where $C_n$ must be chosen so that

$$\sum_{i=1}^{m} \alpha_i^!(t) = 1,$$

and $C_1^{-1}=p(y(t)|y^{t-1})$ is a normalizing factor.

Now we can use $p(\Theta(t)|y^t)$ to compute $p(\Theta(t+1)|y^t)$. The density function for $w(t)$ is:

$$p(w(t))=(1-q)\cdot\delta(w(t))+q\cdot(2\pi)^{-\frac{n}{2}}|R_1|^{-\frac{1}{2}}\cdot e^{-\frac{1}{2}w^T(t)R_1^{-1}w(t)} \tag{2.11}$$

Convolution now gives

$$p(\Theta(t+1)|y^t) = \int_{-\infty}^{\infty} p(w(t))\cdot p(\Theta(t+1)-w(t)|y^t)dw(t) =$$

$$=C_2\cdot \sum_{i=1}^{m} \alpha_i^!(t)[(1-q)\cdot e^{-\frac{1}{2}(\Theta(t+1)-\bar{\Theta}_i^!(t))^T\cdot P_i^!(t)^{-1}(\Theta(t+1)-\bar{\Theta}_i^!(t))}\cdot$$

$$\cdot|P_i^!(t)|^{-\frac{1}{2}} + q\cdot e^{-\frac{1}{2}(\Theta(t+1)-\bar{\Theta}_i^!(t))^T(P_i^!(t)+R_1)^{-1}(\Theta(t+1)-\bar{\Theta}_i^!(t))}\cdot$$

$$\cdot|P_i^!(t) + R_1|^{-\frac{1}{2}}] \tag{2.12}$$

From (2.12) it can be seen that the posterior distribution of $\Theta(t+1)$ given $y^t$ can be described by 2m gaussian density functions. As we do not want the number of gaussian distributions used to approximate the distribution of $\Theta$ to increase, we have to replace the distribution for $\Theta(t+1)$ with one that contains only m gaussian distributions.

One way to do this is to approximate (2.12) by

$$p(\Theta(t+1)|y^t) \simeq$$

$$C_3\Big[\sum_{i\neq i_{min}} \alpha_i'(t) e^{-\frac{1}{2}((\theta(t+1)-\bar{\theta}_i'(t))^T P_i'^{-1}(t)(\theta(t+1)-\bar{\theta}_i'(t))} \cdot |P_i'(t)|^{-\frac{1}{2}}$$

$$+ q\,|R_1|^{-\frac{1}{2}} e^{-\frac{1}{2}((\theta(t+1)-\bar{\theta}'_{i_{max}}(t))^T R_1^{-1}(\theta(t+1)-\bar{\theta}'_{i_{max}}(t))}\Big] \tag{2.13}$$

The estimate, $\hat{\theta}(t)$, of $\theta(t)$ is

$$\hat{\theta}(t) = E[\theta(t)|y^t] =$$

$$= \int_{-\infty}^{\infty} \theta(t)\cdot \sum_{i=1}^{m} \alpha_i(t)\cdot \frac{1}{(2\pi)^{n/2}} \cdot \exp\{-\frac{1}{2}(\theta(t)-\bar{\theta}_i(t))^T\cdot P_i^{-1}(t)\cdot$$

$$\cdot (\theta(t)-\bar{\theta}_i(t))\} \cdot |P_i(t)|^{-\frac{1}{2}} d\theta(t) =$$

$$= \sum_{i=1}^{m} \alpha_i(t)\bar{\theta}_i(t) \tag{2.14}$$

Updating the quantities involved in (2.13)-(2.14) gives the following set of equations:

$$P_i(t) = P_i(t-1) - \frac{P_i(t-1)\varphi(t)\varphi^T(t)P_i(t-1)}{R_2+\varphi^T(t)P_i(t-1)\varphi(t)} \tag{2.15a}$$

$$\bar{\theta}_i(t) = \bar{\theta}_i(t-1) + \frac{1}{R_2}\cdot P_i(t)\varphi(t)\varepsilon_i(t) \tag{2.15b}$$

$$\bar{\alpha}_i(t) = \frac{\alpha_i(t-1)}{\sqrt{R_2+\varphi^T(t)P_i(t-1)\varphi(t)}} \cdot e^{-\frac{1}{2}\frac{\varepsilon_i^2(t)}{R_2+\varphi^T(t)P_i(t-1)\varphi(t)}} \tag{2.15c}$$

$$i_{min}=\arg\min_i \bar{\alpha}_i(t) \qquad i_{max}=\arg\max_i \bar{\alpha}_i(t) \tag{2.15d}$$

$$P_{i_{min}}(t)=R_1 \qquad \bar{\theta}_{i_{min}}(t)=\bar{\theta}_{i_{max}}(t) \qquad \bar{\alpha}_{i_{min}}(t)=q \tag{2.15e}$$

$$\alpha_i(t) = \frac{1}{\sum_{k=1}^{m} \bar{\alpha}_k(t)} \cdot \bar{\alpha}_i(t) \tag{2.15f}$$

$$\hat{\theta}(t) = \sum_{i=1}^{m} \alpha_i(t)\bar{\theta}_i(t) \tag{2.15g}$$

where

$$\varepsilon_i(t) = y(t) - \varphi^T(t)\bar{\theta}_i(t-1) \tag{2.15h}$$

We shall call this approach Adaptive Forgetting through Multiple Models - the AFMM method.

Except for the initial values of $\bar{\theta}_i$, $P_i$, and $\alpha_i$ four choices have to be made to use this method. These design variables are:

$R_1$ covariance matrix for the jumps.

$q$ probability of a jump.

$m$ number of gaussian distributions used to approximate the distribution of the parameter vector.

$R_2$ equation error noise variance.

Simulations indicates that the values of q and $R_1$ are not critical and the only requirement of m is that it is chosen large enough, i.e. of the same order as the number of parameters being estimated. The choice of $R_2$ is however more critical and to obtain an acceptable result it must probably be in the range 0.5-5 times the true value.

The sensitivity with respect to $R_2$ can be a serious problem if it is difficult to estimate this in advance or the noise variance varies too much during the identification. A solution to this problem would be to estimate $R_2$:

$$\hat{R}_2(t)=(1-\mu)\cdot\left(1-\varphi^T(t)P_{i_{max}}(t)\varphi(t)\right)\cdot\varepsilon^2(t)+\mu\cdot\hat{R}_2(t-1) \tag{2.20}$$

where

$$i_{max} = \arg\max_i \alpha_i(t)$$

and $\mu$ is a forgetting factor. This $\hat{R}_2(t)$ then replaces $R_2$ in (2.15abc).

Simulations have shown that augmenting the AFMM method with the above estimation of $R_2$ gives an identification method which also is capable of handling unknown and changing equation error noises.

## 3. SIMULATION EXPERIMENTS: A DIESEL GENERATOR

In this section digital simulations will be used to show how AFMM performs compared to other identification methods. Further results can be found in Andersson (1983).

The AFMM method will be used to identify a simulated diesel-generator, see Goodwin and Sin (1984). A block diagram for the generator is shown in Figure 3.1.

The following values are used:

$I_1=K_1=B_1=K_3=I_2=1 \qquad K_2=5 \qquad B_2=0.1$

The load resistance, L, is assumed to vary. The system was sampled using the sample interval 0.4 sec, giving the sampled

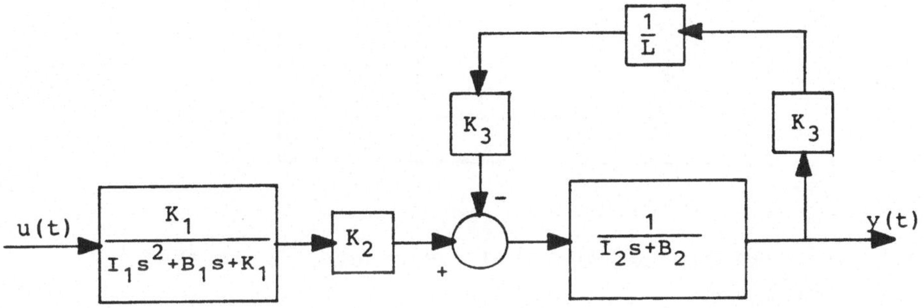


Figure 3.1 Block diagram of the diesel-generator.

| L | $A(q^{-1})$ | $B(q^{-1})$ |
|---|---|---|
| 0.2 | $1-1.670q^{-1}+0.871q^{-2}-0.0872q^{-3}$ | $0.0306+0.0712q^{-1}+0.00918q^{-2}$ |
| 1.0 | $1-2.184q^{-1}+1.662q^{-2}-0.432q^{-3}$ | $0.0431+0.139q^{-1}+0.0283q^{-2}$ |
| 5.0 | $1-2.427q^{-1}+2.036q^{-2}-0.595q^{-3}$ | $0.0466+0.163q^{-1}+0.0359q^{-2}$ |
| 10.0 | $1-2.463q^{-1}+2.092q^{-2}-0.619q^{-3}$ | $0.0471+0.166q^{-1}+0.037q^{-2}$ |

Table 3.1 The sampled diesel-generator for different values of the load resistance, L.

Model: $A(q^{-1})y(t) = q^{-1}B(q^{-1})u(t) + v(t)$

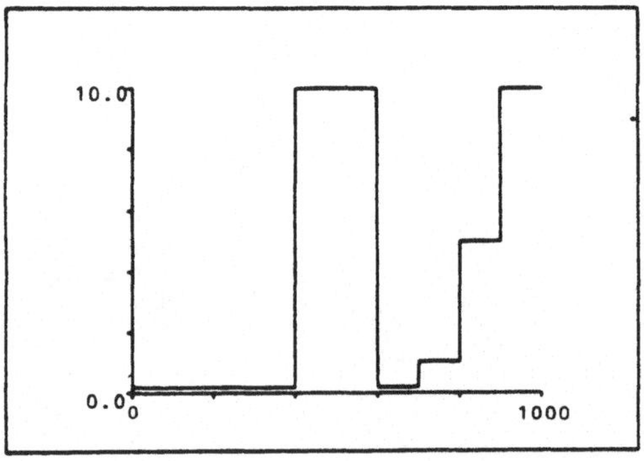


Figure 3.2 The variation of the load resistance, L, in the simulations.

systems listed in Table 3.1 for different values of L. In Figure 3.2 the variation of L during the simulation is shown.

The model used was

$$A(q^{-1})y(t) = q^{-1}B(q^{-1})u(t) + v(t) \tag{3.1}$$

where

$$A(q^{-1}) = 1 + a_1q^{-1} + a_2q^{-2} + a_3q^{-3}$$

$$B(q^{-1}) = b_0 + b_1q^{-1} + b_2q^{-2}$$

$q^{-1}$ is the backward shift operator, i.e. $q^{-1}y(t) = y(t-1)$.

v(t) in (3.1) was gaussian with variance 0.01 and the input signal, u(t), was $\pm$ 0.5 with probability 0.2 to change sign in each sample. The design variables in the AFMM algorithm was chosen as:

$q=0.005 \qquad R_1=1.0 \cdot I \qquad m=5 \qquad \mu=0.98$

The result of the simulation are shown in Figure 3.3 a-c. As comparision, a simulation using the method proposed in Hägglund (1983) is shown in Figure 3.4 a-b. This simulation was performed by T Hägglund. The noise realization used was not the same as in the AFMM simulation and the design variables used in Hägglund's method were

$r_0=0.4 \qquad \gamma_1=0.8 \qquad \gamma_2=0.95 \qquad \lambda=0.99$

$$\beta = \begin{pmatrix} 50 \cdot I_2 & 0 \\ 0 & 5 \cdot I_3 \end{pmatrix}$$

where $I_k$ means the identity matrix of dimension k.

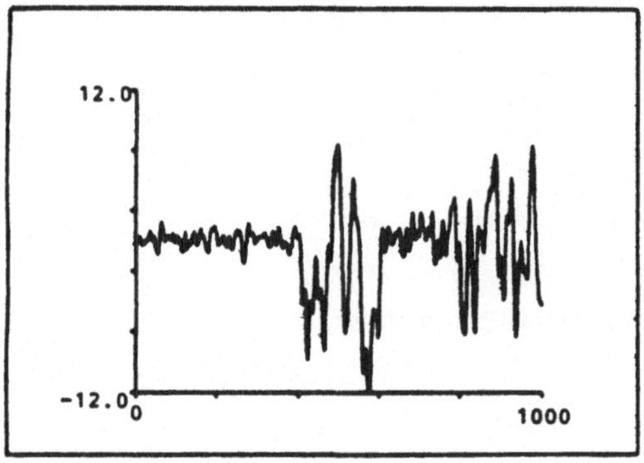


a. The output signal.

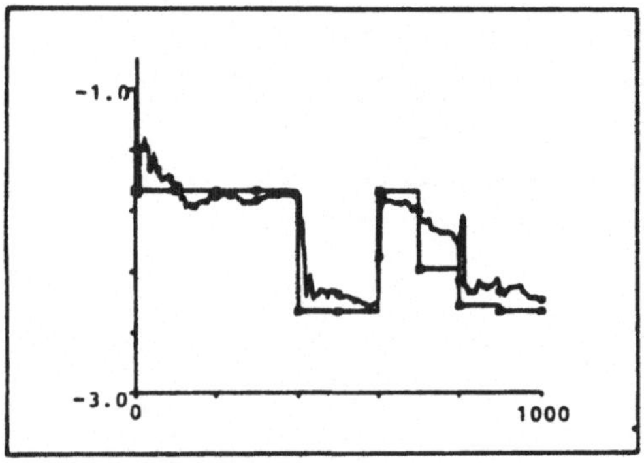


b. The $\hat{\theta}_1$ estimate.

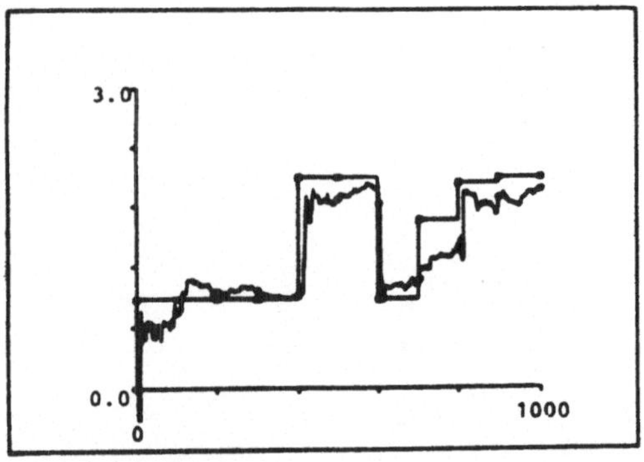


c. The $\hat{\theta}_2$ estimate.

Figure 3.3 The identification of the diesel-generator, with white equation error noise, using the AFMM method.

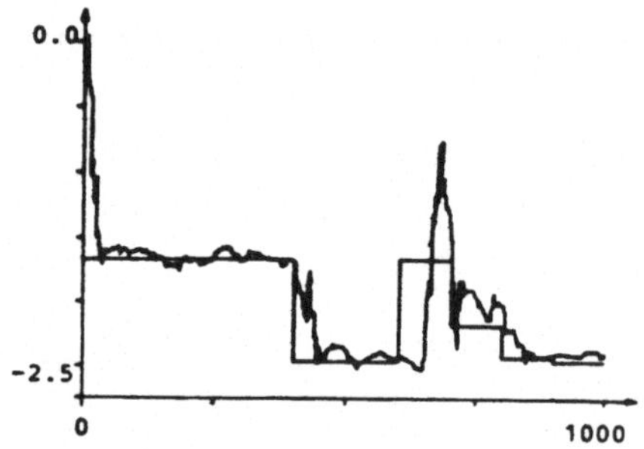


a. The $\hat{\theta}_1$ estimate.

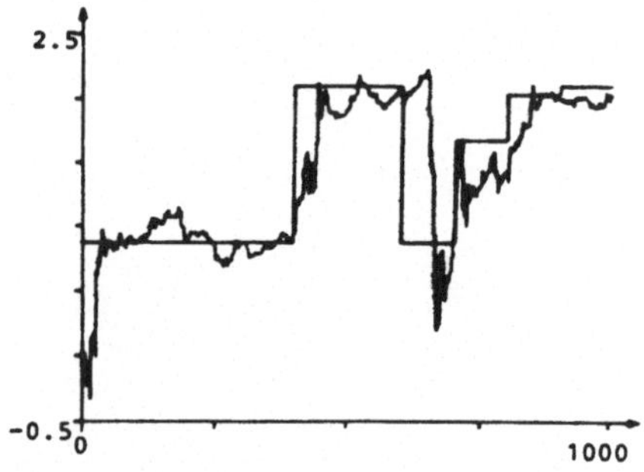


b. The $\hat{\theta}_2$ estimate.

Figure 3.4 The identification of the diesel-generator, with white equation error noise using Hägglund's method.

## 4. CONCLUSION

A new recursive identification method, AFMM, which is capable of identifying systems with changing parameters has been proposed and evaluated, using computer simulations. The approach implements the notion of adaptation of the gain (or the forgetting factor) in a recursive identification algorithm so that both slow, fast and abrupt changes in dynamics may be tracked adequately.

A nice, and probably the most important, feature of the AFMM method is that if it detects a non-existent jump, it is able to correct the mistake without loosing any valuable information. Methods that are based on direct modification of the forgetting factor tend to be more conservative, since a mistake of this sort means that relevant information is destroyed.

In practice it is in most cases not realistic to assume that noise variances, jump probabilities, etc are known. Therefore it is important that the identification method used is not sensitive in this respect. By combining the AFMM method with estimation of the equation error noise variance, the AFMM method is robust with respect to these choices. Also, non-white equation error noises can be dealt with. In this case biased parameter estimates are obtained, but the result describes the system well in the frequency domain.

Although the AFMM method was designed with abruptly changing parameters in mind, simulations have shown that ramp changes in the parameters can also be followed.

No degradation in performance is obtained if the number of models is increased, because the models are updated independently of each other. The only negative effect of increasing the number of models is that the computational burden increases. Since the AFMM method essentially consists of multiple RLS algorithms running in parallel, the computational requirement is approximately the same as for the RLS algorithm times the number of models used.

REFERENCES

Alspach, D. L. and H. W. Sorenson (1972). Nonlinear Bayesian Estimation Using Gaussian Sum Approximations, IEEE Trans Automatic Control, Vol AC-17, No 4, pp 439-448.

Anderson, B. D. O. and J. B. Moore (1979). Optimal Filtering, Prentice Hall.

Andersson, P. (1983). Adaptive forgetting in recursive identification through multiple models. Internal report LiTH-ISY-I-0638, Department of Electrical Engineering, Linköping University, Linköping, Sweden.

Benveniste, A. and G. Ruget (1982). A Measure of the Tracking Capability of Recursive Stochastic Algorithms with Constant Gains, IEEE Trans Automatic Control, Vol. AC-27, No 3, pp 639-649.

Dumont, G. A. (1982). Self-tuning Control of a Chip Refiner Motor Load, Automatica, Vol 18, No 3, pp 307-314.

Fortescue, T. R., L. S. Kershenbaum and B. E. Ydstie (1981). Implementation of Self-tuning Regulators with Variable Forgetting Factors, Automatica, Vol 17, pp 831-835.

Goodwin, G. C. and K. S. Sin (1984). Adaptive Prediction, Filtering and Control, Prentice Hall, to appear.

Hägglund, T. (1982). Adaptive Control with Fault Detection, Report TFRT-7242, Department of Automatic Control,Lund Institute of Technology, Lund, Sweden.

Hägglund, T. (1983). Recursive Least Squares Identification with Forgetting of Old Data, Report TFRT-7254, Department of Automatic Control, Lund Institutet of Technology, Lund, Sweden.

Jazwinski, A. H. (1970). Stochastic Processes and Filtering Theory, Academic Press, New York.

Kesten, H. (1958). Accelerated Stochastic Approximation, Ann Math Stat, Vol 29, pp 41-59.

Ljung, L. and T. Söderström (1983). Theory and Practice of Recursive Identification, MIT Press.

Lo, J. T-H. (1972). Finite-Dimensional Sensor Orbits and Optimal Nonlinear Filtering, IEEE Trans Information Theory, Vol IT-18, No 5, pp 583-588.

Lozano, R. (1983). Convergence Analysis of Recursive Identification Algorithms with Forgetting Factor, Automatica, Vol 19, No 1, pp 95-97.

Sorenson, H. W. and D. L. Alspach (1971). Recursive Bayesian Estimation Using Gaussian Sums, Automatic, Vol 7, pp 465-479.

Trulsson, E. (1983). Adaptive Control Based on Explicit Criterion Minimization, Linköping Studies in Science and Technology. Dissertations. No 106.

Wellstead, P. E. and S. P. Sanoff (1981). Extended self-tuning algorithm, Int J Control, Vol 34, No 3, pp 433-455.

Willsky, A. S. (1976). A Survey of Design Methods for Failure Detection in Dynamic Systems, Automatica, Vol 12, pp 601-611.

# DESCRIPTION D'UN DETECTEUR SEQUENTIEL DE CHANGEMENTS BRUSQUES DE DYNAMIQUES DES MODELES ARMA

D. CANON - C. DONGARLI
LABORATOIRE D'AUTOMATIQUE - Ecole Nationale Supérieure de Mécanique
1, rue de la Noë 44072 NANTES Cédex France

## ABSTRACT

The sequential detection of abrupt changes in ARMA models is an important question in multi sensor/multi target tracking or similar problems (failure detection, E.E.G. analysis ...)

The authors propose an original sequential algorithm both testing the model and estimating its parameters simultaneously. This method is based upon a constant order hypothesis with abrupt changes within the parameters of an ARMA model. This realistic hypothesis is a consequence of the experimental capability of a low order ARMA model to represent correctly any Markovian process.

The identification of the system is sequentially performed by an Extended Kalman Filter which provides both the parameters of the model and a "pseudo-innovation" sequence.

That independant sequence is multiply by itself with a link of one step to define a decision variable and to detect a change of level.

## INTRODUCTION

De nombreux problèmes peuvent se définir comme celui de la segmentation d'un signal en plages possédant certaines caractéristiques statistiques stationnaires. C'est le cas dans l'étude de la parole, comme dans celui des électroencéphalogrammes (E.E.G.), des électrocardiogrammes (E.C.G.) ainsi que dans la poursuite multi-cibles/multi-senseurs.

Une première approche consiste à tester l'ordre du modèle stochastique du signal étudié. Mais, si ces techniques ont théoriquement un caractère séquentiel, elles nécessitent, en fait, un volume important de calculs à chaque étape.

Nous proposons donc une méthode originale visant à conserver effectivement le caractère séquentiel du détecteur tout en limitant le volume de calcul. Pour cela, nous partons de l'hypothèse qu'un processus Markovien peut être modélisé par un modèle ARMA d'ordre 3 ou 4 -sauf cas très particulier où la présence d'harmonique d'ordre élevé est prévisible de par la nature même du processus générateur.

Nous commencerons par rappeler les études déjà réalisées sur ce thème avant de présenter un nouvel algorithme en deux parties :

- le filtre autoadaptatif, qui est un filtre de Kalman étendu, dont les sorties sont les paramètres du système et une séquence pseudo-innovation.
- le détecteur, qui est réalisé à partir de cette séquence (multipliée par elle-même avec un décalage unitaire) avec, ensuite, une comparaison à un seuil autoadaptatif.

Après détection d'une variation du modèle, le filtre de Kalman étendu est réinitialisé et le détecteur est inhibé, le temps d'obtenir une estimation suffisante des paramètres pour éviter les fausses alarmes.

Enfin, nous présenterons le comportement de la méthode sur des simulations.

## I. LE PROBLEME DE LA DETECTION

La détection de ruptures de modèles pour des signaux monodimensionnels a été largement étudiée.

Une comparaison de ces méthodes peut être trouvée dans BASSEVILLE (1981 a, 1982 a) et WILLSKY (1980). Cependant, nous pouvons classer ces méthodes en deux grandes classes :

a. Détection de l'instant du saut en choisissant parmi deux hypothèses ou plus. Ce sont les filtres dits "dérivés et filtrés", ou les tests à sommes cumulatives proposés par SHIRYAEV (1961, 1963, 1965) et développés par HINKLEY (1971). Pour plus de détails, on peut se reporter aussi à BASSEVILLE (1982 b) et CHALMOND (1979).

b. Détection de l'instant et évaluation du saut basées sur un filtrage adapté, et non autoadapté. Ce sont les méthodes de MEHRA et PEESCHON (1971), WILLSKY (1976), et toutes celles basées sur le rapport de vraisemblance généralisé.

Cependant, l'application de ces dernières sur calculateur est généralement lourde. C'est pourquoi, nous présentons une méthode séquentielle réalisant, à la fois, le test et l'estimation des paramètres du modèle sans exiger un volume excessif de calcul.

## II. L'ALGORITHME D'ESTIMATION

Le détecteur proposé dans le quatrième paragraphe est défini à partir de la séquence pseudo-innovation issue d'un filtre de Kalman étendu appliqué à un modèle ARMA. Nous en donnons, ici, le principe.

### II.1. Equations du filtre de Kalman étendu

Nous nous plaçons, avant toutes choses, entre deux ruptures successives du modèle afin de pouvoir faire l'hypothèse de stationnarité.

Le processus discret $\{y_k\}$ est modélisé par l'équation (1) :

$$y_k = \sum_{i=1}^{n} a_i \, y_{k-i} + \varepsilon_k + \sum_{i=1}^{n} c_i \, \varepsilon_{k-i} \qquad (1)$$

les seules données du problème sont la séquence $\{y_k\}$ et l'ordre n du système.

$\{a_i, C_i\ ;\ i = 1,n\}$ sont des paramètres constants et $\{\varepsilon_k\}$ une séquence indépendante gaussienne de moyenne nulle et de variance R.

Du modèle ARMA précédent, nous tirons une représentation sous forme d'équation d'état généralisée. Cela nous donne les équations suivantes :

$x_{k+1} = F \cdot x_k + G\, \varepsilon_k$ (2), équation linéaire

$y_k \ = H(x_k) + \varepsilon_k$ (1), équation non linéaire (produits : $C_i \varepsilon_{k-i}$)

avec

$x_k^T = [a_1 .. a_n\ C_1 .. C_n\ \varepsilon_{k-1} .. \varepsilon_{k-n}]$ vecteur d'état généralisé et

$$F = \left[\begin{array}{c|c} \begin{matrix} 1 & & \\ & \ddots & \\ & & 1 \end{matrix} & 0 \\ \hline 0 & \begin{matrix} 0 & & 0 \\ 1 & \ddots & \\ 0 & 1 & 0 \end{matrix} \end{array}\right] \begin{matrix} \updownarrow 2n \\ \updownarrow n \end{matrix} \qquad \underbrace{\phantom{xxxx}}_{2n}\ \underbrace{\phantom{xx}}_{n}$$

$$G = \big[\underbrace{0 \quad 0}_{2n}\ \underbrace{1\ 0 \quad 0}_{n}\big]$$

Pour pouvoir appliquer les équations du filtre de Kalman, il est nécessaire de linéariser l'équation (1) par un développement au 1er ordre. Si, nous posons :

$$H_k = [y_{k-1} .. y_{k-n}\ |\hat{\varepsilon}_{k-1} .. \hat{\varepsilon}_{k-n}\ |\hat{C}_1 .. \hat{C}_n] \tag{3}$$

$$Z_k = y_k + \sum_{i=1}^{n} \hat{C}_i \cdot \hat{\varepsilon}_{k-i} \tag{4}$$

où $\hat{\varepsilon}$ et $\hat{C}$ représentent les estimations courantes de $\varepsilon$ et C, le système linéarisé se présente comme suit :

$$x_{k+1} = F_k\, x_k + G\, Z_k \quad \text{avec} \quad F_k = F - G\, H_k$$

$$Z_k = H_k\, x_k + \varepsilon_k$$

Alors, les équations du filtre sont aisément développées :

$$\hat{x}_{k+1/k} = F_k\, \hat{x}_{k/k-1} + G\, Z_k + L_k (Z_k - H_k \cdot \hat{x}_{k/k-1})$$

$$\hat{\varepsilon}_k = Z_k - H_k\, \hat{x}_{k/k-1}$$

$$L_k = (F_k\, P_{k/k-1}\, H_k^T)\, (H_k\, P_{k/k-1}\, H_k^T + R_k)^{-1} \text{ , gain du filtre}$$

$$P_{k+1/k} = F_k\, P_{k/k-1}\, F_k^T - L_k (F_k\, P_{k/k-1}\, H_k^T)^T$$

$$R_{k+1} = (1-\alpha)\, R_k + \alpha (Z_k - H_k\, \hat{x}_{k/k-1})^2 \text{ , } 0 < \alpha < 1$$

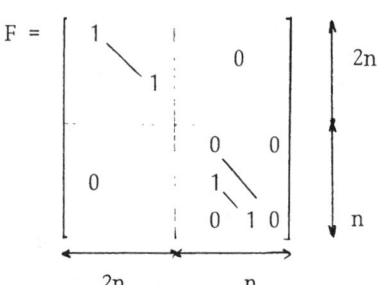

$\hat{x}_{k+1/k}$ est la prédiction de l'état généralisé pour l'instant k+1, connaissant le système jusqu'à l'instant k.

$R_{k+1}$ est l'estimation à l'instant k+1 de la variance R de la séquence pseudo innovation, obtenue par un filtre à oubli exponentiel.

II.2. Stabilisation de l'algorithme

Le système d'équations précédemment présenté est numériquement instable. En effet, $P_{k/k-1}$ peut perdre ses caractères de symétrie et de définie positivité.

Pour stabiliser cet algorithme, nous avons utilisé la forme factorisée de Potter : $P_{k/k-1} = S_k S_k^T$, et nous avons finalement programmé les équations suivantes :

$$F_k = F - G\,H_k$$

$$V_k = S_k^T\,H_k^T$$

$$D_k = V_k^T\,V_k + R_k$$

$$L_k = (F_k S_k\,V_k)/D_k$$

$$\hat{x}_{k+1/k} = F_k\,\hat{x}_{k/k-1} + G\,Z_k + L_k(Z_k - H_k\,\hat{x}_{k/k-1})$$

$$S_{k+1} = F_k\,S_k\,\left[1 - \frac{V_k V_k^T}{\sqrt{D_k}(\sqrt{D_k} + \sqrt{R_k})}\right]$$

$$R_{k+1} = (1 - \alpha)\,R_k + \alpha(Z_k - H_k\,\hat{x}_{k/k-1})^2$$

L'initialisation porte sur $X_{1/0}$, $P_{1/0}$ et $R_1$.

III. EFFET D'UN SAUT SUR LE VECTEUR D'ETAT GENERALISE $x_k$

Le saut sur la valeur des paramètres du modèle est représenté par les équations :

$$x_{k+1} = F_k\,x_k + G\,Z_k + \delta_{\theta,k+1}\mu$$

$$Z_k = H_k\,x_k + \varepsilon_k$$

avec $\delta_{\theta,k+1}$ le symbole de Kronecker, $\mu$ l'amplitude du saut survenant à l'instant $\theta$.

Nous pourrions considérer les effets d'un saut à l'instant $\theta$ sur les variables $x_k$, $Z_k$, $\hat{x}_{k/k-1}$, $\hat{\varepsilon}_k$ définies par les équations du filtre de Kalman (cf. WILLSKY 1976), et en déduire une variable de décision rattachée à la théorie du rapport de vraisemblance généralisé.

En fait, le modèle obtenu et l'emploi du filtre de Kalman étendu pour générer la séquence pseudo-innovation $\{\hat{\varepsilon}_k\}$ permettent de considérer cette dernière comme la sortie d'un système linéaire d'entrée $\{\varepsilon_k\}$. Cela entraîne, en particulier, que la séquence $\{\hat{\varepsilon}_k\}$ est centrée en l'absence, comme en présence, d'un saut. Il n'est donc pas possible d'utiliser directement un détecteur de variation de moyenne sur la séquence $\{\hat{\varepsilon}_k\}$.

Pour implanter une méthode rapide et simple, notre détecteur reposera, en fait, sur le choix entre deux hypothèses :

$H_o : W_k = \hat{\varepsilon}_k \hat{\varepsilon}_{k-1}$ est centré (pas de saut)

$H_1 : W_k = \hat{\varepsilon}_k \hat{\varepsilon}_{k-1}$ n'est pas centré (détection d'un saut)

Ainsi, le problème de la détection d'une modification de spectre sur $y_k$ est ramené à la détection d'une variation de moyenne de la variable $W_k$.

## IV. LA DETECTION

### IV.1. Description du détecteur

Avec les estimations $\hat{\varepsilon}_{k-1}$ et $\hat{\varepsilon}_k$ des valeurs de la séquence $\{\varepsilon_k\}$ pour les instants k-1 et k, nous définissons un nouveau processus stochastique $W_k = \hat{\varepsilon}_k \hat{\varepsilon}_{k-1}$. Nous estimons alors la moyenne de $\{W_k\}$ à l'aide d'un filtre à oubli exponentiel :

$$T_{k+1} = (1 - \beta)\, T_k + \beta W_k \qquad 0 < \beta < 1$$

Quand le filtre estimateur peut être considéré comme optimal, en l'absence de variation brutale des paramètres du système, $\{\hat{\varepsilon}_k\}$ est une séquence indépendante et $T_k$ est proche de zéro.

Mais, quand un saut a eu lieu, l'hypothèse d'indépendance de $\{\hat{\varepsilon}_k\}$ n'est plus vérifiée et $T_k$ devient non nul.

Le détecteur est donc de la forme suivante :

1. Le test est inhibé durant I itérations pour assurer une "bonne" estimation des paramètres du modèle.
2. Nous calculons $|T_k|$ :
   Si $|T(k)| > \lambda$, nous supposons qu'un saut a été détecté et nous réinitialisons le filtre estimateur avant de retourner à l'étape 1.
3. Sinon, le signal $\{y_k\}$ est filtré avant de retourner à l'étape 2.

Ici, nous utilisons un détecteur de variation de moyenne très simple, mais toute autre méthode peut être utilisée.

### IV.2. Réinitialisation du filtre estimateur

Lorsqu'un saut est détecté, la période d'inhibition a pour but de diminuer le risque de fausse alarme.

Un moyen de réinitialiser l'algorithme est de conserver le vecteur d'état généralisé, qui semble le meilleur a priori, mais de corriger la matrice $S_k$ -en augmentant les termes diagonaux, par exemple.

### IV.3. Choix des paramètres β, I et λ

Le facteur d'oubli β détermine ce que nous pouvons appeler la mémoire du détecteur. Si β est proche de 1, le détecteur a peu de mémoire et l'importance est donnée à la dernière valeur de $W_k$. Nous obtenons alors un détecteur nerveux, mais sujet à des fausses alarmes.

Si $\beta$ est proche de 0, le passé du système atténue l'effet de la dernière mesure de $W_k$. Le détecteur est alors plus "mou", car $T_k$ est lissé ce qui diminue le nombre de fausses alarmes tout en n'augmentant pas trop le risque de manque, si l'on choisit convenablement le seuil $\lambda$.

Les paramètres I et $\lambda$ doivent satisfaire des objectifs contradictoires :

- I doit être faible pour ne pas oublier des ruptures du modèle, mais suffisamment élevé pour avoir convergence de l'estimation et diminuer le risque de fausse alarme.
- $\lambda$ ne doit pas être trop grand pour ne pas augmenter le taux de manque, sans, pour autant, élever celui de fausse alarme.

Il est possible d'envisager l'étude en parallèle de l'évolution de la moyenne des variables aléatoires $W_{k,j}$ définies par :

$$W_{k,j} = \hat{\varepsilon}_k \, \hat{\varepsilon}_{k-j} \quad \text{pour } j = 1, 2, \ldots, j_m$$

Nous avons alors $W_k = W_{k,1}$. La densité de probabilité des variables $W_{k,j}$ est de la même forme que celle de $W_k$ et les effets d'une variation des paramètres sont les mêmes sur ces variables. Il est donc légitime d'espérer améliorer les performances du détecteur en multipliant le nombre de variables $W_{k,j}$ et en testant leur ensemble.

Dans le cas général, le taux de non-détection, ou taux de manque, est difficile à évaluer, à chaque instant k, car, lorsqu'un saut a eu lieu, la moyenne de la séquence pseudo-innovation, multipliée par elle-même, à un décalage unitaire près, est fonction à la fois :

- de l'estimation des paramètres,
- de la différence entre les paramètres réels du premier modèle et ceux du second,
- des valeurs mesurées du signal lui-même.

La moyenne est alors une fonction du temps selon un phénomène de propagation d'erreur, amortie par l'utilisation d'un filtre autoadaptatif qui, pour sa part, tend toujours à blanchir la séquence pseudo-innovation.

Nous pouvons représenter l'évolution de la moyenne de $W_k$ en calculant celle du produit de deux variables aléatoires X et Y normales de moyennes respectives $m_1$, $m_2$, d'écarts type respectifs $\sigma_1$, $\sigma_2$ et de coefficient de corrélation $\Omega_{12}$. Alors :

$$E(XY) = m_1 \, m_2 + \Omega_{12} \, \sigma_1 \sigma_2$$

Dans notre cas, $X = \hat{\varepsilon}_{k-1}$ et $Y = \hat{\varepsilon}_k$. En l'absence de saut, $E(XY) = 0$ car $m_1 = m_2 = 0$ $\hat{\varepsilon}_k$, $\hat{\varepsilon}_{k-1}$ étant indépendantes. Par contre, en présence d'un saut si $\hat{\varepsilon}_k$ et $\hat{\varepsilon}_{k-1}$ restent centrées, nous avons : $E(XY) = \Omega_{12} \, \sigma_1 \sigma_2$ et, si $\sigma_1$, $\sigma_2$ peuvent être fonction du temps, c'est surtout $\Omega_{12}$ qui va évoluer et tendre vers zéro tandis que le filtre tendra vers le modèle optimal.

En effet, l'utilisation d'un filtre autoadaptatif permet de pallier la non détection de sauts trop faibles pour être détectés par la méthode choisie, en suivant les non stationnarités lentes.

Le choix de I et $\lambda$ sera donc basé, de préférence, sur le choix d'un taux de fausse alarme acceptable, la valeur de ce taux étant accessible car elle est fonction d'une distribution centrée.

En effet, si nous supposons que la séquence $\{\hat{\varepsilon}_k\}$ est indépendante, gaussienne, centrée, de variance unité, la distribution de $W_k$ est une fonction symétrique $K(\omega)$ :

$$K(\omega) = \int_0^{\infty} e^{-|\omega| \mathrm{ch} t} dt$$

Nous pouvons alors estimer

$$\mathrm{Prob}(\omega < \Omega) = \frac{1}{\pi} \int_{-\infty}^{\Omega} K\left(\frac{|u|}{R}\right) du$$ - voir tableau 1 -

En appliquant ces résultats à notre détecteur et, en supposant que $T_{k-1} \# 0$ et $T_k \# \beta W_k$ la probabilité de fausse alarme PFA est alors :

$$\mathrm{PFA} = 1 - \mathrm{Prob}\,(-\lambda/\beta < W_k < \lambda/\beta)$$

avec

$$\mathrm{Prob}\,(-\lambda/\beta < W_k < \lambda/\beta) = \frac{1}{\pi} \int_{-\lambda/\beta}^{\lambda/\beta} K\left(\frac{|u|}{R}\right) du$$

| $\Omega$ | Prob($\omega < \Omega$) | $\Omega$ | Prob($\omega < \Omega$) |
|---|---|---|---|
| - 0,5 | 0,205 | 0,5 | 0,795 |
| - 1 | 0,104 | 1 | 0,896 |
| - 1,5 | 0,056 | 1,5 | 0,944 |
| - 2 | 0,031 | 2 | 0,969 |
| - 2,5 | 0,017 | 2,5 | 0,983 |
| - 3 | 0,010 | 3 | 0,990 |
| - 3,5 | 0,006 | 3,5 | 0,994 |
| - 4 | 0,003 | 4 | 0,997 |
| - 4,5 | 0,002 | 4,5 | 0,998 |
| - 5 | 0,001 | 5 | 0,999 |

Tableau 1 : Fonction de répartition de $W_k$

## V. SIMULATIONS

Elles s'effectuent sur des échantillons de 700 points avec des changements de paramètres à l'instant k = 350.

Nous commençons par générer le signal $\{y_k\}$ à partir d'une séquence $\{\varepsilon_k\}$ indépendante, centrée de variance R et d'un filtre $A1(Z^{-1})/C1(Z^{-1})$ pour $k \leqslant 350$ et $A2(Z^{-1})/C2(Z^{-1})$ pour $k \leqslant 350$. La simulation se ramène donc au schéma 1.

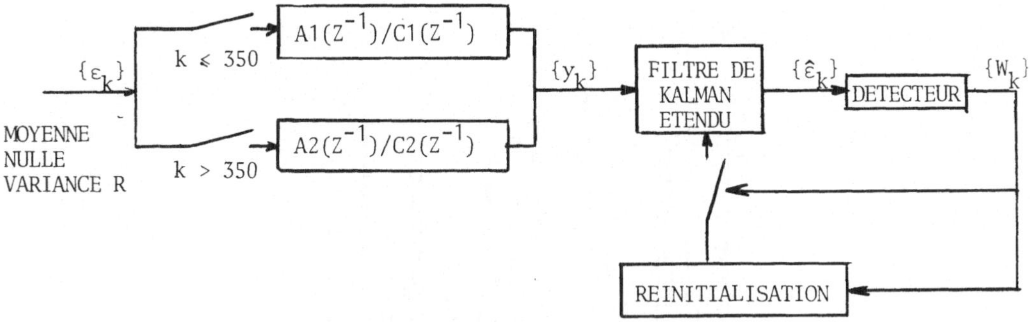


Schéma 1 : Graphe de la simulation

Les modèles choisis sont d'ordre 2 :

1er modèle, $k \leqslant 350$ $\quad A1(Z^{-1}) = 1 - 1,5\ Z^{-1} + 0,56\ Z^{-2}$

$\quad C1(Z^{-1}) = 1 + 1,1\ Z^{-1} + 0,36\ Z^{-2}$

2ème modèle, $k > 350$ $\quad A2(Z^{-1}) = 1 = 1,1\ Z^{-1} + 0,25\ Z^{-2}$

$\quad C2(Z^{-1}) = 1 + 0,5\ Z^{-1} + 0,8\ Z^{-2}$

Afin d'augmenter la difficulté du problème, nous avons essayé de conserver un écart type à peu près constant pour le signal $\{y_k\}$. Pour cela, il a donc été nécessaire de modifier la variance R de la séquence d'entrée $\{\varepsilon_k\}$. Pour $k \leqslant 350$, nous avons pris $R = 3.10^{-3}$ puis, pour $k > 350$, $R = 2,7.10^{-2}$.

Les estimations de la variance, $E\{\hat{\varepsilon}_k^2\}$, et de la moyenne $E\{W_k\} = E\{\hat{\varepsilon}_k\ \hat{\varepsilon}_{k-1}\}$ ont les mêmes facteurs d'oubli : $\alpha = \beta = 0,04$.

Les données ainsi fixées génèrent alors un signal de la forme suivante (voir figure 1).

La modification des dynamiques est visible, mais se limite à une modification du spectre sans diminution de l'amplitude moyenne.

A partir de ce signal, nous avons étudié un détecteur à seuil absolu, puis, devant les premiers résultats, nous avons été amenés à définir un détecteur à seuil relatif.

V.1. Détecteur à seuil $\lambda$ absolu

Les premiers résultats ont été obtenus avec un seuil fixe : $\lambda = 0,015$. L'inconvénient majeur est un risque de fausses alarmes lié à une mauvaise estimation des paramètres du modèle. Cela nous a donc conduits dans un premier temps à augmenter la durée d'inhibition du détecteur : nous avons fixé I à 50.

Comme nous pouvons le constater sur les figures 2 à 6, décrivant les estimations des paramètres et de la variance de $\{\hat{\varepsilon}_k\}$, le détecteur semble peu sensible aux erreurs d'estimation pourvu que I soit suffisant et $\lambda$ bien choisi.

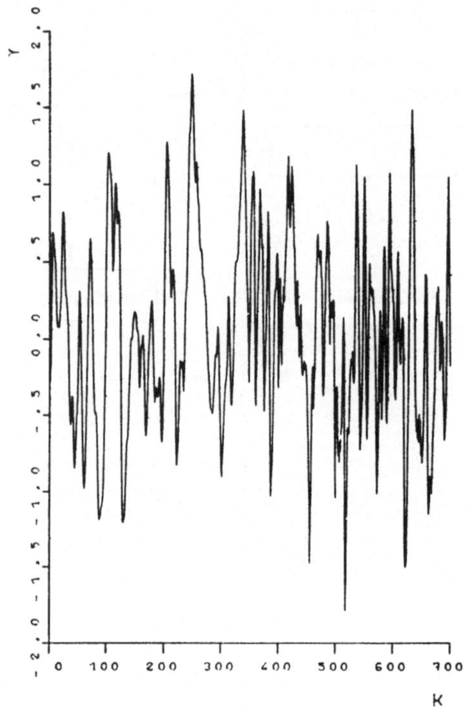


FIG.1: SIGNAL $\{Y_K\}$

MODIFICATION DU SPECTRE POUR K=350

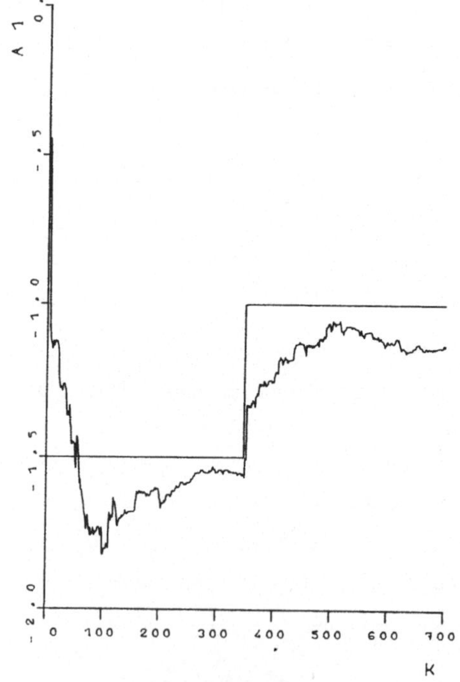


FIG.2: PARAMETRE A1

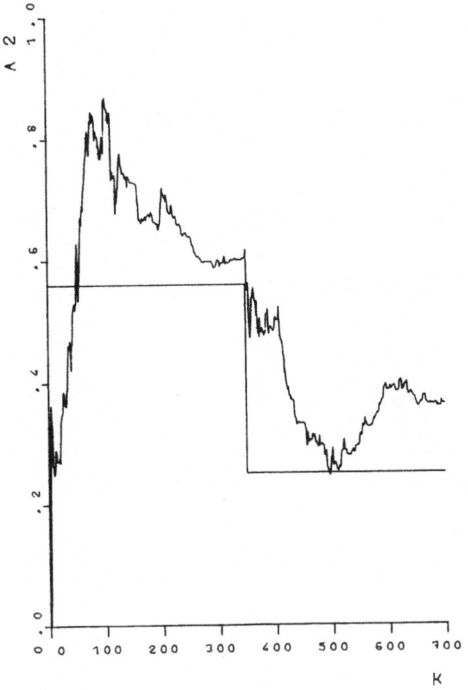


FIG.3: PARAMETRE A2

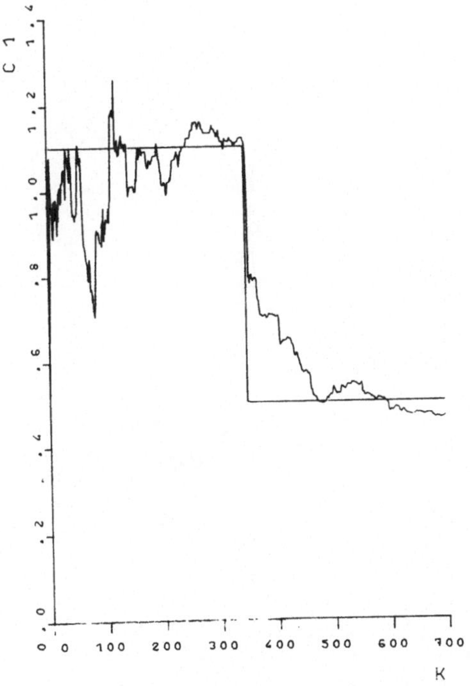


FIG.4: PARAMETRE C1

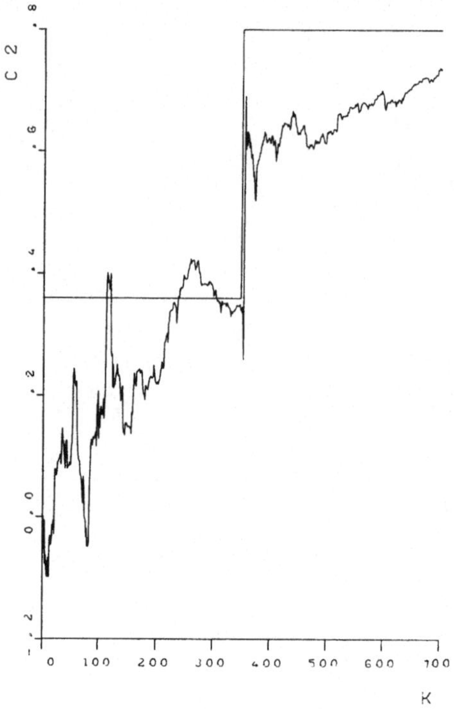


FIG.5: PARAMETRE C2

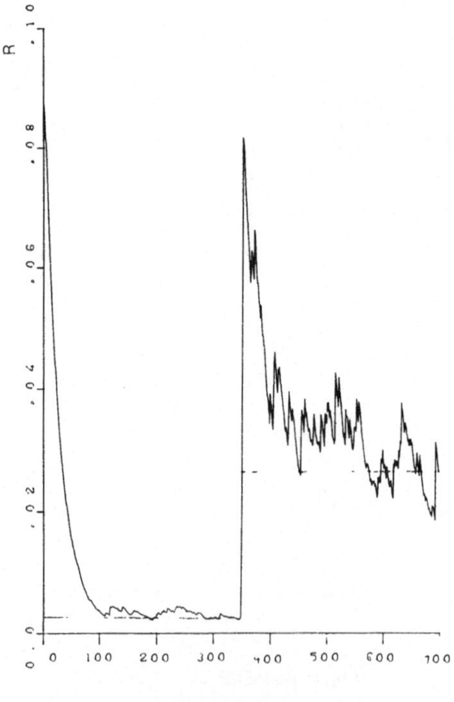


FIG.6:VARIANCE REELLE ET ESTIMEE

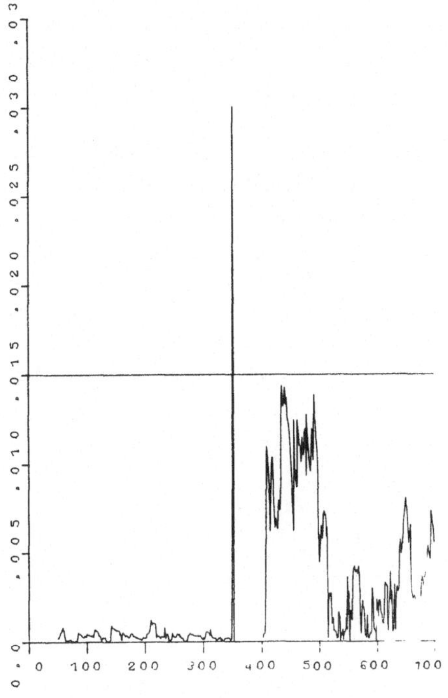


FIG.7: DETECTEUR AVEC SEUIL ABSOLU

(DETECTION K=353)

Nous avons choisi des plages stationnaires sur 350 points. Une première remarque concerne la durée d'inhibition comparée au temps moyen entre chaque changement de paramètres. Il ne faut donc pas que I, trop élevé, nous empêche de détecter des sauts fréquents. Il y a donc intérêt à le diminuer.

Sur la figure 7, nous remarquons la "nervosité" de l'estimateur de $E\{W_k\}$ après le saut. Cela provient du changement de variance de la séquence indépendante d'entrée $\{\varepsilon_k\}$. Le changement est détecté à k = 353.

Cela souligne, en particulier, l'importance du choix de la valeur $\lambda$ du seuil et, surtout, l'inconvénient à la prendre fixe, ce qui l'empêche d'évoluer avec l'étude du système et accroît le risque de fausses alarmes.

Cela, et la nécessité de diminuer la valeur de I, durée d'ihibition, nous a conduits à envisager un détecteur à seuil relatif.

V.2. Détecteur à seuil relatif

Pour nous affranchir des limitations dues au choix d'un seuil fixe, nous avons préféré définir un seuil variable en fonction de la variance estimée de la séquence $\{\hat{\varepsilon}_k\}$. Un premier avantage est la disponibilité de cette valeur, puisqu'elle intervient dans les équations du filtre de Kalman.

Le détecteur s'écrit alors : $|E\{\hat{\varepsilon}_k\hat{\varepsilon}_{k-1}\}| \geq \gamma.E\{\varepsilon_k^2\}$ avec $0 < \gamma < 1$ . La détection revient donc à comparer la pente à l'origine de la fonction d'autocorrélation $\phi(\tau) = E\{\hat{\varepsilon}_k\hat{\varepsilon}_{k-\tau}\}$ à une valeur limite.

Cela nous a permis, dans un premier temps, de ramener la période d'inhibition à I = 10. Nous pouvons ainsi détecter des changements de paramètres plus rapprochés dans le temps, et, nous pouvons aussi utiliser des estimations de paramètres moins bonnes que précédemment.

De plus, comme le montre la figure 8, la détection se fait avec un retard moindre (k = 352 avec $\gamma$ = 0.5).

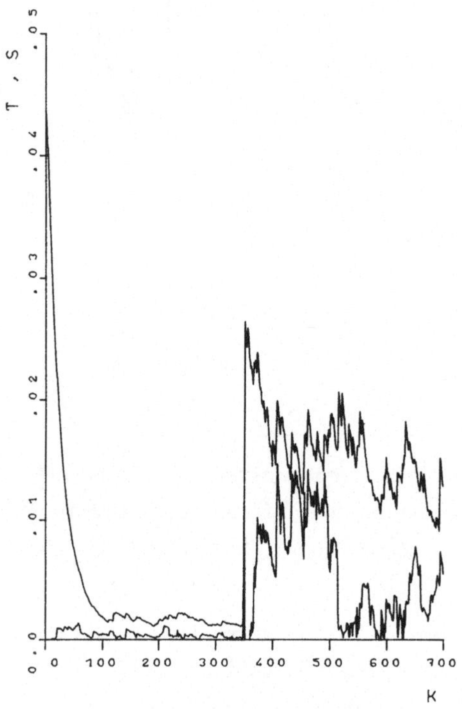


FIG.8: DETECTEUR A SEUIL RELATIF

K=352 $T_K$=0.0131 $R_K$=0.0119

$\gamma$=0.5 I=10 ( $T_K > \gamma.R_K$ )

Cela s'explique par un seuil "instantané" plus faible que lors de l'essai avec un seuil absolu, sans que cela augmente le risque de fausse alarme, puisque seuil et estimateur de $E\{W_k\}$ évoluent en fonction des connaissances acquises à l'instant courant sur le système.

L'étude statistique de ce détecteur -donc l'optimisation de la valeur $\gamma$- est plus complexe que celle du détecteur à seuil fixe.

En effet, l'étude nécessaire à la compréhension des résultats du paragraphe V.1. s'appuie sur celle de la distribution du produit de deux variables laplaciennes indépendantes, centrées, de même variance ; (ce sont les seules hypothèses nécessaires au calcul de PFA, taux de fausses alarmes).

Par contre, en envisageant l'utilisation d'un seuil variable en fonction de $E\{\hat{\varepsilon}_k^2\}$, nous devons étudier la distribution du produit $(\hat{\varepsilon}_{k-1} - \gamma\hat{\varepsilon}_k) \cdot \hat{\varepsilon}_k$, où les variables $(\hat{\varepsilon}_{k-1} - \gamma\hat{\varepsilon}_k)$ et $\hat{\varepsilon}_k$ sont bien laplaciennes centrées, mais non indépendantes et non de même variance.

## CONCLUSION

L'algorithme de détection séquentiel que nous proposons atteint les objectifs que nous nous étions fixés. A savoir :

- de faibles connaissances a priori,
- un volume de calcul réduit,
- de bonnes performances.

L'étude doit cependant se poursuivre, en particulier, pour déterminer les valeurs optimales de $\gamma$, et pour tester la robustesse du détecteur. Nous envisageons, par exemple, les effets d'une modélisation inférieure à l'ordre réel du système, et les effets de sauts rapprochés dans le temps.

Nous pouvons cependant constater que ce détecteur semble robuste. En effet, en cas de fausse alarme, le filtre de Kalman étant réinitialisé, il ne peut y avoir convergence de l'estimation, ce qui entraîne en particulier, une séquence pseudo innovation non indépendante. Dans ce cas, le détecteur est conduit à choisir entre les deux hypothèses, absence ou présence d'une rupture avec un modèle estimé qui, de toute façon, sera erroné. Il devient donc probable de ne pas avoir de détection après une fausse alarme ce qui permet à l'estimateur de converger et au détecteur de rattrapper son erreur. Par contre, face à un manque à la détection, le filtre n'étant pas réactualisé, il s'adaptera de moins en moins au système réel. Le détecteur deviendra alors de plus en plus sensible à une rupture ce qui permet d'envisager un rattrappage dès la rupture suivante.

D'autre part, une comparaison des résultats avec des détecteurs déjà éprouvés devra servir de banc d'essai à cette méthode avant d'en affirmer l'efficacité.

## REFERENCES

BASSEVILLE, M. (1981 a), Edge detection using sequential methods for change in level. Part. II : Sequential detection of change in mean. I.E.E.E. Trans. on A.S.S.P., Vol. ASSP 29, n° 1, pp. 32-50

BASSEVILLE, M. (1982 a), Contribution à la détection séquentielle de ruptures de modèles statistiques. Thèse Docteur ès Sciences, Rennes

BASSEVILLE, M. and A. BENVENISTE (1982 b), Détection séquentielle de changements brusques des caractéristiques spectrales d'un signal numérique. Rapport INRIA, n° 129

BASSEVILLE, M., B. ESPIAU and J. GASNIER (1981 b), Edge detection using sequential method for change in level. Part. I : A sequential edge detection algorithm. IEEE Trans. on ASSP, ASSP 29, n° 1, pp. 24-31

HINKLEY, D.V. (1971), Inference about the change point from cumulative sum tests. Biometrica, 58, n° 3, pp. 509-523

MEHRA, R.K., and J. PEESCHON (1971), An innovation approach to fault detection and diagnosis in dynamic systems. Automatica, 7, p. 637-640

SHIRYAEV, A.N. (1961), The problem of the most rapid detection of a disturbance in a stationnary process. Sov. Math. Dokl, n° 2, pp. 795-799

SHIRYAEV, A.N. (1963), On optimum methods in quickest detection problems. Theory Prob. Appl., 8, n° 1, pp. 22-46

SHIRYAEV, A.N. (1965), Some exact formulas in a disorder problem. Theory Prob. Appl., 10, n° 3, pp. 348-354

WILLSKY, A.S., CHOW, E.Y., GERSCHWIN, S.B., GREENE, C.S., HOUPT, P.K., and KURKJIAN, A.K. (1980), Dynamic model based techniques for the detection of insidients on Freeways. IEEE Trans. A.C., AC 25, n° 3, pp. 347-360

WILLSKY, A.S., and JONES, H.L. (1976), A generalized likelihood ratio approach to the detection and estimation of jumps in linear systems. IEEE Trans. A.C., AC 21, n° 1, pp. 108-112

# DETECTION DES EVOLUTIONS D'UN MOBILE ET ESTIMATION DE SA CINEMATIQUE PAR UNE METHODE DE TESTS D'HYPOTHESES

A. LORENZI - C. BOZZO

CSEE - CETIA

Avenue des Frères Lumière

Z.I. Sainte-Claire

F 83 160 LA VALETTE DU VAR

ABSTRACT : The proposed method deals with the problem of detecting the evolutions of a mobile in motion (aircraft) on the basis of the general theory of significance tests (tests of BAYES, WALD).

The principle is as follows : several possible aircraft flight paths are continuously considered which correspond to different evolutions of the aircraft. Each path is represented as a model in a simple manner ; a 2nd order stationary filter allows the implementation to be performed in real time : considering that the evolutions are performed only at a maximum acceleration the filter is therefore submitted to an acceleration type control, the acceleration being then regarded as a white Gaussian noise with an average value, varying according to the evolution amplitude.

The different filters operate "in parallel" thanks to the measurements provided (Cartesian or polar coordinates). Each filter gives its own estimate of the target kinematics.

The proposed algorithm selects at every moment the estimate given by the filter whose model best corresponds to reality. Taking into account some probabilities of error, such a selection is carried out following a significance test designed on the basis of pseudo-innovations peculiar to each filter (such a test makes allowances for non-stationarities of the measurement noise).

## 1. - INTRODUCTION

La détection des évolutions d'un mobile dont on cherche à estimer la cinématique en temps réel demeure un problème fondamental en poursuite 3D.

Le caractère adaptatif rencontré dans la plupart des estimateurs réside, quand il existe, dans la modification de certaines caractéristiques statistiques des bruits présents dans l'équation d'évolution de la dynamique du mobile.

Ceci suppose bien entendu qu'il est possible de déterminer en temps réel que la valeur de l'un ou de plusieurs de ces paramètres a changé. De cette façon, on réalise, vis-à-vis de la trajectoire décrite par le mobile, une adaptation des erreurs de modélisation associées au modèle moyen de dynamique choisi.

Ainsi, parmi ceux qui sont utilisés dans les problèmes de Tracking, beaucoup font appel à une représentation linéaire où les possibilités d'évolution de la cible sont traduites par la présence dans l'équation d'Etat du système d'un bruit blanc gaussien centré, dont on cherche à identifier les caractéristiques statistiques (variance, écart type...).

Cependant, cette notion assez répandue de l'adaptativité des estimateurs de poursuite implique la recherche d'un compromis relativement difficile à effectuer entre deux objectifs essentiels, et cependant tout à fait contradictoires dans ces conditions.

Il s'agit, d'une part, d'obtenir une estimation de la position du mobile et de ses dérivées qui soit la moins perturbée possible, et d'autre part, de prendre en compte sans traîner les évolutions de la cible.

Même si le modèle adopté par la plupart des auteurs (modèle de SINGER [1] [2]) ne rend compte que de façon très imparfaite des caractéristiques de la trajectoire et des lois de mouvement, l'échec relatif des méthodes basées sur ce principe peut s'expliquer de la façon suivante : on essaie de corriger des erreurs de modélisation du premier ordre (lors d'apparitions de biais correspondant à des évolutions brusques dansle cas d'un modèle non adapté en moyenne) en agissant sur des paramètres qui eux ne peuvent rendre compte que d'erreurs de modélisation du second ordre (petits écarts autour d'un modèle représentant le signal en moyenne).

De plus, sur ces paramètres qui font alors toute l'adaptativité du filtre, on élabore des hypothèses qui ne sont pas forcément en accord avec la physique du problème.

Par exemple, la notion de bruit (brownien, blanc, gaussien, etc...) donne dans ces conditions un caractère aléatoire à la dynamique du mobile qui, s'il peut par exemple représenter des perturbations extérieures (turbulences), ne saurait suffire en aucun cas à la caractérisation d'un problème de poursuite considéré dans un contexte opérationnel.

La méthode proposée dans cet article a pour but une adaptation plus complète du modèle de dynamique (1er et 2è ordre) et de ce fait, permet de concilier les deux principales exigences du Tracking rappelées plus haut.

Dans un premier temps (§ 2), on présentera la méthode de façon générale en cherchant également à la situer par rapport aux méthodes d'estimation Bayésiennes [10], [11], [12], [13], ou à celles d'estimation détection du type GLR [4], [5].

On donnera ensuite (§ 3) les caractéristiques précises de la méthode en accord avec le contexte opérationnel, puis (§ 4) celles du modèle global de dynamique associé. Enfin, (§ 5) on présentera des résultats obtenus en simulation et sur trajectoire réelle d'avion.

## 2. - PRESENTATION GENERALE DE LA METHODE : - SITUATION PAR RAPPORT AUX METHODES D'ESTIMATION BAYESIENNES ET A CELLES DITES D'ESTIMATION-DETECTION (TYPE GLR)

MOOSE et GHOLSON [10] introduisent dans leur modèle de dynamique un terme de commande déterministe (ou une moyenne non nulle dans le bruit de dynamique) qu'il s'agit d'identifier en temps réel par filtrage Bayésien.

Bien que réalisant également une adaptation de la dynamique en moyenne et au second ordre, cette méthode présente deux inconvénients majeurs pour notre application. Elle nécessite un volume de calculs important et d'autre part, la commande identifiée ainsi que les positions et vitesses estimées sont facilement bruitées, en tant que barycentre des sorties de N filtres pondérées par des probabilités qui elles-mêmes sont bruitées.

Enfin, les modèles utilisés par MOOSE et GHOLSON ne permettent pas de commander directement en accélération et de ce fait, ne sont pas à trainage minimal après apparition d'un échelon en accélération.

Nous proposons dans ce papier un système de poursuite consistant à exploiter une banque de filtres fonctionnant en parallèle et qui caractérisent chacun des évolutions différentes (et donc des trajectoires différentes du mobile poursuivi). Tous ces filtres ont les mêmes propriétés en régime permanent (s'il existe) : la méthode adoptée consiste, lors d'une évolution, à choisir le filtre qui conduit à la meilleure estimation des éléments de cinématique au sens d'un certain critère (§ 3).

On choisit ici d'exploiter la théorie générale des tests d'hypothèses (tests de BAYES et de WALD [6], [7], [8].

Pour simplifier la présentation et montrer que la méthode adoptée a de nombreux points communs avec celle de MOOSE et GHOLSON, bien que reposant sur une approche théorique très différente, nous adopterons des modèles linéaires cartésiens très simples (de la famille des modèles de SINGER) (§ 4).

Chaque trajectoire possible du mobile poursuivi se traduira, au niveau du modèle correspondant, par un terme de commande variable suivant l'amplitude de l'évolution et correspondant physiquement à un échelon d'accélération.

L'algorithme proposé sélectionne à chaque instant l'estimée donnée par le filtre dont le modèle correspond le mieux à la "réalité", ce choix étant fait après un test d'hypothèses construit à partir des pseudo-innovations propres à chaque filtre. (Notons que ce test tient compte des non-stationnarités du bruit de mesure).

Comme on le verra dans (§ 3), la détection se fait de façon séquentielle par l'intermédiaire de variables de décision dont le calcul fait intervenir les pseudo-innovations attachées à chaque filtre de façon quadratique.

Or, les méthodes de type GLR [4], [5], introduisent ces mêmes écarts dans le calcul du rapport de vraissemblance généralisé avec aussi, implicitement présente, l'information sur l'instant optimal de la détection et son effet sur l'Etat du système par minimisation du critère quadratique ainsi construit.

La différence fondamentale entre les méthodes de type GLR introduites par WILLSKY [4], et l'algorithme présenté, est la suivante : Outre le fait que ces méthodes sont statistiquement optimales, elles ont pour but une réinitialisation de l'état du système (et parfois [5], des matrices de covariance d'erreur sur l'Etat) à un instant optimal. Les algorithmes correspondants ne sont donc pas à proprement parler adaptatifs vis-à-vis du modèle de dynamique qui est valable en moyenne (matrice $\Phi$) même si ils présentent un caractère adaptatif par rapport à l'Etat du système.

L'algorithme proposé adapte quant à lui le modèle de dynamique au premier ordre et tient compte également des erreurs de modélisation du second ordre.

Indépendamment de ces méthodes d'estimation basées sur la notion de filtrage de KALMAN, on notera l'intérêt revêtu par l'approche de WONHAM [3], qui permet de façon beaucoup plus générale, de caractériser l'évolution de l'état ou des probabilités associées en fonction du temps,et qui justifie a postériori toutes ces approches.

## 3. - DETECTION D'EVOLUTIONS : PRESENTATION DE L'ALGORITHME

### 3.1 - POSITION DU PROBLEME

Il est un fait que toutes les trajectoires d'avion n'ont pas la même probabilité d'existence a priori. On peut envisager autant d'hypothèses que d'évolutions différentes de la cible, à partir de ses capacités propres d'évolution et de sa situation géographique à l'instant présent. On peut considérer que ces hypothèses sont distribuées suivant une densité de probabilité de type Gaussien, fonction du caractère plus ou moins marqué de l'évolution. (Loi centrée sur une évolution nulle). Cependant, afin de minimiser a priori les risques de fausse alarme, en tenant compte des capacités croissantes d'évolutions des mobiles manoeuvrants, nous choisirons dans cette étude d'affecter à chacune des hypothèses la même probabilité a priori. On fait alors l'hypothèse que la densité de probabilité correspondante présente un certain applatissement en son sommet et que donc les hypothèses effectuées concernent cette zone.

En effet, il a été décidé de choisir dans cette étude des modèles de dynamique très simples qui donnent naissance à un ensemble "de signatures de trajectoires" possédant des caractéristiques fondamentales voisines (§ 4). Enfin, dans un contexte opérationnel et dans ces conditions, il devient extrêmement difficile de prédire, même en tenant compte du passé de la trajectoire, l'occurence d'une hypothèse plutôt que d'une autre.

C'est pour cette raison que la "mémoire" du filtre global est volontairement limitée à l'intervalle de temps compris entre deux détections successives. Même si la méthode proposée devient alors sous optimale, elle permet, tout en s'accordant très bien à la réalité du problème posé, de le résoudre de façon simple à partir des résultats obtenus par Wald [6] et Schweppe [7].

De cette façon, toute l'information utile est fournie au test de façon objective par les mesures et leurs prédictions issues du filtrage suivant chaque hypothèse.

D'autre part, il faut faire une remarque d'ordre général quant à la procédure utilisée dans le cas d'un test à N hypothèses. Deux formes (A et B) sont proposées (§ 3.6) pour l'algorithme d'estimation-détection considéré ; dans les deux cas, la structure même du test d'hypothèses utilisée impose la contrainte suivante au niveau du choix des hypothèses : celles-ci étant ordonnées suivant le caractère plus ou moins prononcé des évolutions, on ne peut passer d'une hypothèse à une autre qu'en respectant cet ordre. La détection se fait alors de manière progressive en examinant les hypothèses intermédiaires, ce qui donne au test une structure très simple et indépendante du nombre d'hypothèses considérées.

3.2 - SOLUTION PROPOSEE

On considère ici deux aspects du problème de la détection :

- l'aspect global, qui conduit au test de BAYES
- l'aspect séquentiel, introduit dans le test de WALD [6]

On adoptera dans ce qui suit une solution déduite de celle proposée par SCHWEPPE [7] (et par KLEIN [8] dans sa thèse).

Elle conduit à un algorithme de détection séquentielle étendu à un nombre quelconque d'hypothèses.

3.3 - TEST DE BAYES

On considère un système physique, observable par un processus stochastique z(t). Sachant que deux hypothèses H0 et H1 sont faites sur le système, il s'agit de déterminer de façon optimale l'hypothèse dans laquelle on se trouve, disposant de n observations depuis l'instant initial.

$$Z\ (n) = [\ z\ (0),\ z\ (1),\ ,\ z\ (k),\ ,\ z\ (n-1),\ z\ (n)\ ]$$

est l'ensemble formé de ces observations.

$S_N$ est l'ensemble de toutes ses réalisations. $S_0$ et $S_1$ sont définis de la façon suivante :

Z (n) appartient à $S_0$ dans l'hypothèse H0
Z (n) appartient à $S_1$ dans l'hypothèse H1,

le choix de ces deux sous-espaces déterminant le caractère plus ou moins optimal attaché à la résolution du problème.

Soit Ci,j le coût associé à l'acceptation de l'hypothèse Hi, sachant que Hj est vraie. (Les densités de probabilités de chaque hypothèse sont connues : § 3.1)

Le risque Bayésien est alors défini par :

$$R = \sum_{i=0}^{1} \sum_{j=0}^{1} C_{i,j}\ p(H_j) \int_{S_i} P[Z\ (n)/H_j]\, d\ Z\ (n) \tag{1}$$

qu'il s'agit de minimiser afin d'obtenir $S_0$ et $S_1$.

Le coût associé à une décision correcte (C00 ou C11) est plus faible que celui associé à une erreur (C10 ou C01) et donc :

$$C_\alpha = C_{10} - C_{00} > 0 \text{ et } C_\beta = C_{01} - C_{11} > 0 \tag{2}$$

On peut alors exprimer R en fonction des Cij, $C_\alpha$ et $C_\beta$ de la façon suivante :

$$\begin{aligned} R = {} & C_{00}\ P(H0) + C_{11}\ P(H1) + C_\alpha\ P(H0) \\ & + \int_{So} \left[ C_\beta\ P(H1)\ P\left[Z(n)/H1\right] - C_\alpha\ P(H0)\ P\left[Z(n)/H0\right] \right] \end{aligned} \tag{3}$$

Minimiser R, c'est alors minimiser l'intégrale contenue dans (3). Or,

$$X = C_\beta\, P(H1)\quad P\left[Z(n)/H1\right] > 0 \tag{4}$$

$$Y = C_\alpha\, P(H0)\quad P\left[Z(n)/H0\right] > 0 \tag{5}$$

Les valeurs de Z (n) telles que :

X > Y, font croître R et donc appartiennent à $S_1$

X < Y, font décroître R et donc appartiennent à $S_0$

D'où :

$$Z(n) \in S_\phi \Leftrightarrow \frac{P[Z(n)/H1]}{P[Z(n)/H0]} < L \quad ; \; L = \frac{C\ \alpha\ P(H0)}{C\ \beta\ P(H1)} \tag{6}$$

$$Z(n) \in S_1 \Leftrightarrow \frac{P[Z(n)/H1]}{P[Z(n)/H0]} > L$$

$$T[Z(n)] = \frac{P[Z(n)/H1]}{P[Z(n)/H0]} \tag{7}$$

est appelé rapport de vraisemblance

L est le seuil. Alors, si T[Z (n)] est supérieur au seuil, on choisit l'hypothèse H1, sinon l'hypothèse H0, avec les probabilités d'erreur :

$\alpha$ : probabilité de rejeter H0 quand H0 est vraie
$\beta$ : probabilité d'accepter H0 quand H1 est vraie

et sont dites respectivement de 1ère et 2ème espèce (probabilités de fausse alarme).

$$\alpha = \int_{S_1} P\ [Z\ (n)/H0]\ d\ Z\ (n) \tag{8}$$

$$\beta = \int_{S_\emptyset} P\ [Z\ (n)/H1]\ d\ Z\ (n) \tag{9}$$

### 3.4 - DETECTION SEQUENTIELLE : TEST DE WALD

A chaque hypothèse correspond un modèle du processus envisagé. A l'arrivée d'une observation nouvelle, on divise l'espace $S_N$ en 3 :

$S_N^0$, $S_N^1$ et $S_N^{0,1}$ grâce à 2 bornes A et B précisées ultérieurement à la $p^{ème}$ étape

Si T [Z (p) ] $\geqslant$ A, Z (p) appartient à $S_p^1$, on choisit l'hypothèse H1

Si T [Z (p) ] $\leqslant$ B, Z (p) appartient à $S_p^0$, on choisit l'hypothèse H0

Si B < T [ Z (p) ] < A, Z (p) appartient à $S_p^{0,1}$, on attend la mesure z (p+1)

En considérant les 2 premières hypothèses, on trouve facilement :

$$\frac{1-\beta}{\alpha} \geqslant A \text{ et } \frac{\beta}{1-\alpha} \leqslant B \tag{10}$$

cependant, la période d'échantillonnage utilisée étant faible, on néglige les dépassements de A et B par T [Z (p) ] , d'où :

$$A = \frac{1-\beta}{\alpha} \text{ et } B = \frac{\beta}{1-\alpha} \tag{11}$$

On prendra $\alpha = \beta$ par la suite (cf. 3.1).

### 3.5 - ALGORITHME DE DETECTION SEQUENTIELLE, D'APRES SCHWEPPE

Si $T[Z\ (p)] = \frac{p[Z\ (p)/H1]}{p[Z\ (p)/H0]}$, on obtient la récurence suivante :

$$T[Z\ (p)] = \frac{p[z\ (p)/Z\ (p-1),H1]}{p[z\ (p)/Z\ (p-1),H0]}\, T[Z(p-1)] \tag{12}$$

On considère par la suite la variable :

$$T_{0,1}\ (p) = \mathrm{Log}\ \{T\ [Z(p)]\ \} \tag{13}$$

En effet, c'est SCHWEPPE qui introduit la notion de récurence sur le rapport de vraisemblance. Les équations du filtrage de KALMAN permettent alors d'écrire sous l'hypothèse gaussienne :

$$p\ [z(p)/Z\ (p-1),\ H_l] = \frac{1}{\sqrt{2\Pi}\sigma_l} \exp\left\{-\frac{1}{2}\ \frac{[z(p)\ -\ \hat{z}_l\ p/p-1]^2}{\sigma l^2}\right. \tag{14}$$

avec $l = 0,1$

$\sigma_l$ : covariance de l'erreur de prédiction sur la mesure en régime stationnaire, dans l'hypothèse $l$

$\hat{z}_l p/p-1$ : valeur prédite de la mesure à l'instant p sachant celle à l'instant p-1, dans l'hypothèse 1

d'où :

$$T_{0,1\,(p)} = T_{0,1\,(p-1)} + \mathrm{Log}\ \frac{\sigma_0}{\sigma_1} - \frac{1}{2}\left[\frac{\xi 1 p^2}{\sigma_1{}^2} - \frac{\xi 0 p^2}{\sigma_0{}^2}\right] \tag{15}$$

$\xi_l p$ représentant le processus de pseudo-innovation.

d'où l'algorithme proprement dit, qui permet de tester 2 hypothèses.

Procédure du test à 2 hypothèses

Après s'être fixé au préalable les probabilités de fausse alarme $\alpha$ et $\beta$ qui déterminent les bornes A et B de la variable de décision $T_{o,1}$ on applique la procédure suivante :

* prélèvement d'une mesure z(j)
* prédiction : $\hat{x}_o\ (j/j-1)$ et $\hat{x}_1\ (j/j-1)$
* élaboration des pseudo-innovations : $\xi_o(j)$ et $\xi_1(j)$
* calcul de $T_{o,1}\ (j)$
* comparaison de $T_{o,1}\ (j)$ avec Log B et Log A.
  - Si Log B $\leqslant T_{o,1}(j) \leqslant$ Log A, on ne peut conclure. Il faut alors effectuer une autre mesure z(j+1).
  - Si $T_{o,1}(j) \geqslant$ Log A, on accepte $H_1$.
  - Si $T_{o,1}(j) \leqslant$ Log B, on accepte $H_o$.

Remarques : La période d'échantillonnage étant faible, on suppose qu'aucun changement ne survient entre deux itérations.

Si à l'initialisation, on ne dispose d'aucune information sur le système, on prend $T_{o,1}\ (\emptyset) = \emptyset$ ce qui permet une détection rapide quelle que soit l'hypothèse à choisir.

Il est cependant évident que le choix entre deux modèles n'est pas suffisant pour représenter "l'objet" (Système réel) qui peut correspondre à une hypothèse non envisagée.

Il est donc essentiel d'envisager un nombre n plus important de modèles "possibles" et donc des tests plus complexes à n hypothèses.

* D'autre part, afin d'augmenter la rapidité de la détection sans pour autant remettre en question sa validité, il est nécessaire de réinitialiser la variable de décision après chaque détection. (cf §3.6 Fig 1)

## 3.6 - EXTENSION DU TEST A N HYPOTHESES

Alors que précédemment on avait à choisir entre deux hypothèses, on se propose d'étudier dans cette partie une méthode permettant d'exploiter un modèle global plus fin, c'est-à-dire faisant appel à plusieurs hypothèses. On présente dans un premier temps (3.61) l'algorithme sous une forme A directement extrapolée de la forme étudiée dans la partie (3.5) : elle implique un nombre N d'hypothèses assez faible tout en apportant une grande sécurité dans la détection d'un changement quelle que soit sa nature.

Dans une seconde phase (3.62) on présentera une forme B de l'algorithme duale de la forme A, et qui permet l'utilisation d'un nombre d'hypothèses plus élevé (qui autorise donc une meilleure définition du modèle) tout en conservant au test la même dynamique. Enfin, § (3.63), on définira le mode d'utilisation de ces 2 formes tout en comparant les temps de calcul associés.

### 3.61 - Algorithme d'Estimation-Détection : Forme A

Conformément à ce qui a été fait pour l'algorithme à deux hypothèses, on détermine de façon identique pour n hypothèses n-1 variables de décision et conformément au postulat des hypothèses a priori équiprobables (3.1), on conserve alors les mêmes bornes A et B quelle que soit la variable de décision considérée. Il est à noter que dans cette première forme il n'est pas utile de calculer les n-1 variables de décision envisagées puisque la structure du test permet ici de passer d'une hypothèse Hi à une autre immédiatement voisine dans l'ordre considéré au (§ 3.2) et ceci à chaque période d'échantillonnage. On considère alors à chaque itération 3 hypothèses Hk, Hi, et Hj distribuées dans cet ordre et qui se correspondent par les 2 variables de décision Tk,i et Ti,j suivant le schéma de la figure 1 où les bornes A et B sont remplacées par leur logarithme.

Supposons que l'on soit dans l'hypothèse i à l'itération k

Figure 1

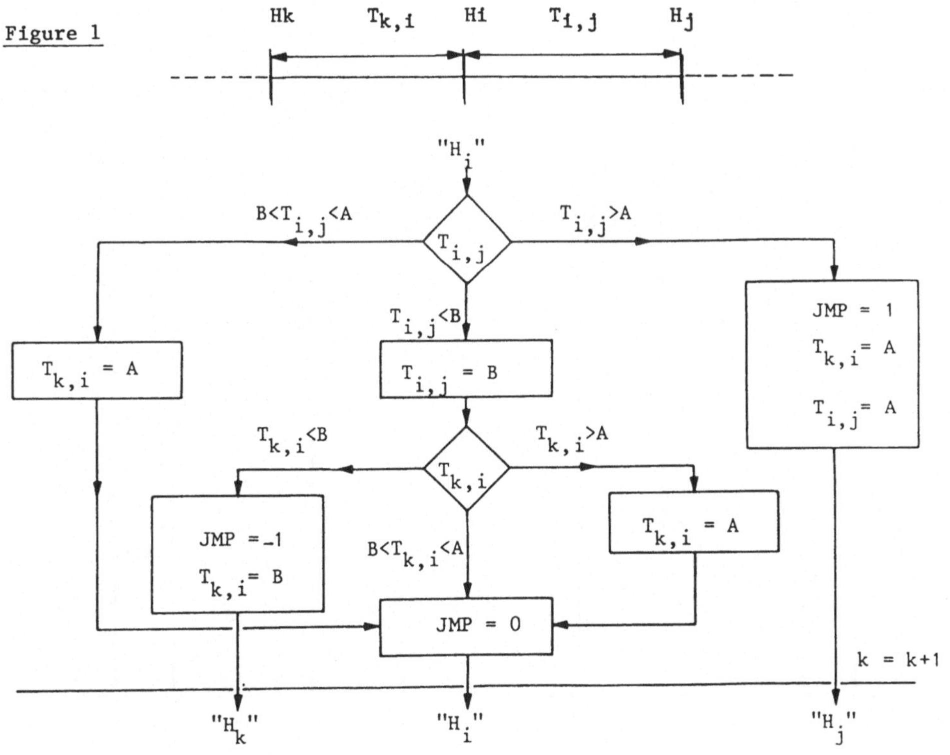

A l'itération k, les variables de décision se situant "avant Hk" sont initialisées à A et celles se situant "après Hj" sont initialisées à B afin de se prévenir des fausses alarmes à l'itération suivante.

Cependant, si cette première formulation du test à N hypothèses offre une grande sécurité dans la détection (une seule détection possible par itération) elle ne présente plus le même intérêt, si, pour une période d'échantillonnage donnée, le nombre d'hypothèses considéré augmente au sein du même domaine des possibilités prévues pour le système. On risque en effet, en voulant augmenter le pouvoir de "définition" de l'algorithme, de ne pas pouvoir enregistrer correctement toutes les évolutions de la dynamique du système : la détection ne se fait alors plus suffisamment vite pour cela, pénalisée par une période d'échantillonnage devenue trop faible en regard du nombre d'hypothèses qu'il faut considérer en temps réel (§ 5). C'est pourquoi on a été amené à mettre en oeuvre une forme duale de l'algorithme précédent qui, bien que légèrement plus coûteuse en temps de calcul, permet cependant de s'affranchir de ce problème.

3.62 - Algorithme d'estimation-détection : Forme B

La caractéristique principale de cet algorithme est la suivante : contrairement à la forme A présentée plus haut, la forme B permet une détection beaucoup plus rapide dans la mesure où sa structure autorise à chaque itération un saut d'une hypothèse Hi à n'importe quelle autre des N-1 hypothèses restantes. Dans le domaine des hypothèses jugées possibles a priori, pour le mobile, on pourra alors considérer un nombre d'hypothèses plus grand sans pour autant augmenter le temps de détection correspondant à l'apparition d'une évolution donnée. Ce nouvel algorithme autorise donc une estimation plus précise de la cinématique du but tout en conservant les mêmes capacités d'adaptation aux évolutions.

Il est à noter que la structure même du test d'hypothèses utilisé devient alors adaptative vis-à-vis des évolutions du mobile : l'identification du modèle de dynamique en moyenne se faisant de façon automatique à chaque itération de la façon suivante :

Par exemple, si on est dans l'hypothèse H0 à l'itération k.

Figure 2

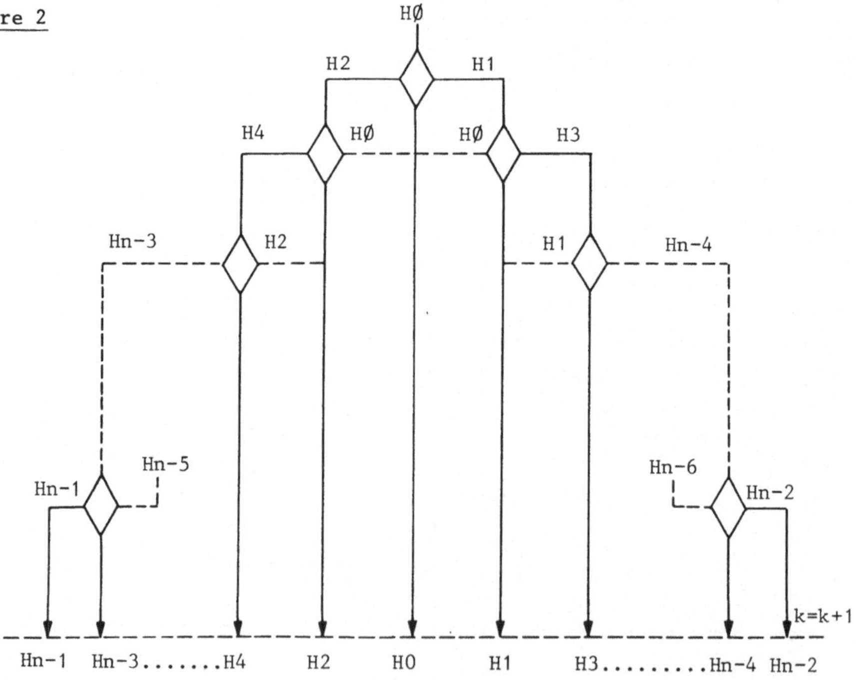

NB : Chaque sous-module correspond à l'algorithme de test présenté figure 1 au § 3.61. Le test global est arrêté lorsqu'un non changement d'hypothèse est détecté par un sous-module ou lors d'une non détection de ce sous-module.

On peut ainsi aboutir à n'importe quelle hypothèse à chaque itération et avec les mêmes probabilités d'erreurs que dans le cas de la forme A, à condition d'adapter les bornes en fonction du nombre d'hypothèses choisies.

### 3.63 - Utilisation et contraintes opérationnelles

Il est à noter que pour un nombre N d'hypothèses donné, les volumes de calcul correspondant à chacune des formes A et B sont très voisins puisque le nombre de filtres mis en oeuvre est le même dans un cas comme dans l'autre, la seule différence à ce niveau se situant dans l'élaboration de N-3 variable de décision supplémentaires pour la forme B à chaque itération.

De la même façon, les temps de calcul associés sont fortement conditionnés par le nombre de filtres mis en jeu et très peu par la forme choisie pour le test d'hypothèses.

On peut donner une estimation du temps de calcul corespondant à une itération dans chacune des formes A et B prévue pour l'algorithme en fonction des temps de base suivants :

- temps mis par un filtre élémentaire pour le calcul à l'instant tk d'une estimée et de la pseudo-innovation correspondante : $\tau f$
- temps de calcul d'une variable de décision : $\tau v$
- temps moyen pour un test à 2 hypothèses (1 variable de décision comme au § 3.5) : $\tau H$

Si $\tau_A^D$ est le temps mis par la partie test de la forme A. On peut écrire alors

$$\tau_H < \tau_A^D < 2\tau_H$$

Si $\tau_B^D$ est le temps mis par la partie test de la forme B. On peut écrire de la même façon :

(voir fig.2) $$\tau_H < \tau_B^D < \frac{N-1}{2}\tau_H$$

N est le nombre d'hypothèses considéré.

D'où le tableau suivant regroupant les performances réalisées par l'algorithme sous les formes A et B pour une itération :

| | Temps Minimum pour 1 itération | Temps Maximum pour 1 itération |
|---|---|---|
| Forme A | $(N)\tau_f + 2\tau_v + \tau_H$ | $(N)\tau_f + 2\tau_v + 2\tau_H$ |
| Forme B | $(N)\tau_f + (N-1)\tau_v + \tau_H$ | $(N)\tau_f + (N-1)\tau_v + \frac{(N-1)}{2}\tau_H$ |

Quand on sait que l'on peut considérer l'inégalité suivante :

$$\tau_H < \tau_v < \tau_f$$

on réalise que la partie adaptative du filtre global n'est pas du tout pénalisante vis-à-vis du temps de calcul total mis en jeu pour les formes A ou B, surtout si les calculs des différents filtres sont effectués de façon successive.

Cependant, il n'en est pas moins vrai que pour obtenir des temps de calcul faibles sans pour autant diminuer le nombre d'hypothèses utilisées, il convient de choisir des modèles de dynamique suffisamment simples (§4).

D'autre part l'intérêt de l'algorithme proposé réside surtout dans le fait que l'on peut réduire considérablement le temps de calcul global en effectuant les calculs de filtrage en parallèle : $(N)\tau_f \longrightarrow \tau_f$

Exemple de temps de calcul obtenus sur microprocesseur 68 000 Motorola :

$$\tau_f = 0{,}7 \text{ ms} \; ; \; \tau_v = 0{,}4 \text{ ms} \; ; \; \tau_H = 0{,}1 \text{ ms}$$

## 4. - MODELISATION CHOISIE : MODELE DOUBLE INTEGRATEUR COMMANDE EN ACCELERATION

Etant donné un processus stochastique z(t) à représentation Gaussienne Markovienne, on dispose de n modèles capables de le représenter à des instants différents.

Chacun d'eux représente l'accélération du mobile comme un bruit blanc Gaussien centré sur une valeur moyenne qui varie d'une hypothèse à l'autre. Il s'agit d'un modèle double intégrateur commandé en accélération suivant le schéma :

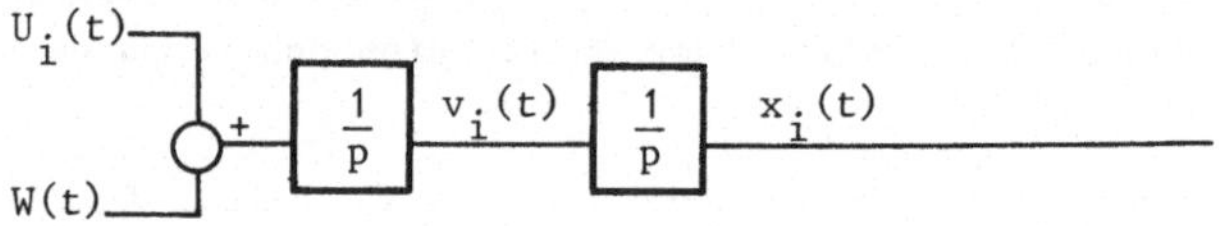

dans l'hypothèse Hi

W(t) est un bruit blanc gaussien centré

Ui(t) est la valeur moyenne de l'accélération du mobile dans l'hypothèse i.

Après discrétisation au 1er ordre, on obtient pour chaque modèle le système d'équations représentatives de l'hypothèse Hi.

$$\text{Hi} \begin{cases} X_i k{+}1 = \Phi X_i k + \Gamma U_i k + \Gamma W k \\ zk = H\, X_i k + V_k \end{cases}$$

$$\text{avec } Xk = \begin{bmatrix} xk \\ vk \end{bmatrix}; \quad \Phi = \begin{bmatrix} 1 & T \\ 0 & 1 \end{bmatrix}; \quad \Gamma = \begin{bmatrix} T^2/2 \\ T \end{bmatrix}$$

zk est la mesure à l'instant k

xk désigne la position

T est la période d'échantillonnage

$U_i k$ est une constante représentant la commande dans l'hypothèse i.

$$E\,[wk\; w\ell] = \sigma_q^2\; \delta k,\ell \; ; \; E[Vk\; V\ell] = \sigma_R^2\; \delta k,\ell \; ; \; E[Wk.\; W\ell] = 0$$

D'autre part, par souci de symétrie, la différence entre des commandes correspondant à des hypothèses consécutives est constante.

De même l'écart-type du bruit d'état, $\sigma Q$, est le même pour chaque modèle et les filtres correspondant possèdent alors le même gain asymptotique.

On remarquera que la méthode de tests d'hypothèses utilisée contribue à adapter le modèle en moyenne alors que les erreurs de modélisation du second ordre sont prises en compte par chacun des filtres constituant l'ensemble du modèle.

## 5. - RESULTATS

### 5.1 - MISE EN EVIDENCE DU FONCTIONNEMENT DE LA FORME A

La simulation est obtenue à partir d'un profil d'accélération utilisant 3 plages correspondant à 3 hypothèses.

$$H_2 : \gamma = -5g \; ; \; H_0 : \gamma = 0 \; ; \; H_1 : \gamma = 5g$$

La mesure est obtenue par addition d'un "bruit blanc gaussien" au signal de position d'écart type $\sigma_R = 10$ m.

Le bruit de dynamique est d'écart-type $\sigma_Q = 10 \text{ ms}^{-2}$ .

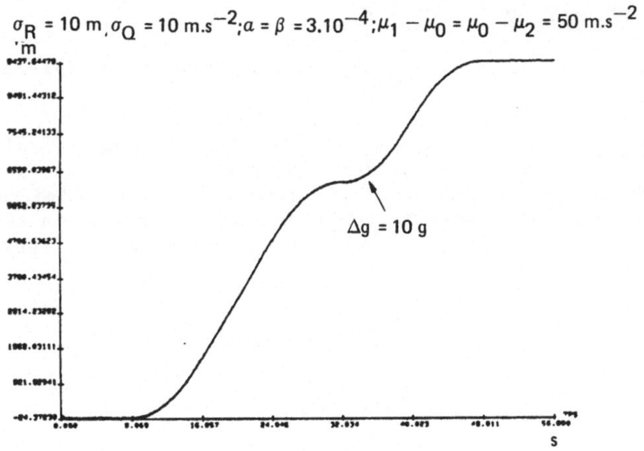


FIGURE 3 : Mesure de la position et, superposée, l'estimation qui en est donnée par l'algorithme dans la forme A (en pointillé).

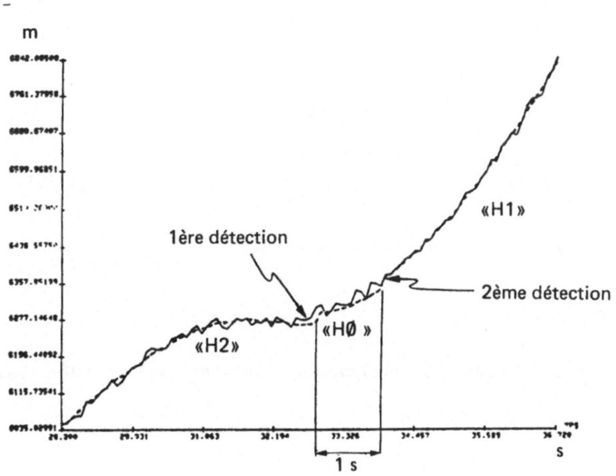


FIGURE 4 : Loupe de la courbe précédente à l'endroit : $\Delta\gamma$ = 10 g

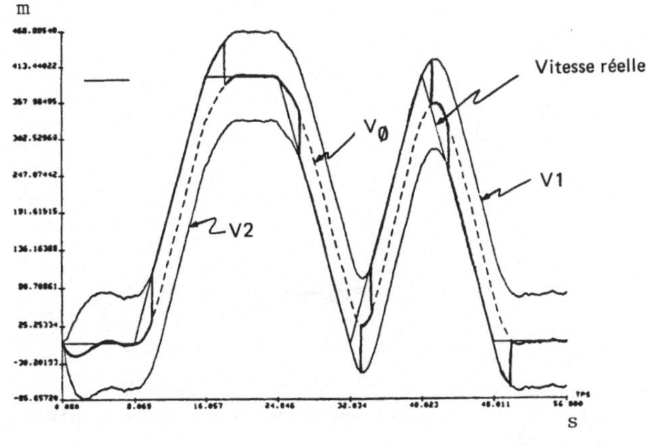


FIGURE 5 : Visualisation du changement d'hypothèse réalisé sur les estimations de vitesse et mise en évidence de la correspondance avec la vitesse réelle.

## 5.2 - MISE EN EVIDENCE DU FONCTIONNEMENT DE LA FORME B

Ici, la simulation est obtenue en accord avec une hypothèse de navigation proportionnelle.

Il s'agit de filtrer une coordonnée d'altitude obtenue par addition d'un bruit blanc gaussien d'écart type $\sigma_R$ = 3 m au signal de position. Le bruit de dynamique est: $\sigma_Q$ = 5 $ms^{-2}$. Pour mettre en évidence les performances de la forme B on a choisi un domaine assez large de trajectoires possibles pour la cible correspondant à une accélération maximale de 10g et une "définition" de 1g en moyenne conduisant à 21 filtres fonctionnant en parallèle.

NB : $\sigma_Q = \frac{\Delta g}{2} = 5 \ ms^{-2}$

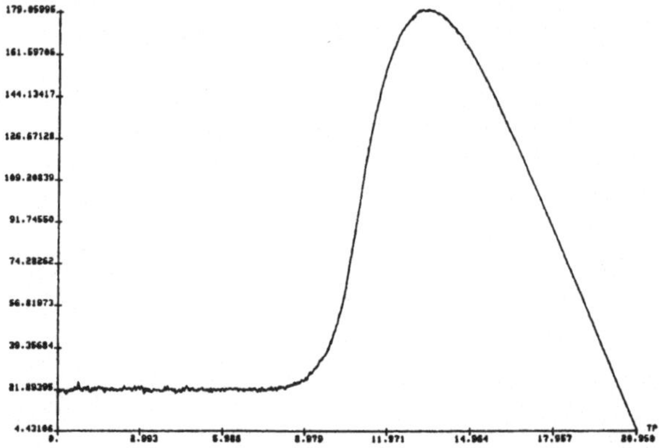

FIGURE 6

FIGURE 6 : Mesure de la position et superposée, l'estimation qui en est donnée par l'algorithme dans la forme B.

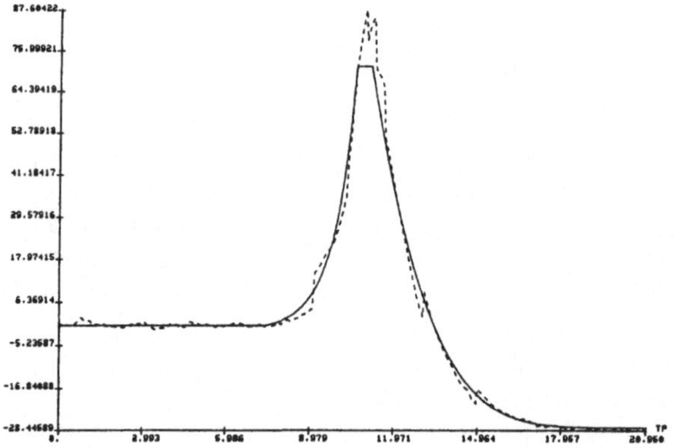

FIGURE 7

FIGURE 7 : Superposée à la valeur vraie de la vitesse en simulation (trait plein), l'estimation de la vitesse donnée par la forme B.

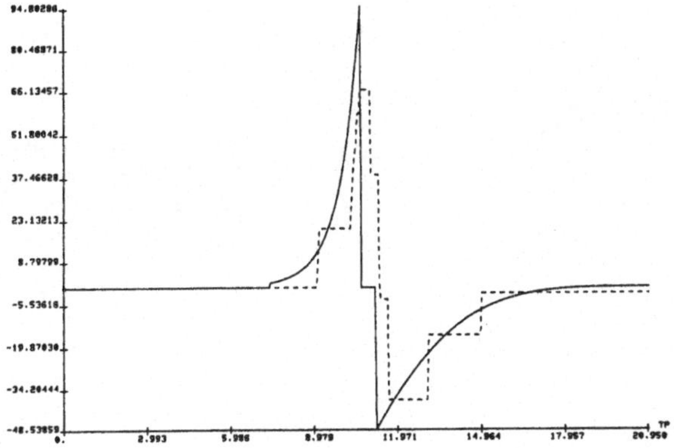

FIGURE 8

FIGURE 8 : Superposée à la valeur vraie de l'accélération en simulation (trait plein), l'estimation de l'accélération avec la forme B.

## 5.3 - TRAJECTOIRE REELLE - COMPARAISON AVEC LE MODELE DE SINGER

Il s'agit d'une "trajectoire baïonnette", dont on étudie la coordonnée "distance". $\sigma_R$ = 5 m. On considère 3 hypothèses.

$\sigma_Q = 5$ m.s$^{-2}$ ; $\sigma_R$  5 m ; $\alpha = \beta = 0,05$ ; $\mu_1 - \mu_0 = \mu_0 - \mu_2 = 2.\sigma_Q$

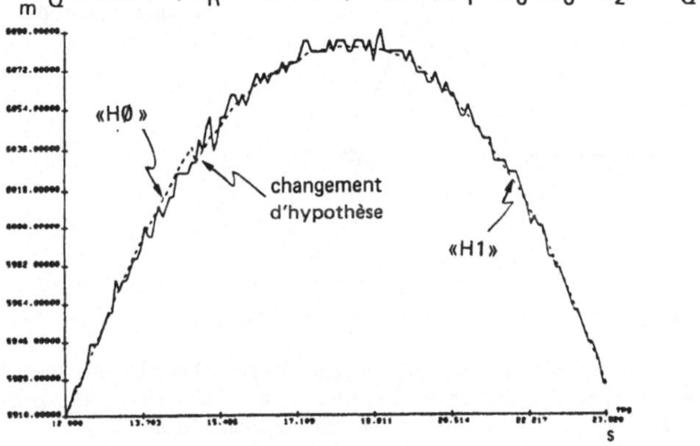

FIGURE 9

FIGURE 9 : Résultat du filtrage lors de la première évolution. (Forme A)

$\sigma_Q = 1 \text{ m.s}^{-3}$ ; $\sigma_R = 5 \text{ m}$ ; $a = \dfrac{1}{10} \text{ s}^{-1}$

en m

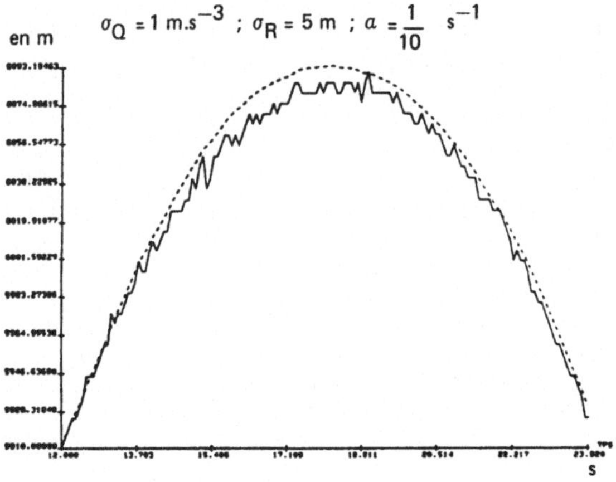

FIGURE 10 : Même passe traitée par un filtre basé sur le modèle de SINGER [6], où on considère un temps de manoeuvre de 10 s. L'estimée est peu perturbée mais le filtre traîne.

$\sigma_Q = 30 \text{ m.s}^{-3}$ ; $\sigma_R = 5 \text{ m}$ ; $a = \dfrac{1}{10} \text{ s}^{-1}$

m

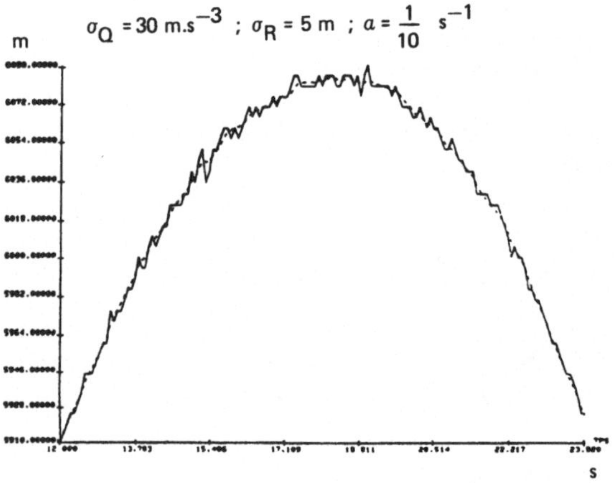

FIGURE 11 : Même traitement que la courbe précédente, mais avec un réglage différent du filtre de SINGER. On a diminué le trainage, mais l'estimée obtenue apparaît nettement plus perturbée que précédemment.

## 6. - CONCLUSION

Le but de cette étude était de réaliser un algorithme conduisant à un filtrage qui tienne compte des évolutions brusques de la cible et qui indépendamment, puisse donner des estimées peu perturbées de sa cinématique.

Nous avons vu au §2 comment se situait cette méthode par rapport à deux grandes classes d'estimateurs qui rendent compte de façon différente du problème d'adaptation de la dynamique (nous avons d'ailleurs indiqué une approche qui permet de décrire de manière beaucoup plus générale et cependant très complètement le problème posé sous la forme d'un processus de diffusion, grâce au générateur associé.[3])

D'autre part, nous avons proposé 2 formes A et B pour l'algorithme suivant le nombre d'hypothèses considérées et la valeur de la période d'échantillonnage.

Enfin, nous avons déterminé un modèle cartésien très simple permettant l'implantation sur calculateur. On étudie actuellement un modèle plus adapté à la nature du test utilisé et qui met en évidence les accélérations normales et tangentielles en tant que terme de commande.

BIBLIOGRAPHIE

- THEORIE DE L'ESTIMATION ET DE LA COMMANDE

[1] R.A. SINGER "Estimating Optimal Tracking Filter Performance For Manned Maneuvering Targets" I.E.E.E. Trans. A.E.S - Vol 6 n°4 - Juillet 1970

[2] C. BOZZO
Le filtrage optimal et ses applications aux problèmes de poursuite Technique et Documentation, Paris 1982

[3] W.M. WONHAM
Probabilistic methods in applied mathematics - BHARUCHA-REID (A.T) Vol. 2,Academic Press - New-York - 1970

- METHODES D'ESTIMATION DETECTION ET TESTS D'HYPOTHESES

[4] A.S WILLSKY and H.L. JONES "A Generalized Likelikood Ratio Approach to State Estimation in Linear Systems Subject to Abrupt Changes".
Report n° 4SL.P.538 - M.I.T. Electronic Systems Laboratory Cambridge - MASS. 1974

[5] C. DONCARLI, A. LORENZI
"Un algorithme d'estimation - détection pour la poursuite de trajectoire" - Huitième collogue GRETSI, NICE, Juin 1981

[6] A. WALD, J. WOLFOWITZ
Optimum character of the sequential probability
Radio Test. Annals of Mathematical Statistics - Vol.19, p. 326 (1948)

[7] F.C. SCHWEPPE
Evaluation of likelikood functions for guaussien signals I.E.E.E. Trans. Information Théory. Vol. I.T.II, pp. 61-70, January 1965.

[8] KLEIN J.P., Iria
"Détection de changement dans les caractéristiques d'un processus à représentation guaussienne markovienne"
Thèse de Docteur-Ingénieur, Paris-Sud, 1973, 117 p.

[9] B. DELLERY, A. LORENZI
C. BOZZO, A. GUILBERT, A. HOULES, L. PASSERON
"Caractérisation de trajectoires - Application à l'estimation et à la prédiction des éléments de cinématique d'un mobile manoeuvrant"
Neuvième collogue GRETSI, NICE, Juin 1983

- METHODES D'ESTIMATION BAYESIENNES

[10] N.H. GHOLSON and R.L. MOOSE
Manoeuvring target tracking using adaptative state estimation I.E.E E Trans. an Aerospace and Electronic Systems - Vol. AES-13, n° 3
May 1977

[11] E. SIFFREDI "Poursuite d'un mobile par filtrage adaptatif" : détermination de la position et du vecteur vitesse - DEA - Université d'Aix Marseille III Centre de St Jérôme 5 Octobre 1979

[12] M. GAUVRIT
"Performances des estimateurs bayesiens en boucle ouverte et Fermée.
Dualité : Identification - Commande".
Thèse d'Etat - Université Paul Sabatier Toulouse (Sciences) - 1982

[13] D.L. ALSPACH
"A parallel fittering Algorithm for Lineair Systems with unknown Time varying Noise Statistics" - I.E.E.E Trans an automatic control, p552-555, october 1974

REMERCIEMENTS

Nous remercions vivement le professeur C. DONCARLI (Université de Nantes),ainsi que monsieur M GAUVRIT (DERA Toulouse) et Monsieur G. SALUT (LAAS Toulouse) pour l'intérêt porté à nos travaux et pour les conseils et encouragements qu'ils ont bien voulu nous prodiguer.

Session 5

# STABILITY II

# STABILITÉ II

# NONLINEAR UNITY-FEEDBACK SYSTEMS AND Q-PARAMETRIZATION

C. A. Desoer and C. A. Lin
Department of Electrical Engineering and Computer Sciences
and the Electronics Research Laboratory
University of California, Berkeley, CA 94720

## ABSTRACT

This paper concerns nonlinear systems, defines a new concept of stability and extends to unity-feedback systems the technique of Q-parametrization introduced by Zames and developed by Desoer, Chen and Gustafson. We specify 1) a global parametrization of all controllers that $\mathcal{S}$-stabilize a given $\mathcal{S}$-stable plant; 2) a parametrization of a class of controllers that stabilize an unstable plant; 3) necessary and sufficient conditions for a nonlinear controller to simultaneously stabilize two nonlinear plants.

## I. INTRODUCTION

The purpose of this paper is to obtain the broadest generalization within the context of <u>nonlinear</u> systems of a number of recent results pertaining to <u>linear</u> feedback systems. For the unity feedback configuration and for a given linear stable plant, Zames (1981) proposed a parametrization of the stabilizing linear controllers in terms of a stable proper transfer function Q. This idea was further developed as a design procedure by Desoer and Chen (1981) and was used for computer aided design by Gustafson and Desoer (1983). In this paper we use also a Q-parametrization but in a nonlinear context. We first generalize the concept of finite-gain stability (incremental stability) to that of $\mathcal{S}$-stability (incremental $\mathcal{S}$-stability, resp.). In Theorem 1, we establish for the <u>nonlinear</u> case, a global parametrization of all I/O maps and of all compensators that result in an $\mathcal{S}$-stable configuration. This theorem generalizes to the nonlinear case, the original linear results of Zames, and in view of the more general stability concept, it also generalizes Desoer and Liu (1981).

In Section IV we consider the case where the plant is unstable. For the linear case, Zames established his "decomposition principle," i.e., stabilize the given linear plant P with a stable linear compensator F, and then proceed with the Q-parametrization as above. Anantharam and Desoer (1982) established a nonlinear version of this result. In Theorem 2 we establish a similar result in the more general concept of $\mathcal{S}$-stability and we weaken the requirement on the stabilizing feedback F: it need not be itself stable but need only lead to a stable feedback configuration of P and F. Note that Theorem 2 generalizes our previous work, first it uses the more general stability concept and, second, the method of proof is greatly improved (Desoer and Lin 1983a).

The problem of simultaneous stability has been formulated and solved in the linear case by Saeks and Murray (1982). Vidyasagar and Viswanadham (1982) also have interesting results along this line. In Section V we consider the nonlinear case: we are given two (possibly unstable) nonlinear plants $\overline{P}_1$ and $\overline{P}_2$ and we derive necessary and sufficient conditions for the existence of a fixed compensator that stabilizes both plants. Theorem 4 is a generalization for nonlinear plants and within the $\mathcal{S}$-stability concept of the linear results of Vidyasagar et al., and of our previous work (Desoer and Lin 1983b).

## II. DEFINITIONS AND NOTATIONS

Let $(\mathcal{L}, \|\cdot\|)$ be a normed space of "time functions": $\mathcal{T} \to \mathcal{V}$ where $\mathcal{T}$ is the time set (typically $\mathbb{R}_+$ or $\mathbb{N}$), $\mathcal{V}$ is a normed space (typically $\mathbb{R}$, $\mathbb{R}^n$, $\mathbb{C}^n$, $\cdots$) and $\|\cdot\|$ is the chosen norm in $\mathcal{L}$. Let $\mathcal{L}_e$ be the corresponding extended space (see e.g. Willems 1971, Desoer and Vidyasagar 1975, Vidyasagar 1978).

A function $\phi : \mathbb{R}_+ \to \mathbb{R}_+$ is said to belong to class K iff $\phi$ is continuous and increasing. $\phi$ is said to belong to class $K_o$ iff $\phi \in K$ and $\phi(0) = 0$. If $\phi_1$ and $\phi_2 \in K_o$, then $\phi_1 + \phi_2$ and $\alpha \mapsto \phi_1(\phi_2(\alpha)) \in K_o$. A nonlinear causal map $H : \mathcal{L}_e^{n_i} \to \mathcal{L}_e^{n_o}$ is said to be $\mathcal{S}$-stable iff $\exists\, \phi \in K$ s.t. $\forall x \in \mathcal{L}_e^{n_i}$, $\forall T \in \mathcal{T}$,

$$\|Hx\|_T \leq \phi(\|x\|_T)$$

H is said to be incrementally $\mathcal{S}$-stable (incr. $\mathcal{S}$-stable) iff (i) H is $\mathcal{S}$-stable, (ii) $\exists\, \tilde{\phi} \in K_o$ s.t. $\forall x, x' \in \mathcal{L}_e^{n_i}$, $\forall T \in \mathcal{T}$,

$$\|Hx-Hx'\|_T \le \tilde{\phi}(\|x-x'\|_T)$$

It can be shown that if the nonlinear causal maps $H_1$ and $H_2$ are $\mathcal{S}$-stable, (incr. $\mathcal{S}$-stable), then $H_1+H_2$ and $H_1 \circ H_2$ are $\mathcal{S}$-stable, (incr. $\mathcal{S}$-stable, resp.). (For simplicity, we drop in the following the symbol "$\circ$" denoting the composition of the maps.)

A feedback system is said to be <u>well-posed</u> iff the relation from the exogenous inputs into each subsystem† variable (i.e., subsystem input and subsystem output) is a well-defined nonlinear causal map between the corresponding extended spaces. More precisely, the system ${}^1S(P,C)$ of Fig. 1, where $P:\mathcal{L}_e^{n_i} \to \mathcal{L}_e^{n_o}$, $C:\mathcal{L}_e^{n_o} \to \mathcal{L}_e^{n_i}$ are causal maps, is said to be <u>well-posed</u> iff $H:(u_1,u_2) \mapsto (e_1,e_2,y_1,y_2)$ is well-defined and causal. Note that ${}^1S(P,C)$ is well-posed implies that†† $(I+PC)^{-1}$ and $(I+CP)^{-1}$ are well-defined and causal. We say that a well-posed nonlinear feedback system is <u>$\mathcal{S}$-stable</u> (incr. $\mathcal{S}$-stable) iff the map from the exogenous inputs to any subsystem variable is $\mathcal{S}$-stable (incre. $\mathcal{S}$-stable, resp.). For the system ${}^1S(P,C)$, since $e_1 = u_1 - y_2$, $e_2 = u_2 + y_1$, we see that $H_{yu}:(u_1,u_2) \mapsto (y_1,y_2)$ is $\mathcal{S}$-stable iff $H_{eu}:(u_1,u_2) \mapsto (e_1,e_2)$ is $\mathcal{S}$-stable iff ${}^1S(P,C)$ is $\mathcal{S}$-stable. The same equivalence holds for <u>incr. $\mathcal{S}$-stability</u>. These concepts of $\mathcal{S}$-stability and incr. $\mathcal{S}$-stability are generalizations of finite-gain stability and incremental stability (Desoer and Vidyasagar 1975); they are in spirit closer to Safonov's work (Safonov 1980). The map $H:\mathcal{L}_e^{n_o} \to \mathcal{L}_e^{n_o}$ is said to be <u>an achievable I/O map</u> of the nonlinear feedback system ${}^1S(P,C)$ iff by some appropriate choice of $C:\mathcal{L}_e^{n_o} \to \mathcal{L}_e^{n_i}$, (i) $H_{y_2u_1} = H$; (ii) ${}^1S(P,C)$ is $\mathcal{S}$-stable.

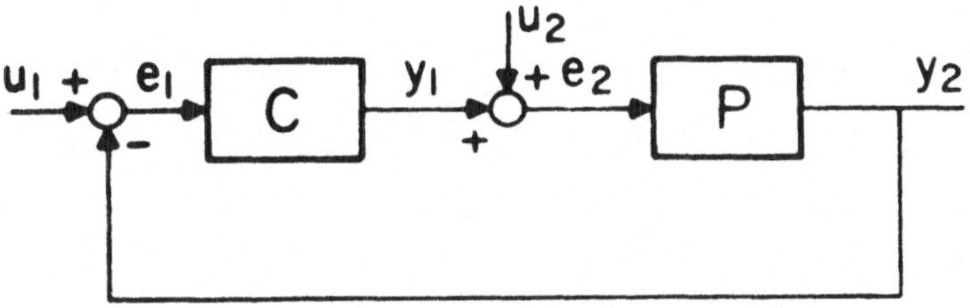


Fig. 1. Shows the system ${}^1S(P,C)$.

†By subsystem we mean any block of the block diagram of the feedback system.

††The meaning of $(I+PC)^{-1}$ deserves clarification: the map C is composed with P then the identity is added, and the resulting map is inverted. Although this formula has the same form as the linear case, it has a completely different interpretation.

We assume throughout this paper that all the nonlinear maps under consideration are causal and that all the nonlinear feedback systems under consideration are well-posed. We use "s.t." to abbreviate "such that," and "u.t.c." to abbreviate "under these conditions."

## III. GLOBAL PARAMETRIZATION OF NONLINEAR $\mathcal{S}$-STABLE I/O MAPS

Consider the well-posed nonlinear unity feedback system ${}^1S(P,C)$ shown in Fig. 1, where $P:\mathcal{L}_e^{n_i}\to\mathcal{L}_e^{n_o}$, $C:\mathcal{L}_e^{n_o}\to\mathcal{L}_e^{n_i}$ are nonlinear causal maps, and $(u_1,u_2)$, $(y_1,y_2)$ and $(e_1,e_2)$ are the "input," "output," and "error" respectively. Theorem 1 is a generalization of a result of Desoer and Liu (1981), it gives a global parametrization of all achievable input-output maps, and of all stabilizing compensators, under the assumption that P is incr. $\mathcal{S}$-stable. This theorem is an extension to the nonlinear case, the well-known linear Q-parametrization result, proved by Zames (1981) in a very general algebraic context.

Theorem 1. (Global parametrization of stable ${}^1S(P,C)$).

Let $P:\mathcal{L}_e^{n_i}\to\mathcal{L}_e^{n_o}$, $C:\mathcal{L}_e^{n_o}\to\mathcal{L}_e^{n_i}$ be nonlinear causal maps. Assume that P is incr. $\mathcal{S}$-stable. Under these conditions (U.t.c.),

(a) $H_{yu}$ is $\mathcal{S}$-stable $\Leftrightarrow \exists$ some $\mathcal{S}$-stable $Q:\mathcal{L}_e^{n_o}\to\mathcal{L}_e^{n_i}$ s.t.

$$C = Q(I-PQ)^{-1} \tag{3.1}$$

(b) $C = Q(I-PQ)^{-1} \Leftrightarrow Q = C(I+PC)^{-1}$ (3.2)

(c) With $u_2 = 0$ and with $C = Q(I-PQ)^{-1}$, the partial map $H_{y_2u_1} : (u_1,0)\mapsto y_2$ is given by

$$H_{y_2u_1} = PQ \tag{3.3}$$

Comments

(i) Equivalence (b) above requires only that ${}^1S(P,C)$ be well-posed.

(ii) Equivalence (a) gives a global parametrization of $\mathcal{C}(P)$, the family of all compensators that result in an $\mathcal{S}$-stable system ${}^1S(P,C)$; more precisely:

$$\mathcal{C}(P) = \{C\,|\,C = Q(I-PQ)^{-1},\ \ Q \text{ is } \mathcal{S}\text{-stable}\}\ .$$

(iii) From (a) and (c), $\mathcal{H}_{y_2u_1}$, the class of all achievable I/O maps is given by

$$\mathcal{H}_{y_2u_1}(P) = \{PQ | Q \text{ is } \mathcal{S}\text{-stable}\} .$$

(iv) Practical design considerations such as robustness of stability, disturbance rejection, plant saturation, etc. impose additional restrictions on Q (see e.g., Desoer and Chen 1981, Gustafson and Desoer 1983).

(v) The equation (3.3), $H_{y_2u_1} = PQ$, raises a number of new problems: given a nonlinear map P, how can one describe the constraints imposed by P on the achievable I/O map $H_{y_2u_1}$? If we have a desired I/O map $H_{y_2u_1}$ and a given P, how does one find a $Q_a$ such that in some appropriate sense, $PQ_a \approx H_{y_2u_1}$? Then having such a $Q_a$ how does one synthesize C?

Proof:

(I) Proof of (b).

We shall prove only the (⇒) implication, since the (⇐) implication can be shown in the same way. By assumption,

$$C = Q(I-PQ)^{-1} .$$

Composing with P and adding identity we obtain successively,

$$I + PC = I + PQ(I-PQ)^{-1} = (I-PQ)^{-1}$$

By taking the inverse, and composing with C, we obtain

$$C(I+PC)^{-1} = Q(I-PQ)^{-1}(I-PQ) = Q$$

Hence, $\quad Q = C(I+PC)^{-1}$ .

(II) Proof of (a).

(⇒) Set $u_2 = 0$, the map $H_{y_1u_1} : u_1 \mapsto y_1$ is given by $H_{y_1u_1} = C(I+PC)^{-1}$ which by assumption is $\mathcal{S}$-stable. Let $Q := C(I+PC)^{-1}$, then Q is $\mathcal{S}$-stable and from (b), we have $C = Q(I-PQ)^{-1}$.

(⇐) Refer to Fig. 1, write the summing node equations

$$e_1 = u_1 - Pe_2 \qquad e_2 = u_2 + Ce_1 \tag{3.4) (3.5}$$

Define

$$\tilde{u}_1 := PC\, e_1 - P(u_2+Ce_1) \tag{3.6}$$

Using (3.5) and (3.6), rewrite (3.4) as

$$e_1 = u_1 + \tilde{u}_1 - PC\, e_1 \tag{3.7}$$

From equation (3.7)

$$e_1 = (I+PC)^{-1}(u_1+\tilde{u}_1) \tag{3.8}$$

$$y_1 = Ce_1 = C(I+PC)^{-1}(u_1+\tilde{u}_1) = Q(u_1+\tilde{u}_1) \tag{3.9}$$

Now, since P is incr. $\mathcal{S}$-stable, $\exists\, \tilde{\phi}_p \in K_o$ s.t. $\forall^\dagger (u_1, u_2) \in \mathcal{L}_e^{n_o} \times \mathcal{L}_e^{n_i}$, $\forall T \in \mathcal{T}$,

$$\|\tilde{u}_1\|_T = \|P(Ce_1)-P(u_2+Ce_1)\|_T \leq \tilde{\phi}_p(\|u_2\|_T) \leq \tilde{\phi}_p(\|u_1\|_T+\|u_2\|_T) \tag{3.10}$$

Hence the map $\tilde{\pi} : (u_1,u_2) \mapsto \tilde{u}_1$ is $\mathcal{S}$-stable. Define the projection map $\pi_i : (u_1,u_2) \mapsto u_i$, $i = 1,2$. From (3.9), the map $H_{y_1u} : (u_1,u_2) \mapsto y_1$ is given by

$$H_{y_1u} = Q(\pi_1+\tilde{\pi}) \tag{3.11}$$

Since $\pi_1$ and $\tilde{\pi}$ are $\mathcal{S}$-stable, and by assumption Q is $\mathcal{S}$-stable, the map $H_{y_1u}$ is $\mathcal{S}$-stable. From Fig. 1, we have

$$y_2 = P(u_2+y_1) \tag{3.12}$$

Hence the map $H_{y_2u} : (u_1,u_2) \mapsto y_2$ is given by

$$H_{y_2u} = P(\pi_2+H_{y_1u}) \tag{3.13}$$

Now $\pi_2$ and $H_{y_1u}$ are $\mathcal{S}$-stable, and by assumption P is $\mathcal{S}$-stable, it follows that $H_{y_2u}$ is $\mathcal{S}$-stable. Therefore $H_{yu}$ is $\mathcal{S}$-stable.

<u>(III) Proof of (c).</u>

Since $C = Q(I-PQ)^{-1}$, from (b) we have $Q = C(I+PC)^{-1}$. With $u_2 = 0$, $H_{y_1u_1} = H_{e_2u_1} = C(I+PC)^{-1} = Q$.

Hence, $H_{y_2u_1} = PQ$ . ◻

---

† For $(u_1,u_2) \in \mathcal{L}^{n_o} \times \mathcal{L}^{n_i}$, we define $\|(u_1,u_2)\| := \|u_1\| + \|u_2\|$.

## IV. Two-Step Stabilization of Nonlinear Plants

The equivalence (a) of Theorem 1 above requires that the plant be incr. $\mathcal{S}$-stable. In practice, unstable plants do occur (e.g., chemical reactors, high performance airplanes, etc.), it is important to extend this method to include unstable plants.

Theorem 2. (Two-step stabilization of nonlinear plants).

Let $P : \mathcal{L}_e^{n_i} \to \mathcal{L}_e^{n_o}$, $F : \mathcal{L}_e^{n_o} \to \mathcal{L}_e^{n_i}$ be nonlinear causal maps such that the system ${}^1S(P,F)$ shown in Fig. 2 is incr. $\mathcal{S}$-stable. Let $P_1 := P[I-F(-P)]^{-1}$.

U.t.c., if

$$C := F + Q(I-P_1Q)^{-1} \text{ for some } \mathcal{S}\text{-stable } Q : \mathcal{L}_e^{n_o} \to \mathcal{L}_e^{n_i} , \tag{4.1}$$

then

(a) the system ${}^1S(P,C)$ is $\mathcal{S}$-stable; and

(b) the system ${}^3S(P,F,C-F)$ shown in Fig. 3 is $\mathcal{S}$-stable.

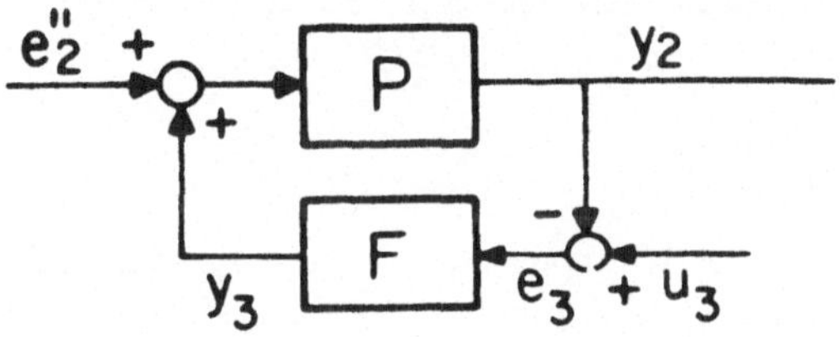


Fig. 2. Shows the system ${}^1S(P,F)$ in which F stabilizes P.

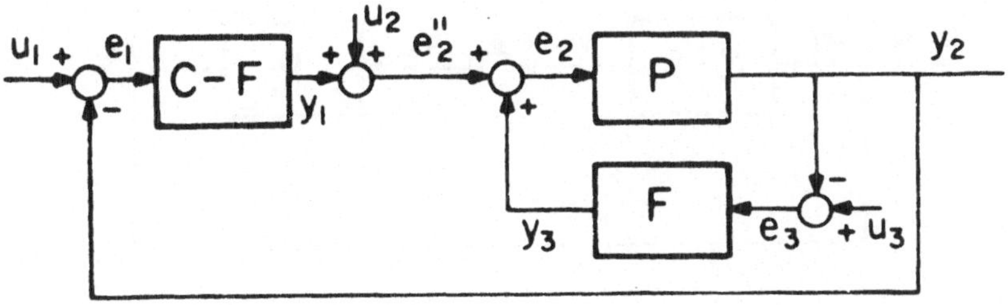


Fig. 3. Shows the system ${}^3S(P,F,C-F)$.

Comments.

(i) None of the maps P, C, F, C-F are required to be stable.

(ii) The key assumptions are (a) well-posedness, (b) ${}^1S(P,F)$ is incr. $\mathcal{S}$-stable, (c) $C = F+Q(I-P_1Q)^{-1}$ where $P_1 = P[I-F(-P)]^{-1}$ and Q is $\mathcal{S}$-stable.

(iii) It can be easily checked (using the summing node equations) that the system ${}^3S(P,F,C-F)$ is $\mathcal{S}$-stable iff the map $(u_1,u_2,u_3) \mapsto (y_1,y_2,y_3)$ is $\mathcal{S}$-stable.

(iv) If P is incr. $\mathcal{S}$-stable, then by choosing F the zero map, we have $P_1 = P$, $C = Q(I-PQ)^{-1}$, and Theorem 2 reduces to Theorem 1.

(v) In the proof we show that (b) implies (a), a simple example shows that (a) does not imply (b). However, if F is incr. $\mathcal{S}$-stable, then (a) and (b) are equivalent (Anantharam and Desoer 1982, Thm. 3).

Proof:

(I) Proof of (b): ${}^3S(P,F,C-F)$ is $\mathcal{S}$-stable.

Consider the system ${}^1S(P,F)$ of Fig. 2, let $\psi = (\psi_2,\psi_3) : (e_2'',u_3) \mapsto (y_2,y_3)$ be its I/O map. Note that $P_1(\cdot) := P[I-F(-P)]^{-1}(\cdot) = \psi_2(\cdot,0)$. By (A.2), $\psi$ is incr. $\mathcal{S}$-stable, hence $P_1$ is incr. $\mathcal{S}$-stable, further from assumption (4.1), Q is $\mathcal{S}$-stable and $C-F = Q(I-P_1Q)^{-1}$; hence, by Theorem 1, these three conclusions imply that the system ${}^1S(P_1,C-F)$ shown in Fig. 4 is $\mathcal{S}$-stable.

Next consider Fig. 3 which shows the system ${}^3S(P,F,C-F)$ with input $(u_1,u_2,u_3)$ and output $(y_1,y_2,e_2'',y_3)$. We claim that the map ${}^3H : (u_1,u_2,u_3) \mapsto (y_1,y_2,e_2'',y_3)$ is $\mathcal{S}$-stable. Let

$$\Delta\tilde{y}_2 := \psi_2(e_2'',u_3) - \psi_2(e_2'',0) \ . \tag{4.2}$$

Drive the system ${}^3S(P,F,C-F)$ with input $(u_1-\Delta\tilde{y}_2,u_2,0)$, call the corresponding output $(\tilde{y}_1,\tilde{y}_2,\tilde{e}_2'',\tilde{y}_3)$, and note that $\tilde{y}_2 = P[I-F(-P)]^{-1}\tilde{e}_2'' = P_1\tilde{e}_2''$; thus if we ignore $\tilde{y}_3$, the system reduces to ${}^1S(P_1,C-F)$, (which has just been shown to be $\mathcal{S}$-stable), with input

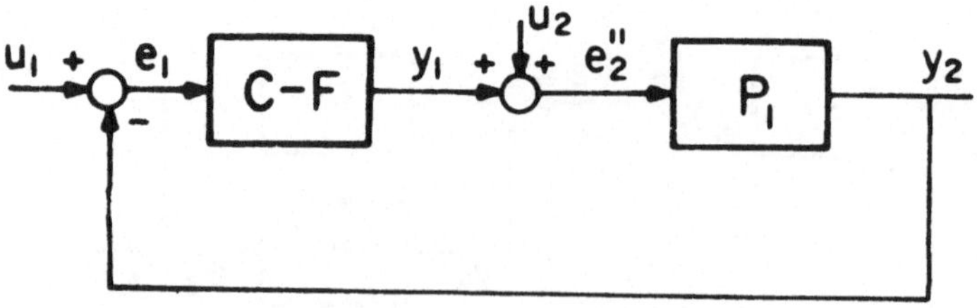


Fig. 4. Shows the system ${}^1S(P_1,C-F)$.

$(u_1-\Delta\tilde{y}_2,u_2)$ and output $(\tilde{y}_1,\tilde{y}_2,\tilde{e}_2'')$. Hence, for ${}^3S(P,F,C-F)$, the partial map (with respect to ${}^3H$), ${}^2\tilde{H}:(u_1-\Delta\tilde{y}_2,u_2,0)\mapsto(\tilde{y}_1,\tilde{y}_2,\tilde{e}_2'')$ is $\mathcal{S}$-stable. Since $\psi_2$ is incr. $\mathcal{S}$-stable, $\exists\,\tilde{\phi}_2\in K_o$ s.t. $\forall e_2''$, $\forall u_3$, $\forall T$,

$$\|\Delta\tilde{y}_2\|_T = \|\psi_2(e_2'',u_3)-\psi_2(e_2'',0)\|_T \le \tilde{\phi}_2(\|u_3\|_T) \le \tilde{\phi}_2(\|u_1\|_T+\|u_2\|_T+\|u_3\|_T) \tag{4.3}$$

Hence the map $(u_1,u_2,u_3)\mapsto\Delta\tilde{y}_2$ is $\mathcal{S}$-stable. Therefore, the map ${}^2\pi:(u_1,u_2,u_3)\mapsto(u_1-\Delta\tilde{y}_2,u_2,0)$ is $\mathcal{S}$-stable. Considering the composition ${}^2\tilde{H}\,{}^2\pi$ we see that, for ${}^3S(P,F,C-F)$, the map $(u_1,u_2,u_3)\mapsto(\tilde{y}_1,\tilde{y}_2,\tilde{e}_2'')$ is $\mathcal{S}$-stable.

Now, we claim that $y_1=\tilde{y}_1$, $y_2=\tilde{y}_2+\Delta\tilde{y}_2$, $e_2''=\tilde{e}_2''$, and hence the map $(u_1,u_2,u_3)\mapsto(y_1,y_2,e_2'')$ is $\mathcal{S}$-stable. To prove this, write the equations for ${}^3S(P,F,C-F)$ with input $(u_1,u_2,u_3)$ and with input $(u_1-\Delta\tilde{y}_2,u_2,0)$, respectively:

| | |
|---|---|
| $y_1 = (C-F)(u_1-y_2)$ (4.4a) | $\tilde{y}_1 = (C-F)(u_1-\Delta\tilde{y}_2-\tilde{y}_2)$ (4.5a) |
| $y_2 = \psi_2(e_2'',u_3)$ (4.4b) | $\tilde{y}_2 = \psi_2(\tilde{e}_2'',0)$ (4.5b) |
| $e_2'' = y_1 + u_2$ (4.4c) | $\tilde{e}_2'' = \tilde{y}_1 + u_2$ (4.5c) |

Using (4.2), rewrite the equations (4.4) as

$$y_1 = (C-F)[u_1-\Delta\tilde{y}_2-(y_2-\Delta\tilde{y}_2)] \tag{4.6a}$$

$$y_2 - \Delta\tilde{y}_2 = \psi_2(e_2'',0) \tag{4.6b}$$

$$e_2'' = y_1 + u_2 \tag{4.6c}$$

From Eqs. (4.5) and (4.6), we see that $(y_1,y_2-\Delta\tilde{y}_2,e_2'')$ and $(\tilde{y}_1,\tilde{y}_2,\tilde{e}_2'')$ satisfy the same equations, by the well-posedness assumption (A.3), Eqs. (4.5) and (4.6) both have a unique solution, hence $y_1=\tilde{y}_1$, $y_2=\tilde{y}_2+\Delta\tilde{y}_2$, $e_2''=\tilde{e}_2''$. Since $y_3=\psi_3(e_2'',u_3)$ and $\psi_3$ is $\mathcal{S}$-stable, the map $(u_1,u_2,u_3)\mapsto y_3$ is $\mathcal{S}$-stable. Consequently, the map ${}^3H:(u_1,u_2,u_3)\mapsto(y_1,y_2,e_2'',y_3)$ is $\mathcal{S}$-stable and (b) is established.

(II) Proof of (a): ${}^1S(P,C)$ is $\mathcal{S}$-stable.

Write the summing node equations for ${}^3S(P,F,C-F)$ in terms of $e_1$, $e_2$, $e_3$, and $e_2''$: (see Fig. 3),

$$e_1 = u_1 - Pe_2 \tag{4.7a}$$

$$e_2'' = u_2 + (C-F)e_1 \tag{4.7b}$$

$$e_2 = e_2'' + Fe_3 \tag{4.7c}$$

$$e_3 = u_3 - Pe_2 \tag{4.7d}$$

Let $u_1 = u_3$, then, by (4.7a) and (4.7d), $e_1 = e_3$; thus by adding (4.7c) and (4.7b) we have

$$e_1 = u_1 - Pe_2 \tag{4.8a}$$

$$e_2 = u_2 + Ce_1 \ . \tag{4.8b}$$

The equations (4.8) describe ${}^1S(P,C)$. Since ${}^3S(P,F,C-F)$ is $\mathcal{S}$-stable, the map $(u_1,u_2,u_1) \to (e_1,e_2)$ defined by (4.8) is $\mathcal{S}$-stable. Hence ${}^1S(P,C)$ is $\mathcal{S}$-stable. ⌑

## V. Simultaneous Stabilization of Nonlinear Plants

The main result is Theorem 4: a necessary and sufficient condition for given two nonlinear plants be simultaneously stabilized by one compensator.

**Theorem 3.**

Let $\bar{P} : \mathcal{L}_e^{n_i} \to \mathcal{L}_e^{n_o}$ and $C, F : \mathcal{L}_e^{n_o} \to \mathcal{L}_e^{n_i}$ be nonlinear causal maps. Let $P := \bar{P}[I-F(-\bar{P})]^{-1}$.

U.t.c., if $F$ is incr. $\mathcal{S}$-stable, then the system ${}^1S(\bar{P},C+F)$ of Fig. 5 is $\mathcal{S}$-stable

$\Leftrightarrow$

the system ${}^1S(P,C)$ is $\mathcal{S}$-stable.

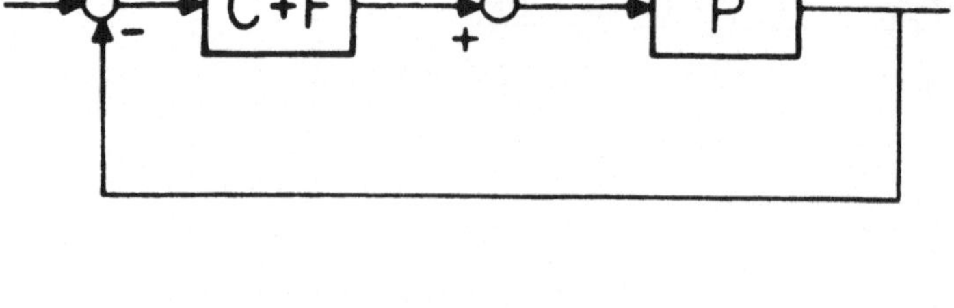

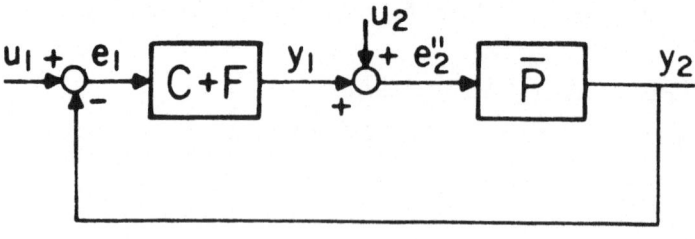

Fig. 5. Shows the system ${}^1S(\bar{P},C+F)$.

Comments.

(i) None of the maps $\overline{P}$, P, and C are required to be stable.

(ii) The Theorem is false if F is not incr. $\mathcal{S}$-stable. Consider the following example: let $\overline{P} = (s-1)/(s+3) =: \overline{n}/\overline{d}$, $F = 3/(s-1)$, and $C = 3/1 =: n_c/d_c$. By calculation, $C + F = 3s/(s-1) =: n_{c+f}/d_{c+f}$, and $P = \overline{P}[I-F(-\overline{P})]^{-1} = (s-1)/(s+6) =: n/d$. The system ${}^1S(P,C)$ is $\mathcal{S}$-stable, since its characteristic polynomial is $nn_c + dd_c = 4s + 3$. However, the system ${}^1S(\overline{P},C+F)$ is unstable, since its characteristic polynomial is $\overline{n}n_{c+f} + \overline{d}d_{c+f} = (s-1)(4s+3)$.

(iii) Traditionally the loop transformation theorem (see e.g., Desoer and Vidyasagar 1975) requires that F be linear, so Theorem 3 is a generalization of the usual stability results obtainable from the loop transformation theorem.

(iv) Roughly speaking, Theorem 3 says given that F is incr. $\mathcal{S}$-stable, and that the system ${}^1S(P,C)$ (${}^1S(\overline{P},C+F)$) is $\mathcal{S}$-stable, if we apply the feedback F (-F, resp.) around the plant and apply the feedforward F (-F, resp.) in parallel with the compensator, then the resulting closed-loop system ${}^1S(\overline{P},C+F)$ (${}^1S(P,C)$, resp.) remains $\mathcal{S}$-stable. This is also the case when the roles of the plant and the compensator are interchanged. The precise statement is given in the following corollary, whose proof can be constructed using the same techniques as those in the proof of Theorem 3.

Corollary 3.1: Let $P : \mathcal{L}_e^{n_i} \to \mathcal{L}_e^{n_o}$, and $\overline{C}$, $F : \mathcal{L}_e^{n_o} \to \mathcal{L}_e^{n_i}$ be nonlinear casual maps. Let $C := \overline{C}(I+F\overline{C})^{-1}$.

U.t.c., if F is incr. $\mathcal{S}$-stable, then

${}^1S(P+F,\overline{C})$ is $\mathcal{S}$-stable $\Leftrightarrow$ ${}^1S(P,C)$ is $\mathcal{S}$-stable.

In order to prove Theorem 3, it is convenient to start by exhibiting the following lemma, whose proof is similar to that of (3.2).

Lemma: Let $\overline{P} : \mathcal{L}_e^{n_i} \mapsto \mathcal{L}_e^{n_o}$ and $F : \mathcal{L}_e^{n_o} \mapsto \mathcal{L}_e^{n_i}$. If $P := \overline{P}[I-F(-\overline{P})]^{-1}$, then $\overline{P} = P[I+F(-P)]^{-1}$.

Comments:

(i) By using relation $P = \overline{P}[I-F(-\overline{P})]^{-1}$, ($\overline{P}=P[I+F(-P)]^{-1}$), the system ${}^1S(P,C)$ of Fig. 1 (${}^1S(\overline{P},C+F)$ of Fig. 5, resp.) can be redrawn as the system of Fig. 6 (Fig. 7, resp.).

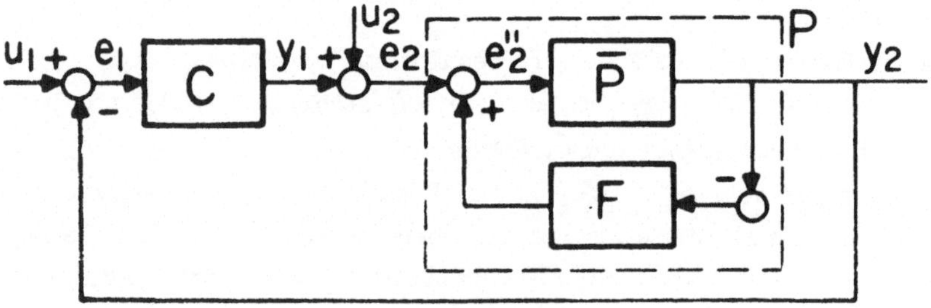


Fig. 6. Shows the system ${}^3S(\bar{P},F,C)$ with $u_3 \equiv 0$.

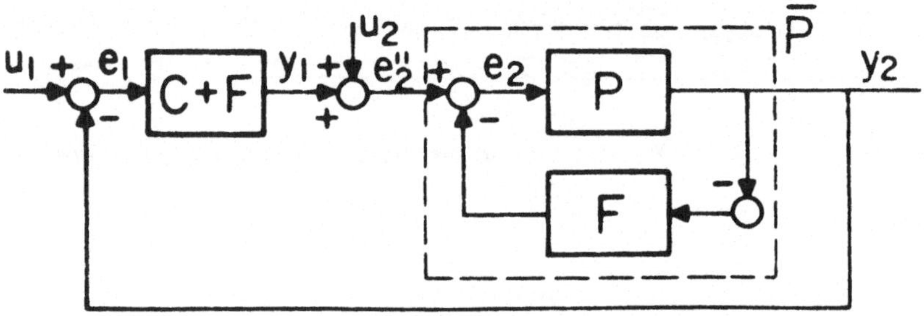


Fig. 7. Shows the system ${}^3S(P,-F,C+F)$ with $u_3 \equiv 0$.

(ii) Note that the system in Fig. 6 (Fig. 7) and the system ${}^1S(P,C)$ (${}^1S(\bar{P},C+F)$, resp.) have the same I/O map $\psi_C : (u_1,u_2) \mapsto (e_1,e_2)$ ($\psi''_{C+F} : (u_1,u_2) \mapsto (e_1,e_2'')$, resp.).

Proof of Theorem 3:

($\Rightarrow$) We show that for the system ${}^1S(P,C)$, the map $\psi_C : (u_1,u_2) \mapsto (e_1,e_2)$ is $\mathcal{S}$-stable. For the system shown in Fig. 6, write the equations defining $e_1$ and $e_2''$:

$$e_1 = u_1 - \bar{P}e_2'' \tag{5.1a}$$

$$e_2'' = u_2 + Ce_1 + F(-\bar{P}e_2'')$$

$$= u_2 + Ce_1 + F(e_1-u_1) \tag{5.1b}$$

Rewrite (5.1b) as

$$e_2'' = u_2 + (C+F)e_1 + [F(e_1-u_1)-Fe_1] \tag{5.2}$$

Let

$$\tilde{u}_1 := u_1 \tag{5.3a}$$

$$\tilde{u}_2 := u_2 + [F(e_1-u_1)-Fe_1] \tag{5.3b}$$

Then, Eqs. (5.1) read

$$e_1 = \tilde{u}_1 - \bar{P}e_2'' \tag{5.4a}$$

$$e_2'' = \tilde{u}_2 + (C+F)e_1 \tag{5.4b}$$

Note that equations (5.4) describe ${}^1S(\bar{P},C+F)$ with input $(\tilde{u}_1,\tilde{u}_2)$; by assumption ${}^1S(\bar{P},C+F)$ is $\mathcal{S}$-stable. Hence the map $\tilde{\psi}_{C+F} : (\tilde{u}_1,\tilde{u}_2) \to (e_1,e_2'')$, specified by (5.4), is $\mathcal{S}$-stable. Since F is incr. $\mathcal{S}$-stable, $\exists\ \tilde{\phi}_F \in K_o$, s.t. $\forall u_1$, $\forall e_1$, $\forall T$,

$$\|F(e_1-u_1)-Fe_1\|_T \le \tilde{\phi}_F(\|u_1\|_T) \le \tilde{\phi}_F(\|u_1\|_T+\|u_2\|_T) \tag{5.5}$$

Hence the map $\tilde{\pi} : (u_1,u_2) \mapsto (\tilde{u}_1,\tilde{u}_2)$ defined by (5.1) and (5.3) is $\mathcal{S}$-stable. Define $\psi_C'' = \tilde{\psi}_{C+F}\tilde{\pi}$, since both $\tilde{\psi}_{C+F}$ and $\tilde{\pi}$ are $\mathcal{S}$-stable, so is $\psi_C'' : (u_1,u_2) \mapsto (e_1,e_2'')$; hence for the system of Fig. 6 the map $(u_1,u_2) \mapsto (e_1,e_2'')$ is $\mathcal{S}$-stable.

Now from Fig. 6,

$$e_2 = e_2'' - F(e_1-u_1) \tag{5.6}$$

Since $\psi_C''$ and F are $\mathcal{S}$-stable, the map $(u_1,u_2) \mapsto e_2$ is $\mathcal{S}$-stable. It then follows that, for the system ${}^1S(P,C)$, $\psi_C : (u_1,u_2) \mapsto (e_1,e_2)$ is $\mathcal{S}$-stable.

($\Leftarrow$) We show that, for the system ${}^1S(\bar{P},C+F)$, the map $\psi_{C+F}'' : (u_1,u_2) \mapsto (e_1,e_2'')$ is $\mathcal{S}$-stable.

Using the Lemma $\bar{P} = P[I+F(-P)]^{-1}$ and redraw ${}^1S(\bar{P},C+F)$ as in Fig. 7. Write the equations defining $(e_1,e_2)$ in Fig. 7.

$$e_1 = u_1 - Pe_2 \tag{5.7a}$$

$$\begin{aligned} e_2 &= u_2 + (C+F)e_1 - F(-Pe_2) \\ &= u_2 + Fe_1 - F(e_1-u_1) + Ce_1 \end{aligned} \tag{5.7b}$$

Let

$$\bar{u}_1 := u_1 \tag{5.8a}$$

$$\bar{u}_2 := u_2 + Fe_1 - F(e_1 - u_1) \tag{5.8b}$$

Since F is incr. $\mathcal{S}$-stable, the map $\bar{\pi} : (u_1,u_2) \to (\bar{u}_1,\bar{u}_2)$ defined by (5.7) and (5.8) is $\mathcal{S}$-stable. Now, with (5.8), equations (5.7) read

$$e_1 = \bar{u}_1 - Pe_2 \tag{5.9a}$$

$$e_2 = \bar{u}_2 + Ce_1 \tag{5.9b}$$

Note that equations (5.9) describe ${}^1S(P,C)$ with input $(\bar{u}_1,\bar{u}_2)$. By assumption ${}^1S(P,C)$ is $\mathcal{S}$-stable, hence the map $\bar{\psi}_C : (\bar{u}_1,\bar{u}_2) \to (e_1,e_2)$, specified by (5.9), is $\mathcal{S}$-stable. Define $\psi_{C+F} = \bar{\psi}_C\bar{\pi}$, since both $\bar{\psi}_C$ and $\bar{\pi}$ are $\mathcal{S}$-stable, so is $\psi_{C+F} : (u_1,u_2) \to (e_1,e_2)$; hence for the system of Fig. 7, the map $(u_1,u_2) \to (e_1,e_2)$ is $\mathcal{S}$-stable.

Now from Fig. 7,

$$e_2'' = e_2 + F(e_1 - u_1) \tag{5.10}$$

Since $\psi_{C+F}$ and F are $\mathcal{S}$-stable, equation (5.10) implies that the map $(u_1,u_2) \to e_2''$ is $\mathcal{S}$-stable. Consequently, we have shown that for the system ${}^1S(\bar{P},C+F)$, $\psi''_{C+F} : (u_1,u_2) \to (e_1,e_2'')$ is $\mathcal{S}$-stable. ◻

<u>Theorem 4</u>. (Simultaneous Stabilization)

Let $\bar{P}_1, \bar{P}_2 : \mathcal{L}_e^{n_i} \to \mathcal{L}_e^{n_o}$ and $F : \mathcal{L}_e^{n_o} \to \mathcal{L}_e^{n_i}$ be nonlinear causal maps. Assume that F is <u>incr. $\mathcal{S}$-stable</u> and is such that $P_1 := \bar{P}_1[I-F(-\bar{P}_1)]^{-1}$ is <u>incr. $\mathcal{S}$-stable</u>. Let $P_2 := \bar{P}_2[I-F(-\bar{P}_2)]^{-1}$. For any $C : \mathcal{L}_e^{n_o} \to \mathcal{L}_e^{n_i}$, let

$$Q := C(I+P_1C)^{-1} \tag{5.11}$$

U.t.c.

${}^1S(\bar{P}_1,C+F)$ and ${}^1S(\bar{P}_2,C+F)$ are $\mathcal{S}$-stable

⇔

Q is $\mathcal{S}$-stable and ${}^1S(P_2-P_1,Q)$ is $\mathcal{S}$-stable (see Fig. 8).

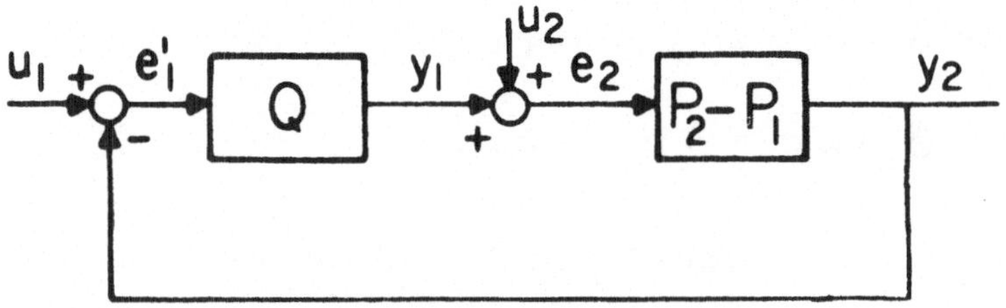


Fig. 8. Shows the system ${}^1S(P_2-P_1,Q)$.

Comments

(i) By Theorem 1, Eq. (5.11) is equivalent to that $C = Q(I-P_1Q)^{-1}$.

(ii) None of the maps $\overline{P}_1$, $\overline{P}_2$, $P_2$, and C are required to be stable.

(iii) The meaning of the theorem is the following: given two nonlinear, not necessarily stable, plants $\overline{P}_1$ and $\overline{P}_2$, if by applying an incr. $\mathcal{S}$-stable feedback F around $\overline{P}_1$ (see Fig. 6), the resulting closed-loop I/O map $P_1 := \overline{P}_1[I-F(-\overline{P}_1)]^{-1}$ is incr. $\mathcal{S}$-stable, then any compensator of the form $Q(I-P_1Q)^{-1} + F$, for some $\mathcal{S}$-stable Q such that ${}^1S(P_2-P_1,Q)$ is $\mathcal{S}$-stable, will stabilize both $\overline{P}_1$ and $\overline{P}_2$.

(iv) If $\overline{P}_1$ is incr. $\mathcal{S}$-stable, take $F = \Theta$, the zero map from $\mathcal{L}_e^{n_o} \leftrightarrow \mathcal{L}_e^{n_i}$, then $P_1 = \overline{P}_1$ and $P_2 = \overline{P}_2$. The theorem shows, for this special case, that given two nonlinear plants $\overline{P}_1$ and $\overline{P}_2$, with $\overline{P}_1$ incr. $\mathcal{S}$-stable, then the problem of finding a compensator to stabilize both $\overline{P}_1$ and $\overline{P}_2$ is equivalent to that of finding an $\mathcal{S}$-stable compensator to stabilize $\overline{P}_2 - \overline{P}_1$. This result was proven for the linear case in (Vidyasagar and Viswanadham 1982, Corollary 3.1.1).

(v) Suppose that we have n nonlinear plants $\overline{P}_1, \overline{P}_2, \cdots, \overline{P}_n$, then we may apply successively the theorem to the pairs $(\overline{P}_i,\overline{P}_1)$, $i = 2, 3, \cdots n$, thus ${}^1S(\overline{P}_i,C+F)$ is $\mathcal{S}$-stable for $i = 1, 2, \cdots, n$ iff $Q := C(I+P_1C)^{-1}$ is $\mathcal{S}$-stable, and ${}^1S(P_i-P_1,Q)$ is $\mathcal{S}$-stable for $i = 2, 3, \cdots, n$.

(vi) To the best of the authors' knowledge, there are no known general conditions under which a general nonlinear plant is stabilizable by a compensator, incr. $\mathcal{S}$-stable or not.

Proof:

Since by assumption F is incr. $\mathcal{S}$-stable, by Theorem 3 we have ${}^1S(\overline{P}_1,C+F)$ and ${}^1S(\overline{P}_2,C+F)$ are $\mathcal{S}$-stable

$\Leftrightarrow$

${}^1S(P_1,C)$ and ${}^1S(P_2,C)$ are $\mathcal{S}$-stable.

Since $P_1$ is incr. $\mathcal{S}$-stable and $Q := C(I+P_1C)^{-1}$, by Theorem 1 we have

$$^{1}S(P_1,C) \text{ is } \mathcal{S}\text{-stable} \Leftrightarrow Q \text{ is } \mathcal{S}\text{-stable.}$$

Using Corollary 3.1 with P replaced by $P_2-P_1$, F replaced by $P_1$, $\bar{C}$ replaced by C, and C replaced by Q, we conclude that

$$^{1}S(P_2,C) \text{ is } \mathcal{S}\text{-stable} \Leftrightarrow {}^{1}S(P_2-P_1,Q) \text{ is } \mathcal{S}\text{-stable.}$$

The assertion follows. ¤

## VI. SUMMARY

In this paper, we introduce a generalized concept of stability: $\mathcal{S}$-stability and incremental $\mathcal{S}$-stability, both applicable to nonlinear systems. Theorem 1 generalizes to the nonlinear case the Q-parametrization results established by Zames (1981). Theorem 2 extends Theorem 1 to include unstable plants. Finally, in Theorem 4, we give a necessary and sufficient condition for the existence of a fixed compensator that stabilizes two given nonlinear plants. It is surprising that these three theorems generalize the linear theory to the nonlinear case and the general formulas of the theory are almost unchanged in form.

Acknowledgement

Research sponsored by National Science Foundation Grant ECS-8119763.

## References

Anantharam, V. and Desoer, C. A., 1982, Proc. 21th IEEE Conf. on Decision and Control, 1199 (to appear IEEE Trans. AC, 1984).
Desoer, C. A. and Chen, M. J., 1981, IEEE Trans. Autom Control, 26, 408.
Desoer, C. A. and Liu, R. W., 1981, System and Control Letters, 1, 249.
Desoer, C. A. and Lin, C. A., 1983a, Ibid., 3, 41.

Desoer, C. A. and Lin, C. A., 1983b, Proc. 1983 Symposium on Circuits and Systems, Newport Beach, California.

Desoer, C. A. and Vidyasagar, M., 1975, Feedback Systems: Input-Output Properties (New York: Academic Press).

Gustafson, C. L. and Desoer, C. A., 1983, Int. J. Control, 37, 881.

Saeks, R. and Murray, J., 1982, IEEE Trans. Autom. Control, 26, 895.

Safonov, M. G., 1980, Stability and Robustness of Multivariable Feedback Systems (Cambridge, MA: MIT Press).

Vidyasagar, M. and Viswanadham, N., 1982, IEEE Trans. Autom. Control, 27, 1085.

Vidyasagar, M., 1978, Nonlinear System Analysis (Englewood Cliffs: Prentice-Hall).

Willems, J. C., 1971, The Analysis of Feedback Systems, (Cambridge, MA: MIT Press).

Zames, G., 1981, IEEE Trans. Autom. Control, 26, 301.

# DECENTRALIZED STABILIZATION OF LARGE-SCALE INTERCONNECTED SYSTEMS

J.C. WILLEMS*
Mathematics Institute
University of Groningen
P.O. Box 800
9700 AV Groningen
The Netherlands

MASAO IKEDA
Department of Systems Engineering
Kobe University
Rokko - Doi, Nada
Kobe 657
Japan

ABSTRACT

In this paper we study the stabilization of interconnected systems by means of decentralized feedback, i.e., in which the control on a given subsystem uses only the measurement on that subsystem. The results obtained use the theory of almost invariant subspaces.

INTRODUCTION

Consider the following interconnected system:

$$\Sigma_i : \dot{x}_i = A_i x_i + B_i u_i + \sum_j A_{ij} x_j \; ; \quad y_i = C_i x_i \tag{1}$$

$i = 1,2\ldots,N$. Here $\Sigma_i$ denotes the i-th subsystem with state $x_i \in \mathbb{R}^{n_i}$, control input $u_i \in \mathbb{R}^{m_i}$, and measured output $y_i \in \mathbb{R}^{p_i}$. The systems $\Sigma_i$ are interconnected through the matrices $A_{ij}$. It is possible to write these equations in a form in which the interconnected structure becomes more apparent. Define therefore $G_i$ and $H_i$ such that $\operatorname{im} G_i = \sum_j \operatorname{im} A_{ij}$ and $\ker H_i = \bigcap_j \ker A_{ji}$. The $A_{ij}$'s may then be written as $A_{ij} = G_i E_{ij} H_j$ for suitable matrices $E_{ij}$. The equations for $\Sigma_i$ become

$$\Sigma_i : \dot{x}_i = A_i x_i + B_i x_i + \sum_j G_i E_{ij} H_j x_j \; ; \quad y_i = C_i x_i \tag{1$'$}$$

These equations can be viewed as the systems:

$$\Sigma_i' : \dot{x}_i = A_i x_i + B_i u_i + G_i v_i \; ; \; y_i = C_i x \; ; \; z_i = H_i x_i \tag{2}$$

(with $u_i$ the control input, $v_i$ the interconnecting input, $y_i$ the measured output, and $z_i$ the interconnecting output), interconnected by the interconnection constraint

$$\Sigma'' : v_i = \sum_j E_{ij} z_j, \quad i = 1,2,\ldots,N \tag{3}$$

---

* The research leading to this paper was made possible by a short term visitor grant from the Japan Society for the Promotion of Science.

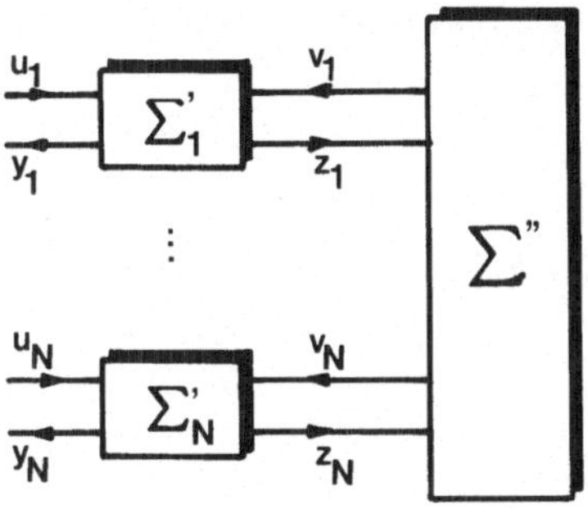


FIGURE 1

We think of a feedback from $y_i$ to $u_i$ as a *local* or *decentralized* feedback control law. This in contradistinction with a feedback from $y_1, y_2, \ldots, y_N$ to $u_i$ which we think of as a *centralized* feedback control law.

The problem addressed in this paper is the following: *Give conditions for the existence of a decentralized control law such that the overall closed loop system is asymptotically stable.*

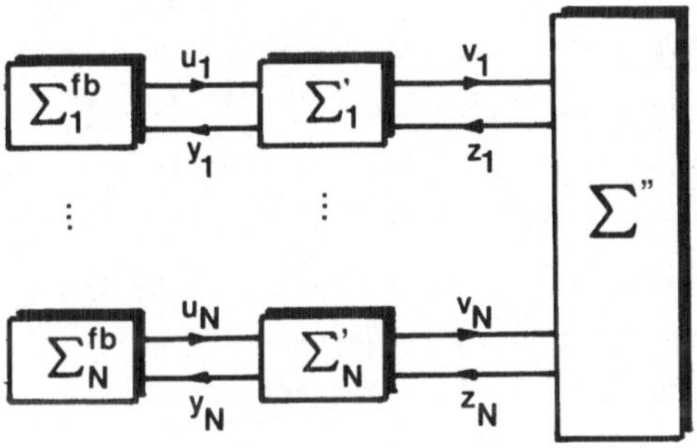


FIGURE 2

The problem of stabilizing interconnected systems has been studied extensively using decentralized *state* feedback, and a number of stabilizability conditions were reported (e.g., [1], [2], [3], and their references). These results, however, do not automatically imply the existence of a decentralized output feedback control law based on observer theory. This is because an observer constructed for an isolated subsystem does not necessarily estimate the state of the subsystem when interconnected with other subsystems, and usually observers for the overall system cannot be obtained in a decentralized form [4].

For this reason, the attempts to use output dynamic feedback for decentralized stabilization have started only recently [5], [6], [7]. The purpose of the present paper is to unify and extend these results using the concept of almost invariant subspaces introduced also recently [8, 9].

A STABILITY LEMMA

Assume that the subsystem $\Sigma_i$ is locally controlled by means of the feedback controller:

$$\Sigma_i^{fb} : \dot{w}_i = K_i w_i + L_i y_i \; ; \; u_i = M_i w_i + N_i y_i \tag{4}$$

Let $$\Sigma_i^{cl} : \begin{bmatrix} \dot{x}_i \\ \dot{w}_i \end{bmatrix} = \begin{bmatrix} A_i + B_i N_i C_i & B_i M_i \\ L_1 C_i & K_i \end{bmatrix} \begin{bmatrix} x_i \\ w_i \end{bmatrix} + \begin{bmatrix} G_i \\ 0 \end{bmatrix} v_i \; ; \; z_i = [H_i \; 0] \begin{bmatrix} x_i \\ w_i \end{bmatrix} \tag{5}$$

denote the i-th closed loop system. This may be written compactly as:

$$\Sigma_i^{cl} : \dot{x}_i^e = A_i^e x_i^e + G_i^e v_i \; ; \; z_i = H_i^e x_i^e \tag{5'}$$

with $x_i^e$, $A_i^e$, $G_i^e$ , and $H_i^e$ defined in the obvious way.

Using the control law (4) yields the following closed loop interconnected system:

$$\Sigma^{cl} : \begin{array}{l} \dot{x}_i = (A_i + B_i N_i C_i) x_i + B_i M_i w_i + G_i \sum_j E_{ij} H_j x_j \\ \dot{w}_i = K_i w_i + L_i C_i x_i , \; i = 1,2,\ldots,N \end{array} \tag{6}$$

This can be written compactly as:

$$\Sigma^{cl} : x_i^e = A_i^e x_i^e + G_i^e \sum_j E_{ij} H_j^e x_j^e , \; i = 1,2,\ldots,N \tag{6'}$$

Let $\sigma(M)$ denote the spectrum, i.e. the set of eigenvalues of the square matrix M, and let Re $\sigma(M) < 0$ signify that all its elements have negative real part. Further let $W_i^e$ denote the impulse response of $\Sigma_i^{cl}$, i.e. $\underline{W}_i^e : t \in \mathbb{R}^+ \rightarrow H_i^e(\exp A_i^e t) G_i^e$. Define $\|\underline{W}_i^e\|_{\mathcal{L}(0,\infty)} := \int_0^\infty \|\underline{W}_i^e(t)\| dt$ (matrix norms are always assumed to be induced norms in this paper). We have:

<u>Lemma 1</u> : Assume that the systems (1) are controlled by the decentralized control laws (4). Then the resulting closed loop system (6) is asymptotically stable if the systems (5) satisfy the following conditions:

(i) Re $\Sigma(A_i^e) < 0$

and (ii) $\|\underline{W}_i^e\|_{\mathcal{L}(0,\infty)} \sum_j \|E_{ij}\| < 1$

for i = 1,2,...,N.

This lemma is an immediate consequence of the 'small gain' stability theorem applied to large scale interconnected systems.

As an immediate consequence of Lemma 1, we have:

<u>Corollary 2</u>: Assume that the systems (2) have the property that for all $\varepsilon > 0$ there exists a feedback controller of the type (4) such that the systems (5) satisfy $\|\underline{W}_i^e\|_{\mathcal{L}(0,\infty)} \leq \varepsilon$. Then the interconnected system (1) can be made asymptotically stable by means of a decentralized feedback control law of the type (4).

ALMOST INVARIANT SUBSPACES

The property of the systems $\Sigma_i'$ required in Corollary 2 ($\|\underline{w}_i^e\|_{\mathcal{L}(0,\infty)} \leq \varepsilon$) is one of the basic questions addressed in the theory of almost invariant subspaces [8, 9]: the problem of *approximate disturbance decoupling* (ADDPM). We will now give the relevant results from there.

Consider the finite dimensional linear time invariant system

$$\Sigma : \dot{x} = Ax + Bu, \; y = Hx$$

with state space $X = \mathbb{R}^n$. Define

$$V^{+}_{b,\ker H} := V^{+}_{\ker H} + R^{*}_{b,\ker H}$$

Here, $V^{+}_{\ker H}$ denotes the supremal asymptotically stable output nulling subspace, formally defined as:

$$V^{+}_{\ker H} := \{x_o \in X \mid \exists \; \underline{x} : \mathbb{R}^* \to X, \; \underline{x} \text{ absolutely continuous, such that } \dot{\underline{x}}(t) - A\underline{x}(t) \in \text{im } B \text{ for almost all } t; \; \underline{x}(0) = x_o, \; \lim_{t\to\infty} \underline{x}(t) = 0, \text{ and } H\underline{x}(t) = 0 \;\; \forall t\}$$

The space $R^{*}_{b,\ker H}$ denotes the largest $\mathcal{L}_1$-almost controllability output nulling subspace defined by

$$R^{*}_{b,\ker H} := \{x_o \in X \mid \exists \; T > 0 \text{ such that, } \forall \varepsilon > 0, \exists \; \underline{x} : \mathbb{R}^+ \to X, \; \underline{x} \text{ absolutely continuous, such that } \dot{\underline{x}}(t) - A\underline{x}(t) \in \text{im } B \text{ for almost all } t; \; \underline{x}(0) = x_o, \; \underline{x}(T) = 0, \text{ and } \int_0^T \|H\underline{x}(t)\| dt \leq \varepsilon\}$$

These subspaces are readily calculated from A,B and H. Indeed, $V^{+}_{\ker H}$ may be calculated using the invariant subspace algorithm (ISA) [10, Section 5.6]. Further, $R^{*}_{b,\ker H}$ may be computed by the almost controllability subspace algorithm (ACSA):

$$R^{k+1}_{\ker H} = \text{im } B + A(\ker H \cap R^{k}_{\ker H}); \; R^{o}_{\ker H} = \{0\} \tag{7}$$

Then

$$R^{n}_{\ker H} = R^{*}_{b,\ker H}.$$

From [8] (see alsO [11]) we have the following result:

Lemma 3: Let (A,B,G,H) be given. Then for all $\varepsilon < 0$ there exists F such that $\text{Re } \sigma(A + BF) < 0$ and $\int_0^\infty \|H \, e^{(A+BF)t} \, G\| \, dt \leq \varepsilon$ if

(i) the pair (A,B) is stabilizable

and (ii) $\text{im } G \subset V^{+}_{b,\ker H}$

The above lemma will give us stabilizability conditions by means of decentralized *state* feedback. In order to obtain conditions for decentralized output feedback we will introduce the dual of $V^{+}_{b,\ker H}$. Intrinsic definitions for these duals are given in [9], but for brevity we will introduce these subspaces here by pure duality. Thus

$$S^{+}_{\mathrm{im}\,G}\ \binom{\text{computed re-}}{\text{lative to (A,C)}} : = (V^{+}_{\ker G^T})^{\perp}\ \binom{\text{computed re-}}{\text{lative to }(A^T,C^T)}$$

$$N^{*}_{b,\mathrm{im}\,G}\ \binom{\text{computed re-}}{\text{lative to (A,C)}} : = (R^{*}_{b,\ker G^T})^{\perp}\binom{\text{computed re-}}{\text{lative to }(A^T,C^T)}$$

and finally

$$S^{+}_{b,\mathrm{im}\,G}\ \binom{\text{computed re-}}{\text{lative to (A,C)}} : = (V^{*}_{b,\ker G^T})^{\perp}\binom{\text{computed re-}}{\text{lative to }(A^T,C^T)}$$

$$= S^{+}_{\mathrm{im}G} \cap N^{*}_{b,\mathrm{im}\,G}$$

Consider now the linear system:

$$\Sigma : \dot{x} = Ax + Bu + Gd \ ; \quad y = Cx \tag{8}$$

and an observer for it:

$$\Sigma^{w} : \dot{w} = Kw + Ly + Pu \ ; \ \hat{x} = Mw + Ny \tag{9}$$

The equation for the error $e := x - \hat{x}$ is given by

$$\begin{bmatrix} \dot{x} \\ \dot{w} \end{bmatrix} = \begin{bmatrix} A & 0 \\ LC & K \end{bmatrix}\begin{bmatrix} x \\ w \end{bmatrix} + \begin{bmatrix} G \\ 0 \end{bmatrix} d + \begin{bmatrix} B \\ P \end{bmatrix} u \ ; \quad e = \begin{bmatrix} I - NC \ \vdots \ -M \end{bmatrix}\begin{bmatrix} x \\ w \end{bmatrix} \tag{10}$$

This may be denoted compactly by

$$\dot{x}^{e} = A^{e}x^{e} + G^{e}d + B^{e}u \ ; \quad e = C^{e}x^{e} \tag{10'}$$

From [9] we obtain the following result:

Lemma 4: Let $\Sigma \approx (A,B,G,C)$ be given. Then for all $\varepsilon > 0$ there exists $\Sigma^{w} \approx (K,L,M,N)$ such that in (10)

1. $\int_0^\infty \| C^{e}(\exp A^{e}t)\, G^{e} \|\, dt \leq \varepsilon$, and $C^{e}(\exp A^{e}t)B^{e} = 0$

and

2. when $\underline{d} = \underline{0}$, then, for all $\underline{x}^{e}(0)$, $\lim_{t\to\infty} \underline{e}(t) = 0$

if (i) the pair (A,C) is detectable

and

(ii) $S^{+}_{b,\mathrm{im}\,G} = \{0\}$ (this subspace should be computed relative to (A,C))

Remark: It is possible to express the condition $S^{+}_{b,\mathrm{im}\,G} = \{0\}$ in terms of the system matrix $\Sigma$ of (10). In fact [12]

$$\{S^{+}_{b,\mathrm{im}\,G} = \{0\}\} \Leftrightarrow \{ \begin{bmatrix} Is-A & G \\ C & 0 \end{bmatrix} \text{ has full column rank for Re } s \geq 0 \}$$

This relation implies that condition (i) in Lemma 4 holds whenever condition (ii) is satisfied.

Consider now the system

$$\Sigma : x = Ax + Bu + Gd \ ; \ y = Cx \ ; \ z = Hx \tag{11}$$

and the following feedback around it :

$$\Sigma^{fb} : \dot{w} = Kw + Ly \; ; \quad u = Mw + Ny \tag{12}$$

This yields the closed loop system

$$\Sigma^{cl} : \begin{bmatrix} \dot{x} \\ \dot{w} \end{bmatrix} = \begin{bmatrix} A+BNC & BM \\ LC & K \end{bmatrix} \begin{bmatrix} x \\ w \end{bmatrix} + \begin{bmatrix} G \\ 0 \end{bmatrix} d \qquad z = \begin{bmatrix} H & 0 \end{bmatrix} \begin{bmatrix} x \\ w \end{bmatrix} \tag{13}$$

This may be written compactly as

$$\Sigma^{cl} : \dot{x}^e = A^e x^e + G^e d \; ; \; z = H^e x^e \tag{13'}$$

Combination of lemmas 3 and 4 yields:

Lemma 5 : Consider the plant (11). Then for all $\varepsilon > 0$, there exists a feedback processor of the form (12) such that the closed loop system (13)' has the properties

1. $\mathrm{Re}\; \sigma(A^e) < 0$

and 2. $\int_0^\infty \|H^e(\exp A^e t)G^e\| \, dt \leq \varepsilon$

if (i) $(A,B)$ is stabilizable

(ii) $(A,C)$ is detectable

(iii) $\mathrm{im}\; G \subset V^+_{b,\ker H}$

(iv) $S^+_{b,\mathrm{im}\, G} = \{0\}$

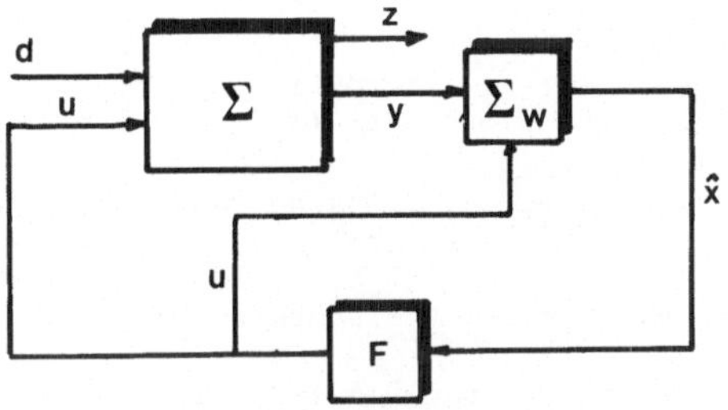


FIGURE 3

Proof: Using (ii) and (iv) yields, be Lemma 4, that there exists an observer with the proporties expressed by Lemma 4. Using (i) and (iii) yields, by Lemma 3, that there exists a feedback matrix F with the properties expressed by Lemma 3. Now use the feedback law $u = F\hat{x}$, with $\hat{x}$ generated as in Lemma 4. Obviously this yields a feedback compensator as (12). We claim that it has the properties 1 and 2 of Lemma 5.

That $\mathrm{Re}\; \sigma(A^e) < o$ follows from the standard controller/observer pole separation of linear systems.

Property (ii) may be shown as follows. The closed loop dynamics of x may be written as

$$\dot{x} = (A + BF)x - BF(x - \hat{x}) + Gd; \; z = Hx$$

Hence, with $\underline{x}(0) = 0$,

$$\underline{z}(t) = \int_0^t H(\exp(A+BF)(t-\tau))G\,\underline{d}(\tau)d\tau - \int_0^t H(\exp(A+BF)(t-\tau)BF(\underline{x}(\tau) - \hat{x}(\tau))d\tau$$

Denote $\underline{W}_e(t) := H(\exp(A + BF)t)G$, and let $\underline{W}_w$ be the impulse response (from d to e) of (10). Now, by (iii) and Lemma 3 there exists, $\forall\varepsilon > 0$, an F such that $\|\underline{W}_c\|_{\mathcal{L}(0,\infty)} \leq \frac{\varepsilon}{2}$. Define $q := \int_0^\infty \|H(\exp(A+BF)t)BF\|\,dt$.

By (iv) and Lemma 4, there exists an observer of the form (9) such that $\|\underline{W}_w\|_{\mathcal{L}(0,\infty)} \leq \varepsilon/2q$. Consequently the impulse response from d to z in (13) will then satisty

$$\int_0^\infty \|H^e(\exp A^e t)G^e\|dt \leq \|\underline{W}_c\|_{\mathcal{L}(0,\infty)} + q\|\underline{W}_w\|_{\mathcal{L}(0,\infty)} \leq \varepsilon$$

This yields the desired result.

□

## STABILIZATION OF INTERCONNECTED SYSTEMS

We now return to the question posed in the introduction. The combination of Corollary 2 and Lemma 5 yields immediately the following results which is the main conclusion of this paper:

Theorem 6: *There exists a decentralized feedback control law* (4) *for the systems* $\Sigma_i$ *of* (1) *such that the closed loop system* (1) *is asymptotically stable if for all* $i = 1,2,\ldots, N$.

(i) $(A_i,B_i)$ *is stabilizable*

(ii) $(A_i,C_i)$ *is detectable*

(iii) $\text{im } G_i \subset \mathcal{V}^+_{b,\ker H_i}$ (*computed relative to* $(A_i,B_i)$)

and (iv) $\mathcal{S}_{b,\text{im}G_i} = \{0\}$ (*computed relative to* $(A_i,C_i)$)

Specializing this to systems in which the state is measured yields:

Corollary 7: Assume $y_i = x_i$ in (1) for $i = 1,2,\ldots, N$.
Then there exists a decentralized state feedback control law $u_i = F_i x_i$ for the systems (1) such that the resulting closed loop system is asymptotically stable if for all $i = 1,2,\ldots, N$

(i) $(A_i,B_i)$ is stabilizable

and (ii) $\text{im } G_i \subset \mathcal{V}^+_{b,\ker H_i}$

An interesting observation, which is easily seen from (7), is that $\text{im } B \subset \mathcal{V}^+_{b,\ker H}$ (and dually $\mathcal{S}^+_{b,\text{im } G} \subset \ker C$). This yields various obvious corollaries of the above results. We state one explicitely:

Corollary 8: There exists a decentralized feedback control law (4) for the systems $\Sigma_i$ of (1) such that the closed loop system (6) is asymptotically stable if for all $i = 1,2,\ldots, N$

(i) $(A_i, B_i)$ is stabilizable

(ii) $(A_i, C_i)$ is detectable

(iii) $\mathrm{im}\, G_i \subset \mathrm{im}\, B_i$

and (iv) $S^+_{b,\, \mathrm{im}\, G_i} = \{0\}$

One final result worth stating is the dual of Theorem 6 which can be obtained by considering

$$\Sigma^*_i : \dot{x}^*_i = A^T_i x^*_i + C^T_i x^*_i + H^T_i v^*_i$$

$$y^*_i = B^T_i x^*_i \ ; \ z^*_i = G^T_i v^*_i$$

and using the control laws

$$\Sigma^{*fb}_i : \begin{array}{l} \dot{w}^*_i = K^T_i w^*_i + M^T_i y^*_i \\ u^*_i = L^T_i w^*_i + F^T_i y^*_i \end{array}$$

After applying Theorem 7 on these systems, computing $(K^T_i, L^T_i, M^T_i, F^T_i)$, and deducing control systems (4) from there , we obtain an asymptotically stable system (6). Stated formally:

Theorem 9: *There exists a decentralized feedback control law (4) for the system* $\Sigma_i$ *of (1) such that the closed loop system (6) is asymptotically stable of for any given* $i = 1,2,\ldots, N$, *either the conditions of Theorem* 6 *are satisfied, or alternatively,*

(i) $(A_i B_i)$ *is stabilizable*

(ii) $(A_i, C_i)$ *is detectable*

(iii) $V^+_{b,\ker H_i} = X$ (*computed relative to* $(A_i, B_i)$)

*and* (iv) $S^+_{b,\mathrm{im}\, G_i} \subset \ker H_i$ (*computed relative to* $(A_i C_i)$)

Of course, the most desirable situation would be to have a theorem, as Theorem 6, which is self-dual. However, this requires a self-dual version of Lemma 5, i.e. the solution of the almost disturbance decoupling problem by measurement feedback and with internal stability (ADDPMS). This, unfortunately, remains an open problem [4].

Remarks 1. It is easy to obtain further refinements of Theorem 6 or 9 to situations where the interconnections are nonlinear, time-varying and/or dynamic. In fact, the results remain true (with asymptotic stability suitably defined) if in (1)' the $E_{ij}$'s are arbitrary input/output bounded systems.

2. It should be noted that the stabilizability condition provided in [5] can readily be derived from Corollary 8 and its dual using the Remark given below Lemma 4. The result of [6] is also included in our results. This can be

understood by the fact that we utilized an *almost* disturbance decoupled observer for the subsystems in which we have the interconnecting inputs as disturbances, while *perfect* disturbance decoupled observers were used in [6].

REFERENCES


[1] E.J. Davison, "The Decentralized Stabilization and Control of a Class of Unknown Nonlinear Time-Varying Systems", *Automatica*, Vol. 10, 1974, pp. 309-316.

[2] M. Ikeda and D.D. Šiljak, "When is a Linear Decentralized Control Optimal?", *Proc. of the Fifth International Conference on Analysis and Optimization of Systems*, New York, Springer Verlag, 1982, pp. 419-431.

[3] M. Ikeda, D.D. Šiljak, and K. Yasuda, "Optimality of Decentralized Control for Large-Scale Systems", *Automatica*, Vol. 19, 1983, pp. 309-316.

[4] D.D. Šiljak and M.B. Vukčević, " On Decentralized Estimation", *Internat. J. Control*, Vol. 27, 1978, pp. 113-131.

[5] Ö. Hüseyin, M.E. Sezer, and D.D. Šiljak, "Robust Decentralized Control Using Output Feedback", *IEEE Proc.*, Vol. 129, Pt. D, 1982, pp. 310-314.

[6] N. Viswanadham and A. Ramakrishna, "Decentralized Estimation and Control for Interconnected Systems", *Large Scale Systems*, Vol. 3, 1982, pp. 255-266.

[7] K. Okuda and M. Ikeda, "Stabilization of Large-Scale Systems Using Dynamic Controllers", *Proc. of the Twelfth SICE Symposium on Control Theory*, 1983, pp. 69-72 (In Japanese).

[8] J.C. Willems, "Almost Invariant Subspace: An Approach to High Feedback Design - Part I: Almost Controlled Invariant Subspaces", *IEEE Trans. on Automatic Control*, Vol. AC-26, 1981, pp. 235-252.

[9] J.C. Willems, "Almost Invariant Subspaces: An Approach to High Gain Feedback Design - Part II: Almost Conditionally Invariant Subspaces", *IEEE Trans. on Automatic Control*, Vol. AC-27, 1982, pp. 1071-1085.

[10] W.M. Wonham, Linear Multivariable Control: A Geometric Approach , New York: Springer Verlag, 1979.

[11] J.L. Willems and J.C. Willems, "Robust Stabilization of Uncertain Systems", *SIAM J. Control and Optimization*, Vol. 21, No. 3, 1983, pp. 352-374.

[12] M.L.J. Hautus and L.M. Silverman, "System Structure and Singular Control", *Linear Algebra and Its Applications*, Vol. 50, 1983, pp. 369-402.

# DEFINITION D'UNE METHODOLOGIE DE CONCEPTION ASSISTEE D'ASSERVISSEMENTS NON LINEAIRES CONTINUS PAR L'UTILISATION DE TECHNIQUES D'AGREGATION PAR NORMES VECTORIELLES

Dominique MEIZEL et Jean-Claude GENTINA (*)

* Laboratoire d'Automatique et d'Informatique Industrielle
Institut Industriel du Nord (I. D. N.)
B.P. 48
59651 VILLENEUVE D'ASCQ CEDEX - FRANCE

## ABSTRACT

We propose to derive a control-system-design-methodology valuable for non-linear continuous-time processes from a stability-analysis-technique based upon the use of vectorial norms (2) (3). The design problem is considered from the stability viewpoint and the design methodology is parameter-optimization. A major-difficulty in non-linear system-stability studies lies in the handling of **sufficient** stability conditions that can be very conservative with respect to the intrinsic stability properties of the system. A key-point of the proposed method consists in the definition of a scalar criterion the negativity of which implies asymptotic stability of the studied control system. The design-method is thus achieved by minimizing the value of this criterion. The arguments of the stability-criterion consists of both the adjustable parameters and a representation basis in the state-space. The values of former arguments are the real synthesis-objective whereas the latter ones are artificial and are used in order to obtain the least conservative stability conditions as possible with respect to the choosen stability-analysis-theorem (2). The minimization of the criterion can be helped by the computation of a steepest-descent direction deduced from the simple-eigenvalues sensitivity theory. We propose then to implement this design-methodology by the use of a special class of state-space representations of the studied systems. A numerical example illustrates this last design-method.

## I - INTRODUCTION

Cette communication propose la définition d'une méthode de synthèse d'asservissements de processus continus non linéaires basée sur l'analyse des propriétés d'asservissements non linéaires par la technique des normes vectorielles (1) (2) (3) (4). L'esprit de la méthode proposée est celui de la synthèse paramétrique (5) (6) (7). On suppose ainsi que la structure (14) de la loi de commande a pu être choisie, et que l'effort de synthèse vise à déterminer la valeur des gains ajustables dans cette structure de manière à stabiliser l'asservissement. L'utilisation de la stabilité appliquée aux processus non-linéaires, pose une difficulté essentielle. En effet, si l'existence d'une fonction de Lyapunov décroissante constitue une condition nécessaire et suffisante de stabilité, la mise en évidence de cette fonction est, en général, pratiquement impossible.

Nous sommes ainsi amenés à utiliser des méthodes d'analyse, d'application plus pratique mais moins générale. On est alors conduit à considérer des conditions **suffisantes** de stabilité. Une méthode de synthèse applicable aux processus non-linéaires doit donc à la fois optimiser les gains ajustables et la méthode d'analyse des propriétés de stabilité de l'asservissement dans le sens de conditions suffisantes de stabilité les moins restrictives possibles. Cette optimisation n'est naturellement envisageable qu'à l'intérieur d'une classe cohérente de méthodes d'analyse. Nous particularisons ici les méthodes déduites de l'utilisation des normes vectorielles [1] [2] [3] [4].

Après avoir défini la classe de processus étudiés dans ce travail, nous rappelons la philosophie des techniques d'agrégation par normes vectorielles ainsi qu'un théorème de stabilité qui s'en déduit. Nous proposons alors d'assimiler le choix judicieux d'une norme vectorielle à celui d'une base de représentation du système étudié.

Ayant de cette manière formulé le problème de synthèse comme la minimisation d'une fonction scalaire, nous proposons une forme remarquable de représentation des systèmes étudiés dite "forme série" qui facilite cette minimisation.

Un exemple illustre la technique proposée.

## II - DEFINITION DES ASSERVISSEMENTS ETUDIES

Nous considérons à priori un problème pour lequel l'objectif de synthèse a pu être ramené à caractériser la stabilité asymptotique de l'origine ($x = 0$) d'un système décrit par l'équation d'état suivante :

$$\overset{\circ}{x}(t) = F(\chi,p) \, . \, x(t) \tag{1}$$

avec : $t \in \mathcal{T} = [t_o, +\infty[$

$x \in \mathcal{S} \subset \mathbb{R}^n$

$\chi \in \mathrm{X}$ représente l'ensemble des arguments des composantes de $F(\chi,p)$ non nécessairement constantes le long des trajectoires possibles du système (1)

$p \in \mathbb{R}^{n_p}$ comporte l'ensemble des gains ajustables dans l'asservissement

On supposera dans ce travail que les éléments non constants de la matrice de régime libre $F(\chi,p)$ peuvent être ramenés par un changement de base dans sa première ligne ou sa première colonne.

$$v_1 \in \mathbb{R}^n \; ; \; v_1^T \triangleq [1, 0, \dots, 0] \; ; \; f(.) : \mathbb{R}^{n_p} \times \mathrm{X} \to \mathbb{R}^n \tag{2}$$

$$F(\chi,p) = F_o(p) + v_1 \, . \, f^T(\chi,p) \tag{2.a}$$

ou

$$F(\chi,p) = F_o(p) + f(\chi,p) \, . \, v_1^T \tag{2.b}$$

L'analyse de ce système par utilisation de normes vectorielles consiste essentiellement : - à choisir à priori une norme vectorielle [2] [3] :

$$q(x) : \mathbb{R}^n \to \mathbb{R}_+^k \tag{3}$$

du vecteur état,

- à former, à partir de q (3) et du système étudié (1), un système de comparaison (4) :

$$\begin{cases} \frac{d}{dt^+}(z(t)) = M(q,\chi,p) \cdot z(t) & (\frac{d}{dt^+} \text{ note la dérivée à droite par rapport au temps}) \\ z(t) \in \mathbb{R}_+^k \end{cases} \quad (4)$$

où les éléments hors diagonaux de la matrice pseudo-majorante [4] $M_{qq}(.)$ sont non-négatifs. La stabilité asymptotique du système de comparaison (4) induit alors la même propriété sur le système initial (1).

La mise en évidence de la stabilité du système de comparaison (lui-même non-linéaire) peut être envisagée à partir du théorème suivant [2] démontré à partir des propriétés des matrices à éléments non négatifs [8].

*THEOREME B - G* [2]

*Considérons un système non linéaire :*

$$\begin{cases} \overset{\circ}{x} = F(\chi,p) \cdot x(t) \\ x(t) \in S \subset \mathbb{R}^n \\ \chi \in X \quad (\text{cf } (1)) \end{cases} \quad (5)$$

*S'il existe une norme vectorielle* q $(\mathbb{R}^n \to \mathbb{R}_+^k)$ *conduisant à la définition d'une matrice majorante* $M(q,\chi,p) \in \mathbb{R}^{k\times k}$ *satisfaisant les propriétés suivantes :*

*i) les éléments non constants de* $M(q,\chi,p)$ *sont isolés dans une seule ligne ou une seule colonne,*

*ii)* $\forall\ \chi \in X$, $M(q,\chi,p)$ *est irréductible,*

*iii)* $\exists\ \varepsilon < 0$ *tels que les mineurs principaux de* $(\varepsilon\, I_k - M(q,\chi,p))$ *soient non négatifs,*

*alors le point d'équilibre* $(x = 0)$ *du système est globalement exponentiellement stable et il existe une fonction de Lyapunov* $v(x(t))$ *satisfaisant l'inégalité :*

$$\overset{\circ}{v}(x(t)) \leq \varepsilon \cdot v(x(t)) \quad (6)$$

*le long des trajectoires du système étudié* (5).

## III - DEFINITION D'UNE METHODE DE SYNTHESE A PARTIR DU THEOREME B - G

L'expression de ce théorème doit être transformée dans un esprit de synthèse. Il s'agit alors de trouver un réglage des composantes de p et une norme vectorielle q telle que l'on puisse conclure à la stabilité par l'utilisation du théorème précédent.

### III.1 - Définition d'un critère optimisable de stabilité asymptotique

Dans cet esprit, nous proposons de définir un critère de stabilité $\varepsilon(q,p)$ comme étant la borne inférieure des valeurs de $\varepsilon$ satisfaisant, pour un choix donné de q et p, à la contrainte (iii) du théorème B - G. On montre [2] que **$\varepsilon(q,p)$** peut être aussi **défini comme la borne supérieure des valeurs propres réelles maximales de l'ensemble des matrices** pseudo-majorantes $\{M(q,\chi,p)$ (4) ; $\chi \in \mathcal{X}\}$.

$$\begin{cases} \varepsilon(q,\chi,p) \triangleq \text{valeur propre réelle maximale de } M(q,\chi,p) \\ \varepsilon(q,p) \triangleq \sup\limits_{\chi \in \mathcal{X}} \{\varepsilon(q,\chi,p)\} \end{cases} \tag{7}$$

L'objectif de synthèse est alors ramené à déterminer une norme vectorielle q satisfaisant à la propriété structurelle (i) ainsi qu'une valeur du vecteur paramètre $p \in \mathbb{R}^{n_p}$ telles que la valeur correspondante du critère $\varepsilon(q,p)$ soit négative. La minimisation de $\varepsilon(q,p)$ par rapport à ses arguments q(.) et p constitue alors une méthodologie de synthèse appropriée à la satisfaction de cet objectif.

Le fait que l'un des arguments du critère soit une norme vectorielle q $(\mathbb{R}^n \to \mathbb{R}^k_+)$, donc un paramètre non-numérique, constitue à priori un obstacle à la mise en œuvre de cette procédure de synthèse. Pour résoudre cette difficulté, nous proposons d'envisager la classe suivante (8) de normes vectorielles déduite de l'utilisation conjointe d'un type standard de normes vectorielles et d'un changement de représentation du système (1) (2.a).

$$q_V : \mathbb{R}^n \to \mathbb{R}^n_+$$

$$\begin{cases} \forall i = 1, \dots, n \quad (q_V)_i(x) = |y_i| = (\bar{y}_i \cdot y_i)^{1/2} \\ \forall x \in \mathbb{R}^n \quad \text{avec : } y \triangleq V^{-1} . x \\ \qquad V \overset{\Delta}{\in} \mathbb{C}^{n\times n} \; ; \; \det(V) \neq 0 \end{cases} \tag{8}$$

$$V \triangleq \left[\begin{array}{c|c} v_{11} & v_{12} \\ \hline 0 & \\ \vdots & V_{22} \\ 0 & \end{array}\right] \; ; \quad v_{11} \in \mathbb{R} \tag{9.a}$$

Remarques

R1) Le choix d'une matrice de changement de base dont la première colonne est colinéaire de $v_1$ (2.a) laisse les éléments non constants de $V^{-1}(F_o(p)+v_1.f^T(x,p))V$ isolés dans la première ligne de cette matrice.

R2) La matrice pseudo-majorante $M(q_v,x,p) \in \mathbb{R}^{n\times n}$ issue du système initial (1) (2.a) et d'une norme vectorielle $q_v(.)$ (8) peut être aisément construite selon la procédure suivante [3] :

- les éléments diagonaux de $M(q_v,x,p)$ sont égaux aux parties réelles des éléments diagonaux de $(V^{-1}.F(x,p).V)$,
- les éléments hors diagonaux de $M(q_v,x,p)$ sont égaux aux modules des éléments correspondants de $(V^{-1}.F(x,p).V)$.

R3) La contrainte (i) d'application du théorème B - G est naturellement satisfaite d'après les remarques précédentes (R1, R2).

R4) La prise en compte de systèmes dont les éléments non constants peuvent être isolés dans une seule colonne (2.b) s'effectue en transposant la matrice V (9.a) définissant les normes vectorielles envisagées (8).

R5) Si l'on considère le cas limite d'un système (1) linéaire stationnaire, il existe toujours un changement de base V (9.a) qui amène un système de comparaison (4) stable asymptotiquement [10] si et seulement si le système linéaire étudié possède cette propriété.

L'utilisation de la classe de normes vectorielles $q_V$ (8) définies à partir d'un changement de base V (9.a) nous permet de noter le critère $\varepsilon(q_V,p)$ (7) sous la forme suivante :

$$\varepsilon(V,p) \triangleq \varepsilon(q_V,p) \tag{10}$$

Les arguments de ce critère sont alors de type numérique et le processus de synthèse défini comme la minimisation de $\varepsilon(V,p)$ jusqu'à l'obtention d'une valeur négative peut alors être mis en œuvre. D'un point de vue pratique, le calcul de $\varepsilon(V,p)$ peut être mené par l'utilisation du lemme de Kotelyanskii [8] par la procédure suivante :

Procédure de calcul de $\varepsilon(V,p)$ (11)

1) Notons $\varepsilon_i(V,p)$ $(i = 2, \ldots, n)$ la valeur propre réelle maximale de la matrice constante obtenue en supprimant les i-1 premières lignes et colonnes de $M(q_V,x,p)$. On calcule $\varepsilon_2(V,p)$ en s'appuyant sur le lemme de Kotelyanskii et l'inégalité suivante qui ordonne les $\varepsilon_i(V,p)$ [8] :

$$\varepsilon_2(V,p) \geq \varepsilon_3(V,p) \geq \ldots\ldots \geq \varepsilon_n(V,p) \tag{12}$$

2) $\varepsilon(V,p) \geqq \varepsilon_2(V,p)$ est racine de l'équation suivante en $\varepsilon \in \mathbb{R}$ :

$$\inf_{x \in \chi} \{\det (\varepsilon I_n - M(q_V, x, p))\} = 0 \qquad (13)$$

ΔΔΔ

Nous proposons maintenant d'expliciter la tendance d'évolution du critère $\varepsilon(V,p)$ sous un nombre restreint d'hypothèses de manière à faciliter cette minimisation.

III.2 - Calcul de la tendance d'évolution de $\varepsilon(V,p)$

Soit $E$ l'espace réel des paramètres de $\varepsilon(V,p)$ constitué des parties réelles et imaginaires des composantes de V (9.a) ainsi que des composantes de p.

Supposons qu'il existe un nombre **fini** $\gamma$ de matrices $\{M(q_V,\hat{x}_i,p)\ ;\ i = 1, \ldots, \gamma\}$ dont $\varepsilon$ (V,p) soit valeur propre. Si ces matrices sont irréductibles $\varepsilon(V,p)$ est alors une valeur propre simple (8). Dans ce cas, on peut calculer la sensibilité des valeurs propres simples de ces matrices par application de la formule suivante (9) (14) :

Soit $\beta$ un argument de $\varepsilon(V,p)$ ($\beta \in E$) et $\delta\beta$ un petit accroissement de ce paramètre. Notons $\Delta_\beta M(q_V,\hat{x},p)$ et $\Delta_\beta \varepsilon(q_V,\hat{x},p)$ les variations d'une matrice pseudo-majorante $M(q_V,\hat{x},p)$ et de sa valeur propre réelle maximale $\varepsilon(V,\hat{x},p)$ correspondantes à l'accroissement $\delta\beta$ du paramètre $\beta$. On a, au premier ordre près (9) :

$$\delta_\beta\, \varepsilon(V,\hat{x},p) = v^T(V,\hat{x},p) \,.\, \Delta_\beta M(q_V,\hat{x},p) \,.\, u(V,\hat{x},p) \qquad (14)$$

où $u(V,\hat{x},p)$ et $v(V,\hat{x},p)$ désignent respectivement les vecteurs propres relatifs à $\varepsilon(V,p)$ de $M(q_V,\hat{x},p)$ et $M^T(q_V,\hat{x},p)$ liés par la relation $u^T(.) \,.\, v(.) = 1$.

ΔΔ

On peut toujours supposer les matrices $F(x,p)$ différentiables par rapport aux composantes de p. Les matrices $M(q_V,x,p)$ (cf R2) sont ainsi toujours différentiables (éventuellement à gauche ou à droite) par rapport aux arguments réels constituant $E$.

A partir des gradients (éventuellement directionnels) des $\gamma$ valeurs propres $\varepsilon(V,\hat{x}_i,p) \equiv \varepsilon(V,p)$ des matrices $M(q_V,\hat{x}_i,p)$, on peut :

- soit définir une direction de plus grande pente du critère $\varepsilon(V,p)$ dans l'espace $E$
- soit caractériser l'existence d'un point stationnaire.

En conclusion, nous avons défini une méthode de synthèse paramétrique basée sur l'utilisation du théorème B - G (4) en définissant un critère numérique $\varepsilon(V,p)$ dépendant à la fois des paramètres ajustables dans l'asservissement étudié ainsi que d'une base de représentation particulière du système étudié.

Le processus de synthèse est alors assimilé à la minimisation de ce critère visant à obtenir une valeur négative de ce dernier.

Le calcul pratique du critère est aisé (procédure (11)) et l'évaluation de sa direction de plus grande pente a pu être explicitée sous un nombre restreint d'hypothèses (§ III.2).

Dans la suite de cette communication, nous proposons de restreindre le choix des bases de représentation à celle conduisant à une forme remarquable [10] de représentation des systèmes étudiés (1) (2.a) et notée "forme série".

Ce mode de représentation est introduit à partir d'une représentation canonique à laquelle on peut ramener les systèmes étudiés et permet une implémentation aisée de la procédure de synthèse proposée.

IV - REPRESENTATION CANONIQUE DES SYSTEMES ETUDIES

Considérons le système décrit par l'équation d'état :

$$\overset{\circ}{x}(t) = (F(p) + v_1 \cdot f^T(\chi,p) \cdot x(t) \tag{15}$$

Si la paire $[F(p), v_1]$ est complètement contrôlable, on peut représenter la matrice de régime libre sous la forme suivante (16) en utilisant deux fois le schéma de l'algorithme de Danilevskii [13] :

$$\exists\, U(p) \in \mathbb{R}^{n\times n} \; ; \; \det(U(p)) \neq 0$$

$$\left\{\begin{array}{l} x' \triangleq U^{-1}(p) \cdot x \\ \overset{\circ}{x}'(t) = F'(\chi,p) \cdot x'(t) \\ F'(\chi,p) = \begin{bmatrix} -g_1(\chi,p) & -g_2(\chi,p) & \cdots & -g_n(\chi,p) \\ 1 & & & \\ & \ddots & & \\ & & 1 & 0 \end{bmatrix} \end{array}\right. \tag{16}$$

Vis à vis de cette forme, on peut définir un polynôme symbolique (17) homologue du polynôme minimal d'une matrice compagnon :

$$P(\lambda,\chi,p) = \lambda^n + g_1(\chi,p) \cdot \lambda^{n-1} + \ldots\ldots + g_n(\chi,p) \tag{17}$$

Dans le cas où $[F(p), v_1]$ n'est pas complètement contrôlable, on peut décomposer le système étudié sous la forme d'un système linéaire stationnaire en série avec un système non linéaire décrit sous forme canonique (16) [11].

Nous pouvons ainsi considérer les systèmes décrits sous forme canonique (16) comme objets de notre étude.

Nous proposons maintenant de définir la forme remarquable "série" de manière à faciliter la mise en œuvre de la procédure de synthèse proposée.

V - FORMES REMARQUABLES "SERIE" [10]

Considérons un vecteur c :

$$c \in \mathbb{C}^{n-1} \qquad c^T \in [\alpha_2, \ldots, \alpha_n] \tag{18}$$

La représentation série du système (16) de paramètre c (18) est alors la suivante :

$$x'' \in \mathbb{C}^n \qquad x'' \triangleq V^{-1}(c) \,.\, x'$$

$$\left\{ \begin{array}{l} \mathring{x}''(t) = S(c,\chi,p) \,.\, x''(t) \\ S(c,\chi,p) = \begin{bmatrix} -b_1(.) & -b_2(.) & \cdots & -b_{n-1}(.) & -b_n(.) \\ 1 & \alpha_2 & & & \\ & \ddots & \ddots & & 0 \\ & & \ddots & \alpha_{n-1} & \\ 0 & & & 1 & \alpha_n \end{bmatrix} \end{array} \right. \tag{19}$$

Les composantes $b_1(.)$ à $b_n(.)$ sont obtenues à partir des différences divisées du polynôme symbolique (17) le long des composantes de c.

Les différences divisées d'une fonction f ($\mathbb{C} \to \mathbb{C}^m$) de classe $C\infty$ le long des composantes de c sont définies de façon récurente selon les expressions suivantes (20) :

$$\left\{ \begin{array}{l} D(\alpha_n, f) = f(\alpha_n) \\ \forall\ i = n-1, \ldots, 2 : \\ D(\alpha_i, \alpha_{i+1}, \ldots, \alpha_n ; f) = \left\{ \begin{array}{ll} \dfrac{\partial}{\partial \alpha_{i+1}} (D(\alpha_{i+1}, \alpha_{i+2}, \ldots, \alpha_n ; f(.)) & \text{si} \quad \alpha_i = \alpha_{i+1} \\ \dfrac{D(\alpha_{i+1}, \alpha_{i+2}, \ldots, \alpha_n ; f(.)) - D(\alpha_i, \alpha_{i+2}, \ldots, \alpha_n ; f(.))}{\alpha_{i+1} - \alpha_i} & \text{si} \quad \alpha_i \neq \alpha_{i+1} \end{array} \right. \end{array} \right. \tag{20}$$

Δ

La première ligne de la matrice S(c,χ,p) (19) est alors exprimée par :

$$\begin{cases} b_n(\chi,p,c) = P(\alpha_n,\chi,p) = D(\alpha_n \; ; \; P(.,\chi,p)) \\ \vdots \\ b_i(\chi,p,c) = D(\alpha_i,\ \alpha_{i+1},\ \ldots,\ \alpha_n \; ; \; P(.,\chi,p)) \qquad i = 2, \ldots, n \\ \vdots \\ b_1(\chi,p,c) = a_1(\chi,p) + \sum_{i=2}^{n} \alpha_i \end{cases} \tag{21}$$

<u>Remarque</u>

Un exemple de cette transformation est donné dans l'exemple d'application (§ VI).
ΔΔ

Vis à vis de ce changement de base, la matrice pseudo-majorante M(c,χ,p) de S(c,χ,p) (19) considérée dans l'étude de stabilité est alors la suivante (cf R2).

$$M(c,\chi,p) = \begin{bmatrix} -\mathbb{R}(b_1(.)) & |b_2(.)| & & |b_n(.)| \\ 1 & \mathbb{R}(\alpha_2) & & 0 \\ & \ddots & \ddots & \\ 0 & & 1 & \mathbb{R}(\alpha_n) \end{bmatrix} \tag{22}$$

La borne supérieure ε(c,p) des valeurs propres réelles maximales ε(c,χ,p) de M(c,χ,p) est alors déterminée par la procédure suivante (cf. (12), (13)) :

$$\begin{cases} \det(\varepsilon I_n - M(c,\chi,p)) \triangleq -|b_n(.)| + (\varepsilon - \mathbb{R}(\alpha_n))(-|b_{n-1}(.)| + (\varepsilon - \mathbb{R}(\alpha_{n-1}))(-|b_{n-2}(.)| + \\ \qquad + \ldots + (\varepsilon - \mathbb{R}(\alpha_3))(-|b_2(.)| + (\varepsilon - \mathbb{R}(\alpha_2))(\varepsilon + \mathbb{R}(b_1(.))) \underbrace{\ldots)}_{n \text{ fois}} \\ \varepsilon_2 = \max_{i=2,\ldots,n} (\mathbb{R}(\alpha_i)) \\ \varepsilon(c,p) \text{ est définie par } \begin{cases} \varepsilon(c,p) \geqq \varepsilon_2 \\ \inf_{\chi \in X} (\det(\varepsilon(c,p) I_n - M(c,\chi,p))) = 0 \end{cases} \end{cases} \tag{23}$$

De même, le calcul des vecteurs propres à droite et à gauche de M(c,χ,p) (22) est immédiat, facilitant ainsi le calcul de la direction de plus grande pente du critère ε(c,p) (14).

Nous proposons maintenant d'illustrer l'utilisation de cette forme remarquable pour effectuer une synthèse paramétrique d'un exemple d'application (§ VI).

## VI - EXEMPLE D'APPLICATION

Considérons le modèle suivant donné pour le fonctionnement d'un système de surchauffe de centrale thermique (4) (12).

$$\mathring{y}(t) = \begin{bmatrix} -b_o(x) & -b_1(x) & -b_2(x) \\ 1 & 0 & 0 \\ 0 & 1 & 0 \end{bmatrix} y(t) + \begin{bmatrix} \ell_1(x) \\ 0 \\ 0 \end{bmatrix} u(t) \tag{24}$$

$$\theta(t) = \begin{bmatrix} 0 & 0 & 1 \end{bmatrix} y(t)$$

Dans ce modèle, $x \in [0\ ,\ 1.3]$ représente la puissance fournie par la centrale du réseau ; $\theta(t)$ la température dans le circuit de surchauffe. Les fonctions définissant les coefficients de la dynamique du processus sont identifiées par :

$$\begin{cases} \ell_1(x) = (4.475\ x^2 - 1.145\ x + 1.278)\ .\ 10^{-1} \\ b_o(x) = (1.705\ x^3 + 0.083\ x + 1.716) \\ b_1(x) = (2.726\ x^{5/2} + 0.029\ x + 1.084) \\ b_2(x) = (3.073\ x^{5/2} + 0.118\ x + 0.228) \end{cases}$$

On souhaite réguler la température de $\theta(t)$ autour d'une valeur $\theta_c$ constante. En conséquence, on a adopté un régulateur P. I. D..

Nous proposons dans cette étude de définir un réglage des paramètres P, I & D tel que l'asservissement constitué soit aussi robuste que possible vis à vis de variations quelconques de $x$ dans un domaine le plus large possible.

Considérons la loi de commande :

$$u(t) = -\ (\beta_1\ \mathring{\theta}(t) + \beta_2\ (\theta(t) - \theta_c) + \beta_3 \int_0^t (\theta(\tau) - \theta_c)\ d\tau) \tag{25}$$

Les gains ajustables sont les composantes de :

$$p = [\beta_1, \beta_2, \beta_3]^T \tag{26}$$

L'asservissement (24) (25) peut être représenté par l'équation en régime libre suivante :

$$\overset{\circ}{x} = A(\chi,p) \,.\, x$$

$$A(\chi,p) = \begin{bmatrix} -a_1(\chi,p) & -a_2(\chi,p) & -a_3(\chi,p) & -a_4(\chi,p) \\ 1 & 0 & 0 & 0 \\ 0 & 1 & 0 & 0 \\ 0 & 0 & 1 & 0 \end{bmatrix} \qquad (27)$$

$$\theta - \theta_c = \begin{bmatrix} 0 & 0 & 1 & 0 \end{bmatrix} x(t)$$

Définissons une forme remarquable série à partir d'un vecteur arbitraire c de paramètres de représentation :

$$c = \begin{bmatrix} \alpha_2, & \alpha_3, & \alpha_4 \end{bmatrix}^T \in \mathbb{C}^3 \qquad (28)$$

Il vient :

$$\begin{cases} S(c,\chi,p) = \begin{bmatrix} -b_1(.) & -b_2(.) & -b_3(.) & -b_4(.) \\ 1 & \alpha_2 & 0 & 0 \\ 0 & 1 & \alpha_3 & 0 \\ 0 & 0 & 1 & \alpha_4 \end{bmatrix} \\ \text{avec} \begin{cases} b_4(\chi,c,p) = \alpha_4^4 + a_1(\chi,p)\,\alpha_4^3 + a_2(\chi,p)\,\alpha_4^2 + a_3(\chi,p)\,\alpha_4 + a_4(\chi,p) \\ b_3(\chi,c,p) = (\alpha_4^3 + \alpha_4^2\,\alpha_3 + \alpha_4\,\alpha_3^2 + \alpha_3^3) + a_1(\chi,p)(\alpha_4^2 + \alpha_4\,\alpha_3 + \alpha_3^2) + \\ \qquad + a_2(\chi,p)(\alpha_4 + \alpha_3) + a_3(\chi,p) \\ b_2(\chi,c,p) = (\alpha_4^2 + \alpha_4\alpha_3 + \alpha_3^2 + \alpha_2\,(\alpha_4 + \alpha_3 + \alpha_2)) + \\ \qquad + a_1(\chi,p)(\alpha_4 + \alpha_3 + \alpha_2) + a_2(\chi,p) \\ b_1(\chi,c,p) = a_1(\chi,p) + \alpha_2 + \alpha_3 + \alpha_4 \end{cases} \end{cases} \qquad (29)$$

Un programme de synthèse paramétrique a été réalisé, il comporte :

- un module de calcul de $\varepsilon(c,p)$ (23),
- un module de calcul de la direction de plus grande pente de $\varepsilon(c,p)$ dans l'espace des paramètres $E = \{p_i\ ;\ \mathbb{R}(\alpha_{i+1}),\ Im(\alpha_{i+1})\ ;\ i = 1, 2, 3\}$. Ce module fait appel au calcul des gradients des valeurs propres simples de matrices irréductibles (14).

La minimisation de $\varepsilon(c,p)$ est effectuée par l'algorithme simple du gradient. Nous avons déterminé un ensemble d'intervalles de variation de $\chi$ recouvrant l'intervalle $[0 ; 1.3]$ et tel que la stabilité exponentielle du processus (27) soit assurée quelles que soient les variations de $\chi$ dans chaque sous intervalle. La stratégie de recherche est initialisée en calculant un réglage de $(\beta_1, \beta_2, \beta_3)$ stabilisant pour le système linéaire déduit de (27) ($\chi \in [\underline{\chi} , \bar{\chi}]$, $\bar{\chi}$ étant initialisé à $\underline{\chi}$). Les paramètres de représentation $(\alpha_2, \alpha_3, \alpha_4)$ sont alors initialisés égaux aux pôles de ce système linéaire.

On effectue alors l'algorithme suivant :

TANT QUE (on peut augmenter $\bar{\chi}$ de $10^{-4}$ en gardant $(\varepsilon(c,p) < 0)$)
  Augmenter $\bar{\chi}$ jusqu'à ce que $\varepsilon(c,p)$ devienne non négatif
  Déterminer $\hat{c}(\bar{\chi})$, $\hat{p}(\bar{\chi})$ qui minimisent $\varepsilon(c,p)$ avec $\chi \in [\underline{\chi} , \bar{\chi}]$
Fin TANT QUE

Les résultats issus de cette stratégie sont contenus dans le tableau en Annexe.

## VII - CONCLUSION

Le travail présenté introduit une méthodologie de synthèse associée à un théorème d'analyse de la stabilité des processus non linéaires [2] déduit des techniques d'agrégation par normes vectorielles. Cette synthèse est assimilée à la minimisation d'un critère numérique qui dépend à la fois des paramètres ajustables dans la loi de commande ainsi que de paramètres caractérisant une forme remarquable de représentation du système.

Ces paramètres de représentation permettent d'optimiser les conditions **suffisantes** de stabilité issues du théorème d'analyse utilisé [2] dans le sens de contraintes les moins restrictives possible.

Cette méthodologie se généralise aisèment aux asservissements dont la matrice de régime libre comporte plusieurs rangées de termes non constants ainsi qu'aux systèmes non linéaires en temps discret [1].

Annexe

| Domaine de variation de $x$ | $u(t) = -\left[\beta_1 \mathring{y}(t) + \beta_2 y(t) + \beta_3 \int_0^t y(\tau)\, d\tau\right]$ Réglage $p^T = [\beta_1, \beta_2, \beta_3]$ | | | Paramètres de représentation $c^T = [\alpha_2, \alpha_3, \alpha_4]$ | | | $\varepsilon(c,p)$ |
|---|---|---|---|---|---|---|---|
| | $\beta_1$ | $\beta_2$ | $\beta_3$ | $\alpha_2$ | $\alpha_3$ | $\alpha_4$ | |
| [ 0 ; 0,3103 ] | 0.2508 | 0.5441 | 0.26751 | - .5647 + j .028665 | - .53028 + j 0.29317 | - 0.096 + j $\varepsilon(10^{-4})$ | $-2.6\ 10^{-4}$ |
| [ 0,3103 ; 0.473 ] | - 0.1827 | - 0.6685 | 0.3517 | - .5682 + j (- .01918) | - .5496 + j (0.2888) | - 0.1271 + j $\varepsilon(10^{-7})$ | $-4\ 10^{-4}$ |
| [ 0.473 ; 0.6731 ] | - 0.5563 | - 1.9095 | 0.30128 | - .65839 + j (- 0.0447) | - .59367 + j (0.36) | - 0.1003 + j $\varepsilon(10^{-4})$ | $-1.5\ 10^{-4}$ |
| [ 0.6731 ; 1.1505 ] | - 0.26639 | - 2.903 | 0.24086 | - .70531 + j (- 0.4805) | - .71484 + j (.5291) | - 0.0625 + j $\varepsilon(10^{-5})$ | $-4.3\ 10^{-5}$ |
| [ 1.1505 ; 2.1199 ] | 3.79607 | 0.92651 | 2.8273 | - 1.28 + j (0.01768) | - .614 + j (0.7189) | - .591 + j (- 0.72679) | $-2.1\ 10^{-5}$ |

BIBLIOGRAPHIE

(1) BORNE P. 1976
"Contribution à l'étude des systèmes discrets non linéaires de grande dimension. Application aux systèmes interconnectés"
Thèse de Doctorat ès Sciences Physiques, n° 346, Université de Lille I, Villeneuve d'Ascq (France).

(2) GENTINA J.C., BORNE P., BURGAT C., BERNUSSOU J., GRUJIĆ Lj.T. 1979
"Sur la stabilité des systèmes de grande dimension. Normes vectorielles"
RAIRO, Vol. 13, n° 1.

(3) GRUJIĆ Lj.T., GENTINA J.C., BORNE P. 1976
"General aggregation of large scale systems by vector-Lyapunov functions and vector norms"
Int. J. of Control, Vol. 24, n° 4.

(4) GENTINA J.C. 1976
"Contribution à l'analyse et à la synthèse des systèmes continus non linéaires de grande dimension"
Thèse de Doctorat ès Sciences Physiques, n° 347, Université de Lille I, Villeneuve d'Ascq (France).

(5) GUARDABASI G., LOCATELLI A., MAFFEZZONI C., SCHIAVONI N. 1982
"A parameter optimization approach to the computer aided design of structurally constrained multivariable regulators"
Congrès IASTED "Modelling, Identification, Control & Robotics", Davos (Suisse).

(6) HÖFLER A.B.
"A software segmentation technique with high control structure flexibility for optimization by gradients"
8ème Congrès IFAC Mondial Triennal, Kyoto, pp. 1611-1616.

(7) SIRISENA H.R., CHOI S.S. 1975
"Pole placement in prescribed regions of the complex plane using output feedback"
IEEE Trans. on A.C., Vol. 1C-20, pp. 810-812.

(8) GANTMACHER F.R. 1966
"Théorie des matrices"
Tome 2, Dunod, Paris.

(9) DEIF A.S. 1982
"Advanced matrix theory for scientists and engineers"
Abacus Press, Halsted Press.

(10) MEIZEL D., GENTINA J.C. 1979
"New aspects on linear single-input single output systems"
Int. J. Control, Vol. 30, n° 6, pp. 1043-1060.

(11) MEIZEL D., GENTINA J.C., DAUPHIN-TANGUY G. 1982
"A parameter optimization design method of robust controllers for large scale non linear processes"
IEEE Int. Large Scale System Symposium, Virginia Beach (U. S. A.), pp. 343-348.

(12) LECOUTURIER J., DUPUY M. 1971
"Centrale de Loire / Rhône Tranche 3, identification des installations de température et de pression du générateur de vapeur"
E. D. F. - D. E. R., Dept Automatique et Moyens de Production, Rapport AMP 84.

(13) FADEEV D.K., FADEEVA V.N. 1963
"Computational methods of linear algebra"
Freeman, San Francisco.

(14) DAVISON A.J., ÖZGÜNER U. 1982
"Synthesis of the decentralized robust servomechanism problem using local models"
IEEE Trans. on A.C., Vol. AC-27, pp. 583-599.

# ON THE STABILIZATION OF POWER SYSTEMS WITH A REDUCED NUMBER OF CONTROLS

Riccardo Marino
Seconda Università di Roma, Tor Vergata
Dipartimento di Elettronica
Via O. Raimondo, I00173, Roma, Italy

Abstract: The well-known nonlinear model for the stabilization problem of a power system network reduced at its generating nodes is considered. If active power controls can be employed at each node, the resulting control problem is not well defined since the location and the number of power controls are to be established: the aim is to use the least number of controls which allows the application of a stabilizing control scheme. We propose the recently developed technique of "feedback linearization".

Preliminary results establish that if the number of controls is equal to the number of generating nodes any power network is globally feedback linearizable; on the other hand if only one control is employed the power network is feedback linearizable if and only if its graph is a straight chain. Subsequently it is shown that, under mild conditions on the structure of a power network with $n$ nodes, n/2 controls guarantee feedback linearizability. In those cases the explicit nonlinear state feedback stabilizing control laws are given.

## 1. INTRODUCTION

In this paper we consider a power system network reduced at its n generating nodes. Each node i is characterized by the voltage $(E_i, \delta_i)$. Each a.c. line ij connecting nodes i and j is characterized by the impedance $(Z_{ij},\phi_{ij})$. Each node i makes available to the rest of the network an amount of active power denoted by $\hat{P}_{mi}$. $\hat{P}_{mi}$ represents the difference between the power produced by generators and the power absorbed by concentrated loads as far as node i is concerned. Given the impedance matrix of the network, for any given realistic set of power injections $(\hat{P}_{m1},\ldots,\hat{P}_{mn})$ there exists a number of stable load flow solutions,

---

This work was partly supported by DOE, Grant No. DE/ACOI/79ET/29367 and MPI (fondi 40%).

denoted by $(E_i^o, \delta_i^o)_j$. The angle $\delta_i$ can be thought as the position of the rotor of a synchronous machine delivering the power $\hat{P}_{mi}$. In correspondence to a stable load flow solution $(E_i^o, \delta_i^o)$ each machine is rotating at the synchronous speed $\omega_s$, which determines the frequency in the a.c. network. If parameter perturbations (changes in the values of power injections or line impedances) or structural perturbations (changes in network structure) occur, the stable load flow solutions are in general no longer the same. The nominal condition $(E_i^o, \delta_i^o)$ at which the unperturbed network was working represent the initial condition for a perturbed trajectory. It is a well-known fact that in this case the dynamics of the mutual angle positions $(\delta_i - \delta_j)$ is much faster than the dynamics of the voltage magnitudes $E_i$. The stabilization problem for power systems is to design control schemes which prevent the mutual angle positions $(\delta_i - \delta_j)$ from reaching certain physical bounds at which circuit breakers are supposed to disconnect the corresponding line ij. A successful control scheme is proposed in [1] where, on the basis of the concept of observation decoupled state space, local state measures are used for bang-bang control laws if n active power controllers are employed, i.e. one for each node.

In this paper a differential geometric approach is followed: assuming full state $(\delta_i, \omega_i)$ measurements, the location and the number of power controllers which induce feedback linearizability are investigated. Feedback linearizability is a recently introduced property for nonlinear systems ([3],[4]) which guarantees the possibility of compensating the nonlinearities by state feedback and, in our case, allows us the construction of state feedback stabilizing control laws.

Let us now introduce the nonlinear control model for power system stabilization problems we will refer to throughout the paper. The following notation is used:

$$\gamma_i = \begin{cases} 1 & \text{if a power control is acting at node i} \\ 0 & \text{otherwise} \end{cases}$$

$$c_{ij} = \begin{cases} 1 & \text{if there is an a.c. line connecting nodes i and j} \\ 0 & \text{otherwise} \end{cases}$$

$$k_{ij} = c_{ij}\, E_i\, E_j / M_i\, Z_{ij}$$

$M_i = J_i \omega_s$ where $J_i$ is the momentum of inertia of the rotor of the synchronous machine i;

$$P_{mi} = \hat{P}_{mi} + \sum_{\substack{j=1 \\ j=i}}^{n} c_{ij} E_i^2 \sin \alpha_{ij} / Z_{ij}$$

$$\alpha_{ij} = \phi_{ij} - 90^\circ$$

$\sum_{\substack{j=1 \\ j=i}}^{n} k_{ij} \sin(\delta_i - \delta_j + \alpha_{ij})$ represents active power which the network can accept from node i.

The model is

$$\begin{bmatrix} \dot{\delta}_1 \\ \cdots \\ \dot{\delta}_n \\ \dot{\omega}_1 \\ \cdots \\ \dot{\omega}_n \end{bmatrix} = \begin{bmatrix} \omega_1 - \omega_s \\ \cdots\cdots \\ \omega_n - \omega_s \\ P_{m1} - \sum_{\substack{j=1 \\ j=1}}^{n} k_{1j} \sin(\delta_1 - \delta_j + \alpha_{1j}) \\ \cdots\cdots\cdots\cdots\cdots \\ P_{mn} - \sum_{\substack{j=1 \\ j=n}}^{n} k_{nj} \sin(\delta_n - \delta_j + \alpha_{nj}) \end{bmatrix} + \begin{bmatrix} 0 \\ \cdots \\ 0 \\ 1 \\ \cdots \\ 0 \end{bmatrix} \gamma_1 u_1(t) +$$

$$+ \begin{bmatrix} 0 \\ \cdots \\ 0 \\ 0 \\ 1 \\ \cdots \\ 0 \end{bmatrix} \gamma_2 u_2(t) + \cdots\cdots + \begin{bmatrix} 0 \\ \cdots \\ 0 \\ 0 \\ 0 \\ \cdots \\ 1 \end{bmatrix} \gamma_n u_n(t) \qquad (1)$$

## 2. MAIN RESULTS

The stabilizing control problem which takes shape from the previous section is: design a stabilizing and hopefully robust control which drives $(\delta_i-\delta_j)$ to $(\delta_i-\delta_j)_c$ and $\omega_i$ to $\omega_s$ for each $i$ ; the state evolves according to the set of nonlinear equations (1); it is desirable to use the least number of controls.

The control theory of smooth nonlinear systems is far from being complete even in the special case of systems linear in control. However recent results ([2],[3],[4],[5]) show that if the system happens to be feedback equivalent to a linear controllable system (see the Appendix for the corresponding definitions and necessary and sufficient conditions which identify this special class of nonlinear systems), then the system can be actually controlled and procedures for constructing nonlinear state feedback control laws are available. In fact feedback equivalence to linear controllable systems implies the existence of a state feedback control law

$$v = a(x) + S(x)\, u$$

which makes the modified system (driven by the new inputs v) linear in a suitable coordinate system.

According to [6] we assume that the mutual angle positions $(\delta_i-\delta_j)$ can be estimated on line on the basis of measurements of currents and voltages and that the angular speeds $\omega_i$ can be directly measured. We also assume the knowledge of structural or parameter changes.

We will therefore pose the question: how many controls are needed and where should they be located in order for the system (1) to be feedback equivalent to a linear controllable system? Geometrically the question becomes: find a subspace G of the control space $\mathbb{R}^n$ such that the distribution

$$M^i = \mathrm{span}\{G,\ ad_f G, \ldots,\ ad_f^i G\}$$

is involutive for each i, $1 \le i \le 2n-1$, and

$$\dim M^{2n-1}(x) = 2n \ \forall x \in U, \text{ open subset in } R^{2n}$$

The answer to this equation will determine how many machines and which ones should be controlled in order to induce the feedback linearizability property and therefore the possibility of stabilizing the system by state feedback.

It turns out that the solution to the problem posed depends on the structure, i.e. the graph theoretic properties, of the a.c. reduced network; more precisely for the power system equations the subspaces G of the control space $R^n$ which make the distribution $M^i$ involutive and the distribution $M^{2n-1}$ of dimension 2n are closely related to the incidence matrix representing the a.c. network.

A first step in the analysis of the relations between the linearizing subspaces G and the a.c. network graph structure is given by the following preliminary proposition which specifies the only a.c. network which is compatible with a one dimensional linearizing subspace of $\mathbb{R}^n$ (for instance $\gamma_1 = \dots = \gamma_{n-1} = 0$). Recall that $k_{ij}$ represents the maximal angular acceleratin which can be transmitted from machine i to machine j through the a.c. line ij.

Define

$$E \triangleq \{x \in \mathbb{R}^{2n}: f(x) \in \text{span}\{\gamma_1 g_1, \dots, \gamma_n g_n\}\}$$

PROPOSITION 1. Given any integer n, system (1) with $\gamma_1 = \dots = \gamma_{n-1} = 0$ is locally feedback equivalent in a sufficiently small neighborhood $U_{x_o}$ of $x_o$, for any $x_o \in E \cap \Omega_n$ where

$$\Omega_n = \{x \in \mathbb{R}^{2n}: |\delta_{i-1} - \delta_i + \alpha_{i-1,i}| \neq l \frac{\pi}{2} ;$$

$$i = 2, \dots, n ; \quad l \text{ odd integer}\}$$

to a single input controllable system $\dot{y} = Ay + bv$ if and only if, up to a machine relabeling, the matrix $[k_{ij}]$ displays the structure

$$K = \begin{bmatrix} 0 & k_{12} & 0 & 0 & \dots & 0 & 0 \\ k_{21} & 0 & k_{23} & 0 & \dots & 0 & 0 \\ 0 & k_{32} & 0 & k_{34} & \dots & 0 & 0 \\ & \cdot & \cdot & \cdot & \cdot & \cdot & \\ 0 & 0 & 0 & 0 & & 0 & k_{n-1,n} \\ 0 & 0 & 0 & 0 & & k_{n,n-1} & 0 \end{bmatrix} \qquad (2)$$

PROOF. By Theorem A (see the Appendix, m=1) we only need to prove that the assumption (2) on the matrix K is equivalent to:

i) span $\{ad_f^{\ell} g_n : 0 \leq \ell \leq j\}$ is an involutive distribution for each $j \leq j \leq 2n-1$, of constant dimension in $\Omega_n$;

ii) span $\{ad_f^{\ell} g_n : 0 \leq \ell \leq 2n-1\}$ has dimension 2n in $\Omega_n$.

We are given an integer n. The Jacobian of f is

$$\frac{df}{dx} = \left[\begin{array}{c|c} 0 & I_{n \times n} \\ \hline \overline{K}(x) & 0 \end{array}\right]$$

where

$$\bar{k}_{ij} = \begin{cases} k_{ji}\cos(\delta_i-\delta_j+\alpha_{ij}) & \text{if } j > i \\ -k_{ij}\cos(\delta_i-\delta_j+\alpha_{ij}) & \text{if } j < i \end{cases}$$

$$\bar{k}_{ii} = \sum_{j\neq i} -k_{ij}\cos(\delta_i-\delta_j+\alpha_{ij})$$

$$g_n = (\overbrace{0,\ldots\ldots,0,0,}^{n}\ \overbrace{0,\ldots\ldots\ldots\ldots\ldots,0,1}^{n})^T.$$

$$ad_f g_n = (0,\ldots\ldots,0,-1,\ 0,\ldots\ldots\ldots\ldots\ldots,0,0)^T$$

$$ad_f^2 g_n = (0,\ldots\ldots,0,0,\ \bar{k}_{1,n},\bar{k}_{2,n},\ldots,\bar{k}_{n-1,n},\bar{k}_{n,n})^T$$

$\mathrm{span}\{g_n, ad_f g_n, ad_f^2 g_n\}$ is involutive if and only if $k_{1,n}=\ldots=k_{n-2,n}=0$, since a relabeling of the machines is allowed; recall that it follows $k_{n,1}=\ldots=k_{n,n-2}=0$. In this case its dimension is 3 in $\Omega_n$: $ad_f^2 g_n$ becomes

$$ad_f^2 g_n = (\overbrace{0,\ldots\ldots,0,\quad 0,\quad 0}^{n},\overbrace{0,\ldots\ldots,0,\bar{k}_{n-1,n},\bar{k}_{n,n}}^{n})^T$$

$ad_f^3 g_n$ becomes

$$ad_f^3 g_n = (\underbrace{0,\ldots\ldots,0,-\bar{k}_{n-1,n},\bar{k}_{n,n}}_{n},\underbrace{0,\ldots,0,\bar{h}_{n-1,n},\bar{h}_{n,n}}_{n})^T$$

where

$$\bar{h}_{i,j} = \begin{cases} (\omega_j-\omega_i)k_{i,j}\sin(\delta_i-\delta_j+\alpha_{ij}) & \text{if } j > i \\ (\omega_i-\omega_j)k_{i,j}\sin(\delta_i-\delta_j+\alpha_{ij}) & \text{if } j < i \end{cases}$$

$$\bar{h}_{i,i} = \sum_{j\neq i} -\bar{h}_{i,j}$$

Under the aforementioned hypothesis $\mathrm{span}\{ad_f^\ell g_n : 0 \le \ell \le 3\}$ is involutive and has dimension 4 in $\Omega_n$.

In order to complete the proof we will use an induction argument. Suppose we proved:

i) $\mathrm{span}\{ad_f g_n : 0 \le \ell \le j\}$ is involutive of dimension $j+1$ for every $x \in \Omega_n$ and for each $j$, $0 \le j \le 2h-1$, if and only if $k_{i,j}=0$ (and therefore $k_{j,i}=0$) for any $j$ and $i$ such that $n-h+2 \le j \le n, 1 \le i\_j \le 2$.

ii) The Lie brackets display the following structure:

$$g_n = (\overbrace{0,\ldots\ldots\ldots\ldots,0,0,}^{n} \quad \overbrace{0,\ldots\ldots\ldots\ldots\ldots,0,1}^{n})^T$$

$$ad_f g_n = (0,\ldots\ldots\ldots\ldots,0,-1, \quad 0,\ldots\ldots\ldots\ldots\ldots,0,0)^T$$

$$ad_f^2 g_n = (0,\ldots\ldots\ldots\ldots,0,\ 0, \quad 0,\ldots\ldots\ldots\ldots,0,*,*)^T$$

$$ad_f^3 g_n = (0,\ldots\ldots\ldots,0,*,*, \quad 0,\ldots\ldots\ldots\ldots,0,*,*)^T$$

$$\ldots\ldots\ldots\ldots\ldots$$

$$ad_f^{2j-2} g_n = (0,\ldots\ldots,0,*,.,*,*, \quad 0,\ldots\ldots,0,\pi^h,\ldots,*,*)^T$$

$$ad_f^{2j-1} g_n = (\underbrace{0,\ldots\ldots,0,}_{n-j}\pi^j,*,.,*,*, \quad \underbrace{0,\ldots\ldots,0,}_{n-j}*,\ldots,*,*)^T$$

$$\ldots\ldots\ldots\ldots\ldots\ldots$$

$$ad_f^{2h-2} g_n = (0,\ldots\ldots,0,*,.,*,*, \quad 0,\ldots,0,\pi^h,*,\ldots\ldots,*,*)^T$$

$$ad_f^{2h-1} g_n = (\underbrace{0,\ldots,0,}_{n-h}\pi^h,*,.,*,*, \quad \underbrace{0,\ldots,0,}_{n-h}*,*,\ldots\ldots,*,*)^T$$

where $\pi^h = \prod_{i=n-h+1}^{n} \bar{k}_{i-1,i}$ and $*$ replaces any nonzero component.

In this case $\mathrm{span}\{ad_f^\ell g_n:\ 0\leq\ell\leq 2h-1\}$ has dimension $2h$ in $\Omega_n$, $h\geq 2$.
If one sets, after an eventual relabeling of the remaining machines, $k_{1,n-h-1}=\ldots=k_{n-h-1,n-h-1}=0$

$$ad_f^{2h} g_n = (\overbrace{0,\ldots\ldots,0,*,\ldots\ldots,*,}^{n} \quad \overbrace{0,\ldots,0,\pi^{h+1},*,\ldots\ldots,*}^{n})^T$$

$$ad_f^{2h+1} g_n = (\underbrace{0,\ldots,0,}_{n-h-1}\pi^{h+1},*,\ldots,*, \quad \underbrace{0,\ldots,0,}_{n-h-1}*, \quad *,\ldots\ldots,*)^T$$

$\mathrm{span}\{ad_f^\ell g_n:\ 0\leq\ell\leq 2h+1\}$ has dimension $2h+2$ in $\Omega_n$ and $\mathrm{span}\{ad_f^\ell g_n:\ 0\leq\ell\leq j\}$ is involutive for each $j$, $0\leq j\leq 2h+1$. On the other hand if there is no relabeling of the remaining machines such that $k_{1,n-h-1}=\ldots=k_{n-h-1,n-h-1}=0$ $\mathrm{span}\{ad_f^\ell g_n:\ 0\leq\ell\leq 2h\}$ is not involutive. Thus properties i) and ii) are established for h+1 if they hold for h. Since we proved that i) and ii) hold for h=2, if followsthat conditions i) and ii) of Theorem A (see the

Appendix) are equivalent to the property (2) for the matrix k.

REMARK 1. The structure (2) of the matrix k corresponds to a network structure of a controlled chain of n machines where the control acts on a machine at the end of the chain.

REMARK 2. The structure of a controlled chain of n machines has a striking resemblance with the Brunovsky canonical form of a chain of 2n integrators controlled at the end and even more to the "linear controllable coupled oscillator" canonical form introduced in [8], which is a chain of n masses connected by linear springs controlled by a linear damping force. As a matter of fact a chain of n machines can be viewed as a chain of n masses connected by nonlinear springs.

REMARK 3. Due to the special structures of the successive Lie brackets, the integral manifold of the involutive distribution $M^{2n-2}=$ $=\mathrm{span}\{ad_f^{\ell} g_n : 0\leq\ell\leq 2n-2\}$ is $\delta_1=$const (or equivalently $F(\delta_1)=$const F is any smooth invertible function): in fact $L_X\delta_1=0$ for every X belonging to $M^{2n-2}$. This is a remarkable property since the actual construction of control laws for feedback linearizable systems depends on the possibility of computing a closed form solution of the system of partial differential equations

$$(dT(x),\ X) = 0 \qquad X \in M^{2n-2}.$$

In fact the feedback-transformation which carries system (1) satisfying the assumptions of Proposition 1 into a linear system with a preassigned characteristic polynomial $a_{2n}+a_{2n-1}s+\ldots+a_1 s^{2n-1}+s^{2n}$

$$\dot{\tilde{x}} = A_c\tilde{x} + b_c v \tag{3}$$

where

$$A_c = \begin{bmatrix} 0 & 1 & 0 & \ldots\ldots & 0 \\ 0 & 0 & 1 & \ldots\ldots & 0 \\ 0 & 0 & 0 & \ldots\ldots & 1 \\ a_1 & a_2 & a_3 & \ldots\ldots & a_{2n} \end{bmatrix}, \quad b_c = \begin{bmatrix} 0 \\ 0 \\ 0 \\ 1 \end{bmatrix}$$

is given by $\tilde{x}=T(x)$, $v(x,u)=a(x)+s(x)u_m$.
The components $(T_1(x)\ldots,T_{2n}(x))$ of the nonlinear change of coordinates are

$$\tilde{x}_1 = T_1(x) = \delta_1 - \delta_1^c$$

$$\tilde{x}_2 = T_2(x) = L_f\delta_1$$

$$\tilde{x}_3 = T_3(x) = L_f^2\delta_1$$

......

$$\tilde{x}_{2n} = T_{2n}(x) = L_f^{2n-1}\delta_1$$

The transformation between the controls $u_m$ and $v$ is given by

$$v(x,u) = L_f^{2n}\delta_1 + L_gL_f^{2n-1}\delta_1 u_m -$$

$$- a_1(\delta_1-\delta_1^c)-\dots - a_{2n}L_f^{2n-1}\delta_1$$

The only network structure which is compatible with the feedback-equivalence to system (3) is determined. The other extreme situation corresponds to the following question: how many controls are needed at most an $n$ machine network of any given structure in order to guarantee the feedback equivalence to a linear controllable system driven by the same number of controls? The following proposition establishes that $n$ controls suffice.

PROPOSITION 2. Suppose $\gamma_1=\dots=\gamma_n=1$ in system (1). Then system (1) is feedback equivalent in $\mathbb{R}^{2n}$ to an n-input linear controllable system with controllability indices (2,...,2).

PROOF. It is a straightforward application of the Theorem A reported in the Appendix.

$$g_n = (0,\dots\dots,0, \quad 0,\dots\dots,1)^T$$

$$g_{n-1} = (0,\dots\dots,0, \quad 0,\dots\dots 1,0)^T$$

......

$$g_2 = (0,\dots\dots,0, \quad 0,1,0,\dots,0)^T$$

$$g_1 = (0,\dots\dots,0, \quad 1,0,\dots\dots,0)^T$$

$$ad_fg_{n-1} = (0,\dots\dots,-1, \quad 0,\dots\dots,0)^T$$

......

$$ad_fg_2 = (0,-1,0,\dots,0, \quad 0,\dots\dots,0)^T$$

$$ad_fg_2 = (\underbrace{-1,0,\dots\dots,0}_{n}, \quad \underbrace{0,\dots\dots,0}_{n})^T$$

Clearly span$\{ad_f^\ell g_i: 0\le\ell\le 1, 1\le i\le n\} = \mathbb{R}^{2n}$ and span$\{g_i: 1\le i\le n\} = \mathbb{R}^n$ are involutive distribution

REMARK 4. From (1) the linearizing feedback transformation can be found by inspection:

$$u_i = -P_i + \sum_{j=1}^{n} k_{ij}\sin(\delta_i - \delta_j + \alpha_{ij}) + v_i \quad l=1,\ldots,n \quad (4)$$

It is apparent that the feedback control law (4) linearizes the system without any need of changes of coordinates. In fact in this case the feedback linearizability property is global. Proposition 2 makes it clear that if each node is controlled there exists a stabilizing control law (v can be designed via linear system control techniques) for any given structure of the a.c. network. It also proves that in this case the power system equations are completely controllable, i.e. given any two states there exists a control which takes the system from one state to the other one in any preassigned fixed positive time.

At this point it is established that, regardless to the graph theoretical properties of the network, if a control action can be performed at each node the power system equations (1) are globally feedback equivalent to a linear controllable system and in particular completely controllable. Thus linear control techniques apply.

On the other hand the only structure which is compatible with the (local) feedback linearizability property is a controlled chain of machines, which is an unrealistic configuration for power system networks reduced to their generation nodes. Clearly one would expect an intermediate result, i.e. that there exist realistic network structures which admit the feedback linearizability property when only p nodes are available for control actions, with $p<n$. The following proposition provides a first answer in this direction.

PROPOSITION 3. Suppose there exists a machine renumbering such that the matrix

$$R(\delta) = \begin{bmatrix} \bar{k}_{12}(\delta) & \bar{k}_{32}(\delta) & \bar{k}_{52}(\delta)\ldots & \bar{k}_{n-1,2}(\delta) \\ \bar{k}_{14}(\delta) & \bar{k}_{34}(\delta) & \bar{k}_{54}(\delta)\ldots & \bar{k}_{n-1,4}(\delta) \\ & \cdot\;\cdot\;\cdot & \cdot\;\cdot\;\cdot & \\ \bar{k}_{1n}(\delta) & \bar{k}_{3n}(\delta) & \bar{k}_{5n}(\delta)\ldots & \bar{k}_{n-1,n}(\delta) \end{bmatrix}$$

is nonsingular in $\Omega$, open and dense submanifold in $\mathbb{R}^{2n}$; then system (1) with $\gamma_i=0$ for any odd $i$, $0<i<n$, is locally feedback equivalent in a neighborhood $U_{x_o} \in \Omega$ for any $x_o \in \Omega$ $\{x \in \mathbb{R}^{2n}: f(x) \in \text{span}\{\gamma_i g_i, i \text{ even}\}$

to an $\frac{n}{2}((\frac{n+1}{2})$ if n is odd) input linear controllable system with controllability indices

$$\underbrace{[4,4,\ldots,4]}_{n/2} \quad (\underbrace{[4,4,\ldots,4,2]}_{n+1/2} \text{ if } n \text{ is odd}).$$

PROOF. One needs to chech directly the conditions i) and ii) of Theorem A (see the Appendix). We only consider the even case. The following successive Lie brackets can be easily computed:

$$g_n = (\overbrace{0,\ldots\ldots,0,0,}^{n} \; \overbrace{0,0,\ldots\ldots,0,1}^{n})^T$$

$$g_{n-2} = (0,\ldots\ldots,0,0, \; 0,0,\ldots,0,1,0,0)^T$$

$$\cdot \quad \cdot \quad \cdot \quad \cdot \quad \cdot$$

$$g_2 = (0,\ldots\ldots,0,0, \; 0,1,\ldots,0,0,0,0)^T$$

$$ad_f g_n = (0,\ldots\ldots,0, \;,0, \; 0,0,\ldots\ldots,0,0,0,0)^T$$

$$ad_f g_{n-2} = (0,\ldots\ldots,0,-1,0,0, \; 0,0,\ldots\ldots,0,0,0,0)^T$$

$$\cdot \quad \cdot \quad \cdot \quad \cdot \quad \cdot$$

$$ad_f g_2 = (0,-1,\ldots\ldots,0,0,0,0, \; 0,0,\ldots\ldots,0,0,0,0)^T$$

$$ad_f^2 g_n = (0,-1,\ldots\ldots,0,0,0,0,\bar{k}_{1,n},\bar{k}_{2,n},\ldots,\bar{k}_{n-1,n},\bar{k}_{n,n})^T$$

$$ad_f^2 g_{n-1} = (0,0,\ldots\ldots,0,0,0,0,\bar{k}_{1,n-2},\bar{k}_{2,n-2},\ldots,\bar{k}_{n-1,n-2},\bar{k}_{n,n-2})^T$$

$$\cdot \quad \cdot \quad \cdot \quad \cdot \quad \cdot$$

$$ad_f^2 g_2 = (0,0,\ldots\ldots,0,0,0,0,\bar{k}_{1,2},\bar{k}_{2,2},\ldots,\bar{k}_{n-1,2},\bar{k}_{n,2})^T$$

$$ad_f^3 g_n = (-\bar{k}_{1,n},-\bar{k}_{2,n},\ldots\ldots,-\bar{k}_{n-1,n},-\bar{k}_{n,n}, \; \bar{h}_{1,n},\bar{h}_{2,n},\ldots,\ldots,\bar{h}_{n-1,n},\bar{h}_{n,n})^T$$

$$ad_f^3 g_{n-2} = (-\bar{k}_{1,n-2},-\bar{k}_{2,n-2},\ldots,-\bar{k}_{n-1,n-2},-\bar{k}_{n,n-2},\bar{h}_{1,n-2},\bar{h}_{2,n-2},\ldots,\bar{h}_{n-1,n-2},\bar{h}_{n,n-2})^T$$

$$\cdot \quad \cdot \quad \cdot \quad \cdot \quad \cdot$$

$$ad_f^3 g_2 = (\underbrace{-\bar{k}_{1,2},-\bar{k}_{2,2},\ldots,\ldots,-\bar{k}_{n-1,2},-\bar{k}_{n,2}}_{n}, \; \underbrace{\bar{h}_{1,2},\bar{h}_{2,2},\ldots,\ldots,\bar{h}_{n-1,2},\bar{h}_{n,2}}_{n})^T$$

span{$ad_f^\ell g_i$: $0 \le \ell \le j$, $1 < i \le n$, i even} is an involutive distribution of constant dimension $\frac{n}{2}(j+1)$ for every $j=0,1,2$ and for every $x \in \Omega$; span{$ad_f^\ell g_i$: $0 \le \ell \le 3$, $1 < i \le n$, i even} has dimension 2n for every $x \in \Omega$. Thus Theorem A applies in $\Omega$ and the Proposition is therefore proved.

REMARK 6. The linearizing feedback transformation to the Brunovsky canonical form can be explicitely and directly constructed if Proposition 3 holds for a suitable machine renumbering

$$\left.\begin{aligned} T_i(x) &= \delta_{2i-1} - \delta_{2i-1}^c \\ T_{2i}(x) &= L_f \delta_{2i-1} \\ T_{3i}(x) &= L_f^2 \delta_{2i-1} \\ T_{4i}(x) &= L_f^3 \delta_{2i-1} \end{aligned}\right\} \quad i = 1, \ldots, \frac{n}{2}$$

$$\begin{pmatrix} v_2 \\ v_4 \\ \cdot \\ \vdots \\ v_n \end{pmatrix} = \begin{pmatrix} L_f^4 \delta 1 \\ L_f^4 \delta_3 \\ \cdot \\ \vdots \\ L_f^4 \delta_{n-1} \end{pmatrix} + \begin{pmatrix} L_{g_2} L_f^3 \delta_1 & L_{g_4} L_f^3 \delta_1 \cdots & L_{g_n} L_f^3 \delta_1 \\ L_{g_2} L_f^3 \delta_3 & L_{g_4} L_f^3 \delta_3 \cdots & L_{g_n} L_f^3 \delta_3 \\ \vdots & \vdots & \vdots \\ L_{g_2} L_f^3 \delta_{n-1} & L_{g_4} L_f^3 \delta_{n-1} & L_{g_n} L_f^3 \delta_{n-1} \end{pmatrix} \begin{pmatrix} u_2 \\ u_4 \\ \cdot \\ \vdots \\ u_n \end{pmatrix}$$

The controls $v_i$, i even, can be designed according to linear control techniques recalling that $y=T(x)$.

REMARK7. Proposition 3 guarantees the feedback linearizability property with $\frac{n}{2}$ controls if n is even ($\frac{n+1}{2}$ if n is odd) if there exists a renumbering of machines such that the matrix $R(\delta)$ is nonsingular in an open and dense submanifold of $\mathbb{R}^{2n}$. This is actually a condition on the graph theoretical properties of the a.c. network. Suppose we divide the n machines into two groups: the controlled (even) ones, whose number is $\frac{n}{2}$ ($\frac{n+1}{2}$ if n is odd), and the $\frac{n}{2}$ uncontrolled (odd) ones. The condition required by Proposition 3 only contains the parameters of all the a.c. lines which connect machines belonging to different groups. It follows that the a.c. lines connecting machines of the same group play no role at all: no requirement is imposed on those lines. The nonsingularity of $R(\delta)$ can then be related to the possibility of transmitting the control action from the controlled set of machines to hte uncontrolled one.

Furthermore the determinant of $R(\delta)$ quantifies in some sense such a capability as a function of the rotor angles.

The capability of transmitting electric power, and therefore control signals, depends on the mutual angle positions: in fact the angular speeds $\omega_i$ do not appear in $R(\delta)$. In view of this interpretation the existence of a machine relabeling which guarantees the nonsingularity of $R(x)$ seems a mild condition: it amounts to the requirement of the existence of two group of machines tightly electrically connected. On the other hand the converse of such an argument shows that the matrix $R(\delta)$ can be an important tool of analysis and design since it measures, on the basis of a rigorous mathematical theory, how connected two group of machines are in terms of the mutual angle displacements.

In conclusion three points emerge from Proposition 3:

- the nonsingularity of $R(\delta)$ for a suitable renumbering of the machines is a very mild and realistic condition on the network structure considering also that the a.c. network (reduced to its generating points) we are referring to is not sparse at all;
- structural perturbations affecting the a.c. lines which are internal to each group do not affect the feedback-linearizability property at all;
- if the renumbering of the machines is such that the determinant of $R(x)$ is maximized over the region of interest, it is reasonable to expect that small perturbations on matrix $R$ parameters could be allowed as well without destroying the feedback linearizability property.

REMARK 8. The number of controls employed determines the number of components of rotor angles which can be arbitrarily assigned in the stable equilibrium point at which the system is supposed to be asymptotically stabilized. In fact any state $x_o$ belonging to the set

$\{x \in \mathbb{R}^{2n}: f(x) \in \text{span}(\gamma_1 g_1, \ldots, \gamma_n g_n)\}$ can be forced by feedback to be an equilibrium point. For instance: $\delta_1^c$ can be arbitrarity chosen in Remark 3; any set $(\delta_1^o, \ldots, \delta_n^o)$ can be chosen if Proposition 2 applies; $\delta_{2i-1}^c$, $i=1,\ldots,n/2$, can be chosen in Remark 6. This is an important issue since the feedback linearizability characterized in the Appendix is a local property and the stable load flow solutions of the perturbed power network are not a priori known. Some freedom in choosing $x_o$ allows to keep the machines together until the original power network is recovered.

## 3. CONCLUSION

It is shown that the concept of feedback linearizability and, in general, the geometric approach to nonlinear control theory are of some interest in the stabilization of power system networks.

In particular, depending on the number of power controls available, feedback linearizability is mirrored by certain structures of the network. It is shown that feedback linearizability is induced: i) in any case if $n$ controls are available where $n$ is the number of nodes; ii) if and only if the network is a chain controlled at the end, whenever only one control is available; iii) under mild conditions on the network if $n/2$ controls are available. Feedback linearizability can be always induced if a sufficient number of controls is used; the trend is: the stronger the conditions on the network structure are the smaller the number of needed controls is. The lower limit is given by the unrealistic structure of a straight chain. Feedback linearizability allows the construction of explicit stabilizing state feedback nonlinear control laws.

## 4. AKNOWLEDGMENTS

The author wishes to thank Professors J. Zaborszky and D.L. Elliott for many stimulating discussions on the topic of this paper.

## 5. APPENDIX ([3],[4],[5],[9])

*Notation*: Let $f(x)$ be a smooth vector field and $T(x)$ a smooth real valued function, $x \in \mathbb{R}^n$

$$L_f T(x) = \langle dT(x), f(x) \rangle = \sum_{i=1}^{n} \frac{\partial T(x)}{\partial x_i} f_i(x)$$

*Notation*: $ad_f g$ denotes the Jacobi bracket of two vector fields $f$ and $g$, which is given in local coordinates by

$$ad_f g = \left[\frac{dg}{dx}\right] f - \left[\frac{df}{dx}\right] g;\ ad_f^0 g = g;\ ad_f^\ell g = ad_f^\ell (ad_f^{\ell-1} g).$$

*Definition*: The system $\frac{dx}{dt} = f(x) + \sum_{i=1}^{m} u_i(t) g_i(x) \triangleq f(x) + G(x)u(t), x \in \mathbb{R}^n$, is said tc be feedback-equivalent in a neighborhood $U_{x_o}$ to the system $\frac{dy}{dt} = Ay + \sum_{i=1}^{m} v_i(t) b_i \triangleq Ay + Bv$, $y \in \mathbb{R}^n$, if there exist a diffeomorphism $T$: $U \to T(U)$, $T(x_o)=0$, and an affine transformation of the control space $\mathbb{R}^m$, $v = a(x) + S(x)u$ (where $a(x_o)=0$ and $S(x)$ is nonsingular in $U_{x_o}$) such that

$$f(x) = \left[\frac{dT}{dx}\right]^{-1} (AT(x) + a(x))$$

$$G(x) = \left[\frac{dT}{dx}\right]^{-1} S(x)$$

*Theorem* A: The system $\frac{dx}{dt}=f(x)+\sum_{i=1}^{m} u_i(t)q_i(x)$, $(f,g_1,\dots,g_m$ smooth vector fields), is feedback equivalent in $U_{x_o}$ to the controllable system $\frac{dy}{dt}=Ay+\sum_{i=1}^{m} v_i(t)b_i$ with controllability indices $(k_1,\dots,k_m)$ if and only if

i) $\text{span}\{ad_f^{\ell} g_i(x)\colon 0\le\ell\le n-m,\ 1\le i\le m\} = \mathbb{R}^n$ sor all $x \in U_{x_o}$

ii) $\text{span}\{ad_f^{\ell} g_i\colon 0\le\ell\le j,\ 1\le i\le m\}$ is an involutivedistribution of constant rank $m_j$ in $U_{x_o}$ for every $j$, $0\le j\le n-m$.

iii) $f(x_o) \in \text{span}\ \{g_1(x_o),\dots,g_m(x_o)\}$

where $k_i$ is equal for every $i$ to the number of $s_j\ge i$, for for $j\ge 0$, where $s_j=m_j-m_{j-1}$ and $s_o=m_o$.

## 6. REFERENCES


[1] J. ZABORSZKY, K.W. WHANG, K.V. PRASAD, *Operation of the large interconnected power system by decision and control in emergencies*, Report SSM7907, Dept. of System Science and Math., Washington University, St. Louis, Mo, 1979.

[2] R.W. BROCKETT, *Feedback invariants for nonlinear systems*, Proc. IFAC Congress, Helsinki, 1978.

[3] B. JAKUBCZYK, W. RESPONDEK, *On linearization of control systems*, Bull. Acad. Pol. Sci. Vol.XXVIII, n.9-10, 517-522.

[4] L.R. HUNT, R. SU, G. MEYER, *Design for multiinput nonlinear systems*, In Differential Geometric Control Theory, R. Brockett,.. ed. 268-298, Birkhäuser 1983.

[5] R. MARINO, W.M. BOOTHBY, D.L. ELLIOTT, *Geometric properties of linearizable control systems*, Submitted to Int. J. of Math. System Theory.

[6] J. ZABORSZKY, *Towards a comprehensive analysis and operating practice of the large compound HV-AC-DC system*, Report n.8203, Dept. of System Science and Math., Washington University, St. Louis, Mo, 1982.

[7] R. MARINO, *Feedback equivalence of nonlinear systems with applications to power system equations*, Doctoral Dissertation, Washington University, St. Louis, Mo, 1982.

[8] R. MARINO, *Stabilization and feedback equivalence to linear coupled oscillators*, To appear in Int. J. of Control.

[9] W.M. BOOTHBY, *Introduction to Differentiable Geometry and Riemannian Manifolds*, Academic Press, N.Y., 1979.

[10] R. MARINO, *A geometric approach for state feedback stabilization of power system networks*, Int. Conf. on Modelling, Identification and Control, IASTED, Innsbruck, 1984.

# STABILITY OF INTERCONNECTED SYSTEMS HAVING SLOPE-BOUNDED NONLINEARITIES*

Michael G. Safonov
Department of Electrical Engineering—Systems
University of Southern California
Los Angeles, CA 90089-0781
U.S.A.

## ABSTRACT

Improved stability criteria are obtained for systems having multiple nonlinearities. The key result (lemma 2) identifies a class of frequency dependent scaling factors d(s) such that, for any time-invariant slope-bounded nonlinearity f(x), the "scaled" operator $dfd^{-1}$ is in the same $L_2$ conic sector as f(x). Previous results admit only constant scaling factors d.

## INTRODUCTION

In the past two decades enumerable papers have appeared dealing with the stability of interconnected systems having multiple nonlinearities, Several very good survey papers describe these results [1], [2], [3]. Recently a number of papers have appeared employing related tools and techniques to address the problem of characterizing the "robustness" (i.e., stability margins) of systems having multiple uncertain components, including uncertain gains, uncertain transfer functions (e.g., [4]-[17]) and, in some cases, uncertain nonlinearities too ([4].[5],[11]). In all cases it is possible to arrange the system equations so that the nonlinearities and other uncertain elements appear as a diagonal operator (e.g., [5],[11],[13],[14]) though when system structure permits some results also permit direct analysis of non-diagonal configurations (e.g., [3],[12]) involving full matrices of nonlinearities or uncertain linear elements. In all cases the stability conclusions are conservative in the sense that sufficient, but not necessary conditions for stability are obtained, and much research has been devoted to the problem of exploiting information regarding the "structure" of the system to reduce this conservativeness (e.g., [13]-[17]). In the nonlinear interconnected system area the numerous results involving "M-matrix" stability conditions (including diagonal dominance results) may be viewed in this vein inasmuch as their proofs typically use the M-matrix condition to guarantee the existence of certain constant scaling factors for various spaces of signals such that standard small gain, positivity or conic sector stability criteria are satisfied or such that a Lyapunov function of a specified form decreases along system trajectories. The advantages of scaling for conservatism reduction are well known in the robustness literature too, but only three papers suggest constructive techniques for selection of these scaling factors. The paper [14] gives an algorithm for computing optimal frequency dependent scaling factors for systems with a diagonal matrix of linear uncertain elements and the results of [16] guarantee the convergence of the algorithm. The papers [12] and [13] suggest methods based on the Perron eigenvectors of certain positive matrices for selecting frequency-dependent scaling factors via a technique closely related to the M-matrix technique employed in the nonlinear literature to guarantee the existence of suitable constant scaling factors. This is discussed in [13] wherein it is pointed out that the added flexibility of using frequency dependent scaling factors provided by the linear case leads to less conservative results.

The starting point for the present work is the closely related work of Zames and Falb [23] and Willems (Theorem 3.15 of [21]) giving a class $\mathcal{M}$ of non-causal frequency dependent multipliers for a memoryless monotone nonlinearity f such that $m$f is a positive operator for each $m \in \mathcal{M}$. In the present paper sector nonlinearities (not necessarily monotonic) are considered. We obtain a useful new result (Lemma 2) characterizing a class $\mathcal{Z}$ of frequency dependent scaling factors d(s) such that the "scaled" operator $dfd^{-1}$ is inside the same conic sector as f. This new result is derived from the above-mentioned multiplier result in [21] and [23] by restricting attention to non-casual multipliers $m$ of the special form $m$ = d*d and using loopshifting transformations and noncausal multiplier factorization techniques. While the "scaling" afforded by the transformation $dfd^{-1}$ offers no benefit in reducing the conservativeness of single-input-single-output nonlinear stability criteria such as the standard circle criterion, diagonal matrices of such scaling factors can vastly reduce the conservativeness of frequency-domain stability tests for systems having multiple nonlinearities. These diagonal matrices of frequency dependent scaling factors are useful with diagonal matrices of slope-bounded nonlinearities (lemma 2), enabling one to greatly reduce the conservativeness of nonlinear interconnected system input-output stability criteria and robustness results such as those described in [1],[2],[5], and [11], for example .

## PROBLEM FORMULATION

Consider the system depicted in Fig.1 consisting of m memoryless

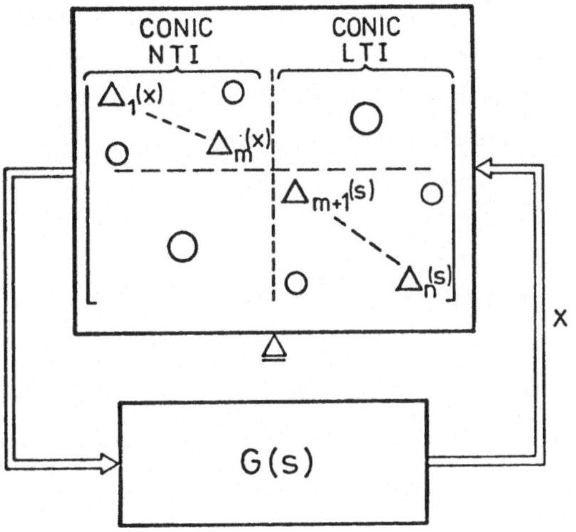


Fig.1 The System

nonlinear time-invariant (NTI) elements denoted $\Delta_i(x_i)$ $(i=1,\ldots,m)$, n-m uncertain linear time-invariant (LTI) elements denoted by their transfer functions $\Delta_i(s)$ $(i=m+1,\ldots,n)$, and an LTI feedback interconnection matrix H(s). We restrict our attention to the case where each NTI element $\Delta_i (i=1,\ldots,m)$ is single-input-single-output (SISO). Each NTI $\Delta_i$ is assumed to satisfy the incremental conicity condition $\Delta_i$ <u>incrementally inside</u> Cone $(c_i,r_i)$ for some real constants $c_i$ and $r_i$ and for all $i=1,\ldots,m$; that is, each NTI $\Delta_i$ satisfies the slope inequalities†

$$c_i - r_i \leq \frac{|\Delta_i(x)-\Delta_i(y)|}{|x-y|} \leq c_i + r_i \qquad \forall\ x-y \neq 0\ ; i=1,\ldots,m. \tag{2}$$

We further assume $\Delta_i(0)=0 \ \forall\ i=1,\ldots,m$.

To keep our notation and theorem statements simple, we also restrict the LTI $\Delta_i$ $(i=m+1,\ldots,n)$ to be SISO. Each LTI $\Delta_i$ is further required to be stable and to satisfy the conicity condition (e.g., [4]) $\Delta_i(s)$ <u>inside</u> Cone$(c_i(s),r_i(s))$ by which we mean

$$|\Delta_i(j\omega) - c_i(j\omega)| \leq r_i(j\omega) \qquad \forall\omega\ . \tag{3}$$

The transfer functions $c_i(s)$ must be analytic in the closed r.h.p.

---

† When $\Delta_i$ is differentiable, this inequality is equivalent to $c_i - r_i < \Delta_i'(x) < c_i + r_i\ \forall x$ where $\Delta_i'(x)$ denotes the slope of $\Delta_i(x)$.

We define two classes of transfer functions useful in scaling for reducing the conservativeness of conic sector stability results.

__Definition 1__: We denote by $Z$ the set of SISO transfer functions $d(s)$ satisfying the following conditions:

$$0 < |d(j0)| \le |d(j\infty)| < \infty \tag{4}$$

$$v(t) \triangleq \frac{1}{2\pi}\int_{-\infty}^{\infty}\left(\left|\frac{d(j\omega)}{d(j\infty)}\right|^2 - 1\right) e^{j\omega t}\, d\omega \le 0 \quad \forall\, t. \tag{5}$$ □

__Definition 2__[††] The set $Z_{RL}$ is the set of rational SISO transfer functions $d(s)$ of the form

$$d(s) = K \prod_{i=1}^{N} \frac{s-z_i}{s-p_i} \tag{6}$$

where $K, z_i$, and $p_i$ are real, $N \ge 0$ is an integer, and the $p_i$ and $z_i$ are ordered such that

$$0 > \ldots > z_i > p_i > z_{i+1} > p_{i+1} \qquad \forall\, i=1,\ldots,N-1 \tag{7}$$ □

The condition (7) of Definition 2 implies for example that each function $d(s)$ in $Z_{RL}$ must have a Bode magnitude plot whose slope is between 0 and 1, i.e.,

$$0 \le \frac{d \ln |d(j\omega)|}{d \ln \omega} \le 1 \qquad \forall\, \omega \tag{8}$$

for every $d(s) \in Z_{RL}$.

The condition (5) in Definition 1 says simply that the strictly proper part of $|d(j\omega)|^2$ must be the Fourier transform of a non-positive function, viz. $v(t)$.

Further insight into the composition of the set $Z$ is provided by the following lemma

__LEMMA 1__:

$$Z_{RL} \subset Z \tag{9}$$

__Proof__:

$$v(t) = \sum_{i=1}^{N} c_i e^{p_i |t|} \qquad \forall\, t \in (-\infty, \infty) \tag{10}$$

[††] That the set $Z_{RL}$ is of significance in SISO nonlinear stability theory is well known [18,23-25]: The elements $d(s) \in Z_{RL}$ are known to be useful as multipliers for systems having a SISO monotonic nonlinearity. The elements of $Z_{RL}$ correspond to driving point impedances of electrical networks consisting solely of resistors and inductors.

where $c_i$ is the residue of the l.h.p pole $p_i$ in the bilateral Laplace transform $V(s) = d(-s)\, d(s)$ of $v(t)$.

Using (6)-(7) it follows that

$$c_i = (s-p_i)\, V(s) \Big|_{s=p_i}$$

$$= K^2 \; \frac{z_i^2 - p_i^2}{-2p_i} \; \prod_{j \neq i} \left( \frac{z_j^2 - p_i^2}{p_j^2 - p_i^2} \right)$$

$$< 0 \;\; \forall\, i = 1,\ldots,N. \tag{11}$$

Hence $v(t) \leq 0 \;\forall t$. □

MAIN RESULT

In analyzing the stability of nonlinear systems such as in Fig.1 it is common practice to employ a diagonal multiplier matrix

$$D(s) = \mathrm{diag}(d_1,\ldots,d_m,d_{m+1}(s),\ldots,d_n(s))$$

as indicated in Fig,2 so as to reduce the conservativeness of conic sector stability results, e.g.[1],[2],[5]. All previous results restrict the terms $d_i$ $(i=1,\ldots,m)$ associated with the nonlinearities $\Delta_i (i=1,\ldots,m)$ to be constants. The following theorem enables further reduction of conservativeness by establishing that frequency-dependent $d_i(s)$ may be used for all $i=1,\ldots,n$ provided that $d_i(s) \in Z$ for the $m$ terms $i=1,\ldots,m$ associated with the nonlinearities.

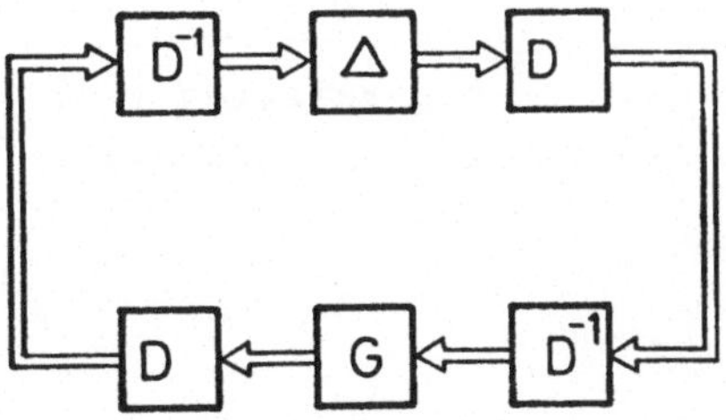


Fig.2 Diagonal scaling $D = \mathrm{diag}(d_1,\ldots d_n)$

<u>Theorem 1 (main result)</u>: Define

$$R(s) \stackrel{\Delta}{=} \mathrm{diag}(r_1,\ldots,r_m,r_{m+1}(s),\ldots,r_n(s)) \tag{12}$$

$$C(s) \stackrel{\Delta}{=} \mathrm{diag}(c_1,\ldots,c_m,c_{m+1}(s),\ldots,c_n(s)) \tag{13}$$

$$M(s) \stackrel{\Delta}{=} R(s)G(s)(I - C(s)G(s))^{-1}. \tag{14}$$

Let D(s) be any diagonal multiplier matrix of the form

$$D(s) = \mathrm{diag}(d_1(s),\ldots\ldots,d_m(s),d_{m+1}(s),\ldots\ldots,d_n(s)) \tag{15}$$

for which the first $m$ terms satisfy

$$d_i(s) \in Z \ , \ i=1,\ldots,m \ . \tag{16}$$

If M(s) is stable and if

$$\sup_{\omega} \bar{\sigma}(D(j\omega)M(j\omega)D^{-1}(j\omega)) < 1 \tag{17}$$

then the system of Fig.1 is stable for every collection of $\Delta_i$ satisfying (2) for $i=1,\ldots,m$ and (3) for $i=m+1,\ldots,n$, respectively. □

<u>Proof</u>: The standard loop-shifting and multiplier techniques (e.g., [19],[20]) enable the system of Fig.1 to be transformed into the equivalent system of Fig.3. To establish stability we employ the small gain theorem [19].

Note that (17) implies for every $\varepsilon$ sufficiently small that

$$g(DMD^{-1}) < 1 \tag{18}$$

where $g(\cdot)$ denotes the $L_2$ gain of the relation $(\cdot)$. To establish stability of the system of Fig.1, it thus suffices to prove that for some $\varepsilon > 0$

$$g(d_i\tilde{\Delta}_i d_i^{-1}) \leq 1 \quad i = 1,\ldots,n \ . \tag{19}$$

where

$$\tilde{\Delta}_i \stackrel{\Delta}{=} (\Delta_i - c_i)\ (1+\varepsilon)^{-1} r_i^{-1} \ . \tag{20}$$

For the LTI terms $i=m+1,\ldots,n$ (19) is clearly satisfied since for $i=m+1,\ldots,n$

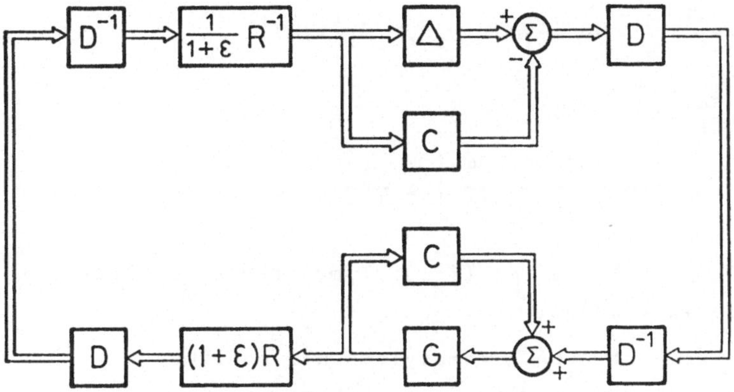


Fig. 3 System with multiplier $(1+\varepsilon)R$, loop shifting C and diagonal scaling $D = \text{diag}(d_1, \ldots, d_n)$

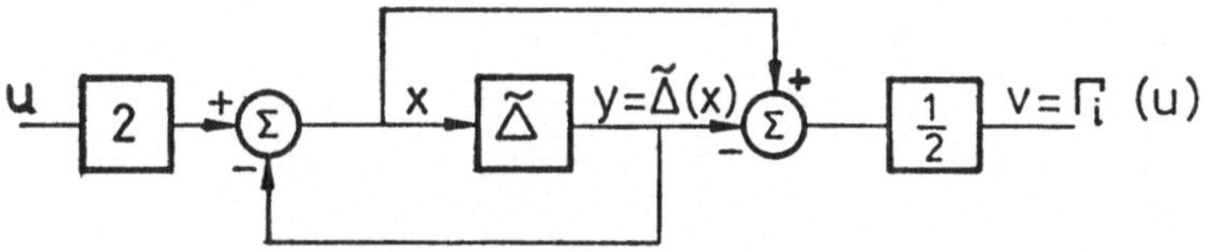


(a) $\Gamma_i(u)$

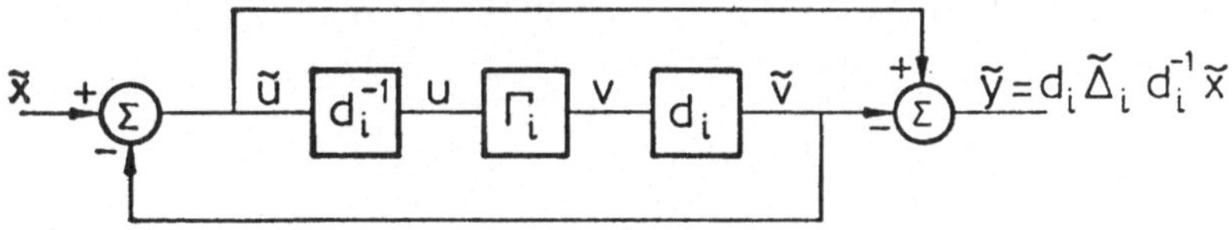


(b) $d_i \tilde{\Delta}_i d_i^{-1}$

Fig. 4

$$g(d_i\tilde{\Delta}_i d_i^{-1}) = \sup_\omega |d_i(j\omega)\tilde{\Delta}_i(j\omega)d_i^{-1}(j\omega)|$$

$$= \sup_\omega |\tilde{\Delta}_i(j\omega)|$$

$$= \sup_\omega \left|\frac{\Delta_i(j\omega) - c_i(j\omega)}{(1+\varepsilon)r_i(j\omega)}\right| \leq \frac{1}{1+\varepsilon} \tag{21}$$

by (3). For the NTI terms ($i=1,\ldots,m$), the inequality (19) follows from the lemma 2 below. □

Lemma 2. Let $\Delta_i$ be a memoryless NTI nonlinearity with $\Delta_i(0) = 0$ satisfying (2) and let $d_i(s) \in Z$ . Then,

$$d_i\Delta_i d_i^{-1} \ \underline{\text{inside}}\ \text{Cone}(c_i, r_i) \ , \tag{22}$$

i.e.,

$$||P_\tau(y-c_i x)||_{L_2} \leq ||P_\tau r_i x||_{L_2} \forall\ y = d_i\Delta_i d_i^{-1}x,\ x\in L_{2e}, \tau\in[0,\infty] \ . \tag{23}$$

Proof of Lemma 2. We assume without loss of generality that $d_i(s)$ and $d_i^{-1}(s)$ are stable and causal (if not, we could substitute the stable causal spectral factor of $d_i(s)d_i(-s)$). It suffices to establish that

$$g(d_i\tilde{\Delta}_i d_i^{-1}) \leq 1 \ \forall\ \varepsilon > 0. \tag{24}$$

where $\tilde{\Delta}_i$ is given by (20). From (2) and the assumption $\Delta_i(0) = 0$ we have

$$|\tilde{\Delta}_i(x) - \tilde{\Delta}_i(y)| \leq (1+\varepsilon)^{-1}|x-y| \forall\ x,y \in R \tag{25a}$$

$$\tilde{\Delta}(0) = 0 \tag{25b}$$

That is, $\tilde{\Delta}_i$ is a contraction with contraction constant $(1+\varepsilon)^{-1}$. Consequently the system of equations (see Fig.4(a))

$$y = \tilde{\Delta}_i x \tag{26a}$$

$$v = \frac{1}{2}(x-y) \tag{26b}$$

$$u = \frac{1}{2}(x+y) \tag{26c}$$

has a unique solution $(x,y,v)$ for any given $u$ and defines a function $v = \Gamma(u)$ . From (25a,b) and (26a,b,c) it follows that $\Gamma(0) = 0$ and that $\Gamma(u)$ is a bounded strictly incrementally positive function, which is to say that for some $\delta > 0$

$$\delta^{-1}|x-y|^2 > (\Gamma_i(x) - \Gamma_i(y))(x-y) > \delta|x-y|^2 \ \forall\ x,y \in R \tag{27}$$

By Theorem 3.15 of [21] it follows that for every $d \in Z$

$$\langle \xi,\eta \rangle_{L_2} \geq 0 \;\forall\; \xi = d^* d \Gamma \eta, \eta \in L_2 \tag{28}$$

where $d^*$ denotes $L_2$ adjoint of the convolution operator $d(s)$. For every $d \in Z$ the operators $d$ and $d^*$ are $L_2$ stable and invertible maps of $L_2$ into $L_2$. The operator $d^*$ and its inverse $d^{*-1}$ are anticausal and have respective bilateral Laplace transforms $d^*(s) = d(-s)$ and $d^{*-1}(s) = d^{-1}(-s)$.
Using the identity

$$\langle d_i^* \xi, \eta \rangle_{L_2} = \langle \xi, d_i \eta \rangle_{L_2} \qquad \forall \xi,\eta \in L_2 \tag{29}$$

and taking $\eta = d_i^{-1} P_\tau \tilde{u}$, it follows from (28) and the causality of $d_i \Gamma_i d_i^{-1}$ that

$$0 \leq \langle P_\tau \tilde{v}, P_\tau \tilde{u} \rangle L_2 \qquad \forall\; \tilde{v} = d_i \Gamma_i d_i^{-1} \tilde{u}, \tilde{u} \in L_{2e} \tag{30}$$

where $P_\tau$ denotes the standard linear truncation operater of input-output stability theory [19]-[21]. That is, $d_i \Gamma_i d_i^{-1}$ is a stable non-negative operator mapping $L_{2e}$ into $L_{2e}$.

Consider now the feedback system (see Fig.4(b)) determined by the equations

$$\tilde{v} = d_i \Gamma_i d_i^{-1} \tilde{u} \tag{31a}$$

$$\tilde{y} = \tilde{u} - \tilde{v} \;. \tag{31b}$$

$$\tilde{x} = \tilde{u} + \tilde{v} \tag{31c}$$

By the positive relation stability criterion, the system is $L_{2e}$ stable. Substitution of (31a,b,c) into (26a,b,c) reveals that every pair $(\tilde{x},\tilde{y})$ satisfying

$$\tilde{y} = d_i \tilde{\Delta}_i d_i^{-1} \tilde{x} \tag{32}$$

is a solution of the system in Fig.4b. Furthermore, observing that (31b,c) implies

$$\tilde{v} = \frac{1}{2}(\tilde{x} - \tilde{y}) \tag{33}$$

$$\tilde{u} = \frac{1}{2}(\tilde{x} + \tilde{y}) \tag{34}$$

it follows from the positivity inequality (30) that for all $\tilde{y} = d_i \tilde{\Delta}_i d_i^{-1} \tilde{x}$, $\tilde{x} \in L_{2e}$ and all $\tau$ that

$$\begin{aligned} 0 &\leq \langle P_\tau \tfrac{1}{2}(\tilde{x}-\tilde{y}),\; P_\tau \tfrac{1}{2}(\tilde{x}+\tilde{y}) \rangle_{L_2} \\ &= \tfrac{1}{4}(\|P_\tau \tilde{x}\|^2_{L_2} - \|P_\tau \tilde{y}\|\rangle^2_{L_2} \end{aligned} \tag{35}$$

and hence

$$g(d_i \Delta_i d_i^{-1}) \leq 1 \qquad (36)$$

which completes the proof of Lemma 2.

## DISCUSSION

Although no special method is proposed by Theorem 1 for selecting the best diagonal D(s) such that (17) is satisfied and $d_i(s) \in Z$ for the values i = 1,...,m corresponding to the nonlinear $\Delta_i$ terms, the results of [13], [14], and [16] provide considerable insight as to how this can be done. The "Perron eigenvector" $\ell_2$-norm scaling given in [13] may be a good choice, but it ignores the constraint $d_i(s) \in Z$ i=1,...,m. The algorithm described in [14] is globally convergent [16] to the optimal D(s), again ignoring the constraint $d_i(s) \in Z \ \forall\ i=1,\ldots,m$. Clearly functions $d_i(j\omega)$ which are "sufficiently smooth" and monotonic (i.e., nondecreasing in magnitude) are in $Z_{RL}$ and hence in Z. Furthermore, the monotonicity property can always be assured without loss of generality since, without altering the value of $\bar{\sigma}(D(j\omega)M(j\omega),D^{-1}(j\omega))$, one may replace D(s) by $d_p^{-1}(s)D(s)$ where $p \in \{1,\ldots,n\}$ is selected to satisfy

$$\frac{d}{d\omega}\,|d_p(j\omega)| \leq \frac{d}{d\omega}|d_i(j\omega)| \ \forall\ i \in \{1,\ldots,m\} \ . \qquad (37)$$

Thus, provided only that D(s) is sufficiently smooth, it is likely that $d_p^{-1}(s)D(s) \in Z$ . The procedures of [13] and [14] both approach a D(s) which minimizes $\bar{\sigma}(D(j\omega)M(j\omega)D^{-1}(j\omega))$ at each frequency $\omega$ and this D(s) will tend to be "smooth" provided that the singular vectors of D(s) very sufficiently smoothly with frequency. Consequently, one may expect the method of [13] and [14] to yield $d_i(s) \in Z$ whenever M(s) has singular vectors which vary sufficiently smoothly with frequency, so the constraint $d_i \in Z$ i=1,...,m would not be an issue. In more difficult situations where the constraint $d_i(s) \in Z$ becomes an active constraint in the minimization

$$\min\{\sup_{\omega} \bar{\sigma}(DMD^{-1}) \,|\, D = \mathrm{diag}(d_1,\ldots,d_n), d_1(s) \in Z \ \forall\ i=1,\ldots,m\} \qquad (38)$$

more sophisticated methods would be invoked to accommodate the functional inequality constraints $d_i(s) \in Z$, i=1,...,m . The facts that the functional $\bar{\sigma}(DMD^{-1})$ is convex in D [16] and that the set $\{d^*d \,|\, d \in Z\}$ is evidently convex suggest that general purpose algorithms for optimization with functional inequality constraints (e.g. [22]) may be useful in optimizing the choice of D(s) in these more difficult situations.

## CONCLUSIONS

The key result is Lemma 2 which establishes that each $d(s) \in Z$ has the property that $dfd^{-1}$ is inside the same conic sector as f for any incrementally-conic memoryless time-invariant nonlinearity f. This property of the class Z is derived via the loop-shifting and multiplier transformations depicted in Figs.3 and 4 from a different, but closely related result of [21,23] which characterizes a class $\mathcal{M}$ of multipliers m(s) for which mf is a positive operator for every memoryless time-invariant monotone nonlinearity f. The class Z and the class $\mathcal{M}$ are related by

$$Z :=\{ d \mid d^*d \in \mathcal{M}\};$$

so, $Z * Z$ is a subset of M.

Lemma 1 establishes that the set $Z_{RL}$ of driving point impedances of resistor inductor networks is a subset of Z. This should not be confused with the related result in [23-25] that $Z_{RL}$ is a subset of $\mathcal{M}$.

Theorem 1 enables improved (i.e., less conservative) sufficient conditions for stability to be obtained for systems containing multiple time-invariant nonlinearities. Like previous results, the present Theorem 1 uses a diagonal "scaling" matrix, viz.

$$D(s) = diag(d_1(s),\ldots,d_m(s),d_{m+1}(s),\ldots,d_m(s)),$$

in conjunction with the small-gain theorem or, more generally, conic relation stability criteria. Whereas in previous criteria the terms $d_i(s)$ for i=1,...,m corresponding to nonlinear elements were required to be constant, theorem 1 expands the class of admissible $d_i(s)$ for time-invariant incrementally conic nonlinearities to the class Z admitting certain frequency dependent $d_i(s)$ including for example $d_i(s)$ terms realizable as the driving point impedance of a resistor-inductor network.

## REFERENCES

[1] N.R. Sandell, P. Varaiya, M. Athans, and M.G. Safonov, "Survey of Decentralized Control Methods for Large Scale Systems", IEEE Trans. on Automatic Control, AC-23, pp.108-128,1978.

[2] M. Araki, "Stability of Large-Scale Nonlinear Systems - Quadratic-Order Theory of Composite System Method Using M-Matrices", IEEE Trans. on Automatic Control, AC-23, pp.129-142, 1978.

[3] A Michel, "On the Status of Stability of Interconnected Systems", IEEE Trans. on Circuits and Systems, CAS-30, pp.326-340, 1983. (Also published in IEEE Trans. on Automatic Control, AC-28, June, 1983 and in IEEE Trans. on Systems Man and Cybernetics, SMC-13, July/August 1983.)

[4] M.G. Safonov, "Robustness and Stability Aspects of Stochastic Multivariable Feedback System Design", Ph.D. Dissertation, Mass. Inst. of Technology, Cambridge, MA, Sept. 1977; also M.G. Safonov, Stability and Robustness of Multivariable Feedback Systems, MIT Press, Cambridge, MA, 1980.

[5] M.G. Safonov and M. Athans, "A Multiloop Generalization of the Circle Stability Criterion for Stability Margin Analysis", IEEE Trans. on Automatic Control. AC-26, pp. 415-422, 1981.

[6] J.C. Doyle, "Robustness of Multiloop Linear Feedback Systems", in Proc. 1978 IEEE Conf. on Decision and Control, San Diego, CA, Jan. 10-12, 1979.

[7] J.C. Doyle and G. Stein, "Multivariable Feedback Design: Concepts for a Classical/Modern Synthesis", IEEE Trans. on Automatic Control, AC-26, pp. 4-16, 1981.

[8] I. Postlethwaite, J.M. Edmunds and A.G.J. MacFarlane, "Principal Gains and Principal Phases in the Analysis of Linear Multivariable Feedback Systems", Ibid., pp. 32-46.

[9] M.G. Safonov, A.J. Laub and G.L. Hartmann, "Feedback Properties of Multivariable Systems: The Role and Use of the Return Difference Matrix", Ibid., pp.47-65.

[10] N.A. Lehtomaki, N.R. Sandell and M. Athans, "Robustness Results in Linear-Quadratic Gaussian Based Multivariable Control Designs, Ibid., pp 75-92.

[11] M.G. Safonov, "Propagation of Conic Model Uncertainty in Hierarchical Systems" IEEE Trans. on Circuits and Systems, CAS-30,pp.388-396, 1983. (Also published in IEEE Trans. on Automatic Control AC-28, June 1983 and in IEEE Trans. on Systems Man and Cybernetics, SMC-13, July/ August 1983.)

[12] D.J.N. Limebeer and Y.S. Hung, "Robust Stability on Inter-connected Systems", Ibid.,pp.397-403.

[13] M.G. Safonov, "Stability Margins of Diagonally Perturbed Multivariable Feedback Systems", IEEE Proc., 129,Pt.D., pp. 251-256, 1982.

[14] J.C. Doyle, "Analysis of Feedback Systems with Structured Uncertainties", Ibid., pp. 242-250.

[15] J.C. Doyle, J.E. Wall and G. Stein, "Performance Robustness Analysis for structured Uncertainty", in Proc. IEEE Conf. on Decision and Control, Orlando, FL, December, 1982.

[16] M.G. Safonov and J.C. Doyle, "Optimal Scaling for Multivariable Stability Margin Singular Value Computation", in Proc. MECO/EES Symposium, Athens, Greece, August 29-September 2, 1983.

[17] M.F. Barratt, "Conservatism with Robustness Tests for Linear Feedback Control Systems", Ph.D. Thesis, University of Minnesota, June 1980; report 80SRC35, Honeywell Systems and Research Center, Minneapolis, MN.

[18] K.S. Narendra and J.H. Taylor, "Frequency Domain Criteria for Absolute Stability, Academic Press, NY, 1973.

[19] G. Zames, "On the Input-Output Stability of Time-Varying Nonlinear Feedback Systems - Part I: Conditions Using Concepts of Loop Gain, Conicity, and Positivity", IEEE Trans. on Automatic Control, AC-11, pp.228-238, 1966.

[20] C.A. Desoer and M. Vidyasagar, "Feedback Systems: Input-Output Properties, Academic Press, NY, 1975.

[21] J.C. Willems "The Analysis of Feedback Systems, MIT Press, Cambridge, MA, 1971.

[22] E. Polak and D.Q. Mayne, "An Algorithm for Optimization Problems with Functional Inequality Constraints", IEEE Trans. on Automatic Control, AC-21, pp.184-193, 1976.

[23] G. Zames and P. Falb, "Stability Conditions for Systems with Monotone and Slope-Restricted Nonlinearities," SIAM J. Control, vol.6, pp.89-108, 1968.

[24] P. Falb and G. Zames, "Multipliers with Real Poles and Zeros: An Application of a Theorem on Stability Conditions," IEEE Trans. on Automatic Control, Vol.AC-13, pp.125-126, 1968.

[25] R.W. Brockett and J.L. Willems, "Frequency-Domain Stability Criteria - Parts I and II," IEEE Trans. on Automatic Control, Vol. Ac-10, pp.255-261 and pp.407-413.

Research supported in part by AFOSR Grant 80-0013, in part by NSF Grant INT-8302754, and in part by Honeywell Systems and Research Center, Minneapolis, MN. This work was completed while the author was an SERC Senior Visiting Fellow at University Engineering Dept., Control and Management Systems Division, Mill Lane, Cambridge CB2 1RX, United Kingdom.

Session 6

# LINEAR SYSTEMS I

# SYSTÈMES LINÉAIRES I

## ON SYMMETRIC EXTRACTION POLYNOMIAL MATRIX SPECTRAL FACTORIZATION

F.M. Callier, Senior Member IEEE
Department of Mathematics
Facultés Universitaires N.-D. de la Paix
8, Rempart de la Vierge
B-5000 Namur BELGIUM

Abstract

We report a revision, [1], of the 1963 Davis algorithm for the spectral factorization of a parahermitian nonnegative polynomial matrix $\Phi$ by symmetric factor extraction : this algorithm is careless about zeros at infinity. By introducing the notion of *diagonal reducedness* of $\Phi$ we obtain an easy sufficient test for the absence of zeros at infinity. We show then i) how to get $\Phi$ diagonally reduced by diagonal excess reduction steps, removing all zeros at infinity and ii) how to remove symmetrically finite zeros while keeping $\Phi$ diagonally reduced, (whence free of zeros at infinity). Didactical examples are given. This results in a revised symmetric extraction spectral factorization algorithm with *monotone degree control.*

### 1. Introduction

It is the objective of this paper to report a revision of the 1963 Davis algorithm for the spectral factorization of a parahermitian nonnegative (p.h.n.n.) polynomial matrix, [1], [2]. This problem (PSF) is as follows :
Let $\Phi(s) \in \mathbb{R}[s]^{m\times m}$ be p.h.n.n., i.e. such that $\Phi(s) = \Phi_*(s) := \Phi^T(-s)$ and $\Phi(j\omega) \geqslant 0$ for all $\omega \in \mathbb{R}$ . (1)
Find a spectral factor $W(s) \in \mathbb{R}[s]^{m\times m}$ such that $\Phi(s) = W_*(s)\, W(s)$ and $W(s)$ has all its finite zeros in $\mathbb{C}_- := \{s : \mathrm{Re}\ s \leqslant 0\}$ . (2)
Due to the recent interest in polynomial matrix fractions in system theory, e.g. [3]-[5], problem (PSF) has turned out to be important in linear quadratic optimal control, e.g. [6], [7].
Based upon the symmetry of the finite spectrum of $\Phi(s)$ , (1) , whereby the set of finite zeros of $\det \Phi(s)$ satisfies

$$Z\,[\det \Phi(s)] = Z_- \cup Z_+ \tag{3}$$

with

$$z \in Z_- = Z\,[\det W(s)] \subset \mathbb{C}_- \quad \text{iff} \quad -z \in Z_+ = Z\,[\det W_*(s)] \subset \mathbb{C}_+ \, , \tag{4}$$

Davis' algorithm uses repetitive symmetric factor extractions

$$\Phi(s) \leftarrow T_{*}(s)^{-1} \cdot \Phi(s) \cdot T(s)^{-1} \tag{5}$$

where $\Phi(s)$ remains a p.h.n.n. polynomial matrix and $T(s) \in \mathbb{R}[s]^{m\times m}$ is the extraction factor, to make $\Phi(s)$ , (1) , first unimodular and then constant removing resp. the finite zeros and zeros at $\infty$ of $\Phi$ .
Using the extraction factors and a final constant factorization $K = U^T U > 0$ with $U \in \mathbb{R}^{m\times m}$ , one can then assemble a spectral factor $W(s)$ , (2) , of $\Phi(s)$.
The trouble however, [1] , is that a careless removal of finite zeros may lead to the introduction of zeros at $\infty$ into $\Phi$ adding unwanted degree content : there is a danger of "degree explosion". Since for a successful computation of the spectral factor it is all important to work at any time with a $\Phi$ of minimal excessive degree content,better first get rid of all zeros at $\infty$ and then apply *optimal* finite zero extractions, i.e. without reintroducing zeros at $\infty$ .... This will be done below using the new notion of *diagonal reducedness of* $\Phi$ . For proofs and additional perspective see [1].

## 2. Infinite Zero Elimination

In this section we report an easy sufficient test for the absence of zeros at $\infty$ of $\Phi$ , (1) , namely whether or not $\Phi$ is diagonally reduced. We report also how any p.h.n.n. $\Phi$ can be made diagonally reduced eliminating hereby its zeros at infinity.
Let $\Phi(s) \in \mathbb{R}[s]^{m\times m}$ be p.h.n.n. , (1) . Let $\delta[\ldots]$ denote the degree of the polynomial between the brackets.
We call *diagonal degree excess* of $\Phi(s)$ the integer

$$\varepsilon[\Phi(s)] = \sum_{i=1}^{m} \delta[\Phi_{ii}(s)] - \delta[\det \Phi(s)] \tag{6}$$

Furthermore, if

$$\delta_i := \frac{1}{2}\delta[\Phi_{ii}(s)] \qquad \text{for } i = 1, 2, \ldots, m \tag{7}$$

denote the *half diagonal degrees* of $\Phi(s)$ , we call (symmetric) *highest degree coefficient matrix* of $\Phi(s)$ the constant matrix

$$\Phi_\infty := \lim_{s\to\infty}\left( \operatorname{diag}[(-s)^{-\delta_i}]_{i=1}^{m} \cdot \Phi(s) \cdot \operatorname{diag}[s^{-\delta_i}]_{i=1}^{m} \right) . \tag{8}$$

We note here that $\Phi_\infty$ can be read from $\Phi(s)$ ; moreover the parameters above play an important role.

Lemma 2.1. Let $\Phi(s) \in \mathbb{R}[s]^{m\times m}$ be p.h.n.n.. Let $W(s) \in \mathbb{R}[s]^{m\times m}$ denote any spectral factor, (2) , with *column degrees* $\gamma_i := \delta_{ci}[W(s)]$ for $i = 1, \ldots, m$. Let $W(s)$ have *column degree excess*

$$\gamma[W(s)] := \sum_{i=1}^{m} \delta_{ci}[W(s)] - \delta[\det W(s)] \tag{9}$$

and a *highest column degree coefficient matrix*

$$W_{c\infty} := \lim_{s\to\infty}\left\{W(s) . \operatorname{diag}[s^{-\gamma_i}]_{i=1}^{m}\right\} . \tag{10}$$

Then the diagonally induced parameters, (6)-(8) of $\Phi(s)$ are such that

a) For all $i = 1, \ldots, m$ and $j = 1, \ldots, m$

$$\delta[\Phi_{ij}(s)] \leqslant \delta_i + \delta_j . \tag{11}$$

b) For all $i = 1, \ldots, m$

$$\delta_i = \gamma_i := \delta_{ci}[W(s)] . \tag{12}$$

c)
$$\varepsilon[\Phi(s)] = 2\gamma[W(s)] \geqslant 0 . \tag{13}$$

d)
$$\Phi_\infty = W_{c\infty}^T . W_{c\infty} \geqslant 0 . \quad \blacksquare \tag{14}$$

Note that the half diagonal degrees of $\Phi$ bound the off diagonal degrees and equal the column degrees of any spectral factor $W$. Moreover the diagonal excess is twice the column excess and the highest degree coefficient matrices are related by the square root relation (14). Hence if one knows, [3]-[5], that the column reducedness of $W$, namely $\gamma[W(s)] = 0$ or equivalently $\det W_{c\infty} \neq 0$, is an easy sufficient test for absence of zeros at $\infty$ of $W$, (e.g. [1]), then the following definition makes sense ... .

Definition 2.1. Let $\Phi(s) \in \mathbb{R}[s]^{m\times m}$ be p.h.n.n.. We say that $\Phi(s)$ is *diagonally reduced* (d.r.) iff $\Phi(s)$ has zero diagonal degree excess, i.e. $\varepsilon[\Phi(s)]$ given by (6) is zero. □

We have then

Theorem 2.1. [Equivalent Definitions]. Let $\Phi(s) \in \mathbb{R}[s]^{m\times m}$ be p.h.n.n. and let $W(s)$ , (2) , be any spectral factor of $\Phi(s)$. Under these conditions

$$\Phi(s) \text{ is d.r.} \tag{15}$$

iff

$$W(s) \quad \text{is} \quad \text{column reduced} \tag{16}$$

or iff

$$\det \ \Phi_\infty \neq 0 \tag{17}$$

where $\Phi_\infty$ is the coefficient matrix (8).

Moreover

$$\Phi(s) \quad \text{is} \quad \text{d.r.} \quad \Rightarrow \quad \Phi(s) \ \text{has no zeros at} \ \infty \ . \qquad \blacksquare \tag{18}$$

We note here that $\Phi(s)$ has no zeros at $\infty$ iff $\Phi(s)^{-1}$ is proper, (bounded at $\infty$), and that a practical sufficient test for this property is (17). Moreover just like any spectral factor can be made column reduced, [3, Th. 2.5.7.], one discovers that any p.h.n.n. polynomial matrix $\Phi$ can be made d.r..

Theorem 2.2. [Getting $\Phi(s)$ d.r.; Infinite Zero Elimination]. Let $\Phi(s) \in \mathbb{R}[s]^{m\times m}$ be p.h.n.n. and not d.r.. Then the diagonal degree excess, (6), of $\Phi(s)$ is reduced by at least two units by the following symmetric extraction procedure called a *diagonal excess reduction step*.

1. Read from $\Phi(s)$ its highest degree coefficient matrix (8).
2. Compute a nonzero vector $k \in \mathbb{R}^m$ such that

   $$\Phi_\infty k = \theta \quad , \tag{19}$$

   where $\theta$ denotes the zero vector.
3. If $k = (k_i)_{i=1}^m$ , compute the active index set

   $$N \ := \ \{i \in \underline{m} \ : \ k_i \neq 0\} \tag{20}$$

   where $\underline{m} = \{1, 2, \ldots, m\}$ , and the *highest active diagonal degree index set*

   $$H \ := \ \{t \in N : \delta_t = \tfrac{1}{2}\,\delta\,[\Phi_{tt}(s)] \geq \delta_i = \tfrac{1}{2}\,\delta\,[\Phi_{ii}(s)] \quad \forall\ i \in N\} \ . \tag{21}$$

4. Denoting by $e_r$ the $r^{th}$ unit vector of $\mathbb{R}^m$ , compute the unimodular polynomial matrix

   $$T(s)^{-1} = I - e_r \cdot e_r^T + \mathrm{diag}\,[s^{\delta_r - \delta_i}]_{i=1}^m \cdot k(k_r)^{-1} \cdot e_r^T \qquad \text{with} \quad r \in H \ . \tag{22}$$

5. Compute

   $$\breve{\Phi}(s) \ = \ T_*(s)^{-1} \cdot \Phi(s) \cdot T(s)^{-1} \quad . \tag{23}$$

STOP : $\varepsilon\,[\check{\Phi}(s)] \leqslant \varepsilon\,[\Phi(s)] - 2$ . (24)

Hence $\Phi(s)$ can be made d.r. by at most $\frac{1}{2}\varepsilon\,[\Phi(s)]$ diagonal excess reduction steps. ◘

Note here especially that $T(s)^{-1}$ , (22), is the unit matrix with column $r$ replaced by a polynomial vector of appropriate monomials with entry $rr$ equal to 1 .
As a consequence in (23), column $r$ of any spectral factor $W$ of $\Phi$ will be replaced by a polynomial combination of columns lowering its degree, [3, Th. 2.5.7.], and similarly so far row $r$ of $W_*$ . Hence (24) follows by (12)-(13).

<u>Example 2.1</u>. We consider the p.h.n.n.

$$\Phi(s) = \begin{bmatrix} 1 & -(1+s)^2 \\ -(1-s)^2 & (1-s^2)(2-s^2) \end{bmatrix} \quad \text{with} \quad \det \Phi(s) = 1 - s^2 .$$

$\Phi$ is not d.r. since $\delta_1 = 0$ , $\delta_2 = 2$ , so $\varepsilon\,[\Phi(s)] = 2$ .
(19) reads

$$\Phi_\infty k = \begin{bmatrix} 1 & -1 \\ -1 & 1 \end{bmatrix} \begin{bmatrix} 1 \\ 1 \end{bmatrix} = \theta$$

with (20) and (21) reading $N = \{1, 2\}$ , $H = \{2\}$ .
As a consequence (22) gives

$$T(s)^{-1} = \begin{bmatrix} 1 & s^2 \\ 0 & 1 \end{bmatrix} ,$$

and so for (23) we get

$$\check{\Phi}(s) = \begin{bmatrix} 1 & -1-2s \\ -1+2s & 2-5s^2 \end{bmatrix} , \quad \det \check{\Phi}(s) = 1 - s^2 ,$$

with $\delta_1 = 0$ , $\delta_1 = 1$ : $\varepsilon\,[\check{\Phi}(s)] = 0$ such that $\check{\Phi}$ is d.r. ◘

Note here the typical loss of excessive degree content when eliminating infinite zeros when $\det \Phi$ does not change : column 2 and row 2 of $\check{\Phi}$ are simpler.

3. Optimal Finite Zero Extraction

In this section we report how to extract symmetrically finite zeros from a d.r. p.h.n.n. polynomial matrix $\Phi$ such that the resulting p.h.n.n. polynomial matrix is again d.r.. As a result no zeros at $\infty$ will be reintroduced by (18).

3.1. Definitions 3.1. Let $\Phi(s) \in \mathbb{R}[s]^{m\times m}$ be p.h.n.n. and d.r. with left and right finite half spectra $Z_-$ resp. $Z_+$, (3)-(4). Let $z \in Z_-$ and $-z \in Z_+$. We call *optimal extraction* of the zeros $z$ and $-z$ the operation described by

$$\check{\Phi}(s) = T_*(s)^{-1} \,.\, \Phi(s) \,.\, T(s)^{-1} \quad , \tag{25}$$

where the following four conditions hold :

a. $T(s) \in \mathbb{R}[s]^{m\times m}$ . (26)

b. $\det T(s) = s - z \quad \text{if} \quad z \in \mathbb{R}$ (27a)
or
$\det T(s) = (s - z)(s - \bar{z}) \quad \text{if} \quad z \in \mathbb{C} \setminus \mathbb{R}$ . (27b)

c. $\check{\Phi}(s) \in \mathbb{R}[s]^{m\times m}$ , (necessarily p.h.n.n.). (28)

d. $\check{\Phi}(s)$ is again d.r.. (29)

We call *optimal inverse standard right factor* (optimal i.s.r.f.) of $\Phi(s)$ any proper rational mxm matrix $T(s)^{-1}$ s.t. (25)-(29) holds and in addition :

e. $T(s)$ is column reduced with highest column degree coefficient matrix

$$T_{c\infty} = I \,. \qquad \square \tag{30}$$

We note here that an optimal i.s.r.f. is associated with a $\mathbb{C}_-$ - zero $z$ (and $\bar{z}$) of $\Phi$ and that it must have real coefficients. It is nice to know that such inverse factor can always be computed from an eigenvector of $\Phi$ at $z$ as indicated below. Since a unimodular d.r. p.h.n.n. matrix is constant, it follows then that a d.r. p.h.n.n. matrix $\Phi$ can be made constant after a finite number of optimal symmetric finite zero extractions. No zeros at $\infty$ will be reintroduced and the diagonal degree excess will be kept zero.

3.2. We describe now the *eigenvector parameters* important for the construction of an optimal i.s.r.f.

Let $\Phi(s) \in \mathbb{R}[s]^{m\times m}$ be p.h.n.n. with a zero

$$z = u + jv \in Z_- \subset \mathbb{C}_- \tag{31}$$

and corresponding possibly complex eigenvector

$$\xi = \eta + j\zeta \in \mathbb{C}^m \tag{32}$$

related by

$$\Phi(z)\,\xi = \theta \quad . \tag{33}$$

Observing that $\xi$ has components

$$\xi = (\xi_i)_{i=1}^m = (\eta_i + j\,\zeta_i)_{i=1}^m \tag{34}$$

and remembering the half diagonal degrees $\delta_i$ , (7), of $\Phi$ , $\xi$ has an *active index set* $N$ and a *highest active diagonal degree index set* $H$ given by

$$N := \{i \in \underline{m} \ : \ \xi_i \neq 0\} \tag{35}$$

$$H := \{t \in N \ : \ \delta_t \geqslant \delta_i \quad \forall \ i \in N\} \subset N \quad . \tag{36}$$

As a consequence the eigenvector $\xi$ has a *highest active diagonal degree subeigenvector*

$$\xi_H := (\xi_t)_{t\in H} = (\eta_t + j\zeta_t)_{t\in H} \in \mathbb{C}^{\# H} \quad , \tag{37}$$

with real and imaginary parts

$$\eta_H := (\eta_t)_{t\in H} \in \mathbb{R}^{\# H} \quad \text{resp.} \quad \zeta_H := (\zeta_t)_{t\in H} \in \mathbb{R}^{\# H} \quad , \tag{38}$$

which generate a *highest active diagonal degree eigenvector submatrix*

$$C_H = \begin{pmatrix}\eta_H & \zeta_H\end{pmatrix} \in \mathbb{R}^{\# H \times 2} \quad , \tag{39}$$

where $\# H$ denotes the number of indices in the set $H$ . Now, observing the scaling freedom on $\xi$ , (it can be multiplied by any nonzero complex constant), we see, [1], that for the zero $z$ and corresponding eigenvector $\xi$ , (31)-(39), we have the following *alternatives* :

a. <u>Case 1</u> : $z \in \mathbb{R}$ and $\xi \in \mathbb{R}^m$ . (40)

b. <u>Case 21</u> : $z \in \mathbb{C} \setminus \mathbb{R}$ and rank $C_H = 1$ , in which case the highest active diagonal degree subeigenvector $\xi_H$ , (37)-(38), can be made real by scaling, i.e. (41)

$$\xi_H = \eta_H + j\,\theta \in \mathbb{R}^{\# H} \quad .$$

c. <u>Case 22</u> : $z \in \mathbb{C} \setminus \mathbb{R}$ and rank $C_H = 2$ , in which case the highest active diagonal degree eigenvector submatrix (39), has a $2 \times 2$ nonsingular submatrix. (42)

It is these alternatives which enable us to construct always an optimal i.s.r.f. as in Definitions 3.1. ... .

3.3. Theorem 3.1. [Optimal Symmetric Finite Zero Extraction]. Let $\Phi(s) \in \mathbb{R}[s]^{m\times m}$ be p.h.n.n. and d.r.. Let $\Phi(s)$ have a zero $z = u + jv \in Z_-$ and corresponding eigenvector $\xi = \eta + j\,\zeta \in \mathbb{C}^m$ with properties (31)-(39) and alternatives (40)-(42). Consider an optimal symmetric extraction of $z \in Z_-$ and $-z \in Z_+$ using an optimal i.s.r.f. $T(s)^{-1}$ as in Definitions 3.1.
Under these considerations such extraction is always possible by satisfying one of the conditions of

Decision Rule 3.1.

a. For case 1, (40), pick $T(s)^{-1}$ of class 1, i.e.

$$T(s)^{-1} = I - e_r \cdot e_r^T + \xi(s-z)^{-1}(\xi_r)^{-1} \cdot e_r^T \tag{43}$$

where one chooses

$$r \in H \quad . \tag{44}$$

b. For case 21, (41), pick $T(s)^{-1}$ of class 21, i.e.

$$T(s)^{-1} = I - e_r \cdot e_r^T + (\eta\, v + \zeta(s-u))((s-u)^2 + v^2)^{-1}(\eta_r\, v)^{-1} \cdot e_r^T \tag{45}$$

where one chooses

$$r \in H \quad . \tag{46}$$

c. For case 22, (42), pick $T(s)^{-1}$ of class 22, i.e.

$$T(s)^{-1} = I - V.V^T + C(sI - A)^{-1}(V^T C)^{-1}.V^T \tag{47}$$

with

$$C = \begin{pmatrix} \eta & \zeta \end{pmatrix} \in \mathbb{R}^{m\times 2}\ , \quad A = \begin{pmatrix} u & +v \\ -v & u \end{pmatrix} \in \mathbb{R}^{2\times 2}\ , \quad V = \begin{pmatrix} e_r & e_q \end{pmatrix} \in \mathbb{R}^{m\times 2}\ , \tag{47a-c}$$

where one chooses the indices

$$r \text{ and } q \text{ in } H \text{ such that } V^T C = \begin{pmatrix} \eta_r & \zeta_r \\ \eta_q & \zeta_q \end{pmatrix} \in \mathbb{R}^{2\times 2} \text{ is nonsingular.} \tag{48}$$

Moreover, under these conditions, denoting by $\delta_i$ and $\breve{\delta}_i$ half diagonal degrees of $\Phi$ and $\breve{\Phi}$ in (25), all half diagonal degrees of $\Phi$ and $\breve{\Phi}$ are identical (49) with the following exceptions :

a. For case 1 , $\breve{\delta}_r = \delta_r - 1$ . (50)

b. For case 21 , $\breve{\delta}_r = \delta_r - 2$ . (51)

c. For case 22 , $\check{\delta}_r = \delta_r - 1$ and $\check{\delta}_q = \delta_q - 1$ . ■ (52)

We note here, [1], that optimal symmetric finite zero extraction is performed by picking (a) diagonal element(s) of highest active degree and transforming the corresponding column(s) and row(s) of $\Phi$ , (see Decision Rule 3.1.); this reflects corresponding transformations on a spectral factor and its parahermitian, ($\Phi = W_*W$), resulting in lower column- resp. row degrees of $W$ resp. $W_*$ :
e.g. for case 1, column $r$ of $W$ is replaced by a polynomial vector having a zero at $z$ , (see (33)), and then divided by $s - z$ lowering the rth column degree of $W$ ... Conclusions (49)-(52) for the half diagonal degrees of $\Phi$ are then natural in view of (12) ... .

Example 3.1. Consider the d.r. $\Phi(s)$, (1), given by

$$\Phi(s) = \begin{bmatrix} -2s^2 + 5 & -5s + 8 \\ 5s + 8 & 13 \end{bmatrix} \qquad \det \Phi(s) = 1 - s^2$$

with $\delta_1 = 1$ and $\delta_2 = 0$ and $Z_- = \{-1\}$. We shall perform an optimal symmetric extraction of $z = -1$ and $-z = 1$ .

According to (33) we have

$$\Phi(-1)\ \xi = \begin{bmatrix} 3 & 13 \\ 3 & 13 \end{bmatrix} \begin{bmatrix} 1 \\ -(3/13) \end{bmatrix} = \theta \quad ,$$

with by (35) and (36), $N = \{1, 2\}$ and $H = \{1\}$ . Since $z = -1$ is real, we are in case 1 with class 1 optimal i.s.r.f.

$$T(s)^{-1} = \begin{bmatrix} (s + 1)^{-1} & 0 \\ -(3/13)(s + 1)^{-1} & 1 \end{bmatrix} ,$$

(see (43)-(44)). Hence according to (25)

$$\check{\Phi}(s) = \begin{bmatrix} 2 & 5 \\ 5 & 13 \end{bmatrix} > 0 \quad , \quad \check{\delta}_1 = 0 \quad \check{\delta}_2 = 0 \quad .$$

Note that (49)-(50) hold. The zeros $-1$ and $1$ have been optimally extracted : $\check{\Phi}$ is constant and d.r..

## 4. A Monotone Degree Control Algorithm for Spectral Factorization

In view of the operations of Theorems 2.2. and 3.1. it is now natural to reformulate Davis' algorithm of the introduction ... .

Algorithm 4.1. [Spectral Factorization]

Data : we are given a p.h.n.n. $\Phi(s) \in \mathbb{R}[s]^{mxm}$ , (1).

1. If $\Phi(s)$ is d.r. , (17), skip. Otherwise get $\Phi(s)$ d.r. by diagonal excess reduction steps, (Theorem 2.2.).

2. If $\Phi(s) \equiv K > 0$ , where $K \in \mathbb{R}^{mxm}$ , skip. Otherwise reduce $\Phi(s)$ to such constant matrix $K > 0$ by optimal symmetric finite zero extractions, (Theorem 3.1. with inverse extraction factor, (43)-(48), associated with (a) zero(s) in $Z_-$ , (3)-(4)).

3. Perform a constant factorization $K = U^T U > 0$ , where $U \in \mathbb{R}^{mxm}$ , e.g. Cholesky factorization.

4. A spectral factor of $\Phi(s)$ is

$$W(s) = U\, T_k(s) . T_{k-1}(s) \ldots T_1(s) \tag{53}$$

where the $T_j(s)$ in $\mathbb{R}[s]^{mxm}$ are the (right) extraction factors detected during stages 1 and 2 . ∎

## Final Comments

α. The extraction factors, (53), are easily computed from their inverses $T(s)^{-1}$ : see [1, formulas (2.37) and (3.35)-(3.37)] .

β. By the comment following Theorem 2.1. and by Theorem 3.1. it follows that in Algorithm 4.1. all half diagonal degrees $\delta_i$ , (7), are driven monotonically to zero. This provides by the dominance formula (11) a monotone degree control for all elements of $\Phi$ .

γ. An example using optimal i.s.r.f.'s of classes 21 and 22 for the extraction of nonreal zeros is given in [1, Sec. 4] .

δ. In contrast to the Davis algorithm of the introduction, Algorithm 4.1., 1) guards against a "degree explosion" by being careful about zeros at $\infty$ : the diagonal degree excess, (an upperbound for the McMillan degree of the zero at $\infty$ of $\Phi$ : see [1, (2.23)]), is immediately reduced to zero in our stage 1 and then kept zero until the end ($\Phi$ remains diagonally reduced ), 2) guarantees a monotone degree control on all elements of $\Phi$ and 3) gives explicit formulas, (45)-(48), for the removal of nonreal zeros of $\Phi$ .

$\varepsilon$. An important contribution of Algorithm 4.1. is *didactical value* enabling the hand calculus of small examples : it is conceptually simple and careful about zeros at $\infty$ through the notion of diagonal reducedness using direct data by (17).

References

[1] F.M. Callier, "On Polynomial Matrix Spectral Factorization by Symmetric Extraction", Report 83/10, Department of Mathematics, Facultés Universitaires de Namur, Namur, Belgium; submitted to the IEEE Transactions on Auto. Control.

[2] M.C. Davis, "Factoring the Spectral Matrix", IEEE Trans. Auto. Control, Vol. AC-8, pp. 296-305, 1963.

[3] W.A. Wolovich, "Linear Multivariable Systems", Springer Verlag, New York, 1974.

[4] T. Kailath, "Linear Systems", Prentice-Hall, Englewood Cliffs, N.J., 1980.

[5] F.M. Callier and C.A. Desoer, "Multivariable Feedback Systems", Springer Verlag, New York, 1982.

[6] V. Kučera, "New Results in State Estimation and Regulation", Automatica, Vol. 17, pp. 745-748, 1981.

[7] F.M. Callier, "Partially Stable LQ-Optimal Control by Spectral Factorization", Int. Jour. Control, to appear, 1984.

# INFINITE ZERO MODULE AND INFINITE POLE MODULE

G. Conte
Ist. Mat. Univ. Genova
via L.B.Alberti 4
16132 Genova - ITALY

A. Perdon
Ist. Mat. Appl. Univ. Padova
via Belzoni 7
35100 Padova - ITALY

## SUMMARY

In this paper we introduce the notion of infinite zero module $Z_\infty(G)$ and infinite pole module $P_\infty(G)$ associated with a transfer function $G(z)$. We show that $Z_\infty(G)$ and $P_\infty(G)$ describe the zero/pole structure at infinity of $G(z)$ and we investigate their dynamical and system theoretic properties. Finally, we apply these concepts to the study of the inverses of $G(z)$.

## INTRODUCTION

Let $G(z)$ denote a rational transfer function matrix of dimensions $p \times m$. In this paper we introduce two abstract algebraic objects, called respectively "infinite zero module" and "infinite pole module" and denoted by $Z_\infty(G)$ and $P_\infty(G)$, which describe the zero/pole structure at infinity of $G(z)$. More precisely, $Z_\infty(G)$ and $P_\infty(G)$ are finitely generated torsion $K[\![z^{-1}]\!]$-modules (and hence finite dimensional K-vector spaces) whose definition in terms of $G(z)$ is based on a dynamical characterization of zeros and poles at infinity. Moreover, when $S(z) = \text{diag}\,\{z^{-\nu_1}, \ldots, z^{-\nu_r}\}$ is the non trivial part of the Smith-MacMillan form at infinity of $G(z)$, the following representations hold : $Z_\infty(G) = \bigoplus_{\nu_i<0} K[\![z^{-1}]\!]/\, z^{\nu_i} K[\![z^{-1}]\!]$ and $P_\infty(G) = \bigoplus_{\nu_i>0} K[\![z^{-1}]\!]/\, z^{-\nu_i} K[\![z^{-1}]\!]$.
Both $Z_\infty(G)$ and $P_\infty(G)$ can be described using special representations of $G(z) : \Lambda U \rightarrow \Lambda Y$ of the form $G(z) = T^{-1}(z)V(z)$, where $T(z)$ and $V(z)$ are matrices with entries from $K[\![z^{-1}]\!]$, coprime in the appropriate sense. In fact, denoting by $\Omega_\infty U$ and $\Omega_\infty Y$ the $K[\![z^{-1}]\!]$-modules of series in $z^{-1}$ with coefficients from $U = K^m$ and $Y = K^p$ respectively, we have that $Z_\infty(G)$ is isomorphic to the torsion submodule of $\Omega_\infty Y/V(z)\Omega_\infty U$ and that $P_\infty(G)$ is isomorphic to $\Omega_\infty Y/T(z)\Omega_\infty Y$.
When $G(z)$ is strictly proper, the connection between $Z_\infty(G)$ and the geometric defini-

tion of zero structure at infinity is given by the existence of a $K[[z^{-1}]]$-isomorphism $f : S^*/R^* \to Z_\infty(G)$. Here, assuming that G(z) has the canonical realization (X,A,B,C), $S^*$ denotes, as usual, the minimum conditionally invariant subspace of X containing Im B and $R^*$ denotes the maximum reachability subspace of X contained in Ker C. The quotient $S^*/R^*$ is endowed with a natural module structure induced from those of $\Lambda U$ and $\Lambda Y$.

Since the zeros at infinity play a fundamental role in many control problems (such as : inversion of linear systems, causal factorization, feedback equivalence), the main motivation to introduce $Z_\infty(G)$ and $P_\infty(G)$ is the interest in having a natural algebraic tool in order to handle such zeros.

Here we apply the notions of infinite zero module and infinite pole module to the study of the inverses of G(z). We obtain, first of all, that given a (right or left) inverse H(z) of G(z) the infinite zero module $Z_\infty(G)$ is contained, in an appropriate sense, in the infinite pole module $P_\infty(H)$ of H(z). The analogous result, concerning the (finite) zero module Z(G) and the (finite) pole module, i.e. the canonical state space, X(H) of H(z) was proved by B.Wyman and M.Sain in [ 8 ].
Then, the fact that MacMillan degree $H(z) = \dim_K X(H) + \dim_K P_\infty(H)$, enables us to state the following result : for any inverse H(z) of G(z), MacMillan degree $H(z) \geq \dim_K Z(G) + \dim_K Z_\infty(G)$. Moreover, recalling that generalized ord. $H(z) = \dim_K X(H) + \dim_K X_\infty(H)$ (see [ 2 ], [ 6 ]), the existence of a canonical projection $\phi : X_\infty(H) \to P_\infty(H)$ and its special properties allow us to state the following result : for any inverse H(z) of G(z), generalized ord. $H(z) \geq \dim_K Z(G) + \dim_K Z_\infty(G)$ + (number of cyclic submodules in direct sum decomposition of $Z_\infty(G)$).
It is possible to find examples in which the lower bounds described above cannot be reached. However, further investigations on this line seem able to give results concerning a constructive characterization of the minimal inverse of a given G(z).

## 1. PRELIMINARIES AND NOTATIONS

Given a field K and a finite dimensional K-vector space S, $\Lambda S$ denotes the set of (formal) Laurent series in $z^{-1}$ with coefficients in S, i.e. series of the form $s = \sum_{t=k}^{\infty} s_t z^{-t}$, $s_t \in S$ and $t \in Z$. The order of s is defined as the index of the first non-zero coefficient $s_t$. A series s will be said proper iff its order is 0 and strictly proper iff its order is greater than 0.

$\Omega S$ denoted the polynomial subset of $\Lambda S$, i.e. the set of elements of the form $\sum_{k}^{0} s_t z^{-t}$, and $\Omega_\infty S$ denotes the power series subset of $\Lambda S$, i.e. the set of elements of the form $\sum_{0}^{\infty} s_t z^{-t}$.

$\Lambda K$ is a field and $\Lambda S$ is a $\Lambda K$-vector space; $\{e_1, e_2, .., e_r\}$ denotes both the canonical basis of S over K and that of $\Lambda S$ over $\Lambda K$. The field $\Lambda K$ contains the ring of polynomials $K[z]$, the ring of power series $K[[z^{-1}]]$ and the field of fractions $K(z)$. As a consequence $\Lambda S$ is, in particular, a $K[z]$-module and, also, a $K[[z^{-1}]]$-module. With respect to these structures $\Omega S$ turns out to be a $K[z]$-submodule and $\Omega_\infty S$ turns out to be a $K[[z^{-1}]]$-submodule. In the following we will consider the quotient modules:

$$\Gamma S := \Lambda S/\Omega S \qquad K[z]\text{-module}$$

$$\Gamma_\infty^* S := \Lambda S/\Omega_\infty S \quad \text{and} \quad \Gamma_\infty S := \Lambda S/z^{-1}\Omega_\infty S \qquad K[[z^{-1}]]\text{-modules.}$$

The elements of $\Gamma S$ can be uniquely represented as strictly proper series, the product by z being the usual product by z in $\Lambda S$ followed by truncation of the polynomial part. The elements of $\Gamma_\infty S$ ($\Gamma_\infty^* S$) can be uniquely represented as polynomials (polynomials without constant term). The product by $z^{-1}$ is the usual product in $\Lambda S$ followed by truncation of the strictly proper part (resp.: the proper part).

Let U and Y be K-vector spaces of dimension m and p respectively. In the following no distinction will be made between $\Lambda K$-linear maps between $\Lambda U$ and $\Lambda Y$ and $p \times m$ matrices associated to them with respect to the canonical basis.

Any $p \times m$ matrix with entries in $\Lambda K$ can be seen as a Laurent series in $z^{-1}$ with matrix coefficients, i.e. $G(z) = \sum_{k}^{\infty} G_t z^{-t}$, $G_t$ $p \times m$ matrix with entries in K, G(z) will be said polynomial, strictly proper, proper iff all its entries are such and it will be said rational iff all its entries belong to K(z).

By a <u>transfer function</u> we mean a $\Lambda K$-linear <u>rational</u> map G(z) between $\Lambda U$ and $\Lambda Y$. Any transfer function G(z) gives rise to the following commutative diagrams: 1.1, where all modules are $K[z]$-modules and all maps are $K[z]$-homomorphisms, and 1.2, where all modules are $K[[z^{-1}]]$-modules and all maps are $K[[z^{-1}]]$-homomorphisms.

1.1

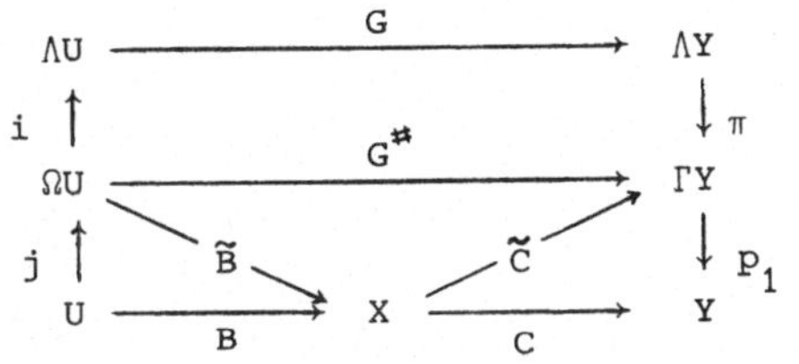


$\pi$ canonical projection

$p_1(\sum_1^{\infty} y_t z^{-t}) = y_1$

i, j canonical inclusion

$X \simeq \mathrm{Im}\, G^{\#} \simeq \Omega U/\mathrm{Ker}\, G^{\#}$ is a f.g. torsion $K[z]$-module. Defining $A: X \to X$ as $Ax = zx$, $(X, A, B, C)$ is a minimal realization of the strictly proper part of $G(z)$ (see[7])

1.2

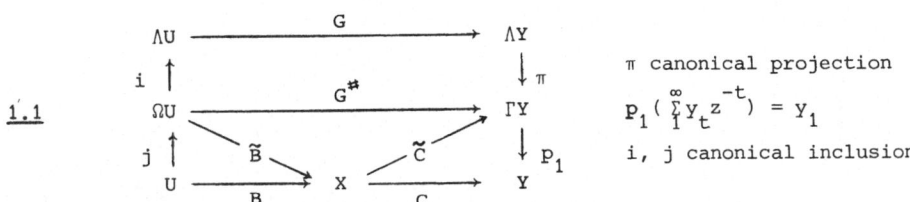

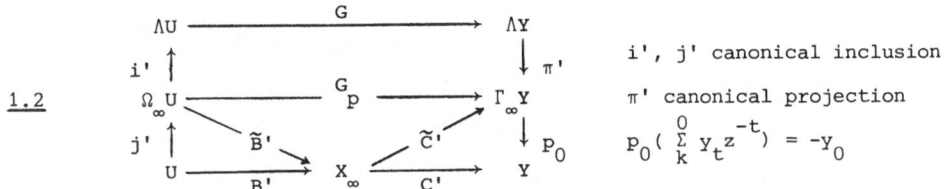

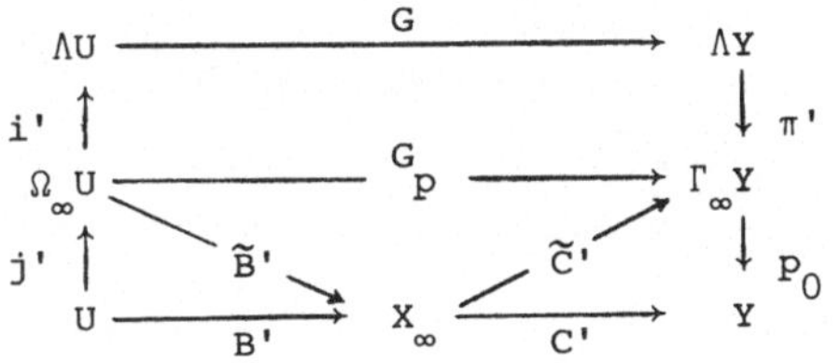


i', j' canonical inclusion

$\pi'$ canonical projection

$p_0(\sum_k^{0} y_t z^{-t}) = -y_0$

$X_\infty \simeq \mathrm{Im}\, G_p \simeq \Omega_\infty U/\mathrm{Ker}\, G_p$ is a f.g. torsion $K[\![z^{-1}]\!]$-module; defining $A': X_\infty \to X_\infty$ as $A'x = z^{-1}x$, $(X_\infty, A', B', C')$ is a minimal generalized state space realization of the polynomial part of $G(z)$ (see[2] ).

Let $H(\omega)$ denote the Smith MacMillan form of $G(1/\omega)$ and let $p_i(\omega) = \omega^{\nu_i} q_i(\omega)$, with $q_i(0) \neq 0$ for $i = 1, .., r = \mathrm{rank}\, G$, be the non-zero (diagonal) elements in $H(\omega)$. Then, the Smith-MacMillan form at infinity of $G(z)$ is the $p \times m$ matrix

$$\begin{pmatrix} S(z) & | & 0 \\ - - - & | & - - \\ 0 & | & 0 \end{pmatrix} \text{ where } S(z) = \mathrm{diag}\,\{z^{-\nu_1}, .., z^{-\nu_r}\} \quad \nu_1 \leq \ldots \leq \nu_r .$$

The r-uple $\{\nu_1, .., \nu_r\}$ is called <u>structure at infinity of G(z)</u> (see [5] 6.5).

The Smith-MacMillan form at infinity can be obtained also by the following procedure. Write $G(z) = z^s \tilde{G}(z)$ where s is the minimum integer such that $\tilde{G}(z)$ is proper and let $M(z)$ be the Smith form of $\tilde{G}(z)$ with respect to the ring $K[\![z^{-1}]\!]$. As $K[\![z^{-1}]\!]$ is a local ring whose maximal ideal is generated by $z^{-1}$, the non-zero (diagonal) elements of $M(z)$ can be assumed to be powers of $z^{-1}$. Then $z^s M(z)$ is the Smith-MacMillan form at infinity of $G(z)$. In particular $G(z) = B_1(z) z^s M(z) B_2(z)$ where $B_1(z)$, $B_2(z)$ are bicausal matrices (i.e. invertible in the ring of proper matrices).

$G(z)$ is said to have a pole at $\infty$ of order $-\nu_i$ for any negative $\nu_i$ in its structure at $\infty$, and, analogously, it is said to have a zero at $\infty$ of order $\nu_j$ for any positive $\nu_j$ in its structure at $\infty$. The total number of poles (zeros) at $\infty$ is then $\sum_{\nu_i < 0} (-\nu_i)$ $(\sum_{\nu_i > 0} \nu_i)$.

Suppose now that $\nu_i \leq 0$ for $i=1,..,k$ and that $\nu_i > 0$ for $i=k+1,..,r$. Denoting by $S^+ = \mathrm{diag}\{z^{\nu_1}, .., .., z^{\nu_k}\}$ and by $S^- = \mathrm{diag}\,\{z^{-\nu_{k+1}}, .., z^{-\nu_r}\}$, a coprime factorization of the Smith-MacMillan form at ∞ of G(z) by proper matrices is given by

$$\begin{pmatrix} S(z) & 0 \\ 0 & 0 \end{pmatrix} = \varepsilon(z)\psi_R^{-1}(z) = \psi_L^{-1}(z)\,\varepsilon(z) \qquad \text{where}$$

$$\varepsilon(z) = \begin{pmatrix} I_k & 0 & 0 \\ 0 & S^- & 0 \\ 0 & 0 & 0 \end{pmatrix} \qquad \psi_R(z) = \begin{pmatrix} S^+ & 0 & 0 \\ 0 & I_{r-k} & 0 \\ 0 & 0 & I_{m-r} \end{pmatrix} \qquad \psi_L(z) = \begin{pmatrix} S^+ & 0 & 0 \\ 0 & I_{r-k} & 0 \\ 0 & 0 & I_{p-r} \end{pmatrix}$$

Coprime factorizations of G(z) by proper matrices are then given by

$$G(z) = [B_1(z)\varepsilon(z)]\,[B_2^{-1}(z)\psi_R(z)]^{-1}$$

$$G(z) = [\psi_L(z)B_1^{-1}(z)]^{-1}[\varepsilon(z)B_2(z)]$$

In the following, for any coprime factorization $G(z) = V(z)T_R^{-1}(z)$ or $G(z) = T_L^{-1}(z)V(z)$ by proper matrices, V(z) will be called proper numerator of G(z) and $T_R(z)$, $T_L(z)$ will be called proper denominators of G(z). It can be proved as in [5] that ε(z) is the Smith form, with respect to $K[\![z^{-1}]\!]$, of every proper numerator of G(z) and that $\psi_R(z)$ (resp. $\psi_L(z)$) has the same nontrivial invariant factors of any proper denominator of G(z).

## 2. INFINITE ZERO MODULE

The aim of this section is to define the module of infinite zeros of a transfer function G(z). Its relations with the classical notion of zeros at infinity and its system theoretic interpretation are investigated.

DEFINITION 2.1 Given a transfer function G(z) its infinite zero module $Z_\infty(G)$ is defined by :

$$Z_\infty(G) = \frac{G^{-1}(\Omega_\infty Y) + \Omega_\infty U}{\mathrm{Ker}\, G + \Omega_\infty U}$$

To motivate the definition given above, let us consider the case m = p = 1. Let u(z) be an element in ΛK and let $k_0$ be its order, then, if $k_0 < 0$, u(z) is said to have $-k_0$ modes at infinity. Certain of these modes may fail to appear in the response of the system y(z) = G(z)u(z), i.e. ord $y(z) = k_1 > k_0$, and this fact is interpreted as the presence of zeros at infinity in G(z). So in defining the abstract module we consider excitations which produce response having no modes at infinity, and we ignore both proper inputs (which have no modes at infinity whose absence can be detected in the output) and Ker G (since identically zero outputs are of little interest).

PROPOSITION 2.2 $Z_\infty(G)$ is a finitely generated torsion $K[\![z^{-1}]\!]$module

Proof. $G(z)$ is $K[\![z^{-1}]\!]$-linear and $\Omega_\infty Y$ is a finitely generated $K[\![z^{-1}]\!]$-module, then $G^{-1}(\Omega_\infty Y)/\ker G$ and $Z_\infty(G)$, which can be viewed as a quotient of the previous one, are finitely generated $K[\![z^{-1}]\!]$-modules.

Every element of $Z_\infty(G)$ is the equivalence class $[u]$ modulo $\mathrm{Ker}\, G + \Omega_\infty U$ of some $u \in \Lambda U$ such that $G(z)u$ is proper. Let k be the degree of the polynomial part of u: then $G(z)(z^{-k}u)$ is proper and $z^{-k}u$ is proper. Therefore $z^{-k}[u] = 0$ and $Z_\infty(G)$ is torsion.

PROPOSITION 2.3 $Z_\infty(G)$ is isomorphic to the torsion submodule of $\Omega_\infty Y/\varepsilon(z)\Omega_\infty U$, where $\varepsilon(z)$ is the Smith form of any proper numerator of $G(z)$.

Proof. We prove the Proposition showing that $Z_\infty(G)$ is isomorphic to the torsion submodule of $\Omega_\infty Y/V(z)\Omega_\infty U$ where $G(z) = T^{-1}(z)V(z)$ is a coprime proper factorization. At this aim, we represent the elements of $Z_\infty(G)$ as $[u]$, for some $u \in \Lambda U$ such that $G(z)u$ is proper. In particular, $V(z)u$ is proper, i.e. $V(z)u \in \Omega_\infty Y$. Moreover, if $[u] = [u']$, i.e. $(u - u') \in (\mathrm{Ker}\, G + \Omega_\infty U)$, we have, since $\mathrm{Ker}\, G(z) = \mathrm{Ker}\, V(z)$, $V(z)(u - u') \in V(z)\Omega_\infty U$. As a consequence, we can define a $K[\![z^{-1}]\!]$-homomorphism $f : Z_\infty(G) \to \Omega_\infty Y/V(z)\Omega_\infty U$ by $f([u]) = [V(z)u]$ (class of $V(z)u$ in $\Omega_\infty Y/V(z)\Omega_\infty U$).
As $Z_\infty(G)$ is torsion, $f(Z_\infty(G))$ is contained in the torsion submodule of $\Omega_\infty Y/V(z)\Omega_\infty U$.
Suppose that $f([u]) = [V(z)u] = 0$: then $V(z)u \in V(z)\Omega_\infty U$ and $u \in \mathrm{Ker}\, V + \Omega_\infty U = \mathrm{Ker}\, G + \Omega_\infty U$. Therefore $[u] = 0$ and f is injective.
Let $[y]$ be a torsion element in $\Omega_\infty Y/V(z)\Omega_\infty U$. By coprimnessof $T(z)$ and $V(z)$, there exist proper matrices $A(z)$ and $B(z)$ such that $y = T(z)A(z)y + V(z)B(z)y$. Then $[y] = [T(z)A(z)y]$ and there exists a positive integer k such that $z^{-k}T(z)A(z)y = V(z)u$ for a suitable u in $\Omega_\infty U$. Let $v = z^k u + B(z)y \in \Lambda U$; then $V(z)v = V(z)z^k u + V(z)B(z)y = T(z)A(z)y + V(z)B(z)y = y$. Now, $z^k u$ belongs to $G^{-1}(\Omega_\infty Y)$ because $G(z)z^k u = T^{-1}(z)V(z)z^k u = T^{-1}(z)V(z)(v - B(z)y) = T^{-1}(z)\ (y - V(z)B(z)y) = T^{-1}(z)(T(z)A(z)y) = A(z)y \in \Omega_\infty Y$. Then $[y] = [V(z)v] = f([u])$ and f is onto the torsion submodule of $\Omega_\infty Y/V(z)\Omega_\infty U$.

The basic property of $Z_\infty(G)$, in connection with the notion of zero at infinity we recalled in the previous section, is pointed out by the following Corollary.

COROLLARY 2.4 The invariant factors of $Z_\infty(G)$ over $K[[z^{-1}]]$ coincide with the non trivial elements of $S^-$.

REMARK 2.5 The above Corollary says that $Z_\infty(G)$ contains all the information about the zero structure at infinity of G(z). More precisely, if $\{\nu_1,\ldots,\nu_r\}$ is the structure at infinity of G(z), the infinite zero module has the following canonical decomposition into a direct sum of cyclic submodules : $Z_\infty(G) = \bigoplus_{\nu_i<0} K[[z^{-1}]]/ z^{\nu_i}K[[z^{-1}]]$.

REMARK 2.6 The $K[[z^{-1}]]$-module $G^{-1}(\Omega_\infty Y)$ which appears in the definition of $Z_\infty(G)$ is the latency kernel of G(z) introduced in [ 4 ]. We will investigate more deeply its relations with the structure at infinity in the following.

## 3. GEOMETRIC CHARACTERIZATION

In this section we assume that G(z) is a $p\times m$ strictly proper transfer function provided with the minimal realization (X,A,B,C). Moreover, we assume, without lost of generality, that $X = \operatorname{Im} G^{\#}$ and, as a consequence, that $\tilde{C}$ is the inclusion (see 1.1).

It is known ([ 1 ]) that the zero structure at infinity of G(z) can be obtained from the quotient $S^*/R^*$, where $S^*$ is the minimum (A,C)-invariant subspace of X containing Im B, and $R^*$ is the maximum controllability subspace of X contained in Ker C. In the following we will characterize $S^*$ and $R^*$ in terms of the transfer function G(z) and then we will prove that $Z_\infty(G)$ is $K[[z^{-1}]]$-isomorphic to $S^*/R^*$.

PROPOSITION 3.1 Define $S = \{s \in \Lambda Y$, s is strictly proper and $s = G(z)u$ for some $u \in \Omega U\} = z^{-1}\Omega_\infty Y \cap G(\Omega U)$ and $R = \{ s \in \Lambda Y$, there exist $u \in \Omega U$ and $\tilde{u}$ strictly proper such that $G(z)u = G(z)\tilde{u} = s\} = G(z^{-1}\Omega_\infty U) \cap G(\Omega U)$. Then $S = S^*$ and $R = R^*$.

Proof. We remark, first of all, that both S and R are contained in $\operatorname{Im} G^{\#} = X$. Since G(z) is strictly proper, we have $\operatorname{Im} B \subset S$. To prove the (A,C)-invariance of S we show that $A(S \cap \operatorname{Ker} C) \subset S$. Any element of S, in fact, is of the form $s = s_1 z^{-1} + \ldots = G(z)u$, with $u \in \Omega U$. As $C(s) = p_1(s) = s_1$, $s \in \operatorname{Ker} C$ iff $s_1 = 0$. For such an element s, $A(s) = zG(z)u = G(z)zu = (s_2 z^{-1} + \ldots)$ belongs clearly to S.

The minimality of S among the (A,C)-invariant subspaces containing Im B will be proved by contradiction. Suppose that V is an (A,C)-invariant subspace of X containing Im B but not containing S, i.e. $G(z)u \in V$ for every constant u and there exist polynomials u(z) such that G(z)u(z) is strictly proper but G(z)u(z) does not belong to V. Let p(z) be such a polynomial of minimum degree : $\deg p(z) \geqslant 1$, as $G(z)u \in V$ for every constant u. Therefore, we have $p(z) = zq(z) + r$, with $r \in U$ and $G(z)p(z) = zG(z)q(z) + G(z)r$. Now, $zG(z)q(z) = G(z)p(z) - G(z)r = y_1z^{-1} + y_2z^{-2} + \ldots\ldots = z(y_1z^{-2} + \ldots\ldots)$ and, since $\deg q(z) \leqslant \deg p(z)$, $G(z)q(z) = y_1z^{-2} + \ldots\ldots$ is an element of $V \cap \operatorname{Ker} C$. By the (A,C)-invariance of V, $AG(z)q(z) = zG(z)q(z) = v \in V$. Thus $G(z)p(z) = v - G(z)r$ belongs to V against the hypothesis.

$R = R^*$ is proved in [ 3 ] § 4.

<u>REMARK 3.2</u> $S^*/R^*$ has a natural $K[\![z^{-1}]\!]$-module structure definied as follows. Let [s] denote an element in $S^*/R^*$, $s = G(z)\tilde{u}(z)$ where $\tilde{u}(z) = zu(z) + u_0 \in \Omega U$. Then $z^{-1}[s] = [G(z)u(z)]$. Definition is consistent, in fact G(z)u(z) is strictly proper and hence $[G(z)u(z)] \in S^*/R^*$, moreover if $[s] = [s']$ and $s' = G(z)\tilde{v}(z)$, $\tilde{v}(z) = zv(z) + v_0$, we have $G(z)\tilde{u}(z) - G(z)\tilde{v}(z) \in R^*$, i.e. $G(z)\tilde{u}(z) - G(z)\tilde{v}(z) = G(z)w(z)$ with w(z) strictly proper. As a consequence $G(z)u(z) - G(z)v(z) = G(z)(z^{-1}w(z)) + G(z)(z^{-1}(v_0 - u_0)) \in R^*$ and $[G(z)u(z)] = [G(z)v(z)]$.

<u>PROPOSITION 3.3</u> $Z_\infty(G)$ and $S^*/R^*$ are isomorphic as $K[\![z^{-1}]\!]$-modules.

Proof. As G(z) is strictly proper, $Z_\infty(G) = G^{-1}(\Omega_\infty Y) / (\operatorname{Ker} G + \Omega_\infty U)$. Let [s], $s = G(z)u(z)$, be an element of $S^*/R^*$. Then s is, in particular, strictly proper and $zu(z) \in G^{-1}(\Omega_\infty Y)$. We define $f : S^*/R^* \to Z_\infty(G)$ as follows : $f([s]) = [zu(z)]$ (class of zu(z) in $Z_\infty(G)$). Definition is consistent, in fact, if $[s] = [s']$, $s' = G(z)v(z)$, then $G(z)(u(z) - v(z)) \in R^*$, i.e. $G(z)(u(z) - v(z)) = G(z)w(z)$ with w(z) strictly proper. As a consequence $zu(z) - zv(z) = zw(z) + p(z)$, with $p(z) \in \operatorname{Ker} G$, and $[zu(z)] = [zv(z)]$ in $Z_\infty(G)$. f is clearly K-linear and, moreover, $f(z^{-1}[s]) - z^{-1}f([s]) = 0$. in fact, if $s = G(z)(zu(z) + u_0)$, $f(z^{-1}[s]) = f([G(z)u(z)]) = [zu(z)]$ in $Z_\infty(G)$. On the other hand, $z^{-1}f([s]) = z^{-1}[z^2u(z) + zu_0] = [zu(z) + u_0] = [zu(z)]$ in $Z_\infty(G)$ since $u_0 \in \Omega_\infty U$. Hence f is $K[\![z^{-1}]\!]$-linear as both $S^*/R^*$ and $Z_\infty(G)$ are torsion.

To show that f is injective, assume that, for $s = G(z)u(z)$, $f([s]) = 0$. Then $zu(z) \in \operatorname{Ker} G + \Omega_\infty U$, i.e. $zu(z) = v(z) + w(z)$, $v(z) \in \operatorname{Ker} G$ and $w(z)$ proper. Multiplying by $z^{-1}$ and applying $G(z)$ we have $G(z)u(z) = G(z)(z^{-1}w(z))$, hence $G(z)u(z) \in$ $\in R^*$ and $[s] = 0$ in $S^*/R^*$.

To show that f is surjective, let us recall that any element in $Z_\infty(G)$ is the equivalence class, modulo $\operatorname{Ker} G + \Omega_\infty U$, of an element in $G^{-1}(\Omega_\infty Y)$. Then any element can be represented as $[zu(z)]$, where $u(z)$ is a polynomial such that $G(z)u(z)$ is strictly proper, and it follows that $[zu(z)] = f([s])$ with $s = G(z)u(z) \in S^*$.

COROLLARY 3.4 The invariant factors of $S^*/R^*$ over $K[[z^{-1}]]$ describe the zero structure at infinity of $G(z)$.

Proof. Trivial by 3.3 and 2.4.

## 4. INFINITE POLE MODULE

To apply the notion of infinite zero module to the study of inverse transfer functions, we need the dual notion of infinite pole module. It has been remarked in [ 8 ] that the finite pole module of a rational $G(z)$ is essentially the state space of a minimal realization of the strictly proper part of $G(z)$. Clearly, the generalized state space of a minimal realization of the polynomial part of $G(z)$ cannot be chosen to represent the infinite pole module we need since it may contain a nondynamical component (see [ 2 ]).

In the same way as in section 2, where we considered the definition of $Z_\infty(G)$, the case $m = p = 1$ suggests to us the following abstract definition :

DEFINITION 4.1 Given a transfer function $G(z)$ its infinite pole module $P_\infty(G)$ is defined by :

$$P_\infty(G) = \frac{G(\Omega_\infty U) + \Omega_\infty Y}{\Omega_\infty Y}$$

PROPOSITION 4.2 $P_\infty(G)$ is a finitely generated torsion $K[[z^{-1}]]$-module whose nontrivial invariant factors over $K[[z^{-1}]]$ coincide with the nontrivial elements of $S^+$.

Proof. $G(\Omega_\infty U)$ and $\Omega_\infty Y$ are finitely generated $K[[z^{-1}]]$-modules, then $P_\infty(G)$ is finitely generated. Any element in $P_\infty(G)$ is the equivalence class, modulo $\Omega_\infty Y$, of some $y = G(z)u$, with $u \in \Omega_\infty U$. If $y$ is proper, $[y] = 0$. If $y$ has a polynomial part of degree $k$, $z^{-k}y$ is proper and $z^{-k}[y] = 0$; hence $P_\infty(G)$ is torsion.

To prove the second part of the proposition, we show that $P_\infty(G)$ is isomorphic to the torsion $K[\![z^{-1}]\!]$-module $\Omega_\infty Y/T(z)\Omega_\infty Y$ where $G(z) = T^{-1}(z)V(z)$ is a coprime proper factorization. Remark, first of all, that for any $u \in \Omega_\infty U$, $T(z)G(z)u = V(z)u$ belongs to $\Omega_\infty Y$. Therefore, $T(G(\Omega_\infty U) + \Omega_\infty Y) \subset \Omega_\infty Y$ and there exists $h : P_\infty(G) \to \Omega_\infty Y/T(z)\Omega_\infty Y$ such that the following diagram, where the upper vertical maps are canonical inclusions and the lower ones are canonical projections, commutes :

4.3
$$\begin{array}{ccc} \Omega_\infty Y & \xrightarrow{\quad T \quad} & T(z)\Omega_\infty Y \\ \downarrow & & \downarrow \\ G(\Omega_\infty U) + \Omega_\infty Y & \xrightarrow{\quad T \quad} & \Omega_\infty Y \\ p \downarrow & & \downarrow q \\ P_\infty(G) & \xrightarrow{\quad h \quad} & \Omega_\infty Y/T(z)\Omega_\infty Y \end{array}$$

Assume that $h(y) = 0$, with $y = p(G(z)u)$ and $u \in \Omega_\infty U$. Then $qTG(z)u = hpG(z)u = 0$ and $TG(z)u \in T(z)\Omega_\infty Y$. Let $TG(z)u = Tv$, $v \in \Omega_\infty Y$, then $G(z)u = v \in \Omega_\infty Y$ and $y = pv = 0$ in $P_\infty(G)$; hence $h$ is injective.

Let $v$ be an element in $\Omega_\infty Y/T(z)\Omega_\infty Y$, i.e. $v = qy$, $y \in \Omega_\infty Y$. By coprimness of $T(z)$ and $V(z)$ there exist proper matrices $A(z)$ and $B(z)$ such that $y = T(z)A(z)y + V(z)B(z)y$ and $qy = q(V(z)B(z)y)$. Take $u = B(z)y$ in $\Omega_\infty U$, then $hpG(z)u = qTG(z)B(z)y = qV(z)B(z)y = qy = v$ and $h$ is onto.

<u>REMARK 4.4</u> A consequence of 4.2 is that $P_\infty(G)$ contains all the information about the pole structure at infinity of $G(z)$. More precisely, if $\{\nu_1,\ldots\ldots,\nu_r\}$ is the structure at infinity of $G(z)$, the infinite pole module has the following canonical decomposition into a direct sum of cyclic submodules : $P_\infty(G) = \bigoplus_{\nu_i>0} K[\![z^{-1}]\!]/\, z^{-\nu_i}K[\![z^{-1}]\!]$ .

In conclusion, the decomposition into direct sums of cyclic submodules of $Z_\infty(G)$ and $P_\infty(G)$ determines the non zero indices of the structure at infinity of $G(z)$. Moreover, the structure at infinity contains a number of zeros equal to the difference (rank G - (number of cyclic submodules in direct sum decompositions of $Z_\infty(G)$ and $P_\infty(G)$)).

<u>REMARK 4.5</u> It is easy to see that $P_\infty(G)$ is isomorphic to the quotient module

$$\frac{\Omega_\infty U}{G^{-1}(\Omega_\infty Y) \cap \Omega_\infty U}$$

. This alternative representation points out the relation between $P_\infty(G)$ and the latency kernel $G^{-1}(\Omega_\infty Y)$ (see [ 4 ]). This, together with 2.6, gives an insight into the connection between the concept of latency and the structure at infinity. In particular, it appears that the latency kernel contains information on both the infinite zeros and the infinite poles of $G(z)$. However, as $G^{-1}(\Omega_\infty Y)$ is not finitely generated unless $G(z)$ is injective, ([ 4 ] 6.16), $Z_\infty(G)$ and $P_\infty(G)$ are more handable algebraic objects.

In case $G(z)$ is injective and strictly proper, the latency indices $\{\lambda_1,\ldots\ldots,\lambda_m\}$ are defined in [ 4 ] in the following way : let $\{d_1,\ldots\ldots,d_m\}$ be an ordered proper basis of $G^{-1}(\Omega_\infty Y)$; then $\mathrm{ord}\, d_i \leqslant -1$ and $\lambda_i = -\mathrm{ord}\, d_i - 1$. Remarking that the polynomial part of any $d_i$ generates a cyclic submodule of order equal to $-\mathrm{ord}\, d_i$ in $Z_\infty(G)$, we have that the latency indices coincide with the order of the infinite zeros decreased by 1. As a consequence, $G(z)$ is non latent iff all its infinite zeros have order 1.

When $G(z)$ is proper, obviously $G(\Omega_\infty U) \subset \Omega_\infty Y$ and $P_\infty(G) = 0$. Let now $G(z)$ be a $p \times m$ transfer function of order $k < 0$. To clarify the relation between $P_\infty(G)$ and $X_\infty(G)$, the generalized state space of the minimal realization of the polynomial part of $G(z)$, let us consider the following diagram (see also 1.2) :

4.6

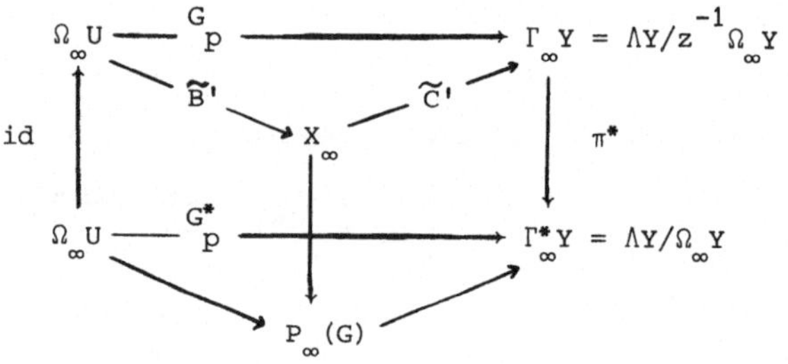


Where $\pi^* : \Gamma_\infty Y \to \Gamma^*_\infty Y$ is the projection $\pi^*(u_n z^n + \ldots\ldots + u_1 z + u_0) = u_n z^n + \ldots + u_1 z$ and $\phi$ is the restriction of $\pi^*$ (remark that $\phi$ is well defined since $\pi^* G_p = G^*_p$).

<u>PROPOSITION 4.7</u> The morphism $\phi : X_\infty(G) \to P_\infty(G)$ is surjective. The cyclic submodules of order k+1 of $X_\infty(G)$ are mapped onto cyclic submodules of order k of $P_\infty(G)$.

Proof. The surjectivity of $\phi$ follows by the commutativity of 4.6.
Let $\{x\}$ be a cyclic submodule of order k+1 of $X_\infty(G)$, i.e. $z^{-i}x \neq 0$ in $\Gamma_\infty Y$ for $i \leqslant k$ and $z^{-k-1}x = 0$ in $\Gamma_\infty Y$. In other words, $z^{-k-1}x$ is strictly proper, $z^{-k}x$ is proper and $z^{-i}x$ has negative order for $i < k$. Then $z^{-i}\phi(x) = z^{-i}\pi^* x = \pi^*(z^{-i}x) \neq 0$ for $i \leqslant k-1$, $z^{-k}\phi(x) = \pi^*(z^{-k}x) = 0$. Hence $\{\phi(x)\}$ is a cyclic submodule of $P_\infty(G)$ of order k.

REMARK 4.8 Let $X_\infty(G) = \bigoplus_i K[\![z^{-1}]\!]/\,z^{-\mu_i}K[\![z^{-1}]\!]$ be the canonical decomposition of $X_\infty(G)$ into a direct sum of cyclic submodules. Then $P_\infty(G) = \bigoplus_i K[\![z^{-1}]\!]/\,z^{-\mu_i+1}K[\![z^{-1}]\!]$ and the indices $\mu_i-1$ coincide with the indices $\nu_i$ of the pole structure at infinity of G(z). Moreover, denoting by $G_{pol}(z)$ the polynomial part of G(z), we have by 4.7 and [ 2 ] : generalized ord. $G_{pol}(z) = \dim_K X_\infty(G) = \sum_i \mu_i \geqslant \sum_{\nu_i>0} (\nu_i+1)$ and (number of indipendent impulsive motions of G(z)) $= \dim_K P_\infty(G) = \sum_{\nu_i>0} \nu_i$.

Hence, the difference between $\dim_K X_\infty(G)$ and $\dim_K P_\infty(G)$ is equal to the number of cyclic submodules in the direct sum decomposition of $X_\infty(G)$ or, equivalently, to the number of cyclic submodules in the direct sum decomposition of $P_\infty(G)$ plus the number of (non dynamical) cyclic submodules of order 1 of $X_\infty(G)$.

## 5. INVERSE TRANSFER FUNCTIONS

In this section we investigate the connection between the infinite zero module of G(z) and the infinite pole module of a (right or left) inverse H(z) of G(z). In the case m = p = 1 any G(z) has a unique inverse H(z) whose number of poles at infinity is equal to the number of zeros at infinity of G(z). In the multivariable case, it will be proved that $Z_\infty(G)$ is a sort of lower bound, in an module theoretic sense, for $P_\infty(H)$. More precisely, we have the following two propositions.

PROPOSITION 5.1 Let $G(z) : \Lambda U \to \Lambda Y$ be an injective transfer function and let $H(z) : \Lambda Y \to \Lambda U$ be a left inverse of G(z), i.e. $H(z)G(z) = 1_{\Lambda U}$. Then there exists an injective $K[\![z^{-1}]\!]$-morphism $j : Z_\infty(G) \to P_\infty(H)$.

Proof. For any $u \in \Omega_\infty U$ such that $G(z)u = y$ belongs to $\Omega_\infty Y$, we have $H(z)G(z)u = u = H(z)y$, hence $G^{-1}(\Omega_\infty Y) \subset H(\Omega_\infty Y)$. This assure the existence of $j : Z_\infty(G) \to P_\infty(G)$ such that the following diagram commutes :

$$\begin{array}{ccccccc}
0 & \longrightarrow & \Omega_\infty U & \xrightarrow{\text{id}} & \Omega_\infty U & \longrightarrow & 0 \\
 & & \downarrow & & \downarrow & & \\
0 & \longrightarrow & G^{-1}(\Omega_\infty Y) + \Omega_\infty U & \xrightarrow{\text{incl}} & H(\Omega_\infty Y) + \Omega_\infty U & & \\
 & & \downarrow & & \downarrow & & \\
 & & Z_\infty(G) & \xrightarrow{j} & P_\infty(H) & &
\end{array}$$

Moreover, j is uniquely determined by the above property and it is easily seen, using the snake lemma, to be injective.

PROPOSITION 5.2 Let $G(z) : \Lambda U \to \Lambda Y$ be a surjective transfer function and let $H(z) : \Lambda Y \to \Lambda U$ be a right inverse of $G(z)$, i.e. $G(z)H(z) = 1_{\Lambda Y}$. Then there exists a surjective $K[\![z^{-1}]\!]$-morphism $p : P_\infty(H) \to Z_\infty(G)$.

Proof. Let $u = H(z)y$ be an element of $H(\Omega_\infty Y)$, then $G(z)u = G(z)H(z)y = y$ belongs to $\Omega_\infty Y$ and $H(\Omega_\infty Y) \subset G^{-1}(\Omega_\infty Y)$. This assure the existence of $p : P_\infty(H) \to Z_\infty(G)$ such that the following diagram commutes :

$$\begin{array}{ccc}
\Omega_\infty U & \xrightarrow{\text{incl}} & \mathrm{Ker}\, G + \Omega_\infty U \\
\downarrow & & \downarrow \\
H(\Omega_\infty Y) + \Omega_\infty U & \xrightarrow{\text{incl}} & G^{-1}(\Omega_\infty Y) + \Omega_\infty U \\
p_1 \downarrow & & \downarrow p_2 \\
P_\infty(H) & \xrightarrow{p} & Z_\infty(G)
\end{array}$$

p is uniquely determined by the above property.

Let x be an element of $Z_\infty(G)$, $x = p_2 u$ with $G(z)u = y \in \Omega_\infty Y$. We have $y = G(z)H(z)y$ and therefore $G(z)(u - H(z)y) = 0$. This implies that $(u - H(z)y)$ belongs to $\mathrm{Ker}\, G \subset \subset \mathrm{Ker}\, p_2$ and that $pp_1 H(z)y - x = p_2(H(z)y - u) = 0$. As a consequence, $x = pp_1 H(z)y$ and p is surjective.

Now, as MacMillan degree $H(z) = \dim_K X(H) + \dim_K P_\infty(H)$ and generalized ord. $H(z) = = \dim_K X(H) + \dim_K X_\infty(H)$ (see [ 2 ], [ 5 ], [ 6 ]), we have the following corollary :

COROLLARY 5.3 Let $H(z)$ be a (right or left) inverse of the transfer function $G(z)$. Then MacMillan degree $H(z) = \dim_K X(H) + \dim_K P_\infty(H) \geqslant \dim_K Z(G) + \dim_K Z_\infty(G)$ and generalized ord. $H(z) = \dim_K X(H) + \dim_K X_\infty(H) = \dim_K X(H) + \dim_K P_\infty(H) +$ (number of cyclic submodules in direct sum decomposition of $X_\infty(H)$) $\geqslant \dim_K Z(G) + \dim_K Z_\infty(G) +$ (number of cyclic submodules in direct sum decomposition of $Z_\infty(G)$).

Proof. By [ 2 ] and by 4.6, 5.1, 5.2.

REMARK 5.4 We remark that using the same techniques, with the obvious modifications, as in [ 3 ] 3.6 and 3.9 it is possible to construct right or left inverses such that j or, respectively, p are isomorphism.

## CONCLUSION

Two abstract algebraic objects associated with any transfer function G(z), namely the infinite zero module $Z_\infty(G)$ and the infinite pole module $P_\infty(G)$, have been introduced. It has been shown that they describe the zero/pole structure at infinity of G(z) and that there exists a canonical relation between $Z_\infty(G)$ and $P_\infty(H)$ where H(z) is a (right or left) inverse of G(z). More precisely, $Z_\infty(G)$ is contained, in a suitable sense, in $P_\infty(H)$.

These results complete the algebraic theory of the (finite) zero and pole module in the sense of B.Wyman and M.Sain [ 8].

Moreover, together with the realization theory for non proper rational transfer functions developed in [ 2], they give a better understanding of the problems involved in the construction of the minimal inverse of a given G(z), as shown in 5.3. Further investigations on this subject with the aid of the algebraic tools described here will be the argument of a forthcoming paper.

## REFERENCES

[ 1] C.Commault and J.M.Dion - Structure at infinity of linear multivariable systems : a geometric approach - 20th IEEE Conf. on Decision and Control (1981)

[ 2] G.Conte and A.Perdon - Generalized state space realization of non proper rational transfer functions - System & Control Letters 1 (1982)

[ 3] G.Conte and A.Perdon - An algebraic notion of zeros for systems over rings - MTNS 1983 Conf., Beer Sheva (1983)

[ 4] J.Hammer and M.Heymann - Causal factorization an linear feedback - SIAM J. Control Opt. 19 (1981)

[ 5] T.Kailath - Linear Systems - Prentice Hall (1980)

[ 6] H.Rosenbrock - Structural properties of linear dynamical systems - Int. J. Control 20 (1974)

[ 7] B.Wyman - Linear systems over commutative rings - Lecture Notes, Stanford Univ. (1972)

[ 8] B.Wyman and M.Sain - The zero module and essential inverse systems - IEEE Trans. Circuit and Systems CAS-28 (1981)

# ON LINEAR SYSTEMS AND PARTIAL REALIZATIONS

A. C. Antoulas

Department of Electrical Engineering
Rice University
Houston, Texas 77251, U.S.A.

ABSTRACT. The new approach to synthesis of linear feedback systems recently proposed by the author is extended to include model matching problems by dynamic output feedback.

## 1. INTRODUCTION.

The general feedback synthesis problem in linear systems is the following.

(1.1)

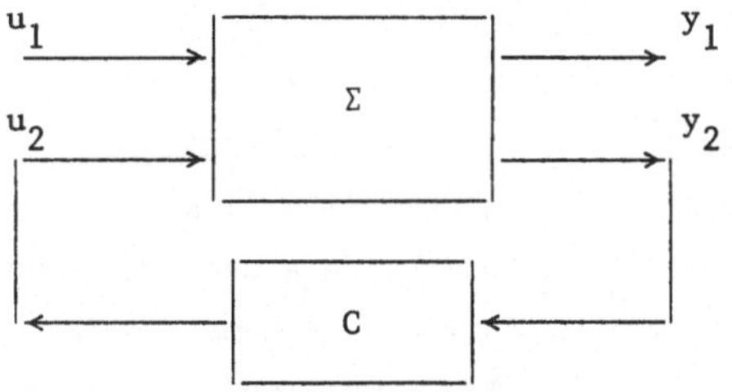


$$Z_\Sigma = \begin{pmatrix} Z_{11} & Z_{12} \\ Z_{21} & Z_{22} \end{pmatrix}.$$

Given is the system $\Sigma$; $u_1$ is the disturbance input, $u_2$ is the control input, $y_1$ the output-to-be-controlled, and $y_2$ the measured output. The transfer function of $\Sigma$ is $Z_\Sigma$, where $Z_{11}$, $Z_{12}$, $Z_{21}$ are $r \times q$, $r \times m$, $p \times q$ proper rational matrices and $Z_{22}$ is a $p \times m$ strictly proper rational matrix.

The goal is to find all compensators $C$, in particular the low-order ones, which achieve certain objectives, e.g. internal stabilization, regulation, placement of the poles, matching of a desired closed-loop transfer function e.t.c.

The equation relating the four given transfer functions and the transfer functions $Z_C$ of $C$ and $Z_y$ of the resulting closed-loop system is:

(1.2) $$Z_{11} = Z_{12}Z_C(I + Z_{22}Z_C)^{-1}Z_{21} + Z_y,$$

which is non-linear in $Z_C$. We are looking for solutions $Z_C$, $Z_y$ of (1.2) which satisfy the following fundamental requirements:

(1.3) $Z_C$: proper rational (which implies the properness of $Z_y$),

(1.4) $Z_C$: internally stabilizing,

(1.5) $Z_C$: regulating, i.e. $Z_y$: stable.

In order to be able to look for low order compensators fulfilling the above as well as further constraints, we need a parametrization of the solutions of (1.2) which keeps track of the MacMillan degree of $Z_C$. Thus if $\Delta$ is the matrix parameter which parametrizes the solutions, we need a relationship of the sort:

(1.6) $$\delta(Z_C) - \delta(\Delta) = \text{constant},$$

where $\delta(\cdot)$ denotes the MacMillan degree. If instead, we want to keep track of the MacMillan degree of $Z_y$, we need a parametrization which satisfies:

(1.7) $\delta(Z_y) - \delta(\Delta) = \text{constant}.$

Problem (1.1) and numerous special cases thereof, have been investigated by many researchers. The first successful approach was in state space by WONHAM and PEARSON [1974] using the so-called geometric theory. Along the same lines we also have WILLEMS and COMMAULT [1981], SCHUMACHER [1982]. In the frequency domain, there exist various solutions of more or less restricted versions of (1.1): BENGTSSON [1977], CHENG and PEARSON [1978], WOLOVICH and FERREIRA [1979], DESOER, LIU, MURRAY, and SAEKS [1980], KHARGONEKAR and OZGULER [1982], and many others. The main characteristic of the above approaches is to provide a necessary and sufficient condition for solvability of the problem, followed by a method for obtaining one or some of the solutions satisfying (1.3-5). The first parametrization of all solutions of equation (1.2) subject to (1.3-5) was obtained by PERNEBO [1981] and was later refined by CHENG and PEARSON [1981].

The main drawback of all those treatments is that the derived parametrizations do not satisfy relationships like (1.6) or (1.7). Consequently, nothing can be said about the order of the compensators or of the closed-loop transfer functions which are computed.

Using a new approach ANTOULAS [1983] was able to remedy this situation and obtain a parametrization of all solutions of equation (1.2) subject to conditions (1.3-6). The essense of the new theory is the theory of partial realizations.

The purpose of this paper is to show that it is possible to find a parametrization which satisfies (1.7), i.e. keeps track of the MacMillan degree of $Z_y$, and at the same time parametrizes all admissible closed-loop transfer functions $Z_y$. As a byproduct, we obtain the solution of the model matching problem.

In the next section we present a summary of the main results of ANTOULAS [1983]. Section three develops some preliminary results, used to prove the main results in section four.

## 2. SUMMARY OF PREVIOUS RESULTS.

The parametrization of all solutions of equation (1.2) subject to (1.3-6) is summarized below. The first step is to apply the so-called Youla parametrization, which linearizes (1.2). Let

(2.1) $Z_{22} = LM^{-1} = T^{-1}U, \quad TA + UB = I,$

where $L, M, T, U, A, B$ are polynomial matrices, with $L, M$ right coprime and $T, U$ left coprime. The Youla parameter $Z_x$ is defined as follows:

(2.2) $Z_C(I + Z_{22}Z_C)^{-1} = (B + MZ_x)T.$

Equation (1.2) thus becomes

(2.3) $Z_1 = Z_2 Z_x Z_3 + Z_y$, where: $Z_1 = Z_{11} - Z_{12}BTZ_{21}$, $Z_2 = Z_{12}M$, $Z_3 = TZ_{21}$,

which is linear in $Z_x$ and $Z_y$. Moreover the internal stability requirement (1.4) is equivalent to

(2.4) $Z_x$: stable.

Moreover,

(2.5) $\delta(Z_x) = \delta(Z_C) + \delta(Z_{22})$.

We are thus looking for stable and proper rational solutions of equation (2.3), which satisfy (1.6). The following result can be shown: equation (2.3) has stable solutions $Z_x$, $Z_y$ if and only if there exist a polynomial matrix $X$ and a stable rational matrix $Z$ such that

(2.6) $Z_1 = Z_2 X Z_3 + Z$.

The proof of this result in ANTOULAS [1983] contains a constructive procedure for finding such an $X$ and a $Z$, if they exist. Let

(2.7) $Z_2 = ND^{-1}$, $Z_3 = Q^{-1}P$,

be coprime polynomial factorizations. We also write

$$D = D_+ D_- \ , \quad Q = Q_- Q_+ \ ,$$

where $\det D_-$, $\det Q_-$ are stable polynomials and $\det D_+$, $\det Q_+$ are completely unstable polynomials. We are now ready to write down the parametrization of all rational solutions of (1.2) which is suitable for our purposes.

(2.8) LEMMA. <u>The rational matrices</u> $Z_C$, $Z_y$ <u>satisfy equation</u> (1.2) <u>if and only if</u>

$$Z_C = (B + MZ_x)(A - LZ_x)^{-1}, \quad Z_x = D_+ \Delta Q_+ + X,$$

$$Z_y = - ND_-^{-1} \Delta Q_-^{-1} P + Z,$$

<u>for some rational matrix of appropriate dimensions</u> $\Delta$.

(2.9) COROLLARY. <u>If</u> $\Delta$ <u>is stable it follows that</u> $\delta(Z_x) = \delta(\Delta)$.

From the corollary it follows together with (2.5) that $\Delta$ satisfies (1.6) if it is stable.

The crucial quantity in this theory is:

(2.10) $\Theta \neq D_+^{-1}(M^{-1}B + X)Q_+^{-1}$.

Without loss of generality, X, Z in (2.6) can be chosen so that $\Theta$ is strictly proper rational. Before we proceed let us say a few words about the uniquness of $\Theta$. Recall the definition (2.7) of the polynomial matrices N, D, P, Q. Let R be the greatest common right divisor of NoP' and DoQ' ( o denotes Kronecker product and prime denotes transpose). We say that the problem data exhibit unstable cross cancellations if R is unstable. The following result can be proved.

(2.11) PROPOSITION. The strictly proper rational matrix $\Theta$ defined by (2.10) is unique if and only if the problem data exhibit no unstable cross cancellations.

In the present paper, we assume for simplicity that the problem data do no exhibit unstable cross cancellations. The case where they do is investigated in ANTOULAS [1984].

We can thus write the formal power series expansion of $\Theta$:

$$\Theta = \Lambda_1 z^{-1} + \Lambda_2 z^{-2} + \Lambda_3 z^{-3} + \ldots,$$

where $\Lambda_t$, $t > 0$, are constant matrices of appropriate dimensions. We can also assume without loss of generality that $MD_+$ is a column reduced polynomial matrix with column degrees $\kappa_i$, and that $Q_+T$ is a row reduced polynomial matrix with row degrees $\nu_j$. From the $\Lambda_t$'s, the $\kappa_i$'s and the $\nu_j$'s we define the central quantity of this theory, which is a finite sequence of constant matrices

(2.12) $S = (A_1, \ldots, A_\mu)$, $\mu = \max\{\kappa_i + \nu_j - 1\}$,

so that its (i,j)-th elements are

$$(A_t)_{ij} = (\Lambda_t)_{ij}, \text{ if } t < \kappa_i + \nu_j, \text{ for all } i \text{ and } j,$$

and free otherwise. (Notice, that $(M)_{ij}$ denotes the (i.j)-th element of M.)

The main result can be stated as follows.

(2.13) THEOREM. The solutions $Z_C$, $Z_y$ of equation (1.2) given by lemma (2.8) are:

(a) proper rational if and only if the parameter $\Delta$ is a partial realization of the sequence S defined above.

(b) proper rational and satisfy the stability requirements (1.4,5) if and only if the parameter $\Delta$ is a stable partial realization of the sequence S.

(c) In addition to (b), $\delta(Z_C)$ is minimal if and only if the parameter $\Delta$ is a minimal partial stable realization of the sequence S.

## 3. SOME PRELIMINARY RESULTS.

3.1. The first part of this section is devoted to the proof of

(3.1) LEMMA. Given are the rational matrices $Z_1$, $Z_2$, $Z_3$. Assume that the equation $Z_1 = Z_2 Z_x Z_3 + Z_y$ has stable rational solutions $Z_x$, $Z_y$. If $Z_2$ has full row rank over $k[z]$ and $Z_3$ has full column rank over $k[z]$, there exist a stable rational matrix $\bar{Z}$ and a polynomial matrix $\bar{Y}$ such that

$$Z_1 = Z_2 \bar{Z} Z_3 + \bar{Y}.$$

For the proof we need the

(3.2) PROPOSITION. Given the stable rational matrix $Z$ and the completely unstable and invertible polynomial matrices $N_+$, $P_+$ there exist a stable rational matrix $\Delta$ and a polynomial matrix $Y$ such that

$$Z = N_+ \Delta P_+ + Y.$$

PROOF. Using the Kronecker product technique, the result can be reduced to the case $P_+ = I$. Let $Z = R^{-1}S$ be a polynomial left coprime representation. By assumption $R$ is a stable polynomial matrix. Since $N_+$ is completely unstable, we can interchange $R$ and $N_+$, i.e. we can write $RN_+ = N_+'R'$, where $\det R = \det R'$, and $\det N_+ = \det N_+'$. Let $X$, $Y$ be polynomial matrices such that $S = RY + N_+'X$. Then with $\Delta = (R')^{-1}X$, which is a stable rational matrix, $Z = N_+\Delta + Y$.

PROOF OF (3.1). Let $Z_2 = ND^{-1}$, $Z_3 = Q^{-1}P$ be coprime polynomial representations. Let also $N = N_+N_-$, $P = P_-P_+$, where $N_+$, $P_+$ are non-singular and completely unstable polynomial matrices, while $N_-$, $P_-$ have stable invariant factors and full row, column rank respectively. By (2.6) $Z_1 = Z_2 X Z_3 + Z$. Since $Z$ is stable, by proposition (3.2) we can write $Z = N_+\Delta P_+ + Y$. Let $N_-^{\#}$, $P_-^{\#}$ denote the right, left inverse of $N_-$, $P_-$; they exist because of the full rank assumptions and are stable. It follows that $\bar{Z} := X + DN_-^{\#}\Delta P_-^{\#}Q$, and $\bar{Y} := Y$, are the required matrices.

3.2. Let $N_-$ be an $r \times m$ $(r \leq m)$ row reduced polynomial matrix, and $P_-$ a $p \times q$ $(p \geq q)$ column reduced polynomial matrix, both with stable invariant factors. Given is also the sequence $S$ of $m \times p$ constant matrices, like the one in (2.12). The purpose of this second part of the section is to discuss briefly the equation

(3.3) $N_-\Delta P_- = \bar{\Delta}$,

where $\Delta$ is required to be a partial realization of $S$. In particular, what restriction does this condition impose on the $\bar{\Delta}$'s? What can be said about the stability of $\Delta$, and $\bar{\Delta}$ ?

Before stating the general result, let us consider the following example:

$$N_- = \begin{pmatrix} z^2+z+1 & 0 & z \\ z & z+1 & 0 \end{pmatrix}; \quad S = (A_1, A_2), \quad A_1 = \begin{pmatrix} 1 & 1 \\ 0 & 2 \\ -1 & 0 \end{pmatrix}, \quad A_2 = \begin{pmatrix} 0 & x \\ 0 & 1 \\ y & 0 \end{pmatrix},$$

and $P_- = I$, while $x$, $y$ are free parameters. $\Delta$ has the general form $\Delta = A_1 z^{-1} + A_2 z^{-2} + A_3 z^{-3} + \ldots$, where $A_1$, $A_2$ are as above and the rest of the $A_t$'s are free. A straightforward computation yields:

$$N_- \Delta = \bar{A}_{-1} z + \bar{A}_0 + \bar{A}_1 z^{-1} + \bar{A}_2 z^{-2} + \ldots = \bar{\Delta}, \quad \text{where}$$

$$\bar{A}_{-1} = \begin{pmatrix} 1 & 1 \\ 0 & 0 \end{pmatrix}, \quad \bar{A}_0 = \begin{pmatrix} 0 & x+1 \\ 1 & 3 \end{pmatrix}, \quad \bar{A}_1 = \begin{pmatrix} ? & ?? \\ 0 & x+3 \end{pmatrix}.$$

Because $N_-$ is row reduced ? and ?? are free parameters; so are all $\bar{A}_t$, $t > 1$. The converse is also true again because of the row-reducedness of $N_-$, i.e. given any $\bar{\Delta}$ with $\bar{A}_{-1}$, $\bar{A}_0$, $\bar{A}_1$ as above, there exists a partial realization $\Delta$ of $S$, such that $N_-\Delta = \bar{\Delta}$. Thus if we define the sequence

$$\bar{S} = (\bar{A}_{-1}, \bar{A}_0, \bar{A}_1),$$

we can restate the result as follows: $\bar{\Delta} = N_-\Delta$ for some partial realization $\Delta$ of $S$, if and only if $\bar{\Delta}$ is a partial realization of $\bar{S}$.

Clearly,

$$\bar{\Delta}^* = \begin{pmatrix} z & z-2 \\ 1 & 3 \end{pmatrix},$$

is a partial realization (of zero MacMillan degree) of $\bar{S}$, for $x = -3$, $? = ?? = 0$. We want to find a stable partial realization $\Delta^*$ of $S$ such that (3.3) is satisfied. Notice first that (3.3) says that part of the left denominator polynomial matrix of the required $\Delta^*$ is fixed. There remains to determine a row polynomial vector $n = (n_1 \;\; n_2 \;\; n_3)$ such that

(3.4a) $n\Delta^* = \gamma$ = polynomial row vector, and

(3.4b) $\det \begin{pmatrix} N_- \\ n \end{pmatrix}$ = stable polynomial.

From the theory of partial realizations it follows that (3.4a) is satisfied for every row polynomial vector $n$ of high enough degree. For such an $n$ the coefficients are determined so that (3.4b) is satisfied. This is always possible since the invariant factors of $N_-$ are assumed stable. In our example, if we choose $y = 1$ and $n = (0 \;\; 0 \;\; z+1)$ we obtain $\gamma = (-1 \;\; 0)$. Thus a desired solution of (3.3) is

$$\Delta^* = \begin{pmatrix} z^2+z+1 & 0 & z \\ z & z+1 & 0 \\ 0 & 0 & z+1 \end{pmatrix}^{-1} \begin{pmatrix} z & z-2 \\ 1 & 3 \\ -1 & 0 \end{pmatrix}.$$

The procedure illustrated in the above example concerning the construction of the sequence $\bar{S}$ and of stable solutions to stable partial realizations of $\bar{S}$, can be carried out in general. It also follows that if (3.3) is solvable it has a stable solution as well. Moreover the general case can be reduced to the case $P_- = I$ using Kronecker products.

We close this section by stating the main result. The proof of this result which goes along the same lines as the considerations above, is omitted.

(3.5) LEMMA. Let $N_-$, $P_-$ be row, column reduced with stable invariant factors. The sequence $S$ of constant matrices is given and the sequence $\bar{S}$ is constucted as in the example above.

(a) If $\Delta$ is a stable partial realization of $S$, $\bar{\Delta} = N_-\Delta P_-$ is a stable partial realization of $\bar{S}$.

(b) Conversely, if $\bar{\Delta}$ is a stable partial realization of $\bar{S}$, there exists a stable partial realization $\Delta$ of $S$, such that $\bar{\Delta} = N_-\Delta P_-$.

## 4. THE RESULTS.

The proper rational function $Z^*$ is an admissible closed-loop transfer function for the system $\Sigma$ if it stable and there exists an internally stabilizing compensator $C$ such that

$$Z^* = Z_{11} - Z_{12}Z_C(I + Z_{22}Z_C)^{-1}Z_{21}.$$

If $Z_{12}$ does not have full row rank over $k[z]$, there exists an $r \times r$ polynomial unimodular matrix $U_1$ such that

$$U_1Z_{12} = \begin{pmatrix} \bar{Z}_{12} \\ 0 \end{pmatrix},$$

where $\bar{Z}_{12}$ has full row rank, say $\bar{r}$. Similarly, if $Z_{21}$ does not have full column rank over $k[z]$, there exists a polynomial unimodular $q \times q$ matrix $U_2$ such that

$$Z_{21}U_2 = (\bar{Z}_{21} \quad 0),$$

where $\bar{Z}_{21}$ has full column rank, say $\bar{q}$. If either or both of the above cases occur, a necessary condition for $Z^*$ to be admissible is derived easily, namely:

$$U_1(Z_{11} - Z^*)U_2 = \begin{pmatrix} \bar{Z}_{11} - \bar{Z}^* & 0 \\ 0 & 0 \end{pmatrix},$$

i.e. the last $r - \bar{r}$ rows and the last $q - \bar{q}$ columns of $U_1(Z_{11} - Z^*)U_2$ must be zero. The above considarations imply that for our purposes, we can assume without loss of generality:

(4.1) row $\text{rank}_{k[z]} Z_{12} = r$, i.e. full; column $\text{rank}_{k[z]} Z_{21} = q$, i.e. full.

In this section instead of the parametrization of the solutions of equation (1.2) given in lemma (2.8), we will use a slightly modified one. Recall lemma (3.1) and the definition (2.7) of the polynomial matrices N, D, Q, P. The following result is readily proved.

(4.2) LEMMA. <u>Under the assumption</u> (4.1), <u>the rational matrices</u> $Z_C$, $Z_y$ <u>satisfy equation</u> (1.2) <u>if and only if</u>

$$Z_C = (B + MZ_x)(A - LZ_x)^{-1}, \quad Z_x = D\Delta Q + \bar{Z},$$

$$Z_y = - N\Delta P + \bar{Y},$$

<u>for some rational matrix of appropriate dimensions</u> $\Delta$.

All the results of section two, except for part (c) of theorem (2.13), hold true after replacing $D_+$ by D, $P_+$ by P, $D_-$ and $P_-$ by the identity matrix I, X by $\bar{Z}$ and Z by $\bar{Y}$. The definition (2.12) of the sequence S has to be modified accordingly.

Using the second relationship in lemma (4.2) we can characterize the family of admissible closed-loop transfer functions as follows:

(4.3) $\underline{Z}_y := \{Z^* = - N\Delta P + \bar{Y}$, for some stable partial realization $\Delta$ of $S\}$.

In the sequel we will give an explicit criterion for checking whether a given $Z^*$ belongs to $\underline{Z}_y$. The difficulty arises of course, from the fact that in general, N, P are rectangular (see remark (4. 7a)). In case $Z^*$ is admissible, the method leads to the computation of the corresponding parameter $\Delta$. Thus using the first relationship in lemma (4.2) the compensators C which give rise to $Z^*$ are obtained. The main result is theorem (4.5) followed by a necessary and sufficient condition for 0 to be admissible, i.e. for the disturbance rejection problem to be solvable.

As before, we decompose the polynomial matrices N, P in the following way:

(4.4) $N = N_+N_-$, $N_+$ completely unstable, invertible; invariant factors of $N_-$ stable;
$P = P_-P_+$, $P_+$ completely unstable, invertible; invariant factors of $P_-$ stable.

Moreover, because of assumption (4.1), $N_+$ and $P_+$ can be chosen so that $N_-$ and $P_-$ are row and column reduced respectively.

We can now state the main result which is based on lemma (3.5). Recall (2.13).

(4.5) THEOREM. Let (4.1) hold true. From the sequence S, and $N_-$, $P_-$ we construct the sequence $\bar{S}$ as indicated in section 3.

(a) $Z^*$ is admissible if and only if $Z^* = -N_+\bar{\Delta}P_+ + \bar{Y}$, for some stable partial realization $\bar{\Delta}$ of $\bar{S}$.

(b) In addition, $\delta(Z^*)$ is minimal among all admissible transfer functions if and only if $\bar{\Delta}$ is a minimal stable partial realization of $\bar{S}$.

(4.6) COROLLARY. Solvability of the disturbance rejection problem. 0 is admissible if and only if $N_+^{-1}\bar{Y}P_+^{-1}$ is a stable partial realization of $\bar{S}$.

(4.7) REMARKS. (a). The purpose of the theorem above is to answer the question of admissibility of a given transfer function, by means of a formula like the one for $Z_x$ in lemma (4.2), in which the parameter is multiplied on each side be a square non-singular matrix.

The fact that $N_+$, $P_+$ are completely unstable polynomial matrices and $\bar{Y}$ a polynomial matrix is crucial for the minimality result (4.5b).

In order to check whether $Z^*$ is admissible we only need to check that $N_+^{-1}(\bar{Y} - Z^*)P_+^{-1}$ is a stable partial realization of $\bar{S}$. Notice that a necessary condition for admissibility is: $N_+$, $P_+$ must be zeros of $\bar{Y} - Z^*$.

(b). We have assumed that $Z_{12}$, $Z_{21}$ have full row, column rank. If one of these assumptions is not satisfied, according to the discussion at the beginning of this section, part of the closed-loop transfer function is apriori fixed and unaffected by all choices of the compensator C. In this case (4.5b) tells us how to minimize the MacMillan degree of that part of the closed-loop transfer function which depends on C.

## 5. AN EXAMPLE.

We conclude with the discussion of an example from OHM, HOWZE, and BHATTACHARYYA [1983]. Consider

$$Z_{11} = 1, \qquad Z_{12} = -(z-1)/z(z-2),$$

$$Z_{21} = (z-1)/(z+1), \qquad Z_{22} = -(z-1)^2/z(z+1)(z-2).$$

We are looking for a parametrization of all minimal-MacMillan-degree admissible closed-loop transfer functions & the corresponding compensators.

According to the procedure given at the beginning of section 2, we write: $Z_{22} = LM^{-1} = T^{-1}U$, $TA + UB = I$, where: $L = U = -(z-1)^2$,

$M = T = z(z+1)(z-2)$, $A = (z-3)/4$, $B = (z^2 - 2z - 4)/4$. From (2.3) we get

$$Z_1 = 1 + (z-1)^2(z^2 - 2z - 4)/4,$$

$$Z_2 = -(z-1)(z+1),$$

$$Z_3 = z(z-2)(z-1).$$

It is readily checked that since $Z_1$ has no poles $X = 0$, $Y = Z = Z_1$. Because of the special form of $Z_1$, $Z_2$, $Z_3$ the two parametrizations in Lemmata (2.8) and (4.2) of $Z_x$, $Z_y$ coincide:

$$Z_x = \Delta,$$
$$Z_y = z(z+1)(z-1)^2(z-2)\Delta + 1 + (z-1)^2(z^2 - 2z - 4)/4.$$

Thus by (2.12)

$$S = (1/4, -1/4, -3/4, -5/4, -11/4).$$

In this case $N_+ = z(z-1)^2(z-2)$, $N_- = (z+1)$, while $P_+ = P_- = 1$. By applying the procedure in section 3 we thus obtain

$$\bar{S} = (\bar{A}_0, \bar{A}_1, \bar{A}_2, \bar{A}_3, \bar{A}_4), \text{ where } \bar{A}_0 = 1/4, \bar{A}_1 = 0, \bar{A}_2 = -1, \bar{A}_3 = -2,$$
$$\bar{A}_4 = -4.$$

According to Theorem (4.5) we can rewrite $Z_y$ as follows:

$$Z_y = z(z-1)^2(z-2)\bar{\Delta} + 1 + (z-1)^2(z^2 - 2z - 4)/4.$$

Using the theory of partial realizations, we conclude that the minimal stable partial realizations $\bar{\Delta}_{min}$ of $\bar{S}$ have MacMillan degree 4; a parametrization is:

$$\bar{\Delta}_{min} = 1 - 4\{z^2 + (\alpha+2)z + 2\alpha + \beta + 4\}/\chi(z),$$

where $\chi(z) = z^4 + \alpha z^3 + \beta z^2 + \gamma z + \delta$, is a stable polynomial. By Theorem (4.5b) all resulting minimal-Macmillan-degree closed loop transfer functions are:

$$Z_y^{min} = 1 - (z-1)^2\{(4\alpha + 2\beta + \gamma + 8)z + \delta\}/\chi(z).$$

The compensators which give rise to the above closed-loop transfer functions are, according to the formulae in Lemma (4.2):

$$C = (z + 1)\{(4\alpha + 2\beta + \gamma + 8)z + \delta\}/\{z^2 - (3\alpha + 2\beta + \gamma + 6)z$$
$$+ 2\alpha + \beta - \delta + 4\}.$$

Thus, the corresponding compensators are of second order. In addition no choice of the parameters $\alpha,\beta,\gamma,\delta$ yields a stable compensator $C$.

Once the family $Z_y^{min}$ is obtained, the parameters $\alpha,\beta,\gamma,\delta$ can be determined by imposing further requirements, like minimal sensitivity, etc.

REFERENCES.

A. C. ANTOULAS

[1983] "A new approach to synthesis problems in linear system theory", Report 8302, Dept. of Elec. Eng., Rice University.

[1984] "A unifying approach to the design of linear systems", IFAC congress, Budapest.

G. BENGTSSON

[1977] "Output regulation and internal models - A frequency domain approach", Automatica, 13: 333-345.

L. CHENG AND J. B. PEARSON

[1978] "Frequency domain synthesis of linear multivariable regulators", IEEE Transactions on Automatic Control, AC-23: 3-15.

[1981] "Synthesis of linear multivariable regulators", IEEE Transactions on Automatic Control, AC-26: 194-202.

C. A. DESOER, R. W. LIU, J. MURRAY, and R. SAEKS

[1980] "Feedback system design: the fractional representation approach to analysis and synthesis", IEEE AC-25: 399-412.

P. P. KHARGONEKAR and A. B. ÖZGÜLER

[1982] "Regulator problem with internal stability", Technical report, Electrical Engineering Department, University of Florida.

D. Y. OHM, T. W. HOWZE, and S. P. BHATTACHARYYA

[1983] "Structural synthesis of multivariable controllers", Report #8308, Texas A&M University, Dept. of Electrical Engineering.

L. PERNEBO

[1981] "An algebraic theory for the design of controllers for linear multivarible systems; Part I: Structure matrices and feedforward design; Part II: Feedback realizations and feedback design", IEEE Transactions on Automatic Control, AC-26: 171-194.

J. M. SCHUMACHER

[1982] "Regulator synthesis using (C,A,B )-pairs", IEEE Transactions on Automatic Control, AC-27: 1211-1221.

J. C. WILLEMS and C. COMMAULT

[1981] "Disturbance decoupling by measurement feedback with stability or pole placement", SIAM J. Control Opt., 19: 490-504.

W. A. WOLOVICH and P. FERREIRA

[1979] "Output regulation and tracking in linear multivariable systems", IEEE Transactions on Automatic Control, AC-24: 460-465.

W. M. WONHAM and J. B. PEARSON

[1974] "Regulation and internal stabilization in linear multivariable systems", SIAM J. Control and Opt., 12: 5-18.

D. C. YOULA, J. J. BONGIORNO, and H. A. JABR

[1976] "Modern Wiener-Hopf design of optimal controllers, Part I: the single input-output case", IEEE Transactions on Automatic Control, AC-21: 3-13.

D. C. YOULA, H. A. JABR, and J. J. BONGIORNO

[1976] "Modern Wiener-Hopf design of optimal controllers, Part II: the multivariable case", IEEE Transactions on Automatic Control, AC-21: 318-338.

# POURSUITE DE MODELE A ENTREE BORNEE

J.M. DION et C. COMMAULT
Laboratoire d'Automatique de Grenoble
E.N.S.I.E.G. - I.N.P.G.
B.P. 46, 38402 - SAINT-MARTIN-D'HERES

=:=:=:=:=:=:=:=:=:=

RESUME.

Dans cet article on étudie l'équation AX = B où A et B sont des matrices sur le corps des fractions rationnelles. On s'intéressera aux solutions X à éléments appartenant à des anneaux particuliers tels que l'anneau des fractions rationnelles propres stables. La solution obtenue généralise les résultats connus où A, X et B appartiennent au même corps ou au même anneau. Quelques applications à divers problèmes de commande sont présentées. On s'intéressera plus particulièrement à la poursuite de modèle à entrée bornée.

ABSTRACT.

In this paper specific solutions to the equation AX=B are considered where A and B are matrices with rational entries. We focus our interest on solutions X with entries in particular rings such as the ring of proper rational stable functions. The given solution generalizes previous results where A,B and X belong to the same field or ring. Applications to some control problems are presented. The bounded input model following problem will be emphasised.

## I. INTRODUCTION.

L'étude du problème de la poursuite de modèle par retour d'état revient à l'étude de l'équation matricielle AX = B où A est la matrice de transfert du procédé à commander, B celle du modèle et X celle du précompensateur équivalent à l'action du retour d'état. L'obtention de solutions propres (resp. propres et stables) revient à l'étude de l'équation AX = B sur l'anneau des fractions rationnelles propres (resp. propres et stables). Une littératuree abondante a été consacrée à ces problèmes voir par exemple [1] - [8]. La solubilité de cette équation s'exprime simplement en termes de formes de Hermite ou de formes de Smith sur l'anneau considéré.

Dans cet article on étudie l'équation $AX = B$ où A et B sont des matrices sur le corps des fractions rationnelles. On s'interessera aux solutions X à éléments dans des anneaux particuliers tels que l'anneau des fractions rationnelles propres stables. Cette formulation permet de considérer des systèmes ayant des parties non propres ou non stables. Des transferts non propres peuvent apparaitre comme limites de systèmes singulièrement perturbés [9], ces transferts ont été largement étudiés dans [10], [5]. D'autre part, l'étude du rejet de perturbations conduit également à ce type d'équation, et dans ce cas A et B ne sont pas toujours stables, mais nous sommes intéressés par des commandes bornées et donc par des solutions X stables.

Ce type d'équation assez général permet d'appréhender divers problèmes de commande tels que :

+ Poursuite ou rejet de perturbations à entrées bornées par retour d'état ou de sortie.
+ Presque-poursuite ou presque rejet de perturbations à entrées bornées.

## II. NOTATIONS ET PRELIMINAIRES.

Soient $R_p(s)$ l'anneau des fractions rationnelles propres et $R_{ps}(s)$ l'anneau des fractions rationnelles propres stables. La stabilité étant définie comme suit :

Soit $C_b$ une partie symétrique du plan complexe coupant l'axe réel en au moins un point, une fonction de transfert sera dite stable si tous ses pôles finis sont dans $C_b$.

Soient $R_p^{pxm}(s)$ (resp. $R_{ps}^{pxm}(s)$) l'ensemble des matrices pxm à coefficients dans $R_p(s)$ (resp. $R_{ps}(s)$).

Un élément inversible (unité) B(s) de $R_p^{nxn}(s)$ est appelé une matrice bicausale et est caractérisé comme suit : $\det(\lim_{s\to\infty} B(s)) \neq 0$
Une telle matrice bicausale ne possède ni zéros, ni pôles à l'infini.

Un élément inversible (unité) B(s) de $R_{ps}^{nxn}(s)$ est appelé une matrice bicausale bistable, une telle matrice n'a ni zéros, ni pôles à l'infini ou hors du domaine de stabilité.

Les anneaux $R_p(s)$ et $R_{ps}(s)$ sont Euclidiens, on a pour ces anneaux un algorithme de division qui permet de calculer de façon relativement simple les formes de Smith et de Hermite de matrices à coefficients dans ces anneaux, pour plus de détails voir [11].

Soit T(s) une matrice (pxm) rationnelle propre de rang r et (s) sa forme de Smith sur l'anneau $R_p(s)$

on a $T(s) = B_1(s)\,\Lambda(s)\,B_2(s)$ où $B_1(s)$ et $B_2(s)$ sont bicausales

et $\Lambda(s) = \begin{bmatrix} \Delta(s) & o \\ o & o \end{bmatrix}$ avec $\Delta(s) = \mathrm{diag}(s^{-n_1} \ldots\ldots s^{-n_r})$.

Les $n_i$ sont donc les ordres des zéros à l'infini de T(s).

Notons que la forme de Smith de T(s) sur $R_{ps}(s)$ contiendrait en plus les pôles et les zéros instables de T(s).

Considérons l'équation AX= B où A et B sont des matrices à éléments dans $R_p(s)$ (resp. $R_{ps}(s)$) on a le résultat important suivant :

Théorème 1 : [4].

Soit l'équation AX = B où A et B sont des matrices (pxm) et (pxn) à coefficients dans $R_p(s)$ (resp. $R_{ps}(s)$). Cette équation a une solution X dans $R_p^{mxn}(s)$ (resp. $R_{ps}^{mxn}(s)$) si et seulement si (A,B) et (A,0) ont la même forme de Smith sur l'anneau $R_p(s)$ (resp. $R_{ps}(s)$).

Remarque : En d'autres termes l'équation est soluble sur $R_p(s)$ si et seulement si (A,B) et (A,0) ont la même structure à l'infini. Cette équation est soluble sur $R_{ps}(s)$ si et seulement si (A,B) et (A,0) ont les mêmes zéros infinis et les mêmes zéros instables.

## III. POURSUITE DE MODELE A ENTREE BORNEE.

### III.1. Poursuite de modèle.

L'étude du problème de la poursuite de modèle revient à trouver un précompensateur propre de transfert X solution de l'équation AX = B où A et B sont respectivement le transfert du procédé et du modèle à poursuivre. Si A et B sont des transferts propres, le théorème 1 s'applique directement.

Considérons maintenant le problème de la poursuite de modèle avec stabilité. On désire préserver la stabilité interne et externe, c'est-à-dire que le système compensé soit stable et qu'il n'y ait pas de simplifications "poles-zéros" instables. Il faut donc que le modèle choisi soit stable. De même le précompensateur doit être stable de façon à garantir la stabilité interne. Si le procédé est instable, on le stabilise par un retour d'état et on se retrouve dans les conditions d'application du théorème 1 sur $R_{ps}(s)$.

### III.2. Rejet de perturbations.

Le problème du rejet de perturbations a été défini dans [12] de la manière suivante :

Soit le système $\dot{x} = \bar{A}x + \bar{B}u + \bar{E}q$ , $z = \bar{D}x$ déterminer une commande par retour d'état $u = F x$ telle que la sortie z ne dépende pas de la perturbation q. Le problème sera résolu avec stabilité si le spectre de $A + BF$ est stable.

Considérons les transferts suivants :

$$A = \bar{D}(sI-\bar{A})^{-1}\,\bar{B}$$

$$B = \bar{D}(sI-\bar{A})^{-1}\,\bar{E}$$

On a le résultat suivant [13].

Théorème 2 :

Le problème de rejet de perturbations a une solution si et seulement si il existe une solution strictement propre X à l'équation AX = B.

Le problème est soluble avec stabilité si et seulement si

(i) il existe une solution strictement propre stable X à AX = B

(ii) $(\bar{A},\bar{B})$ est stabilisable.

Remarques :

+ Dans le cas où la perturbation est mesurable, il suffit de chercher une solution X propre.
+ Le théorème 2 reste valable dans le cas où A et B sont des transferts propres.
+ Ces résultats montrent "l'équivalence" des problèmes de rejet de perturbations et de poursuite de modèle.

## III.3. Poursuite de modèle ou rejet de perturbations à entrées bornées.

Il peut être intéressant de ne pas se restreindre à des transferts stables comme le montre l'exemple suivant :

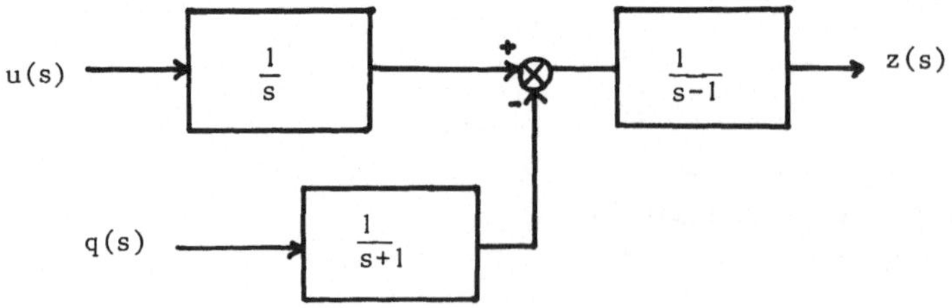

Dans ce cas, le problème du rejet de la perturbation mesurable q ne peut être résolu avec stabilité. En effet le transfert perturbation sortie étant instable, dans toute représentation d'état $(\bar{A},\bar{B},\bar{E},\bar{D})$ la paire $(\bar{A},\bar{B})$ n'est pas stabilisable. Il existe pourtant une commande bornée qui permet d'annuler le transfert entre q et z après bouclage :

$$u(s) = \frac{s}{s+1}\, q(s)$$

Dans ce paragraphe on s'intéressera à la résolution du problème de rejet de perturbations avec entrée bornée défini comme suit :

Définition 1 :

Soit $A \in R_p^{pxm}(s)$ le transfert entre la commande u et la sortie z. Soit $B \in R_p^{pxn}(s)$ le transfert entre la perturbation q et la sortie z. Le problème du rejet de perturbations à entrées bornées consiste à trouver un précompensateur $X \in R_{ps}^{mxn}(s)$ donc propre et stable tel que z soit indépendant de q.

On définirait de même le problème de la poursuite de modèle à entrées bornées qui lui est équivalent.

Pour ces deux problèmes le théorème 1 ne s'applique pas, A et B n'étant pas forcément stables.

On va donc présenter une généralisation du théorème 1 au cas où A et B ne sont pas stables (ni même forcément propres).

## III.4. Résultat général.

On a le théorème suivant qui permet de résoudre le problème défini au paragraphe (III.3.).

Théorème 3 :

Soient $A \in R^{pxm}(s)$ et $B \in R^{pxn}(s)$, il existe une solution $X \in R_{ps}^{mxn}(s)$ à l'équation $AX = B$ si et seulement si il existe U unité de $R_{ps}^{(m+n)x(m+n)}(s)$ (bicausale et bistable) telle que :

$$(A,B) = (A,0)\,U$$

Démonstration :

CN : Il existe une solution propre stable X à $AX = B$ donc

$$(A,B) = (A,AX) = (A,0)\begin{bmatrix} I & X \\ 0 & I \end{bmatrix} = (A,0)\,U$$

X étant propre et stable, U est propre et stable, il en est de même pour $U^{-1} = \begin{bmatrix} I & -X \\ 0 & I \end{bmatrix}$ , U est donc une unité de l'anneau.

CS : Soit U unité de l'anneau considéré telle que :

$$(A,B) = (A,0)U = (A,0) \begin{bmatrix} U_{11} & U_{12} \\ U_{21} & U_{22} \end{bmatrix} = (A\,U_{11},\ A\,U_{12})$$

donc $B = A\,U_{12}$ avec $U_{12}$ propre et stable.

Dans le cas où A et B sont des transferts propres stables, ce théorème est équivalent au théorème 1, en effet, (A,B) = (A,0) U implique que (A,B) et (A,0) ont la même forme de Smith sur $R_{ps}(s)$. Réciproquement si (A,B) et (A,0) ont la même forme de Smith sur $R_{ps}(s)$, il existe X solution propre stable à AX = B ce qui implique

$$(A,B) = (A,0) \begin{bmatrix} I & X \\ 0 & I \end{bmatrix} = (A,0)U$$

Le théorème 3 nous donne une CNS de solubilité du problème de rejet de perturbations (ou de poursuite de modèle) à entrées bornées, ceci pour des systèmes dont les transferts ne sont pas forcément propres ou stables.

On présente maintenant une caractérisation équivalente en termes de factorisations des transferts considérés.

**Théorème 4 :**

Soient $A \in R^{p \times m}(s)$ et $B \in R^{p \times n}(s)$, il existe une solution $X \in R_{ps}{}^{m \times n}(s)$ à l'équation AX = B si et seulement si il existe U,T et V telles que

A = TU avec U bicausale et $U^{-1}$ stable.

B = TV avec V propre et $U^{-1}V$ stable.

**Démonstration :**

CN : X est une solution propre et stable à AX = B

donc $(A,B) = (A,0) \begin{bmatrix} I & X \\ 0 & I \end{bmatrix}$

d'autre part, on peut toujours factoriser A en A = TU avec U bicausale et $U^{-1}$ stable. En effet, il existe un entier n et un polynôme p tels que $A = \frac{s^n}{p} A'$ où A' est propre et stable. Soit H' la forme de Hermite de A' sur $R_{ps}(s)$ on a A' = H'U avec U bicausale bistable, $T = \frac{s^n}{p} H'$ donne le résultat. Il vient :

$$(TU,B) = (TU,0)\begin{bmatrix} I & X \\ 0 & I \end{bmatrix} \quad \text{donc} \quad B = T\,U\,X = TV \text{ avec } V = UX$$

$V$ est propre et $U^{-1}V = X$ est stable par hypothèse.

CS :

$$\text{On a} \quad (A,B) = (A,0)\begin{bmatrix} I & U^{-1}V \\ 0 & I \end{bmatrix}$$

où $U^{-1}V$ est propre et stable par hypothèse, l'application du théorème 3 donne le résultat.

N.B. : Dans ce théorème on aurait pu de manière équivalente factoriser A en $A = TU$ avec $U$ bicausale et bistable et B en $B = TV$ avec V propre et stable.

Sur l'exemple précédent on peut choisir

$$T = \frac{1}{s(s-1)} , \quad U = 1, \quad V = \frac{s}{s+1}$$

Remarques :

- i) Si on se restreint à des matrices A et B propres les conditions du théorème 4 apparaissent dans un théorème de [14], mais sont appliquées de manière incorrecte.

  ii) Si on se restreint à des matrices A et B propres et stables, on retrouve le fait que B doit s'écrire $B = HV$ où H est la forme de Hermite de A sur l'anneau des fractions rationnelles propres stables et où V est propre et stable.

  iii) Pour que le problème soit soluble, il faut que les zéros instables de A soient "contenus" dans B et que les pôles instables de B soient "contenus" dans A. Ceci est bien clair en monovariable.

  iv) Le théorème 3 n'utilise pas les caractéristiques particulières de $R_{ps}(s)$, il pourrait donc être généralisé à d'autres anneaux.

## CONCLUSIONS.

On a présenté ici une application du théorème 3 aux problèmes de rejet de perturbations et de poursuite de modèle à entrée bornée. D'autres applications de ce théorème sont possibles. Par exemple, une C.N.S. de solubilité du problème de presque rejet de perturbations avec "entrées bornées", ceci en considérant des solutions stables à $AX = B$.

Les problèmes traités précédemment conduisent à une équation du type $AXB = C$ dans le cas de commandes par retour de mesures. Le produit de Kronecker permet de se ramener au cas précédent [6].

Le problème du presque rejet de perturbations par retour de sorties a été posé et résolu dans [15]. Il revient à résoudre l'équation $AXB = C$ sur les fractions rationnelles avec A,B et C strictement propres, où A est le transfert entrée-sortie, B le transfert perturbation-mesure et C le transfert perturbation-sortie. L'extension au presque rejet de perturbations par retour de sortie avec stabilité se résoud avec les outils développés dans [16]. Le presque rejet de perturbations par retour de sorties à entrées bornées se résoud en cherchant une solution rationnelle stable à $AXB = C$, donc en utilisant le produit de Kronecker et le théorème 3.

REFERENCES.

[1] S.H. WANG and E.J. DAVISON. "A minimization algorithm for the design of linear multivariable systems". I.E.E.E. T.A.C. A.C-18, pp. 220-225, 1973.

[2] G.D. FORNEY,Jr. "Minimal bases of rational vector spaces with applications to multivariable linear systems". SIAM J. on Control, 13, pp. 493-520. 1975.

[3] S. KUNG and T. KAILATH. "Some notes on valuation theory in linear systems" Proc. I.E.E.E. C.D.C. San Diego 1978. pp. 515-517.

[4] A.S. MORSE. "System invariants under feedback and cascade control". Proc. Int. Symp. Udine Springer Verlag. 1975.

[5] G. VERGHESE. "Infinite frequency behaviour in generalized dynamical systems" Ph.D. dissertation, Dept. of Elect. Engg. Stanford University. 1978.

[6] L. PERNEBO. "Algebraic control theory for linear multivariable systems. Ph.D. dissertation. Lund Institute of Technology Sweden, 1978.

[7] W.A. WOLOWICH and P.L. FALB. "Invariants and canonical forms under dynamic compensation" SIAM J. on Control, 14, pp. 996-1008, 1976.

[8] M. MALABRE. "Structure à l'infini des triplets invariants, application à la poursuite parfaite de modèle" 5ème Conf. Int. Analysis and Opt. of Systems. INRIA. Springer-Verlag. 1982.

[9] B. FRANCIS. "Convergence in the boundary layer for singularly perturbed equations" Automatica, 18, pp. 57-62, 1982.

[10] H.H. ROSENBROCK. "State space and multivariable theory". Nelson Wiley.1970.

[11] M. MARCUS. "Introduction to modern algebra". Marcel Dekker. New York. 1978.

[12] W.M. WONHAM and A.S. MORSE. "Decoupling and pole assignment in linear multivariable systems : a geometric approach". SIAM J. on Control, 8, pp. 1-18, 1970.

[13] M.L.J. HAUTUS. "(A,B)-invariant and stabilizability subspaces, a frequency domain description" Automatica, 16, pp. 703-707, 1980.

[14] BHATTACHARYYA, A.C. del NERO GOMES and J.W. HOWZE. "The structure of robust disturbance rejection control". I.E.E.E. T.A.C. AC-28, pp. 874-881, 1983.

[15] J.C. WILLEMS. "Almost invariant subspaces : An approach to high gain feedback design. Part.II : Almost conditionally invariant subspaces". I.E.E.E. T.A.C., AC-27, pp. 1071-1096. 1982.

[16] C. COMMAULT, J.M. DION and S. PEREZ. "Transfer matrix approach to the disturbance decoupling problem". I.F.A.C. Congress - Budapest - July 1984.

Session 7

# DISTRIBUTED PARAMETER SYSTEMS

# SYSTÈMES À PARAMÈTRES DISTRIBUÉS

## OPTIMAL CONTROL FOR LINEAR SYSTEMS WITH RETARDED STATE AND OBSERVATION AND QUADRATIC COST

Elena M. Fernandez-Berdaguer† and E. Bruce Lee
Department of Electrical Engineering
University of Minnesota
123 Church Street
Minneapolis, Minnesota 55455

†Now at the Department of Mathematics of the Virginia Polytechnic and State University, Blacksburg, VA 24060.

### Abstract

Stabilization (feedback control) of linear finite dimensional systems has long been based on the use of input-output models and ideas associated with equivalent systems. When such ideas are applied to the linear hereditary systems or other infinite dimensional systems new results appear which are significant in terms of controller synthesis and analysis.

Recently the input-output model in the hereditary situation was generalized by admitting delays in the observation (or reconstruction) of state type data. Also considerations of various categories of equivalent linear hereditary systems under, for example, the action of the feedback group has been undertaken. Each of these studies has led to new insights into the formulation of optimal control questions and questions of stabilizability.

Here we extend previous results on optimal control (quadratic formulation) of the linear hereditary systems. The main contribution is the extension of the optimal control theory to the infinite horizon case (with delayed observation) where questions of stability with the optimal controller become a significant issue in optimal synthesis.

The setting is the standard semigroup formulation in the Hilbert space $M^2$. The main result is a characterization of the optimal stabilizing feedback controller in terms of the solution of a certain system of operator Riccati equations in the delayed cost formulation.

### §1. Preliminaries

In what follows, R will denote the field of real numbers. Given two real numbers a and b with $a < b$ and E a real Banach space, $C(a,b;E)$ will denote the Banach space of all continuous maps $[a,b] \to E$ endowed with the sup norm, $L^p(a,b;E)$ for $1 \leq p < \infty$ will denote the Banach space of all measurable maps $[a,b] \to E$ which are p-integrable, and $W^{1,p}(a,b;E)$ will denote the Sobolev space of all maps x in $L^p(a,b;E)$ with a distributional derivative Dx in $L^p(a,b;E)$.

Given two Banach spaces X and Y, $\mathcal{L}(X,Y)$ will denote the Banach space of all continuous linear maps $X \to Y$ endowed with the natural norm. When $X = Y$, $\mathcal{L}(X,Y)$ will be abbreviated by $\mathcal{L}(X)$. Now if E is a Hilbert space we define the space $M^2$: $M^2(a,b;E)$ will denote the product space $E \times L^2(a,b;E)$, the elements of $M^2(a,b;E)$

being pairs $h = (h^0,h^1)$ with $h^0 \varepsilon E$ and $h^1 \varepsilon L^2(a,b;E)$. With inner product given by

$$(h,k)_{M^2(a,b;E)} = (h^0,k^0)_E + (h^1,k^1)_{L^2(a,b;E)}$$

$M^2(a,b;E)$ becomes a Hilbert space.

$W^2$ will denote the subspace of $M^2$ defined as

$$W^2 = \{(x^0,x^1) \varepsilon M^2 \mid x^0 = x^1(0),\ x^1 \varepsilon W^{1,2}\}.$$

As the model of the controlled process we will consider the functional differential equation:

$$(1.1)\qquad \dot{x}(t) = \sum_{i=0}^{N} A_i x(t-h_i) + \int_{-\gamma}^{0} A(\theta)x(t+\theta)d\theta + Bu(t)$$

where $0=h_0<h_1<\ldots<h_N=\gamma$, $A_i \varepsilon \mathcal{L}(R^n)$, $A(\cdot)\varepsilon L^2([-\gamma,0];\mathcal{L}(R^n))$, and $u \varepsilon L^2([0,T];R^P)$.

The initial condition for (1.1) is

$$(1.2)\qquad x(0)=x_0;\ x(t)=\phi(t),\ t\varepsilon[-\gamma,0).$$

It is well known [7] that for any $x^0 \varepsilon R^n$ and $\phi(t)\varepsilon L^2([-\gamma,0];R^n)$ there is a unique solution to equation (1.1) with initial condition (1.2) and that the solution belongs to $W^{1,2}_{loc}([0,\infty);R^n)$.

Also if $\tilde{\phi}\varepsilon W^2$ and $u \varepsilon C^1$ then equation (1.1) with initial condition (1.2) can be written as an evolution equation in the space $M^2([-\gamma,0];R^n)$ in the following way (see [2],[8])

$$(1.3)\qquad \frac{d\tilde{x}(t)}{dt} = \tilde{A}\,\tilde{x}(t) + \tilde{B}u(t)$$

$$\tilde{x}(0) = \tilde{x}_0 = (x_0,\phi)$$

where $\tilde{x}(t) = (x(t),x^1(t))$, $x^1(t)(\theta) = x(t+\theta)$.

$\tilde{B}\varepsilon\mathcal{L}(R^P, M^2)$ is defined by $\tilde{B}u=(Bu,0)$ and $\tilde{A}$ is the generator of the strongly continuous semigroup $T(t)$. Then the integral version of (1.3) is

$$(1.4)\qquad \tilde{x}(t) = T(t)\,\tilde{x}_0 + \int_0^t T(t-s)\,\tilde{B}u(s)ds\,.$$

In §2 we will use the following perturbation result:

<u>Theorem 1.1</u>: Let H denote a Hilbert space, let $T(\cdot,\cdot)$ be a strongly continuous semigroup in $M^2$ which is uniformly bounded and let C be in $B_\infty(t_0,T;\mathcal{L}(H))$. Then the operator integral equation

$$(1.5)\qquad S(t,s)\tilde{x} \equiv T(t,s)\tilde{x} + \int_s^t T(t,\eta)\,C(\eta)\,S(\eta,s)\tilde{x}d\eta,\ \tilde{x}\varepsilon H$$

has a unique solution $S(.,.)$ in the class of strongly continuous bounded linear operators on $H$. $S(.,.)$ is an evolution operator called the perturbed evolution operator of $T(.,.)$ by $C(.)$, $S(.,.)$ is also the unique solution of:

$$(1.6) \qquad S(t,s)\tilde{x} = T(t,s)\tilde{x} + \int_s^t S(t,\eta)\, C(\eta)\, T(\eta,s)\tilde{x}d\eta \text{ for } \tilde{x}\varepsilon H$$

## §2. Generalized quadratic optimal control on a finite interval

In this section we consider the question of minimizing the functional

$$(2.1) \qquad J(t_0,\tilde{x}(t_0),u) = \int_{t_0}^{T} \Big[\sum_{i=0}^{N} | P_i\, x(t-\tau_i) |^2 + (u(t),\, Uu(t))\Big]\, dt$$

with respect to $u$ where $0=\tau_0<\tau_1<\ldots<\tau_N=b$; $P_i\varepsilon R^{n\times n}$; and $U\varepsilon R^{p\times p}$ is symmetric and positive definite (that is $\| U \|_{R^{p\times p}}>\alpha$ for some $\alpha>0$). If $t>t_0$, $x(t)$ in (2.1) is the solution of (1.1); if $t\varepsilon[t_0-b,t_0]$, $x(t)=\phi(t)$; and in case $b>\gamma$ we define $x(t)=\phi(t)=0$ for $t\varepsilon[t_0-b,\, t_0-\gamma]$.

The state $\tilde{x}(t)$ in the above case will be considered in the space $M^2([-\beta,0];R^n)$ with $\beta = \max(\gamma,b)$ (this is done in order to include for each $t$ all the values of $x(t-\tau_i)$ in the function part of $x(t)$ in $M^2$.)

From now on $M^2$ will mean $M^2([-\beta,0];R^n)$.

We define the (unbounded) operator $\Pi$:domain$(\tilde{A})\to M^2$ by

$$(2.2) \quad \Pi(\phi^0,\phi^1)=(P_0\phi(0)+\Sigma P_i\phi(-\tau_i),0)$$

to write 2.1 in the form

$$(2.3) \qquad J(t_0,\tilde{x}(t_0),u) = \int_{t_0}^{T} [\|\Pi\tilde{x}(t)\|^2_{M^2} + (u(t),\, Uu(t))]dt.$$

Note that even in the case $\tilde{\phi}$ doesn't belong to domain$(\tilde{A})$ $\|\Pi\tilde{x}(t)\|_{M^2} \varepsilon L^2([t_0,T])$.

Denote $L^2(t_0,T;R^p)$ by $U_{t_0}$, and $L^2(t_0,T;M^2)$ by $\Xi_{t_0}$. Now we can write

$$(2.4) \qquad J(t_0,\tilde{x}(t_0),\, u) = \int_{t_0}^{T} | \overline{H}_{t_0}\, \tilde{x}(t) |^2\, dt + a_{t_0}(u,u) + 2\Lambda_{t_0}(u).$$

We define the operator $\overline{H}_{t_0}\varepsilon\mathcal{L}(M^2,\Xi_{t_0})$, the continuous bilinear form $a_{t_0}$ on $U_{t_0}$ and $\Lambda_{t_0}\varepsilon\mathcal{L}(U_{t_0})$ in the following way:

$$(2.5) \qquad (\overline{H}_{t_0}\tilde{x})(t) = \Pi T(t-t_0)\tilde{x}$$

$$(2.6)\qquad a_{t_0}(u,v)=\int_{t_0}^{T}[((\tau_{t_0}u)(t),(\tau_{t_0}v)(t))_{M^2}+(Uu(t),v(t))]dt$$

where $\tau_{t_0}\varepsilon\mathcal{L}(U_{t_0},\Xi_{t_0})$ is defined by $(\tilde{B}u(s)\varepsilon\Xi_{t_0})$

$$(2.7)\qquad (\tau_{t_0}u)(t) = \Pi\int_{t_0}^{t}T(t-s)\,\tilde{B}\,u(s)\,ds$$

$$(2.8)\qquad \Lambda_{t_0}u = \int_{t_o}^{T}((\tau_{t_0}u)(t),(\overline{H}_{t_0}\tilde{x})(t))_{M^2}dt\,.$$

Also we can write (2.6) as

$$(2.9)\ a_{t_0}(u,v) = (A_{t_0}u,v)_{U_{t_0}}$$

where $A_{t_0}\varepsilon\mathcal{L}(U_{t_0})$ and $A_{t_0}$ is Hermitian (this follows from $a_{t_0}$ being continuous and symmetric). In fact $A_{t_0}$ can be computed as

$$(2.10)\qquad (A_{t_0}u)(t) = (\tau^{*}_{t_0}\tau_{t_0}u)(t) + U\,u(t).$$

Also $\Lambda_{t_0}(u)$ can be expressed as

$$(2.11)\qquad \Lambda_{t_0}(u) = (u,g_{t_0})_{U_{t_0}}$$

with $g_{t_0}\varepsilon U_{t_0}$ given by

$$(2.12)\qquad g_{t_0}(s) = (\tau^{*}_{t_0}\overline{H}_{t_0}\tilde{x})(s).$$

It is known that under our hypotheses there is a unique solution to the minimization question (see [16]) and that the optimal u is characterized by

$$(2.13)\ a_{t_0}(u,v) + \Lambda_{t_0}(v) = 0 \qquad \forall\ v\varepsilon U_{t_0}\,.$$

Hence u must satisfy

$$(2.14)\qquad (A_{t_0}u)(t) + (\tau^{*}_{t_0}\overline{H}_{t_0}\,\tilde{x})(t) = 0 \qquad \text{a.e. in } [t_0,T].$$

From our previous observation on $A_{t_0}$ and the fact that $\| A_{t_0} \| > \alpha$ it follows that $A_{t_0}^{-1}$ exists, $A_{t_0}^{-1}\varepsilon\mathcal{L}(U_{t_0})$ and $\| A_{t_0}^{-1} \| < \alpha$.

Thus from (2.14) it follows that

$$(2.15)\qquad u(t) = -A_{t_0}^{-1}\,(\tau^{*}_{t_0}\,\overline{H}_{t_0}\,\tilde{x})(t) \qquad \text{a.e. in } [t_0,T]\,.$$

It is clear that the map $\tilde{x}\mapsto u$ is continuous from $M^2$ into $U_{t_0}$.

All of the above reasoning can be applied to minimizing $J(s,\tilde{x},u)$ in $[s,T]$, $s\varepsilon[t_0,T]$. Therefore if u is the optimal solution u is given by

$$(2.16)\qquad u(t) = -A_s^{-1}(\tau^{*}_s\overline{H}_s\ \tilde{x})(t)\ ,\ t\varepsilon[s,T].$$

If $\tilde{x}(t)$ is the corresponding optimal trajectory we have

(2.17) $\tilde{x}(t) = T(t-s)\tilde{x} - \int_s^t T(t-\sigma)\tilde{B}A_s^{-1}(\tau_s^*\overline{H}_s\tilde{x}))(\sigma)d\sigma.$

From the form of $\tau_s^*$ and $\overline{H}_s$ and the fact that $\|T(t-s)\| \leq M$ for some M and $t_0 \leq s \leq t \leq T$ it follows that

$$\|(\tau_s^*\overline{H}_s\tilde{x})(t)\|_{U_{t_0}} \leq K\|\tilde{x}\|_{M^2}.$$

Then from above inequality and $\|A_{t_0}^{-1}\|_{\mathcal{L}(U_{t_0})} \leq 1/\alpha$, we have that

$$\|(A_s^{-1}\tau_s^*\overline{H}_s)(t)\| \leq K, \quad t_0 \leq s \leq t \leq T .$$

Therefore it can be proved that the operator defined as

(2.18) $S(t,s)\tilde{x} = T(t-s)\tilde{x} - \int_s^t T(t-\sigma)\tilde{B}(A_s^{-1}\tau_s^*\overline{H}_s\tilde{x})(\sigma)d\sigma$

is an evolution operator on $M^2$ (For a proof see [15]).

Now we will find the feedback operator of(2.15) in a form for which we will find a Riccati equation type realization.

By using (2.10), (2.5) and (2.7) we can write equation (2.14) as

$$0=\tau_{t_0}^*[\tau_{t_0}u + \overline{H}_{t_0}\tilde{x}](t)+Uu(t)=\tau_{t_0}^*[\Pi\tilde{x}(\cdot)](t) + Uu(t).$$

Hence

(2.19) $u(t) = U^{-1}[\tau_{t_0}^*\Pi\tilde{x}(\cdot)](t).$

Let H be the operator $H\in\mathcal{L}(\Xi_\sigma)$ ;$t_0 \leq \sigma \leq T$

(2.20) $(\underset{\sigma}{H}\tilde{x})(t) = \Pi\int_\sigma^t T(t-s)\tilde{x}(s)ds .$

It is clear that

(2.21) $(\underset{\sigma}{H}\tilde{x})(t) = \int_\sigma^t \Pi T(t-\sigma)\tilde{x}(\sigma)d\sigma .$

We define $P \in \mathcal{L}(M^2, \Xi_{t_0})$ as

(2.22) $P(t)\tilde{x} = (H_t^*\Pi S(\cdot,t)\tilde{x})(t) .$

Now if $\tilde{x}(\cdot)$ is the optimal trajectory

(2.23) $\tilde{u}(t) = -U^{-1}\tilde{B}^* P(t)\tilde{x}(t).$

<u>Lemma 2.1</u>: $P(t)\in\mathcal{L}(M^2)$ and $P\mapsto P(t)x$ $[t_0,T]\to M^2$ is continuous.

Proof: see appendix.

Now $\tilde{B}U^{-1}\tilde{B}^*P(t)\varepsilon B_\infty([t_0,T);M^2)$ and replacing $u(\sigma)$ $(=A_S\tau_S H_S\tilde{x})(\sigma))$ in (2.18) by (2.23) we can use theorem 1.1 to conclude that S(t,s) is the perturbed evolution operator of T(t) by $-\tilde{B}U^{-1}\tilde{B}^*P(t)$.

Theorem 2.1: P(t) satisfies the Riccati equation:

$$(2.24) \quad P(t)\tilde{x} = (H^*\Pi T(\cdot -t)\tilde{x})(t) - \int_t^T T^*(\sigma-t)\ P(\sigma)\tilde{B}U^{-1}\tilde{B}^*P(\sigma)T(\sigma-t)\tilde{x}d\sigma.$$

Proof:

From (2.22) we have for $\tilde{x}\varepsilon M^2;\tilde{y}\varepsilon\Xi_{t_0}$

$$(2.25)\ (P(\cdot)\tilde{x},\tilde{y}(\cdot))_{\Xi_{t_0}} = \int_{t_0}^T (\Pi \int_{t_0}^\sigma T(\sigma-s)\tilde{y}(s),\ \Pi S(\sigma,s)\tilde{x})_{M^2} ds d\sigma.$$

Replacing $S(\sigma,s)$ in (2.25) by (1.6) we obtain

$$(P(\cdot)\tilde{x},\tilde{y}(\cdot))_{\Xi_{t_0}} = \int_{t_0}^T (\Pi \int_{t_0}^\sigma T(\sigma-s)\tilde{y}(s),\ \Pi T(\sigma-s)\tilde{x}$$

$$- \int_s^\sigma \Pi\ S(\sigma,\eta)\ \tilde{B}U^{-1}\tilde{B}^*\ P(\eta)\ T(\eta-s)\ \tilde{x}\ d\eta)dsd\sigma$$

$$= I_1 - \int_{t_0}^T (\Pi \int_{t_0}^\sigma T(\sigma-s)\tilde{y}(s),\ \int_s^\sigma \Pi\ S(\sigma,\eta)\tilde{B}U^{-1}\tilde{B}^*\ P(\eta)T(\eta-s)\tilde{x}d\eta ds)d\sigma$$

$$(2.26) = I_1 - \int_{t_0}^T \int_{t_0}^\sigma (\Pi T(\sigma-\eta) \int_{t_0}^\eta (T(\eta-s)\tilde{y}(s),\ \Pi S(\sigma,\eta)\tilde{B}U^{-1}\tilde{B}^*P(\eta)T(\eta-s)\tilde{x}dsd\eta)d\sigma.$$

The last formula was obtained by using that $T(\sigma-\eta)T(\eta-s)=T(\sigma-s)(s<\eta<\sigma)$ and interchanging the order of integration. From (2.25) and (2.26) we obtain

$$(P(\cdot)\tilde{x},\tilde{y}(\cdot))_{\Xi_{t_0}} = I_1 - \int_{t_0}^T \int_{t_0}^\eta (T(\eta-s)\tilde{y}(s),\ P(\eta)\tilde{B}U^{-1}\tilde{B}^*(\eta)T(\eta-s)\tilde{x})dsd\eta$$

$$= I_1 - \int_{t_0}^T (\tilde{y}(s),\ \int_\eta^T T^*(\eta-s)\ P(\eta)\ \tilde{B}U^{-1}\tilde{B}^*P(\eta)T(\eta-s)\tilde{x}d\eta)ds.$$

Clearly

$$I_1 = \int_{t_0}^T (\tilde{y}(s),(H^*\Pi T(\cdot-s)\tilde{x})(s))ds.$$

Theorem 2.2: The optimization question

$$\min\{J(t_0,\tilde{x}(t_0),\upsilon)\,|\,\upsilon\varepsilon\ L^2([t_0,T];R^P]\}$$

has a unique solution u(t) given by (2.23) where P(t) is the unique solution of (2.24).

Moreover the optimal cost is given by

$$(2.27)\quad J(t_0,\tilde{x}(t_0),u) = (P(t_0)\tilde{x}(t_0),\tilde{x}(t_0))_{M_2}\ .$$

Proof:

Uniqueness of solution to (2.24): If $P_1$, $P_2$ are solutions of (2.24) then

$$P_1(\eta)-P_2(\eta) = \int_\eta^T [T^*(\eta-s)[P_1(\eta)-P_2(\eta)]\tilde{B}Q^{-1}\tilde{B}^*P_2(\eta)\ T(\eta-s)\ d\eta$$

$$+\int_\eta^T T^*(\eta-s)\ P_1(\eta)\ \tilde{B}Q^{-1}\tilde{B}^*[P_1(\eta)-P_2(\eta)]T(\eta-s)d\eta\ .$$

Now from Gronwall's lemma it follows that $P_1=P_2$.

To prove (2.27) we use (2.25) with $\tilde{y}(s) = \tilde{y}\chi_{[t_0 T]}(s)$, $\tilde{y}\varepsilon M$.

Then

$$(2.28)\quad \int_{t_0}^T (P(t)\tilde{x},\tilde{y})_{M^2}dt = \int_{t_0}^T\int_{t_0}^\sigma (\Pi T(\sigma-s)\tilde{y}(s),\ \Pi S(\sigma,s)\tilde{x})_{M^2}dsd\sigma$$

$$= \int_{t_0}^T\int_{t_0}^\sigma (\Pi S(\sigma,s)\tilde{y}(s),\ \Pi S(\sigma,s)\tilde{x})dsd\sigma$$

$$-\int_{t_0}^T\int_{t_0}^\sigma (\ \Pi\int_\sigma^s T(s-\eta)\ \tilde{B}U^{-1}\tilde{B}^*P(\eta)S(\eta,s)\tilde{y}d\eta,\ \Pi S(\sigma,s)\tilde{x})dsd\sigma$$

$$= \int_{t_0}^T\int_s^T (\Pi S(\sigma,s)\tilde{y},\Pi S(\sigma,s)\tilde{x})d\sigma\chi_{[t_0,T]}(s)ds$$

$$-\int_{t_0}^T\int_{t_0}^\sigma (\tilde{B}U^{-1}\tilde{B}^*P(\sigma)S(\sigma,s)\tilde{y},\ P(\sigma)\ S(\sigma,s)\tilde{x})dsd\sigma$$

$$= \int_{t_0}^T\int_s^T (\Pi S(\sigma,s)\tilde{y},\Pi S(\sigma,s)\tilde{x})d\sigma\ \chi_{[t_0,T]}(s)\ ds$$

$$-\int_{t_0}^T\int_s^T (\tilde{B}U^{-1}\tilde{B}^*P(\sigma)\ S(\sigma,s)\tilde{y},\ P(\sigma)S(\sigma,s)\tilde{x}d\sigma)\chi_{[t_0,T]}(s)\ ds.$$

Now (2.27) follows from lemma (2.1) and (2.28).

<u>Corollary 2.1</u>: P(t) satisfies the equation:

$$(2.29)\quad P(t)\tilde{x}=(\hat{H}^*\Pi S(\cdot,t)\tilde{x})(t) - \int_t^T S^*(\sigma,t)\ P(\sigma)(\tilde{B}U^{-1}\tilde{B}^*P(\sigma)S(\sigma,t)\tilde{x}d\sigma$$

where $\hat{H}\in\mathfrak{L}(\Xi_{to})$, $(\hat{H}\tilde{x})(t) = \Pi\int_{t_0}^t S(t,s)\tilde{x}(s)ds.$

## §3. Optimal control on the infinite interval

In this section we consider the question of §2 for $T=\infty$, that is minimizing

$$(3.1)\qquad J_\infty(\tilde{x},u) = \int_0^\infty [\,|\sum_{i=1}^{N} P_i\ x(t-\tau_i)\,|^2 + (u(t),\ Uu(t))]dt,\ u\varepsilon L^2([0,\infty);R^P)$$

where $P_i$ and $U$ are as in §2.

We will apply the theory developed by Gibson in [16].

Definition 3.1: A function $u\colon [0,\infty)\to R^P$ is an admissible control for $\tilde{x}$ if $u$ is strongly measurable on $[0,\infty)$ and $J(x,u)$ is finite. $U_{ad}$ will denote the set of admissible controls.

Definition 3.2: The controlled system (1.1) (1.3) is said to be stabilizable if there exists some operator $G\varepsilon\mathcal{L}(W^2,R^P)$ of the form

$$(3.2)\qquad G\tilde{x} = \sum_{i=0}^{H} G_i\ x\ (-\sigma_i) + \int_{-\beta}^{0} G_{01}(\theta)\ x(\theta)d\theta \quad .$$

where $0=\sigma_0<\ldots<\sigma_H = \beta$, $G_{01}\varepsilon\ L^2[-\beta,0]$ such that the resulting closed loop system

$$(3.3)\qquad \begin{aligned}\dot{\tilde{x}}(t) &= (\tilde{A}+\tilde{B}G)\tilde{x}(t) \qquad\qquad \text{a.e.in } [0,\infty)\\ \tilde{x}(0) &= \tilde{x}_0\end{aligned}$$

is $L^2$-stable.

Lemma 3.1: If (1.1) is stabilizable and $P_T(s)$ is the operator solution of (2.24) on the interval $[0,T]$ then

$$\lim_{T\to\infty} (P_T(s)\tilde{x},\tilde{x})_{M^2} = \lim_{T\to\infty} \int_s^T [\,|\Pi\tilde{x}(t)|^2 + (u(t),Uu(t))]dt=c<\infty$$

Proof:

This proof follows the lines of the proof of theorem 5.7 in [9].

If (1.1) is stabilizable there is a feedback operator $G$ such that $\tilde{A}+\tilde{B}G$ is stable.

Let $\tilde{x}_G(t)$ be the solution of

$$\dot{\tilde{x}}(t)=(\tilde{A}+\tilde{B}G)\tilde{x}(t)\ .$$

Then

$$(3.4)\qquad (P_T(s)\tilde{x},\tilde{x})\le \inf\{J_T^{\,s}(\upsilon,\tilde{x})\,|\,\upsilon\varepsilon U_s\} \le \int_s^T [\,|\Pi\tilde{x}_G(t-s)|^2 + (UG\tilde{x}_G(t-s),\ \tilde{x}_G(t-s))]ds$$

and

$$(3.5)\qquad \int_0^T |\Pi\tilde{x}_G(t)|^2 dt = \int_0^T |\sum P_i\; x(t-\tau_i)|^2$$
$$\leq \sum \| P_i \|_{R^{n\times n}} [\|\tilde{x}'\|^2_{L^2[-\beta,0];R^r} + \|x(t)\|^2_{L^2(0,T)} \leq \sum k\|P_i\|\|\tilde{x}\|_{M^2}\;.$$

Also

$$(3.6)\qquad \int_0^T (UG\tilde{x}_G(t), G\tilde{x}_G(t))_{M^2} dt \leq K\,\|\tilde{x}\|^2_{M^2}\;.$$

Thus

$$(3.7)\qquad (P_T(s)\tilde{x},\tilde{x})_{M^2} \leq C\|\tilde{x}\|^2_{M^2} \quad \forall\tilde{x} \text{ and } \forall\; T>s>0\;.$$

Therefore

$$(3.8)\qquad \|P_T(s)\|_{\mathcal{L}(M^2)} \leq c, \qquad \forall\; T>\sigma\;.$$

From $P_{T_2}(s) \geq P_{T_1}(s)$ if $T_2 > T_1 > s$ and (3.8) the limit $\lim_{T\to\infty} P_T(s)\tilde{x}$ exists for every $\tilde{x}\in M^2$ and

$$\lim_{T\to\infty} (P_T(s)\tilde{x},\tilde{x}) \leq C\,\|\tilde{x}\|^2_{M^2}\;.$$

<u>Theorem 3.1</u> If (1.1) is stabilizable then there is a unique non-negative self adjoint operator $P_\infty\in\mathcal{L}(M^2)$ that satisfies the Riccati equation:

$$(3.9)\qquad P_\infty\tilde{x} = H^*(\Pi T(\cdot)\tilde{x}) - \int_0^\infty T^*(\eta)P_\infty\tilde{B}U^{-1}\tilde{B}^*P_\infty\; T(\eta)\tilde{x}d\eta\;.$$

Also we have

$$(3.10)\qquad \min_{u\in U_0} J_\infty(\tilde{x},u) = (P_\infty\tilde{x},\tilde{x})_{M^2}$$

Moreover $P_\infty$ satisfies the equation

$$(3.11)\qquad P_\infty\tilde{x} = H^*(\Pi S_\infty(\cdot)\tilde{x}) - \int_0^\infty S_\infty^*(\eta)P_\infty\tilde{B}U^{-1}\tilde{B}^*P_\infty S_\infty(\eta)\tilde{x}d\eta\,.$$

where $S_\infty$ is the solution of

$$(3.12)\qquad S_\infty(t)\tilde{x} = T(t)\tilde{x} - \int_0^t T(t-\eta)\tilde{B}U^{-1}\tilde{B}^*P_\infty\; S_\infty(\eta)\tilde{x}d\eta$$

$$(3.12)'\qquad\qquad = T(t)\tilde{x} - \int_0^t S_\infty(t-\eta)BU^{-1}\tilde{B}^*P_\infty\; T(\eta)\tilde{x}d\eta\;.$$

Finally, the optimal control $u(t)$ and the optimal trajectory $\tilde{x}(t)$ are given by

$$(3.13)\qquad u(t) = -U^{-1}\tilde{B}^*P_\infty\tilde{x}(t)$$

$$(3.14)\qquad \tilde{x}(t) = S_\infty(t)\tilde{x}\qquad\qquad .$$

Proof: The difference between our formulation and the one considered in [15] is that the weight of $\tilde{x}$ in our cost is not a bounded operator as it is in [15].

But the proofs are still valid since $\Pi\tilde{x}(t)$ is an element of $L^2_{loc}([s,\infty],M^2)$.
From lemma 3.1 there is a sequence $T_n \to \infty$ such that $\lim_{T_n\to\infty} (P_{T_n}(s)\tilde{x},\tilde{x}) = c < \infty$. The results of theorem 4.1 in [16] still hold in our case by the previous observation, then there is an admissible control for any initial time s and state $\tilde{x}$.

From theorem 4.2 in [15] there is a unique non-negative self adjoint operator $P_\infty(s) \varepsilon \mathcal{L}(M^2)$ such that

$$\min_{\nu \varepsilon U_{ad}} J_\infty(s,\tilde{x},\nu) = (P_\infty(s)\tilde{x},\tilde{x})_{M^2}$$

and for any sequence $\{T_n\}$ such that $T_n \to \infty$

$$P_{T_n}(s)\tilde{x} \to P_\infty(s)\tilde{x} \text{ strongly in } M^2 .$$

Also from Theorem 4.1 in [15]:

$$u_n(\cdot) \to u(\cdot) \text{ strongly in } U_0$$

$$x_n(\cdot) \to x(\cdot) \text{ strongly in } \Xi_0.$$

Then we have

$$(3.15) \qquad \tilde{x}(t) = T(t)\tilde{x} - \int_0^t T(t-s)BU^{-1}\tilde{B}^*P_\infty\tilde{x}(s)ds .$$

Therefore $\tilde{x}(t)$ is continuous and $\tilde{x}(t)=S_\infty(t)\tilde{x}$ where $S_\infty(t)$ is as in (3.12). Moreover $\lim_{n\to\infty} S_{T_n}(t)\tilde{x} = S_\infty(t)\tilde{x}$ . This follows from (3.12) corollary 2.1 and Gronwall's lemma.

Now we have

$$(3.16) \qquad (P_\infty x,y) = \int_0^\infty (\Pi T(\sigma)\tilde{x},\Pi S_\infty(\sigma)\tilde{y})\, d\sigma .$$

By replacing $S_\infty$ in (3.16) with (3.12)' we obtain

$$(P_\infty x,y) = (H^*\Pi T(\cdot)\tilde{x},\tilde{y}) - \left(\int_0^\infty T^*(\eta)P_\infty\tilde{B}U^{-1}\tilde{B}^*P_\infty T(\eta)\tilde{x}d\eta ,y\right)$$

$\forall y \varepsilon M^2$ which proves (3.9).

Finally, since $\tilde{x}$ is continuous and

$$(3.17) \qquad u(t) = -U^{-1}\tilde{B}^*P_\infty\tilde{x}(t) \quad \text{in } U_0$$

(3.17) holds pointwise.

<u>Theorem 3.2</u> If there is a positive definite solution to the Riccati equation (3.9) then there is a solution to the minimization question which is given by (3.13) where $P_\infty$ is some solution to the Riccati equation.

Proof.

If there is a solution to the Riccati equation (3.9) then there is an admissible control (Theorem 4.5 in [15]) for any s and $\tilde{x} \varepsilon M^2$, therefore there is a $P_\infty$ satisfying

$$(P_\infty \tilde{x}, \tilde{x}) = \min_{\nu \varepsilon U_{ad}} J_\infty(\tilde{x}, \nu).$$

Now the proof follows as in theorem 3.1.

The solution to the Riccati equations (3.9) and (3.11) is unique if any admissible control drives the state to zero. This condition is satisfied in our case for example if

$$(3.18) \quad \| x(t) \|_{L^2[0,T]} \leq k \int_0^T \| \Pi \tilde{x}(t) \|_{M^2} dt \qquad \forall\, T \varepsilon [0,\infty).$$

A necessary and sufficient condition for (3.18) to hold is that the Laplace transform $\hat{L}$ of the Green's function of the operator

$$(Ly)(t) = \sum P_i\, y(t-h_i) = \sum (P_i \delta_{h_i} * y)(t) \qquad y \varepsilon C_0$$

satisfies: (see [1])

1) $\hat{L}(P)$ is holomorphic

2) $\hat{L}(\overline{P}) = \overline{\hat{L}(P)}$

3) $\kappa \overline{\hat{L}^T(P)}\; \hat{L}(P) - 1_n$ is a non negative definite matrix

Finally we will prove a result about the spectrum of the optimal system:

<u>Theorem 3.3</u>: If a complex number $\lambda < 0$ is an eigenvalue of $\tilde{A} - \tilde{B}U^{-1}\tilde{B}^* P_\infty$ then it is a zero of:

$$\det \begin{vmatrix} A(\lambda) & -BU^{-1}B^T \\ \Pi^T(\lambda)\Pi(\lambda) & -A^T(\lambda) \end{vmatrix}$$

where

$$A(\lambda) = \lambda I - \sum_{i=0}^{N} A_i e^{-\lambda h_i}$$

and

$$\Pi(\lambda) = \sum_{i=0}^{N} P_i\, e^{-\lambda h_i}$$

Proof:

$$(3.19)\quad P_\infty \tilde{x} = (H^*\Pi S)\tilde{x} = \sum_{i=0}^{N} \int_{\tau_i}^{\infty} T^*(\eta-\tau_i)(P_i^T \Pi S(\eta)\tilde{x},0)d\eta$$

$$+\sum_{i=0}^{N}\int_{-\tau_i}^{0} P_i^T\Pi S(\eta+\tau_i)\tilde{x}d\eta$$

(See proof of Lemma 2.1 for the expression for H*.)

Let's denote by $P_\infty'$ the operator

$$(3.20)\qquad P_\infty'\tilde{x} = \sum \int_{\tau_i}^{\infty} T^*(\eta-\tau_i)(P_i^T\Pi S(\eta)\tilde{x},0)d\eta .$$

Now if $\lambda<0$ is an eigenvalue of $\tilde{A}-\tilde{B}U^{-1}\tilde{B}^*P_\infty$ with eigenvector $\tilde{x}$, $e^{\lambda t}$ is an eigenvalue of $S(t)$ with corresponding eigenvector $\tilde{x}$. Hence

$$(3.21)\qquad P_\infty'\ S(t)\ \tilde{x} = \sum_{i=0}^{N}\int_0^{\infty} T^*(\eta)(P_i^T\Pi S(\eta+t+\tau_i)\tilde{x},0)d\eta$$

$$= \sum_{i=0}^{N}\int_0^{\infty} T^*(\eta)(P_i^T P_j\ e^{\lambda(\eta+t+\tau_i-\tau_j)}\tilde{x}^\circ,0)d\eta$$

Now

(3.22)

$$\lambda P_\infty' S(t)\tilde{x} = (-\sum_{i=0}^{N} e^{-\lambda\tau_i}P_i^T P_j\ e^{\lambda(\tau_j+t)}x^\circ,0) - A^*\sum_{i=0}^{N}\int_0^{\infty} T^*(\eta)(P_i^T P_j e^{\lambda(\eta+t+\tau_i+\tau_j)}x^\circ,0)d\eta$$

$$= (\sum_{i=0}^{N} P_i^T\ P_j\ e^{\lambda(\tau_i-\tau_j+t)}x^\circ,0) - A^*P_\infty' S(t)\tilde{x}.$$

Let $\Pi(\lambda)\ \varepsilon\ \mathcal{L}(R^n)$ be defined as

$$\Pi(\lambda) = \Sigma\ P_i e^{\lambda\tau_i} .$$

Then

$$\lambda(P_\infty' S(t)\tilde{x})^\circ = -\Pi^T(\lambda)\Pi(\lambda)\ e^{\lambda t}\tilde{x}^0 - A_0^T(P_\infty' S(t)\tilde{x})^\circ - (P_\infty' S(t)\tilde{x})'(0).$$

Since $P_\infty' S(t)\tilde{x}\varepsilon$ domain$(\tilde{A}^*)$

$$(P_\infty' S(t)\tilde{x})'(\theta) = \sum\ e^{\lambda(\theta+\tau_i)}A_i^T\ \chi_{(-h_i,0]}(\theta)(P'S\ (t)\tilde{x})^\circ.$$

Hence

$$(3.23)\qquad \lambda(P_\infty' S(t)\tilde{x})^\circ = -\Pi^T(\lambda)\ \Pi(\lambda)e^{\lambda t}\tilde{x}^\circ - A_0^T\ (P_\infty' S(t)\tilde{x})^\circ - \sum A_i^T e^{\lambda\tau_i}(P_\infty' S(t)\tilde{x})^\circ .$$

Setting $(P_\infty'\tilde{x})^\circ = y$ (Then $(P_\infty' S(t)\tilde{x})^\circ = e^{\lambda t}(P_\infty'\tilde{x})^\circ$ from (3.20) and (3.21)) we have

$$\lambda y = -\Pi^T(\lambda)\Pi(\lambda)\tilde{x}^\circ - A^T(\lambda)y.$$

Acknowledgement The reseach reported on here has been supported by the National Science Foundation under Grant Number ECS82 17375.

References

1. Beltrami & Wholers Distributions and the Boundary Values of Analytic Functions, Academic Press, New York and London, 1966.

2. C. Bernier and A. Manitius, "On semigroups in $R^n x L^p$ corresponding to differential equations with delays", Canadian Journal of Mathematics, 30 No. 5 (1978), pp. 897-914.

3. A. V. Balakrishnan, Applied Functional Analysis, Springer-Verlag, New York, Heidelberg, Berlin, 1976.

4. R. F. Curtain and A. J. Pritchard, "The finite dimensional Riccati equation for systems defined by evolution operators", SIAM J. Control & Opt., 14(1976), pp. 951-983.

5. R. F. Curtain and A. J. Pritchard, "An abstract theory for unbounded control action for distributed parameter systems", SIAM J. Control & Opt., 15 (1977), pp. 566-611.

6. M. C. Delfour and S.K, Mitter, "Controllability, observability and optimal feedback control of affine hereditary differential systems", SIAM J. Control, 10, (1972), pp. 298-327.

7. M. C. Delfour and S. K. Mitter, "Hereditary differential systems with constant delays, I. general case". J. Differential Equations, 12 (1972), pp. 231-235.

8. M. C. Delfour and S. K. Mitter, "Hereditary differential systems with constant delays, II. A class of finite systems and the adjoint problem", J. Differential Equations, 18 (1975), pp. 18-28.

9. M. C. Delfour, C. McCalla and S. K. Mitter, "Stability and the infinite-time quadratic cost problem for linear hereditary differential systems", SIAM J. Control, 13 (1975), pp. 48-88.

10. M. C. Delfour, "State theory of linear hereditary differential systems", J. Math. Anal. Applic. 60 (1977) pp. 8-35.

11. M. C. Delfour, "The product space approach in the state space theory of linear time-invariant differentail systems with delays in state and control variables", Report CRMA1013, Centre de Recherches de Mathematiques Appliquees. February 1981.

12. M. C. Delfour, "The linear quadratic optimal control problem with delays in state and control variables: A state space approach", CRMA 1012 Centre de Recherches de Mathematiques Appliquees. March 1981. (Revised as CRMA-118 Sept. 1982.)

13. R. Datko, "Extending a theorem of A.M. Lyapunov to Hilbert spaces", J. Math. Anal. Appl., 32 (1970), pp. 610-616.

14. R. E. Edwards, Functional Analysis, Holt, Rinehart and Winston, Inc. 1965.

15. J. S. Gibson "The Riccati Integral Equations for Optimal Control Problems on Hilbert Spaces", SIAM J. Control and Opt. Vol. 17, (1979)

16. ------------ "Linear Quadradtic Optimal Control of Hereditary Differential Systems: Infinite dimensional Riccati Equations and Numerical Approximations", Siam J. Control and Opt., 21 (1983), pp. 95-139.

17. J. Hale, Theory of Functional Differential Equations, Springer-Verlag, New York, 1977.

18. A. Ichikawa, "Evolution equations with delay", Control Theory Centre. Reprot, No. 52 (revised), University of Warwick, January 1977.

19. A. Ichikawa, "Optimal control and filtering of evolution with delay in control and observation", Control Theory Centre. Report, No. 53, June 1977.

20. E. W. Kamen, An operator theory of linear functional differential equations", Journal of Differential Equations, 27 (1978), pp. 274-297.

21. H. Koivo and E. B. Lee, "Controller synthesis for linear systems with retarded state and control variables and quadratic cost", Automatica 8, (1972), pp. 203-208.

22. E. B. Lee, "Generalized quadratic optimal controllers for linear hereditary systems", IEEE Trans. on Automat. Contr., AC-25 (1980), pp. 528-531.

23. E. B. Lee and S. H. Żak, "On spectrum placement for linear time invariant delay systems." IEEE Trans. on Automat. Contr. AC--27 (1982 pp. 446-449.

24. E. B. Lee and L. Markus, Foundations Optimal Control Theory, John Wiley and Sons, Inc. New York 1967.

25. J. L. Lions, Optimal Control of Systems governed by Partial Differential Equations, Springer-Verlag, New York, Heidelberg, Berlin, 1971.

26. A. Manitius and A. W. Olbrot, "Finite spectrum assignment problem for systems with delays", IEEE Trans. Automat. Contr., Vol. AC-24 (1979), pp. 541-553.

27. A. Manitius, "Optimal control of hereditary systems", In "Control Theory and Topics in Functional Analysis". Vol. II, pp. 43-178, International Atomic Energy Agency, Vienna 1976.

28. A. S. Morse, "Ring models for delay differential systems", Automatica, 12 (1976), pp. 529-531.

29. A. W. Olbrot, "Stabilizability, detectability and spectrum assignment for linear systems with general time delays", IEEE Trans. Automat. Contr. Vol. AC-23 (1978( pp. 887-890.

30. A. W. Olbrot, "On non-degeneracy and related problems for linear constant time-lag systems", Ricerche di Automatica, Vol. 3, No. 3 (1972), pp. 203-220.

31. V. N. Popov, "Pointwise degeneracy of linar, time-invariant, delay-differential equations", Journal of Differential equations, 11 (1972), pp. 541-561.

32. W. Rudin, Functional Analysis, McGraw-Hill Book Company, 1973.

33. R. B. Vinter and R. H. Kwong, "The infinite time quadratic control problem for linear systems with state and control delays: an evolution equation approach", SIAM J. Control and Opt. Vol 19 (1981) pp. 139-153.

Appendix

Proof of Lemma 2.1: We define

$$H_0\tilde{x}(t) = \int_{t_0}^{t} \sum_{i=0}^{N} P_i\, T(t-s-\tau_i)\, \chi_{[t_0,t-\tau_i]}(s)\, \tilde{x}(s)ds = \int_{t_0}^{t} \hat{H}_0(s)\tilde{x}(s)ds.$$

$$H_1\tilde{x}(t) = \int_{t_0}^{t} \sum_{i=0}^{N} P_i\, \tilde{x}'(s)(t-s-\tau_i)\, \chi_{(t-\tau_i,t]}(s)\, ds.$$

$$(H\tilde{x}(\cdot),(y^\circ(\cdot),0))_{\Xi_{t_0}} = \int_{t_0}^{T} ((H_0\tilde{x})(t),\, y^\circ(t))_{M^2}dt + \int_{t_0}^{T} (H'\tilde{x}(t),\tilde{y}(t))dt$$

$$= \int_{t_0}^{T} (\tilde{x}(s),\, \int_{s}^{T} (\hat{H}_0^* y)(t)dt)_{M^2}ds + \int_{t_0}^{T} (\sum P_i \int_{s}^{T} \tilde{x}(s)(t-s-\tau_i)\chi(s)_{(t-\tau_i,t]}\tilde{y}(t))dt\, ds$$

$$+ \int_{t_0}^{T} \sum \int_{-\gamma}^{0} (x'(s)(\theta)\, ,\, P_i^T\tilde{y}(s+\theta+\tau_i)\, \chi_{[-\tau_i,0)}(\theta)d\theta ds.$$

Hence we have:

$$(H^*y)^\circ(s) = (H^*_0 y)^\circ(s)$$

$$(H^*y)^1(s)(\theta) = (H^*y)^1(s)(\theta) + \sum P_i\tilde{y}(s+\theta+\tau_i)\chi_{[-\tau_i,0)}(\theta)$$

Now

$$P(t)\tilde{x} = (H_0^* \Pi S(\cdot,t)\tilde{x})(t) + (0,\sum P_i^T(\Pi S(t+\theta+\tau_i,t)\tilde{x}\chi_{[-\tau_i,0)}(\theta))$$

It is clear that

$$\| (H_0^*\Pi S(\cdot,t)\tilde{x})(t) \|_{M^2} \leq k \|\tilde{x}\|_{M^2}$$

and

$$\| (0,\sum P_i^T\, S(s+\tau_i,t)\tilde{x}\, \chi_{[-\tau_i,0)}(\cdot) \|_{M^2} \leq \sum \|P_i^T\|_{R^{n\times n}} \|S\|_{M^2} \|\tilde{x}^1\|_{L^2[-\beta,0];R^n)}$$

$$\leq k \|\tilde{x}\|_{M^2}\, .$$

Continuity of P on t: Clearly $(H_0^*(\Pi S(\cdot,t)\tilde{x})(t)$ is continuous in t

From

$$S(t+\theta+\tau_i,t)\tilde{x}\, \chi_{[-\tau_i,0)}(\theta) = [T(\theta+\tau_i)\tilde{x} - \int_{t}^{t+\theta+\tau_i} T(t+\theta+\tau_i-\eta)\tilde{B}U^{-1}\tilde{B}^*S(\eta,t)\tilde{x}d\eta]\, \chi_{[-\tau_i,0)}(\theta)$$

and $\tilde{x}\to\Pi T(t)\tilde{x}$ continuous from $M^2$ into $\Xi_{t_0}$ it follows that

$$\int_{-\gamma}^{0} |\sum P_i^T\, \Pi[S(t+\theta+\tau_i,t) - S(t+\theta+\tau_i,t_1)]\tilde{x}\, \chi_{[-\tau_i,0)}(\theta)\, |^2\, d\theta \underset{t\to t_1}{\to} 0\, .$$

# ON THE FINITE ELEMENT APPROXIMATION OF THE BOUNDARY CONTROL FOR TWO-PHASE STEFAN PROBLEMS

P. Neittaanmäki* and D. Tiba**

*Lappeenranta University of Technology, Department of Physics and Mathematics, Box 20, SF-53851 Lappeenranta 85, Finland

**INCREST, Department of Mathematics, Bd. Păcii 220, 79622 Bucuresti, Romania

Abstract A boundary control for a two-phase Stefan problem is considered. The problem is regularized by utilizing the Yosida approximation and the Friedrichs mollifier. Next it is discretized by finite elements in space and finite differences in time. The solution of these auxiliary problems are shown to be minimizing sequences for the original problem when certain parameters approach to zero. A gradient algorithm is presented for the discretized problem. Numerical test example illustrates the efficiency of the methods.

## 1. INTRODUCTION

We shall be concerned with the approximation of the boundary control problem:

(*P*) Minimize $\int_0^T \{\frac{1}{2}\|y-d\|^2_{L^2(\Omega)} + \frac{1}{2}\|u\|^2_{L^2(\partial\Omega)}\}dt$

subject to:

$$(1.1)\qquad \frac{\partial}{\partial t}v(t,x) - \Delta y(t,x) = f(t,x) \quad \text{in } Q$$
$$v(t,x) \in \beta(y(t,x)) \quad \text{in } Q$$

$$(1.2)\qquad \frac{\partial}{\partial n}y(t,x) = u(t,x) \quad \text{in } \Sigma$$

$$(1.3)\qquad v(0,x) = v_0(x) \quad \text{in } \Omega$$

Above $\Omega \subset \mathbb{R}^n$ , $n \geq 1$, is a bounded domain with a smooth boundary $\partial\Omega$ and $Q = \Omega \times ]0,T[$ is a cylinder with lateral face $\Sigma$. We assume that $v_0 \in L^2(\Omega)$, $d,f \in L^2(Q)$ and that $\beta$ is a strongly maximal monotone graph in $\mathbb{R} \times \mathbb{R}$ :

$$(1.4)\qquad (\beta(y)-\beta(z))\,(y-z) \geq \alpha|y-z|^2 , \quad \alpha > 0 ,$$

bounded on bounded sets. When $\beta$ is given by

$$(1.5)\qquad \beta(y) = \begin{cases} y - r_0 & , \ y > r_0 \\ [-\sigma,0] & , \ y = r_0 \\ \kappa(y-r_0)-\sigma & , \ y < r_0 \end{cases},$$

where $\kappa$, $\sigma > 0$, we obtain a two-phase Stefan problem (J.L. Lions [18], p. 196).

For the definition and the existence of the solution of (1.1)-(1.3) via the Duvaut-Frémond freezing index formulation see [9], [12] and also the recent works [27, 22].

Consider the regularized problem:

$$(P_\varepsilon)\qquad \text{Minimize} \int_0^T \{\tfrac{1}{2}\|y-d\|^2_{L^2(\Omega)} + \tfrac{1}{2}\|u\|^2_{L^2(\partial\Omega)}\}dt$$

subject to:

$$(1.6)\qquad \frac{\partial}{\partial t}\beta^\varepsilon(y(t,x)) - \Delta y(t,x) = f(t,x) \quad \text{in } Q$$

and (1.2), (1.3) with

$$(1.7)\qquad \beta^\varepsilon(y) = y + \int_{-\infty}^{\infty} \gamma_\varepsilon(y-\varepsilon^2\xi)\rho(\xi)d\xi .$$

Here $\gamma_\varepsilon$ is the Yosida approximation of the maximal monotone graph $\gamma(y) = \beta(y)-y$ (it is assumed for convenience that $\alpha \geq 1$ in (1.4)) and $\rho$ is the Friedrichs mollifier.

This smoothing technique is mainly due to V. Barbu [3], [4] in the study of necessary optimality conditions for control problems governed by variational inequalities.

In the previous paper [28] the existence of at least one optimal control $u^*$ in $L^2(\Sigma)$ is established for Problem $(P)$. Denote by $u_\varepsilon$ any optimal control for Problem $(P_\varepsilon)$ and by $\pi(u^*)$, $\pi_\varepsilon(u_\varepsilon)$ the optimal values for Problem $(P)$ and for Problem $(P_\varepsilon)$, respectively. The following approximation result is known from [28] and it shows that $\{u_\varepsilon\}$ is a minimizing sequence for Problem $(P)$:

<u>Theorem 1.1</u> <u>When</u> $\varepsilon \to 0$ <u>we</u> <u>have</u>:

$$(1.8)\qquad \pi(u_\varepsilon) \to \pi(u^*) ,$$

$$(1.9)\qquad \pi_\varepsilon(u_\varepsilon) \to \pi(u^*) . \qquad \square$$

In section 2 we define a discretized control problem $(P_{h,k})$ and we show a similar approximation relation between $(P_{h,k})$ and $(P_\varepsilon)$. Therefore, in order to obtain a suboptimal control for Problem $(P)$, one can solve $(P_\varepsilon)$ or $(P_{h,k})$ for sufficiently small parameters. Due to the differentiability properties, gradient methods can be utilized efficiently. We emphasize the descent property of the algorithm, not the convergence to a minimum point which may not be true since we have not convexity. The gradient algorithm is presented in section 3 with a numerical example (section 4).

The main result of this paper is Theorem 2.4 which states that the finite dimensional problem $(P_{h,k})$ provides a minimizing sequence for $(P_\varepsilon)$.

Results available for the Stefan problems and their derivatives have been comprehensively reviewed in [11, 15, 25]. For the use of finite element method in Stefan like problems we refer to [5, 8, 10, 11, 16, 21, 23, 26, 29]; in this connection we also refer to monograph [13], where numerical analysis of variational inequalities is considered. For related problems or methods in connection of control problems for Stefan type processes, both from the theoretical and the numerical point of view, we also quote [1, 5, 19, 20, 21, 23, 26, 28].

## 2. THE DISCRETIZED PROBLEM

For the sake of simplicity we assume that $\Omega$ is a convex polygonal domain in $R^2$. Throughout the paper let the symbols $L^2(\Omega)$, $L^2(\partial\Omega)$, $H^1(\Omega)$, $L^2(0,T;\, H^1(\Omega))$, $\|\cdot\|_{0,\Omega}$, $(\cdot,\cdot)_{0,\Omega}$, $\|\cdot\|_{L^2(\partial\Omega)}$, $\|\cdot\|_{1,\Omega}$ etc. have the usual meaning (the same as in [7, 13, 18]).

Let $T_h$ be a regular triangulation of $\Omega : \overline{\Omega} = \cup\{K \mid K \in T_h\}$ ([7]) and let $V_h$ be the space of continuous functions which are linear on each triangle $K \in T_h$ equipped with the norm $|\cdot|_h$ induced by the modified $L^2(\Omega)$ inner product:

$$(2.1) \qquad (u,v)_h = \frac{1}{3} \sum_{i \le I} W_i u_i v_i \quad , \qquad u,v \in V_h \quad .$$

Here I (the dimension of $V_h$) is the number of vertices associated with $T_h$, $W_i$ is the sum of the areas of the triangles with a vertex in i and $u_i$, $v_i$ are the values of u and v at node i, respectively. It is well known that norms $\|\cdot\|_{0,\Omega}$ and $|\cdot|_h$ are equivalent in $V_h$ and
$|(u,v)_h-(u,v)_{0,\Omega}| \le Ch^2\|u\|_{1,\Omega}\|v\|_{1,\Omega} \le Ch\|u\|_{0,\Omega}\|v\|_{1,\Omega} \quad \forall u,v \in V_h$ ,

where $C > 0$ is a generic constant which may vary with the context ([8]). By $L_h$ we denote the space of traces of functions from $V_h$, endowed with the $L^2(\partial\Omega)$ norm. Let the dimension of $L_h$ be J; $J < I$.

Assume that the interval $[0,T]$ is divided into N equal subintervals $[t_n, t_{n+1}]$ of length $k > 0$: $t_n = nk$, $k = T/N$, $n = 0,\ldots, N-1$. We consider the following approximation of $(P_\varepsilon)$:

$$(2.2) \qquad (P_{h,k}) \quad \text{Minimize } \frac{k}{2} \sum_{n=1}^{N} \{|y^n - d^n|_h^2 + \|u^n\|^2_{L^2(\partial\Omega)}\}$$

subject to $u \in L_h^N$ $(:=L_h \times\cdots\times L_h)$ and $y \in V_h^N$ such that

$$(2.3) \qquad \left(\frac{v^{n+1}-v^n}{k}, v\right)_h + \int_\Omega \nabla y^{n+1}\cdot \nabla v - \int_{\partial\Omega} u^{n+1} v = (f^{n+1}, v)_h \quad \forall v \in V_h, \quad n \leq N-1$$

$$(2.4) \qquad v_i^n = \beta^\varepsilon(y_i^n), \ i \leq I, \quad n \leq N, \quad v^0 = v_0 .$$

Here $d, f \in V_h^N$ and in the sequel they will be appropriate discretizations of d and $f \in L^2(Q)$.

It holds (see [21]).

<u>Proposition 2.1</u> <u>Problem $(P_{h,k})$ has at least one optimal pair $[y_{h,k}, u_{h,k}]$.</u> □

Let $\theta : L_h^N \to V_h^N$ be the mapping $u \to y$ defined by (2.3), (2.4). It is a nonlinear mapping and moreover it is Gateaux differentiable([21]):

<u>Proposition 2.2</u> <u>The mapping $\theta$ is Gateaux differentiable and for every</u> $u, w \in L_h^N$, $r = \nabla\theta(u)w$ <u>satisfies</u>

$$(2.5) \qquad \frac{1}{k}\left(\nabla\beta^\varepsilon(y^{n+1})r^{n+1} - \nabla\beta^\varepsilon(y^n)r^n, v\right)_h + \int_\Omega \nabla r^{n+1}\nabla v - \int_{\partial\Omega} w^{n+1} v = 0$$

$$\forall\, v \in V_h, \quad n \leq N-1,$$

$$(2.6) \qquad r^0 = 0 .$$

□

Let $\pi_{h,k}: L_h^N \to ]-\infty, \infty[$ be a cost functional associated with $(P_{h,k})$. Then $u_{h,k}$ is a minimum point of $\pi_{h,k}$ and the Gateaux differential vanishes:

$$(2.7) \qquad \nabla\pi_{h,k}(u_{h,k})w = 0 \qquad \forall\, w \in L_h^N .$$

We have:

$$\lim_{\lambda\to 0} \frac{\pi_{h,k}(u+\lambda w)-\pi_{h,k}(u)}{\lambda} = k\sum_{n=1}^{N}\{(y^n-d^n,r^n)_h + (u^n,w^n)_{L^2(\partial\Omega)}\}$$

$$= k\sum_{n=1}^{N}\{([\nabla\theta(u)^*(y-d)]^n,w^n)_h + (u^n,w^n)_{L^2(\partial\Omega)}\} .$$

By (2.7) and the above equality we get

(2.8) $$\nabla\theta(u_{h,k})^*(y_{h,k}-d) = -u_{h,k} .$$

We define <u>the adjoint state</u> $p_{h,k}\in V_h$ by

(2.9) $$(\nabla\beta^\varepsilon(y^n_{h,k})\cdot\frac{p^n-p^{n+1}}{k}, v)_h + \int_\Omega \nabla p^n\cdot\nabla v = -(y^n_{h,k}-d^n, v)_h ,$$

$$\forall\, v\in V_h ,\quad n\le N ,$$

(2.10) $$p^N = 0 .$$

This is a linear system in implicit form and obviously there is a unique solution $p_{h,k}\in V_h$. We have ([21]) .

<u>Proposition 2.3</u> <u>Relation</u> (2.8) <u>is equivalent with</u>

(2.11) $$u^n_{h,k} = p^n_{h,k}|_{\partial\Omega} ,\quad n\le N .$$

□

Now we can state and prove the main result of this paper:

<u>Theorem 2.4</u> <u>Under the monotonicity and boundedness assumptions for</u> $\beta$ <u>we have</u>

(2.12) $$\lim_{k\to 0}\lim_{h\to 0}\pi_{h,k}(u_{h,k}) = \pi_\varepsilon(u_\varepsilon) ,$$

<u>and therefore</u> $u_{h,k}$ <u>is a minimizing sequence for Problem</u> $(P_\varepsilon)$.

<u>Proof.</u> For every $h,k>0$, we obtain:

(2.13) $$\frac{k}{2}\sum_{n=1}^{N}\{|y^n_{h,k}-d^n|^2_h + \|u^n_{h,k}\|^2_{L^2(\partial\Omega)}\}$$

$$\le \frac{k}{2}\sum_{n=1}^{N}\{|y^n_0-d^n|^2_h + \|u^n_0\|^2_{L^2(\partial\Omega)}\}$$

for all $u_0\in L^N_h$ , $y_0=\theta(u_0)\in V^N_h$ .

Let $u_0 \in H^1(0,T; L^2(\partial\Omega))$ be fixed, $u_0^{h,k} \in L_h^N$ be the discretization of $u_0$ and $y_0^{h,k} = \theta(u_0^{h,k})$. It is known that $y_0^{h,k} \to y_0$ strongly in $L^2(Q)$ as $h,k \to 0$, see [11], Ch III or the recent work [29].

By (2.13) we infer that

$$\frac{k}{2} \sum_{n=1}^{N} \{ |y_{h,k}^n - d^n|_h^2 + \|u_{h,k}^n\|_{L^2(\partial\Omega)}^2 \} \tag{2.14}$$

is bounded with respect to $h,k > 0$.

For other estimates we put $v = y_{h,k}^{n+1}$ in (2.3):

$$\frac{1}{k}(y^{n+1} - y^n, y^{n+1})_h + \frac{1}{k}(w^{n+1} - w^n, y^{n+1})_h + \|\nabla y^{n+1}\|_{0,\Omega}^2 \tag{2.15}$$

$$= \int_{\partial\Omega} y^{n+1} u^{n+1} + (f^{n+1}, y^{n+1})_h .$$

We omit the subscripts $h,k$ and we write $w^n = \beta^\varepsilon(y^n) - y^n = \gamma^\varepsilon(y^n)$ according to (1.7). By a device due to O. Grange and F. Mignot [14] it is known that

$$\sum_{n=0}^{p-1} (w^{n+1} - w^n, y^{n+1}) \geq C , \tag{2.16}$$

for appropriate $p$, $0 < p \leq N$ .

From (2.15) and (2.16) we get

$$\frac{1}{k} \sum_{n=0}^{p-1} (y^{n+1} - y^n, y^{n+1})_h + \frac{C}{k} + \sum_{n=0}^{p-1} \|\nabla y^{n+1}\|_{0,\Omega}^2$$

$$\leq \sum_{n=0}^{p-1} \int_{\partial\Omega} y^{n+1} u^{n+1} + \sum_{n=0}^{p-1} (f^{n+1}, y^{n+1})_h .$$

Then

$$\frac{1}{2k} |y^p|_h^2 + \frac{1}{2} \sum_{n=0}^{p-1} \|\nabla y^{n+1}\|_{0,\Omega}^2 \leq \frac{C}{k} + C \sum_{n=0}^{p-1} \|u^n\|_{L^2(\partial\Omega)}^2 + \frac{1}{2} \sum_{n=0}^{p-1} \|y^{n+1}\|_{0,\Omega}^2 .$$

Finally, (2.14) yields

$$\|y^n\|_{0,\Omega} \leq C , \quad \forall\, n,h,k \tag{2.17}$$

$$\sum_{n=0}^{N-1} k \, \|y^{n+1}\|_{1,\Omega}^2 \leq C , \quad \forall\, h,k . \tag{2.18}$$

Let $k > 0$ be fixed. By taking subsequences we get:

$$u_{h,k}^n \xrightarrow{h \to 0} u_k^n \quad \text{weakly in } L^2(\partial\Omega) , \tag{2.19}$$

$$(2.20) \qquad y^n_{h,k} \xrightarrow{h\to 0} y^n_k \quad \text{strongly in } L^2(\Omega) .$$

Since $\beta^\varepsilon$ is Lipschitz continuous, we have

$$(2.21) \qquad v^n_{h,k} \xrightarrow{h\to 0} v^n_k = \beta^\varepsilon(y^n_k) \quad \text{strongly in } L^2(\Omega).$$

We pass to the limit with respect to $h\to 0$ on (2.3) to obtain:

$$(2.22) \qquad (\frac{v^{n+1}_k - v^n_k}{k}, v) + \int_\Omega \nabla y^{n+1}_k \cdot \nabla v - \int_{\partial\Omega} u^{n+1}_k v = \int_\Omega f^{n+1}_k v, \ \forall v \in H^1(\Omega).$$

The weak lower semicontinuity of the norm and (2.17)-(2.21) give:

$$(2.23) \qquad \sum_{n=1}^{N} \{k\|u^n_k\|^2_{L^2(\partial\Omega)} + k\|y^n_k\|^2_{1,\Omega}\} + \|y^n_k\|^2_{0,\Omega} \leq C$$

for every $k > 0$. Moreover, from (2.22) we obtain

$$\left\| \frac{v^{n+1}_k - v^n_k}{k} \right\|^2_{H^1(\Omega)^*} \leq C(\|y^{n+1}_k\|^2_{1,\Omega} + \|u^{n+1}_k\|^2_{L^2(\partial\Omega)} + \|f^{n+1}_k\|^2_{0,\Omega}) ,$$

that is

$$(2.24) \qquad \sum_{n=0}^{N-1} k \left\| \frac{v^{n+1}_k - v^n_k}{k} \right\|^2_{H^1(\Omega)^*} \leq C , \quad \forall k > 0 .$$

Here $H^1(\Omega)^*$ is the dual of $H^1(\Omega)$. Denote by $u_k$, $y_k$, $v_k$ the mesh functions defined on $[0, T]$ from the vectors $u^n_k$, $y^n_k$, $v^n_k$ as usual.

The above estimates show that for subsequences we have

$$v_k \to \tilde{v} \quad \text{strongly in } L^2(0, T; H^1(\Omega)^*) ,$$

$$(2.25) \qquad y_k \to \tilde{y} \quad \text{weakly in } L^2(0, T; H^1(\Omega)) .$$

Since the maximal monotone operator induced by $\beta^\varepsilon$ in $L^2(0, T; H^1(\Omega)) \times L^2(0, T; H^1(\Omega)^*)$ is demiclosed, by (2.21) we have

$$(2.26) \qquad \tilde{v} = \beta^\varepsilon(\tilde{y}) \quad \text{a.e.} \quad \text{in } Q .$$

Moreover,

$$(2.27) \qquad u_k \to \tilde{u} \qquad \text{weakly in } L^2(\Sigma) .$$

We shall show that $[\tilde{y}, \tilde{u}]$ is an optimal pair for Problem $(P_\varepsilon)$. First by (2.26), (2.22) a standard argument proves that $\tilde{y}$, $\tilde{u}$ satisfy the state equation (1.6), (1.2), (1.3) for instance in a weak sense ([11]) or in a variational sense ([9]). Moreover, let $u_0$ in (2.13) be sufficiently regular such as $u_0^{h,k} \to u_0$ strongly in $L^2(\Sigma)$ ([7]).

Because $y_0^{h,k} \to y_0 = \theta(u_0)$ strongly in $L^2(Q)$ we can pass to the limit in (2.13) by (2.27), (2.20), (2,25) and the weak lower semicontinuity of the norm:

$$(2.28)\qquad \int_0^T \{\frac{1}{2}\,\|\tilde{y} - d\|^2_{0,\Omega} + \frac{1}{2}\|\tilde{u}\|^2_{L^2(\partial\Omega)}\}dt$$

$$\le \int_0^T \{\frac{1}{2}\,\|y_0 - d\|^2_{0,\Omega} + \frac{1}{2}\,\|u_0\|^2_{L^2(\partial\Omega)}\}dt \;.$$

Now, by a density argument and the Lipschitz dependence of the solution in (1.6), (1.2), (1.3) on the boundary data (see [29]) we obtain the optimality of the pair $[\tilde{y}, \tilde{u}]$ and we denote it by $[y_\varepsilon, u_\varepsilon]$.

To complete the proof we use the adjoint system (2.9)-(2.11) in order to get strong convergence in (2.27), (2.19). We take $v = p^n_{h,k} - p^{n+1}_{h,k}$ in (2.9):

$$k\,\left|\frac{p^n - p^{n+1}}{k}\right|^2_h + \int_\Omega \nabla p^n(\nabla p^n - \nabla p^{n+1}) = -(y^n - d^n,\; p^n - p^{n+1})_h \;.$$

Summing with respect to n, after an easy computation, we obtain:

$$(2.29)\qquad \|p^n\|_{1,\Omega} \le C\;, \qquad \forall\; n,\; h,\; k$$

$$(2.30)\qquad \sum_{n=0}^{n-1} k\,\left\|\frac{p^n - p^{n+1}}{k}\right\|_h \le C\;, \qquad \forall\; h,\; k\;.$$

Let k be fixed. On a subsequence (which can be the same as in (2.19) we have

$$\lim_{h\to 0} p^n_{h,k} = p^n_k \qquad \text{strongly in } H^{3/4}(\Omega)\;.$$

The trace theorem and (2.11) give strong convergence in (2.19).

As $\{p^n_k\}$ also satisfies (2.29), (2.30) with $|\cdot|_h$ replaced by $|\cdot|_{0,\Omega}$, the Aubin theorem [2] yields:

$$\lim_{k\to 0} p^n_k = \tilde{p} \qquad \text{strongly in } L^2(0,\, T;\, H^{3/4}(\Omega))$$

on the same subsequence as in (2.27) (by taking further subsequences, for instance). Again, the trace theorem and (2.11) give strong convergence in (2.27). Therefore on a subsequence of the iterated limit, we have:

(2.31) $$\lim_{k\to 0}\lim_{h\to 0} u_{h,k} = u_\varepsilon \qquad \text{strongly in } L^2(\Sigma) .$$

We underline that for $\{p_{h,k}\}$ it is not necessary to consider iterated limits, but for $\{u_{h,k}\}$ it is.

Next, the Lipschitz dependence on the boundary data in (1.6), (1.2), (1.3) implies:

(2.32) $$\lim_{k\to 0}\lim_{h\to 0} \theta_\varepsilon(u_{h,k}) = y_\varepsilon \qquad \text{strongly in } L^2(Q) .$$

In (2.32), $\theta_\varepsilon(u_{h,k})$ denotes the solution of (1.6), (1.2), (1.3) corresponding to $u_{h,k}$ and may be different from $y_{h,k}$. From (2.31) and (2.32) follows that we may pass to the limit in the cost functional. Since the optimal value is unique, the convergence is true on the initial sequence and (2.12) is proved. □

Remark 2.5 By Theorem 1.1 and Theorem 2.4 the solution of the Problem $(\mathcal{P})$ is reduced to the solution of the Problem $(\mathcal{P}_{h,k})$.

## 3. AN ALGORITHM

Due to the Proposition 2.2 we can construct a gradient algorithm for solving the Problem $(\mathcal{P}_{h,k})$.

Algorithm 3.1 (for the discretized Problem $(\mathcal{P}_{h,k})$).

Step 1 - choose any $u_0$ and set $n:=0$.

Step 2 - compute $y_n$ by solving (2.5), (2.6).

Step 3 - test if the pair $[y_n, u_n]$ is satisfactory;
if YES then STOP; otherwise GO TO step 4.

Step 4 - compute $p_n$ by solving (2.9), (2.10).

Step 5 - compute $u_{n+1}$ by the equation.

(3.1) $$u_{n+1} = u_n - \rho_n(u_n - p_n|_\Sigma) ,$$

where $\rho_n$ is a real parameter .

Step 6 - set $n:=n+1$ and GO TO step 2 . □

The convergence test involved in Step 3 is the difference $|\pi_{h,k}(u_n) - \pi_{h,k}(u_{n+1})|$ to be smaller than a given parameter. In step 5 $\rho_n$ can be selected by utilizing the line search.

It is known that without convexity properties above algorithm may not converge to a minimum point of $\pi_{h,k}$ (see [6] for example). Since $\pi_{h,k}$ is not convex, our result emphasizes the descent property of the Algorithm 3.1:

Proposition 3.2 (i) *Let* $h, k, > 0$ *be fixed. The sequence* $\pi_{h,k}(u_n)$ *is convergent when* $n \to \infty$ .

(ii) *Assume that the initial approximation* $u_0$ *is sufficiently regular and let* $\tilde{u}_{h,k}$ *be the value computed by Algorithm 3.1 for* $u_{h,k}$. *The sequence* $\pi_{h,k}(\tilde{u}_{h,k})$ *is bounded with respect to* $h,k$ *and every cluster point* $\tilde{\pi}$ *satisfies*

$$\pi_\varepsilon(u_\varepsilon) \leq \tilde{\pi} \leq \pi_\varepsilon(u_0) . \tag{3.2}$$

Proof. (i) The sequence decreases and it is bounded below.

(ii) We assume that $u_0$ is sufficiently regular such as $u_0^{h,k}$ (the discretization of $u_0$) approaches $u_0$ in $L^2(\Sigma)$ when $h,k \to 0$ .
We have

$$\pi_{h,k}(u_{h,k}) \leq \pi_{h,k}(\tilde{u}_{h,k}) < \pi_{h,k}(u_0^{h,k}) . \tag{3.3}$$

By Theorem 2.4 and the properties of $u_0^{h,k}$ we can pass to the limit in (3.3) and finish the proof. □

Remark 3.3 The significance of the Proposition 3.2 is that in real problems we do not search optimal performance since the computed $\pi_{h,k}(\tilde{u}_{h,k})$ may be different from $\pi_{h,k}(u_{h,k})$. We start the algorithm with the control $u_0$ already used in practice and we improve the performance given by it. If $u_0$ is not sufficiently regular we may replace it by a regular approximation $\tilde{u}_0$ due to the Lipschitz properties of the correspondence defined by the state system (1.6), (1.2), (1.3).

Remark 3.4 In our attempt to justify mathematically the numerical computations we deal with $\tilde{u}_{h,k}$ , the computed values. However it must be pointed out that a similar assertion would be useful for the sequence $\pi_\varepsilon(\tilde{u}_{h,k})$.

## 4. A NUMERICAL EXAMPLE

To illustrate our theoretical results, the following numerical example is considered:

$$\Omega = ]0,1[ \times ]0,1[$$
$$T = 1$$

$$(4.1) \qquad \beta(y) = \begin{cases} y & , y < 0 \\ [0,\ 2] & , y = 0 \\ 4y + 2 & , y > 0 \end{cases}$$

$$(4.2) \qquad f(t,x_1,x_2) = \begin{cases} 8(2e^{-2t} - 1), & x_1^2 + x_2^2 > e^{-2t} \\ 2(e^{-2t} - 2), & x_1^2 + x_2^2 \le e^{-2t} \end{cases}$$

$$(4.3) \qquad v_0 = \beta(y_0)$$

$$(4.4) \qquad y_0(x_1,x_2) = \begin{cases} x_1^2 + x_2^2 - 1\ , & x_1^2 + x_2^2 \le 1 \\ 2(x_1^2 + x_2^2 - 1), & x_1^2 + x_2^2 > 1\ . \end{cases}$$

For boundary control u,

$$(4.5) \qquad u(t,x_1,x_2) = \begin{cases} 0\ , & \text{if } x_1 = 0 \text{ or } x_2 = 0 \\ 4 & \text{on the remaining of } \partial\Omega \end{cases}$$

the classical solution of (1.1)-(1.3) with above data is

$$(4.6) \qquad y(t,x_1,x_2) = \begin{cases} 2(x_1^2 + x_2^2 - e^{-2t}), & \text{if } x_1^2 + x_2^2 > e^{-2t} \\ x_1^2 + x_2^2 - e^{-2t}\ , & \text{if } x_1^2 + x_2^2 \le e^{-2t} \end{cases}$$

In [21, 29] the accuracy of finite element approximation for y is considered.

Let $d \equiv 0$ in Problem ($P$), and we want to minimize the functional

$$(4.7) \qquad \Pi(u) = \frac{1}{2} \int_0^1 \int_\Omega y^2\, dxdt + \frac{1}{2} \int_0^1 \int_{\partial\Omega} u^2 d\sigma dt\ .$$

For $u_0$ given by (4.5) the value of $\Pi(u_0)$ is about 16.1 .

In practical computations we have replaced $\beta$ by the piecewise linear approximation $\beta_\varepsilon$:

$$\beta_\varepsilon(y) = \begin{cases} y & , \quad y < 0 \\ \frac{2+4\varepsilon}{\varepsilon} y & , \quad y \in [0,\varepsilon] \\ 2y + 4, & \quad y > \varepsilon . \end{cases}$$

The relation between parameters h,k and $\varepsilon$ has been chosen as follows: $k \approx h^{4/3}$ and $\varepsilon \approx h^{4/3}$.

In numerical tests a bundle algorithm for nonsmooth optimization ([17]) and a conjugate gradient algorithm (Algorithm 5 in [24]) has been applied. The first one seems to be very efficient. In Table 4.1 we can see the diminution of $\Pi_{h,k}(u_n^{h,k})$ versus iteration obtained by the bundle algorithm ([17]). The discretization parameters are h = 1/8 and k = 1/16 resulting 544 unknowns in minimization; $\varepsilon$ = 1/16.

| Number of iteration | Value of $\Pi_{h,k}(u_n^{h,k})$ |
|---|---|
| 0 | 16.003 |
| 1 | .596 |
| 2 | .271 |
| 3 | .251 |
| 4 | .228 |

Table 4.1 Value of $\Pi_{h,k}(u_n^{h,k})$ versus iteration.

The total CPU-time was 372 seconds. The algebraic equations for solving the state problem and the adjoint state problem were solved iteratively (a variant of S.O.R.); the CPU-time was about 17 seconds for each of the problems (Univac 1100/70, single precision).

In Figures 4.2 and 4.3 we can see the values of the boundary control and the corresponding finite element solution of the state problem on two different time levels (iteration 4).

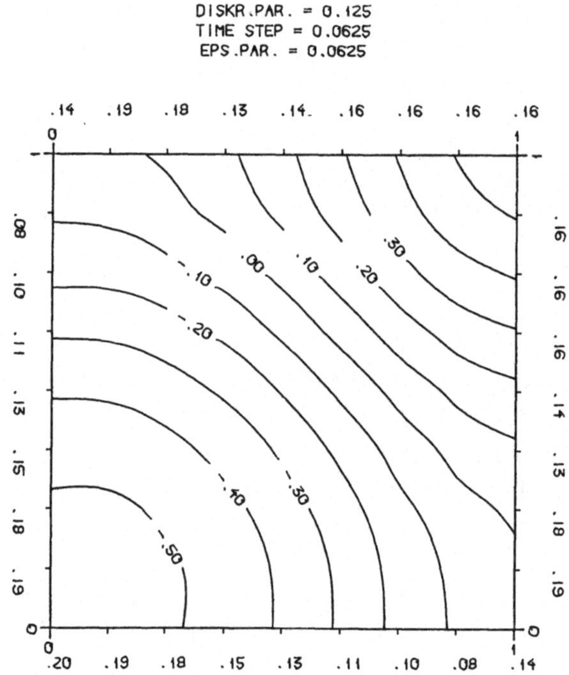


<u>Figure 4.2</u> Boundary control and temperature distribution; time t = .3125 .

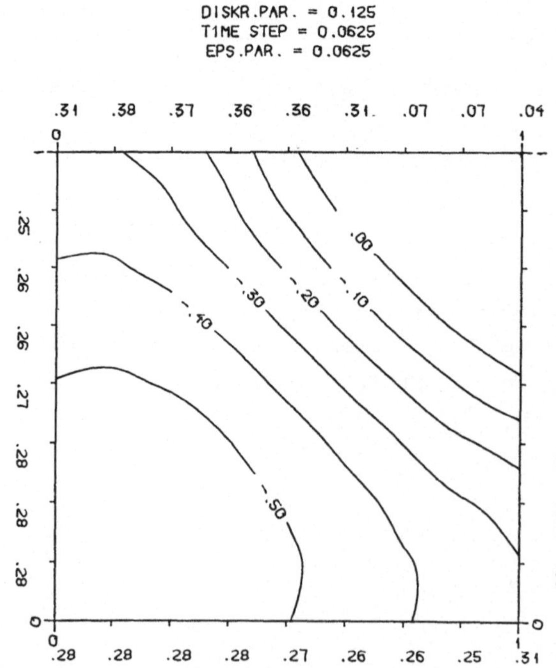


<u>Figure 4.3</u> Boundary control and temperature distribution; time t = .6875 .

ACKNOWLEDGEMENTS

We wish to thank Professor C. Lemaréchal for providing us with his program code M2FC1 (nonsmooth optimization with bounds, by a bundle method). We are also indebted to Mr. E. Laitinen for his help in carrying out numerical tests.

REFERENCES

[1] V. Arnăutu: Approximation of optimal distributed control problems governed by variational inequalities, Numer. Math., 38, 1982, 393-416

[2] J.P. Aubin: Un théorème de compacite, C.R.A.S, Paris, 256, 1963

[3] V. Barbu: Necessary conditions for distributed control problems governed by parabolic variational inequalities, SIAM, J. Control Optim. 19, 1981, 64-86

[4] V. Barbu: Boundary control problems with nonlinear state equation, SIAM, J. Control Optim. 20, 1982, 125-143

[5] F. Bonnans: Application de methodes lagrangiennes en controle, analyse et contrôle de systèmes distributes semi-lineares instables, These, Univ. de Technologie de Compiègne, 1982

[6] J. Cea: Optimization - Theory and Algorithms, Tata Institute of Fundamental Research, Springer, Bombay, 1978

[7] P.G. Ciarlet: The finite element method for elliptic problems. North Holland Publishing Company, Amsterdam, 1978

[8] J.F. Ciavaldini: Analyse numérique d'un problème de Stefan á deux phases par une méthode d'elements finis. SIAM J. Numer. Anal. 12, 1975, 464-487

[9] G. Duvaut: The solution of a two-phase Stefan problem by variational inequality, Proc. Symposium on moving boundary problems in heat flow and diffusion, J.R. Ockendon-W.R. Hodgkins (ed.), Clarendon Press, Oxford, 1975, 173-181

[10] C.M. Elliot: On the finite element approximation of an elliptic variational inequality arising from an implicit time discretization of the Stefan problem, IMA J. Numer. Anal., 1, 1981, 115-125

[11] C.M. Elliot and J.R. Ockendon: Weak and variational methods for moving boundary problems. Research Notes in Mathematics 59, Pitman, Boston, 1982

[12] M. Fremond: Diffusion problems with free boundaries, Autumn Course on Applications of Analysis to Mechanics, I.C.T.P. Trieste, 1976

[13] R. Glowinski, J.L. Lions and R. Tremolières: Numerical analysis of variational inequalities, Studies in Mathematics and its Applications 8. North Holland, Amsterdam, 1981

[14] O. Grange and F. Mignot: Sur la résolution d'une équation et d'une inéquation paraboliques non lineares, J. Funct. Anal., 11, 1972, 77-93

[15] K.-H. Hoffmann and M. Niezgodka: Control of parabolic systems involving free boundaries, in Free Boundary Problems - Theory and Applications (ed. by A. Fasano and M. Primicerio) Research Notes in Mathematics, 79, Pitman, Boston, 1983, 431-462

[16] J.W. Jerome and M.E. Rose: Error estimates for the multidimensional two-phase Stefan problem, Math. Comp., 39, 1982, 377-414

[17] C. Lemaréchal, J.J. Strodiot and A. Bihain: On a bundle algorithm for nonsmooth obtimization, in Nonlinear Programming 4, Academic Press, New York, 1981, 245-281

[18] J.L. Lions: Quelques méthodes de résolution des problèmes aux limites non linéaires, Dunod, Paris, 1969

[19] Z. Meike and D. Tiba: Optimal control for a Stefan problem, in Analysis and Optimization of Systems (ed. by A. Bensoussan and J.L. Lions), Lecture Notes in Control and Information Sciences 44, Springer-Verlag, Berlin, 1982, 776-787

[20] P. Neittaanmäki and D. Tiba: A descent method for the boundary control of a two-phase Stefan problem, in Proc. of the 3nd Int. Workshop on Control Theory and Differential Equations, Buckarest, August 26-30, 1983, INCREST, 1983, 111-127

[21] P. Neittaanmäki and D. Tiba: A finite element approximation of the boundary control of two-phase Stefan problems, Lappeenranta University of Technology, Dept. Physics and Mathematics, Research Report 3/83, 1983

[22] I. Pawlow: A variational inequality approach to generalized two-phase Stefan problem in several space variables, Ann. Mat. Pura Appl., vol. CXXXI, 1982, 333-373

[23] I. Pawlow: Error estimates for Galerkin approximation of boundary control problems for two-phase Stefan type processes, see [20], 129-147

[24] O. Pironneau: Optimal shape design for elliptic systems, Lecture Notes in Comp. Physics, Springer Verlag, New York, 1983

[25] L. Rubinstein: The Stefan problem, Transl. Math. Monographs, Vol. 27, Amer. Math. Soc., Providence, R.I., 1971

[26] C. Saguez: Contrôle optimal de systèmes à frontière libre, Thèse, Univ. Technologique Compiegne, 1980

[27] D.A. Tarzia: Etude de l'inèquation variationelle proposée par Duvaut pour le problème de Stefan á deux phases I, Boll U.M.I. 6, 1982, 865-883

[28] D. Tiba: Boundary control for a Stefan problem, in Optimale Kontrolle partieller Differentialgleichungen mit Schwerpunkt auf numerischen Verfahren, (ed. by K.-H. Hoffmann and W. Krabs), ISNM, Birkhauser, Basel, 1983

[29] D. Tiba and M. Tiba: Regularity of the boundary data and the finite element discretization in two-phase Stefan problems, Int. J. Eng. Sciences (submitted) also available as INCREST preprint, no 41, 1983 .

# Spectrally Canonical Distributed Parameter Systems

L. Pandolfi
Dipartimento di Matematica
Politecnico di Torino - Torino, ITALIA

Abstract: We consider the transfer function $T(z)$ of a distributed parameter system. Under suitable assumptions, we study the relationships which occur between the structures of the poles and zeros of $T(z)$ and the properties of the system.

## 1. Introduction

A great number of non equivalent controllability and observability definitions for distributed parameter systems have been studied in recent times ([1, 2, 3]). The number of these definitions is very large, but only few of them have been applied to the study of problems in systems theory. In particular:

1. Spectral controllability has been applied to the stabilization problem ([4]) and also to several problems in the theory of delay systems ([5, 6, 7]).
2. Approximate controllability. Applications have been made to the realization theory ([8]), high gain feedback ([9]) and stabilization ([10]).
3. Continuous controllability was introduced in the study of the realization theory ([11]).
4. Uniform zero controllability was applied to the stabilization of reversible systems ([12]).

The conditions 3, 4 are very restrictive and imply condition 2, which implies 1. For a large class of systems (not systems of hyperbolic type, however) spectral controllability is equivalent to the full stabilization property; hence, it seems interesting to investigate more properties of spectrally controllable (and observable) systems. In this paper, we study the relationships between the properties of the system and those of its transfer function for the class of systems that we describe now.

We consider the system

$$\dot{x} = Ax + Bu$$
$$y = Cx \tag{1}$$

where $x \in X$, a Hilbert space, $u \in R^m$, $y \in R^p$. The operator $A$ is the infinitesimal generator of a strongly continuous semigroup of bounded operators $E(t)$. The operator $B$ is linear and bounded (its domain is finite dimensional). The operator $C$ is linear, but we do not assume that it is bounded (this assumption would leave out delay systems); instead, we assume that:

A.1. $\mathcal{D}(C) \supseteq \mathcal{D}(A)$.

Assumption A.1. implies that the <u>transfer function</u>

$$T(z) = C(zI-A)^{-1}B$$

is a $p \times m$ matrix function of the complex variable $z \in \rho(A)$, the resolvent set of $A$.

We shall consider a fixed point $z_o$ of the complex plane, and we assume that:

A.2. The point $z_o$ belongs to $\rho(A)$ or is an isolated point of $\sigma(A)$, which is an eigenvalue of finite multiplicity.

We shall say that system (1) is (<u>spectrally</u>) <u>controllable</u> at $z_o$ when

$$\text{Im}\,(z_oI-A) + \text{Im}\,B = X \tag{2}$$

and that it is (<u>spectrally</u>) <u>observable</u> at $z_o$ when

$$\{\text{Ker}(z_oI-A)\} \cap \text{Ker}\,C = \{0\}. \tag{3}$$

When both conditions (2) and (3) hold, we say that the system is (<u>spectrally</u>) <u>canonical</u> at $z_o$. We shall say that system (1) is (spectrally) controllable, observable, canonical when A.2 and conditions (2), (3) hold at every complex point $z_o$.

We observe that, under our assumptions, the codimension of $\text{Im}(z_oI-A)$ is finite and that the elements of the matrix function $T(z)$ are holomorphic in a neighboorhood of $z_o$, with the possible exception of $z_o$, where they may have a pole. Hence ([6]) we can find two square holomorphic matrices $F(z)$, $G(z)$ of suitable dimensions, such that

1. $\text{Det}\,F(z) \equiv 1$, $\det G(z) \equiv 1$.

$$2.\ M(z) = F(z)T(z)G(z) = \begin{cases} [\bar{M}(z),0] & p < m \\ \bar{M}(z) & p = m \\ \begin{bmatrix}\bar{M}(z)\\ 0\end{bmatrix} & p > m \end{cases}$$

and the matrix $\bar{M}(z)$ has the form

$$\bar{M}(z) = \operatorname{diag}\left[M_i(z)(z-z_o)^{s_i}\right]_{1 \leq i \leq k}.$$

The exponents $s_i$ are integer numbers ($s_k$, the last one, may be $+\infty$), ordered increasingly. The matrices $M_i(z)$ have holomorphic elements and are square $d \times d$ matrices, and $\det M_i(z_o) \neq 0$ for any index $i$, with the possible exception of $i=k$. If $\det M_k(z_o) = 0$ then the matrix $M_k(z)$ is identically zero and we assume, by definition, that $s_k = +\infty$.

The matrix $T(z)$ has a pole at $z_o$ when at least one of the exponents $s_i$ is negative. In this case the matrix

$$\begin{pmatrix} s_1,\ldots,s_r \\ d_1,\ldots,d_r \end{pmatrix},$$

where $r$ is the last index such that $s_r$ is negative, is called the structure of the pole. We call order of the pole $z_o$ the number

$$m_T(z_o) = \sum_{1}^{r} s_i d_i$$

(this number is also called the Mac Millan degree of the principal part of $T(z)$ at $z_o$. We use the shortest word "order" for simplicity and also for the sake of simmetry with the zero case. Since we never study the single components of the matrix $T(z)$, no confusion can arise with the order of the poles of the components).

We say that $T(z)$ has a zero at $z_o$ when at least one of the exponets $s_i$ is positive and finite. In this case, the structure of the zero is given by the matrix

$$\begin{pmatrix} s_q,\ldots,s_p \\ d_q,\ldots,d_p \end{pmatrix}$$

where $q$ is the least index such that $s_q$ is positive and $p = k-1$ if $s_k = +\infty$, $p = k$ otherwise. The order of the zero is the number

$$\bar{m}_T(z_o) = \sum_{q}^{p} s_i d_i.$$

In the following, we shall call system matrix the operator matrix

$$S(z) = \begin{bmatrix} zI-A & -B \\ C & 0 \end{bmatrix}.$$

## 2. A Remark on Canonical Realizations

The realization problem require to construct system (1) once that its transfer function $T(z)$ is known. This problem, for distributed parameter systems, is largely unsolved. However, it is clear that if (1) is a realization of $T(z)$, the spectrum of $A$ must contain the singularities of $T(z)$. It is easy to prove the following result:

Theorem 2.1. *Let us assume that system (1) is canonical at* $z_o$, *a pole of* $T(z)$ *whose multiplicity is* $r$. *Then, the generalized eigenspace of* $A$ *which corresponds to the eigenvalue* $z_o$ *has dimension* $r$.

*Proof.* Let $N$ be the generalized eigenspace of $z_o$. The operator $A$ can be reduced by $N(z_o)$ and $\mathrm{Im}\,(z_oI-A)^m$ for some number $m$ ([13], pag. 306). So, we can write $A = \mathrm{diag}(A_1,A_2)$, $B = \mathrm{Col}(B_1,B_2)$, $C = \mathrm{row}(C_1,C_2)$; moreover, $z_o$ is in the resolvent set of $A_2$. Now,

$$T(z) = C_1(zI-A_1)^{-1}B_1 + C_2(zI-A_2)^{-1}B_2 = T_1(z) + T_2(z).$$

From ([6]) we know that the order of the pole of $T(z)$ and of $T_1(z)$ are the same. This last function is the transfer function of a finite dimensional system, for which Theorem 2.1. is known to hold. This completes the proof, since $z_o \in \rho(A_2)$.

In fact, we can give a sharper version of the above theorem. Let $u(z)$ be a function which is regular at $z_o$, $u(z_o) \neq 0$, and such that $T(z)u(z)$ has a pole of order $s$. If we write the matrix of the finite dimensional operator $A_1$ in Jordan form, we see that at least one of the blocks has dimension $s$. Hence:

Lemma 2.2. *Let the structure of the pole* $z_o$ *of* $T(z)$ *be*

$$\begin{pmatrix} d_1, \dots, d_r \\ s_1, \dots, s_r \end{pmatrix}. \tag{4}$$

*For each number* $s_i$ *we can find* $d_i$ *blocks of* $A_1$, *of dimension* $s_i$, *if the system is canonicat at* $z_o$.

We can note that the structure of the poles of $T(z)$ and $T_1(z)$ coincide. Hence:

**Theorem 2.3.** *Let the matrix (4) be the structure of the pole of T(z) at* $z_o$. *For each index* $i$, $1 \le i \le r$, *we can find* $d_i$ *maximal Jordan chains of the operator A, of lenght* $s_i$, *if system (S) is canonical at* $z_o$.

The converse result holds:

**Theorem 2.4.** *If system (1) is not canonical at* $z_o$, *then the multiplicity of* $z_o$ *as a pole of T(z) is strictly minor then the dimension of the generalized eigenspace of the operator A at* $z_o$.

*Proof.* In fact, this is true for the finite dimensional system identified by the matrices $C_1$, $A_1$, $B_1$.

The above simple observations show that the knowledge of the behaviour of the transfer function near the point $z_o$ completely specifies the spectral properties of the operator A of any canonical realization at the eigenvalue $z_o$, exactly as in the case of finite dimensional systems. In this last case, more information on the system can be recovered by the study of the structure of the zeros. This explains the reason why we study the properties of the zeros of the transfer matrix of a distributed parameter system.

## 3. The Zeros of T(z): Analytic Properties

The arguments of this section rest on the following results from [6]:

**Theorem 3.1.** *Let H(z) be a pxm matrix of meromorphic functions in a given region of the complex plane. If* $z_o$ *is a zero of H(z) of multiplicity* $r > 0$, *then it is possible to find holomorphic functions* $\zeta_i(z)$, *defined in a neigboorhood of* $z_o$, *such that:*

1. *The vectors* $\zeta_i(z_o)$ *are independent;*
2. *The function* $H(z)\zeta_i(z)$ *has a zero of order* $r_i$ *in* $z_o$;
3. $r = \sum r_i$.

**Theorem 3.2.** *Let H(z) be as in theorem 3.1. Let us assume that* $p \ge m$ *and that the matrix H(z) has rank m for at least one point z. Let us assume that we can find holomorphic functions* $\zeta_i(z)$, *defined in a* *neigboorhood of* $z_o$*, such that the properties 1, 2 of the above theorem hold. Then, the point* $z_o$ *is a zero of* $H(z)$ *of multiplicity at least* $\sum r_i$.

Now, in order to simplify the exposition of the results of this paper, we introduce the following convenctions:

1. The functions designed as $u(z)$, $\varphi(z)$, $x(z)$, $\xi(z)$, $\psi(z)$ take values respectively in the spaces $R^m$, $R^p$, $X$, $X \times R^m$, $X \times R^p$.
2. The above functions are holomorphic in a neigboorhood of the point $z_o$ that we are considering.

We have the following result:

Proposition 3.3. *Let us assume that system (S) is canonical at* $z_o$ *and let* $u(z)$ *be a function such that* $T(z)u(z)$ *is bounded near* $z_o$*. Then, the function* $x(z) = (zI-A)^{-1}Bu(z)$ *is bounded near* $z_o$ *(i.e. to denote it* $x(z)$ *is consistent with the convenctions above).*

*Proof*. The result is obvious if $z_o \notin \sigma(A)$. Hence, we assume that $z_o$ is an eigenvalue of the operator $A$ and we write $X = K \oplus K'$, where $K$ is the generalized eigenspace of the eigenvalue $z_o$

Thus:

$$A = \begin{bmatrix} A_1 & 0 \\ 0 & A_3 \end{bmatrix} \qquad B = \begin{bmatrix} B_1 \\ B_2 \end{bmatrix} \qquad C = [C_1, C_2].$$

$$T(z) = C_1(zI-A_1)^{-1}B_1u(z) + $$
$$+ C_2(zI-A_3)^{-1}B_2u(z).$$

(The operator $I$ denotes the identity operator. The space on which it acts is clear from the context).

In the above expression for $T(z)$, only the first addendum contains unbounded functions. However, it is bounded, since we assumed that $T(z)$ is bounded. This implies that the function $(zI-A_1)^{-1}B_1u(z)$ is bounded, since it corresponds to the transfer function of a finite dimensional system ([5], theorem 4.3.).

Theorem 3.4. *Let (S) be a canonical system. Let us assume that it is possible to find functions* $u_i(z)$, $\varphi_i(z)$ *such that*

$T(z)u_i(z) = (z-z_o)^{r_i}\varphi_i(z) \qquad r_i > 0$

*with the following properties:*

*1. The vectors* $u_i(z_o)$ *are independent*

*2. the vectors* $\varphi_i(z_o)$ *are not zero.*

*Then, we can find functions* $\xi_i(z)$, $\varphi_i(z)$ *such that*

*i. The vectors* $\xi_i(z_o)$ *are independent*

*ii.* $S(z)\xi_i(z) = (z-z_o)^{r_i}\varphi_i(z)$

*Proof.* Let us consider the functions

$$\xi_i(z) = \begin{bmatrix} (zI-A)^{-1}Bu_i(z) \\ \\ u_i(z) \end{bmatrix}.$$

The above proposition implies that these functions are bounded in $z_o$ (so that they can be extended to holomorphic functions). The vectors $u_i(z_o)$ are independent, so that condition i. is satisfied. Condition ii. is obvious.

REMARK. Let us observe that the above two results depend only on the observability of system (S).

Now, we investigate the converse of the above result. Let us assume that we can find functions $\xi(z)$ such that

$$\xi(z) = \begin{bmatrix} x(z) \\ \\ u(z) \end{bmatrix}, \quad S(z)\xi(z) = (z-z_o)^s\varphi(z). \tag{5}$$

Then:

Theorem 3.5. *If condition (5) holds, and if (S) is canonical at* $z_o$*, then it is possible to find a function* $u'(z)$ *such that*

$T(z)\left[u(z) + (z-z_o)^s u'(z)\right] = (z-z_o)^s\varphi(z).$

*Proof.* In order to show a different trick, which may be useful, we write now $X = K + K$ , where $K$ is the generalized eigenspace of $z_o$. The matrix $S(z)$ can be put in the form

$$S(z) = \begin{bmatrix} zI-A_1 & -A_2 & -B_1 \\ 0 & zI-A_3 & -B_2 \\ C_1 & C_2 & 0 \end{bmatrix}.$$

This form of $S(z)$ can be obtained after a changment of coordinates in the space $X$, and this operation leaves the transfer function $T(z)$ unchanged.

The operator

$$\begin{bmatrix} I & A_2(zI- A_3)^{-1} \\ 0 & I \end{bmatrix}$$

is boundedly invertible. Moreover,

$$S'(z) = \begin{bmatrix} I & A_2(zI-A_3)^{-1} & 0 \\ 0 & I & 0 \\ 0 & 0 & I \end{bmatrix} \begin{bmatrix} zI-A_1 & -A_2 & -B_1 \\ 0 & zI-A_3 & -B_2 \\ C_1 & C_2 & 0 \end{bmatrix} =$$

$$= \begin{bmatrix} zI-A & 0 & -B'(z) \\ 0 & zI-A_3 & -B_2 \\ C_1 & C_2 & 0 \end{bmatrix},$$

where $B'(z) = B_1 + A_2(zI-A_3)^{-1}B_2$. Of course, in a neigboorhood of $z_o$, $z_o$ included, we have that

$$\text{Im} \begin{bmatrix} zI-A_1 & 0 \\ 0 & zI-A_3 \end{bmatrix} + \text{Im} \begin{bmatrix} B'(z) \\ B_2 \end{bmatrix} = X.$$

Hence, if $X \neq K^\perp$, $B'(z_o)$ covers **a complement** of the image of $(zI-A_1)$. Now, we **call** $\psi_i(z)$ the block components of $\psi(z)$, according with the block structure of $S(z)$ (so that i = 1, 2, 3). We see that

$$S'(z) \zeta(z) = (z-z_o)^s \begin{bmatrix} \psi_1(z) + A_2(zI-A_3)^{-1}\psi_2(z) \\ \psi_2(z) \\ \psi_3(z) \end{bmatrix} = (z-z_o)^s \psi'(z).$$

Now, let us call $x_1(z)$, $x_2(z)$, $u(z)$ the block components of $\zeta(z)$ according with the block structure of $S'(z)$. **For every** $u'(z)$ we have

$$S'(z) = \begin{bmatrix} x_1(z) \\ x_2(z) \\ u(z) + (z-z_o)^s u'(z) \end{bmatrix} = (z-z_o)^s \begin{bmatrix} \psi'_1(z) - B'(z)u'(z) \\ \psi'_2(z) - B_2u'(z) \\ \psi'_3(z) \end{bmatrix}$$

where the functions $\psi'_i(z)$ are the components of $\psi'(z)$. From the above equality we have that

$$x_1(z) = (zI-A_1)^{-1}B'(z)\left[u(z) + (z-z_o)^s u'(z)\right] + (z-z_o)^s(zI-A_1)^{-1}\left[\psi'_1(z) - B'(z)u'(z)\right]$$

$$x_2(z) = (zI-A_3)^{-1}B_2\left[u(z) + (z-z_o)^s u'(z)\right] + (z-z_o)^s(zI-A_3)^{-1}\left[\psi'_2(z) - B_2u'(z)\right].$$

Hence,

$$T(z)\left[u(z) + (z-z_o)^s u'(z)\right] = (z-z_o)^s\left\{\psi_3(z) - C_1(zI-A_1)^{-1}\left[\psi_1(z) - B(z)u'(z)\right]\right\} - (z-z_o)^s C_2(zI-A_3)^{-1}\left[\psi_2(z) - B_2u'(z)\right].$$

The term

$(z - z_o)^s \{\psi_3(z) - C_2(zI - A_3)^{-1} [\psi_2(z) - B_2 u'(z)]\}$

has a zero of order s at least for every function u'(z). We do not have information on the other term, unless a special choiche of u'(z) is made. We observed already that $B'(z_o)$ covers a complement of the image of $(zI-A_1)$. Hence, it is possible to find a function u'(z) such that $\psi_1(z) - B(z)u'(z)$ belongs to Im $(zI-A_1)$. With this choice of the function u'(z), also the first addendum has a zero of multiplicity at least s for $z=z_o$. This completes the proof.

REMARK. With the above choice of u'(z), the vectors $\psi_1(z) - B'(z)u'(z)$ belong to Im $(zI-A_1)$

Now, let us try to test the order of the zero of the system in this way: we consider the operator matrix S(z). We operate with S(z) on functions $\xi_i(z)$ such that the vectors $\xi_i(z_o)$ are independent. Then, we read the order of the zero of the functions $S(z)\xi_i(z)$, for $z = z_o$. We have the following result:

Theorem 3.6. *Let*

$$\xi_i(z) = \begin{pmatrix} x_i(z) \\ u_i(z) \end{pmatrix}, \qquad S(z)\xi_i(z) = (z-z_o)^{r_i}\psi_i(z), \quad \psi_i(z_o) \neq 0, \quad r_i > 0.$$

*If the vectors $\xi_i(z_o)$ are independent, then the vectors $u_i(z_o)$ are independent (of course, when the system is a canonical one).*

*Proof*. Again, we can prove the theorem with respect to the matrix S'(z). Let us assume that we can find numbers $\alpha_i$ such that $\sum \alpha_i u_i(z_o) = 0$. We consider the function $\xi(z) = \sum \alpha_i \xi_i(z)$. Of course, $S'(z_o)\xi(z_o) = 0$, so that

$C_1 \sum \alpha_i x_{i_1}(z_o) + C_2 \sum \alpha_i x_{i_2}(z_o) = 0.$

Of course,

$C_2 \sum \alpha_i x_{i_2}(z_o) = C_2(zI-A_2)^{-1} B_2 \sum \alpha_i u_i(z_o)$

is zero since $z_o \notin \sigma(A_2)$. Hence,

$C_1 \sum \alpha_i x_{i_1}(z_o) = 0.$

The problem is now reduced to a finite dimensional one, and the positive answer is in Lemma 4.5 in [5].

An important consequence is the following:

Corollary 3.7. *If the system (S) is canonical at* $z_o$, *we can find at most* $m$ *functions* $\xi_i(z)$ *such that*

*i.* $S(z_o)\xi_i(z_o) = 0$

*ii. the vectors* $\xi_i(z_o)$ *are independent.*

## 3. The Transmission Properties of the Zeros

Let $x(t)$ be a solution of the differential equation in (1), which corresponds to the input $u(t)$. We say that the pair $(x(t),u(t))$ is a conjugate pair when the corresponding input $y(t)$ is identically zero. Of course, the conjugate pairs are the solutions of the equation

$$\frac{d}{dt}\begin{bmatrix} I & 0 \\ 0 & 0 \end{bmatrix}\begin{bmatrix} x(t) \\ u(t) \end{bmatrix} = \begin{bmatrix} A & B \\ C & 0 \end{bmatrix}\begin{bmatrix} x(t) \\ u(t) \end{bmatrix}.$$

We put $X(t) = \mathrm{col}(x(t),u(t))$ so that the above equation can be written

$$\left(\frac{d}{dt}E - L\right)X(t) = 0 \qquad (6)$$

with the obvious meanings of the operators $E$ and $L$. Let us observe explicitly that $S(z) = zE-L$.

Let us assume now that we can find a function $\xi(z)$ such that $S(z)\xi(z)$ has a zero of order $(r+1)$, $r \geq 0$. Then, we have the following lemma:

Lemma 4.1. *If* $S(z)\xi(z) = (z-z_o)^{r+1}\psi(z)$, $r \geq 0$, *then we can find a conjugate pair of the form*

$x(t) = \exp(z_o t)\sum_0^r x_i t^i, \quad u(t) = \exp(z_o t)\sum_0^r u_i t^i.$

*Proof*. Of course, we need only to show that there exists a function $X(t) = \exp(z_o t)\sum_0^r X_i t^i$ which is solution of Eq. (6). In order to simplify the notation we assume (without any real restriction) that $z_o = 0$.

Let us write

$$\xi(z) = \sum_0^\infty \xi_i z^i.$$

The function

$$(zE-L)\xi(z) = \sum_1^\infty E\xi_{j-1}z^j - \sum_0^\infty L\xi_j z^j = -L\xi_o + \sum_1^\infty (E\xi_{j-1} - L\xi_j)z^j$$

has a zero of order $(r+1)$. Hence,

$$L\xi_o = 0 \qquad (7)$$

$$E\xi_{j-1} = L\xi_j \quad 1 \leq j \leq r. \qquad (8)$$

Let us assume now that we can find a solution $X(t)$ of Eq. (6), $X(t) = \sum_0^r X_i t^i$. Then we have

$$E \sum_0^r X_i i t^{i-1} - L \sum_0^r X_i t^i = 0$$

i.e.

$$0 = \sum_0^{r-1} \left[EX_{i+1}(i+1) - LX_i\right] t^i - LX_r t^r.$$

From (7), (8) we see that this condition is satisfied if we chose $X_i = (\zeta_{r-i}/i!)$. This finishes the proof.

Let now $z_o$ be a zero of $T(z)$. From the preceeding section we know that it is possible to find several functions $\zeta_i(z)$, $\zeta_i(z) = \mathrm{col}(x_i(z), u_i(z))$ such that the vectors $u_i(z_o)$ are independent, and $S(z)\zeta_i(z)$ has a zero of order $(r+1)$. Hence:

Theorem 4.2. *Let* $z_o$ *be a zero of* $T(z)$. *We can find a finite number of conjugate pairs* $(x_j(t), u_j(t))$,
$x_j(t) = exp(z_o t) \sum_0^{r_j} x^j_i t^i$, $\quad u_j(t) = exp(z_o t) \sum_0^{r_j} u^j_i t^i$,
*such that the leading coefficients of the polynomials which appear in the expressions of the functions* $u_j(t)$ *are linearly independent.*

REMARK. The above theorem must be compared with the blocking property of the zeros described in [14].

Now we can read back the proof of lemma 4.1. We have the following result:

Theorem 4.3. *Let*

$$X(t) = exp(z_o t) \sum_0^r X_i t^i \qquad (X_i = col\ (x_i, u_i)\ ) \tag{9}$$

*be a conjugate pair. Then, we can find functions* $\zeta(z)$ *such that* $S(z)\zeta(z)$ *has a zero, for* $z = z_o$, *of order* $(r+1)$ *at least.*

A conjugate pair of the form (9) will be denoted, from now on, with the simbol $X(t;z_o,r)$.

Using again Lemma 4.1. we see that, when $S(z)\zeta(z)$ has a zero of order $(r+1)$ at $z_o$, we can associate with $\zeta(z)$ a conjugate pair $X(t;z_o,s)$ for every $s$, $0 \le s \le r$. The coefficients of $X(t;z_o,s)$ are calculated by the relations (7), (8). A sequence of functions of this type will be called a chain of conjugate pairs (associated to $\xi(z)$). We say that a chain of conjugate pairs is maximal when there exists no longer chain, associated with the same function $\xi(z)$, which contains the given one. Of course, it is possible that a chain be infinite. This is the case if (and only if by the next result) the function $S(z)\xi(z)$ is identically zero.

Theorem 4.4. *The function* $S(z)\xi(z)$ *has a zero of order* $k$ *for* $z = z_o$ *(or, it is identically zero) if and only if the longest chain associated with* $\xi(z)$ *has* $k$ *elements (or, is infinite).*

*Proof*. Obvious, from Lemma 4.1. and Theorem 4.3.

REMARK. The last results of this section should be compared with the results in [6]. In particular, they could be used to extend the results on S-invariant subspaces contained in that paper.

## 5. Conclusions

In this paper we proved that the very weak assumption of spectral observability and controllability is strong enough to extend to an important class of distributed parameter systems the main results which are known on the pole-zero structure of lumped systems.

The results of this paper were suggested by the theorem proved in [5] for delay systems (the results of sect. 4 may be compared with [15]). The following example shows an application to a problem of different nature.

Let us consider an insulated bar (of lenght one) which is been heating by a source distributed on a segment of the bar itself. We observe the temperature on the left extremum. If $S(t,s)$ is the temperature, we have the problem

$S(t,.) \in L^2(0,1)$

$S_t = S_{ss} - b(s)u(t)$

$S_s(t,0) = S_s(t,1) = 0.$

Let $\hat{}$ denote Laplace transform. Few calculations show that, for $z \neq 0$,

$\hat{S}(z,s) = \{ (\exp(-\sqrt{z}s) + \exp(\sqrt{z}s))/(\exp(\sqrt{z}) - \exp(-\sqrt{z}))\} \cdot$
$\{e^{-\sqrt{z}} \int_0^1 e^{2\sqrt{z}r} \int_0^r e^{-\sqrt{z}t} b(t)dtds - (e^{\sqrt{z}}/z) \int_0^1 e^{-\sqrt{z}t} b(t)dt\} \quad \hat{u}(z) +$
$+ e^{\sqrt{z}x} \int_0^x e^{-\sqrt{z}t} b(t)dt \quad \hat{u}(z)$

Let us assume now that $b(s) = 1 \quad 0 \leq s \leq 1$. It is easy to see that in this case we have a canonical system and that the Laplace transform of the output $y(t)$ is given by

$$\hat{y}(z) = \frac{e^{\sqrt{z}} + e^{-\sqrt{z}}}{e^{\sqrt{z}} - e^{-\sqrt{z}}} \left\{ \frac{e^{\sqrt{z}} - e^{-\sqrt{z}}}{2z} - \frac{\sqrt{z} + 1}{z} + \frac{1 - \sqrt{z}}{z} e^{-\sqrt{z}} \right\} \hat{u}(z).$$

Hence, the points $z = (4k\pi + \pi)^2$ are simple zeros of the transfer function. We can illustrate the material of sect. 4. We look for a conjugate pair which corresponds to these zeros. We must have
$u(t) = \exp((4k+1)^2\pi^2 t)u_o$, $\quad S(t,s) = \exp((4k+1)^2\pi^2 t)\varphi(x)u_o$. If $u_o = 1$, we see by direct substitution that $\varphi(x)$ must satisfy the equation

$$\varphi'' = (4k+1)^2\pi^2\varphi - 1$$

so that

$$\varphi(s) = c\{\exp(-4k+1)\pi s) - \exp((4k+1)\pi s)\} - \{\exp((4k+1)^2\pi^2 s) - 1\}/\{(4k+1)^2\pi^2\}$$

and $c$ is an arbitrary constant.

## R E F E R E N C E S


1. Dolecki, S., A Classification of Controllability Concepts for Infinite Dimensional Linear Systems, Control and Cybernetics, ''-44, 1976.

2. Dolecki, S., Russel, D.L., A General Theory of Observation and Control, SIAM J. Control Opt., 15, 185-220, 1977.

3. Bacciotti, A., Sedici modi di definire la completa controllabilità negli spazi di Banach, To appear: Rend. del Sem. Matem. Univ. Pol. Torino

4. Bhat, K.P.M., Wonham, W.M., Stabilizability and Detectability for Evolution Systems in Banach Spaces, 1976 IEEE Conf. on Decision and Control.

5. Pandolfi, L., The transmission Zeros of Systems with Delays, International J. Control, 36, 959-976, 1982.

6. Pandolfi, L., Some Observations about the Structure of Systems with Delays, in "Analysis and Optimization of Systems" Bensoussan, A., Lions

J.L. Ed., Lecture Notes in Control and Inf. Sci. 44, Springer Verlag, Berlin, 1982.

7. Pandolfi, L., Canonical Realizations of Systems with Delays, SIAM J. Control Opt., 21, 598-613, 1983.

8. Fuhran, P.A., Linear Systems and Operators in Hilbert Spaces, Mc Grow Hill Internatioanl Book Co., New York, 1981

9. Pojolainen, S., Computation of Transmission Zeros for Distributed Parameter Systems, Int. J. Control, 33, 199-212, 1981.

10. Helton, J.W., Systems with Infinite Dimensional State Space: a Hilbert Space approach, Proc. of the IEEE, 64, 145-160, 1976.

11. Benchimol, C.D., A Note on Weak Stabilizability of Contraction Semigroups, SIAM J. Control Opt., 16, 373-379, 1978.

12. Zabczyk, J., Complete Stabilizability Implies Exact Controllability, Seminarul de Ecuatii Functionale, Univ. din Timishoara, 38, 1-8, 1976.

13. Taylor, A., Introduction to Functional Analysis, Chapman & Hall, Ltd, London, 1958.

14. Callier, F.M., Cheng, V.H.L., Desoer, C.A., Dynamic Interpretation of Poles and Transmission Zeros for Distributed Multivariable Systems, IEEE Trans. Cyrcuit and Systems, CAS-28, 300-306, 1981.

15. Przyłuski, K.M., Zeros of Linear Distributed Parameter Systems with application to the theory of Delay Systems, Technical Report, Institute of Electronics Fundamentals, Warsaw Technical University, Warsaw, 1979.

ACKNOWLEDGMENT: This paper has been written according with the research programs of the GNAFA-CNR and with the financial support of the Ministero della Pubblica Istruzione.

# BOUNDARY FEEDBACK STABILIZATION OF A PARABOLIC EQUATION*

Thomas I. Seidman
Department of Mathematics and Computer Science
University of Maryland Baltimore County
Baltimore, Maryland 21228 U.S.A.
(301) 455-2438

ABSTRACT The equation (*) $u_t = u_{xx} + qu + f$ will, in general, be unstable for positive $q$. We consider control through the boundary conditions $u(\cdot,0) = 0$, $u_x(\cdot,1) = \phi$ with observation available of $\psi := u(\cdot,1)$ and no knowledge of the initial state or of the input $f$. It is shown that one can construct a linear feedback law of the form (**) $\phi(t) = \langle\lambda,\psi^t\rangle + \langle\mu,\phi^t\rangle$ ($\phi^t$, $\psi^t$ are intervals of past history) which stabilizes (*).

1. INTRODUCTION We consider stabilization by boundary feedback of the parabolic equation (for $t > 0$, $0 < x < 1$):

$$u_t = u_{xx} + q(x)u + f(t,x,u) \tag{1.1}$$

with the boundary conditions

$$u(\cdot,0) = 0, \qquad u_x(\cdot,1) = \phi = \text{control}. \tag{1.2}$$

The function $f$ is supposed to consist of a quasilinear perturbation $f_0 = o(u)$ and an 'external' input $f_1$. Thus, for example,

$$f(t,x,r) = f_0(r) + f_1(t,x) \tag{1.3}$$

$$\text{(1.4)}\qquad \text{(i)}\quad |f_0(r)| \le C_0(1+|r|^\alpha) \qquad (\alpha < 1)$$

$$\text{(ii)}\quad \|f_1(t,\cdot)\| \le C_1$$

where the norm in (ii) is that of $X := L^2(0,1)$. We assume that the coefficient $q = q(x)$ is, e.g., continuous on $[0,1]$ although this condition will not be used explicitly and can be weakened considerably.

The strategic organization of the paper is to consider the linear, homogeneous problem ($f \equiv 0$) from the point of view of semigroup theory and obtain a stabilizing feedback law in the form

---

*This research has been partially supported by the U.S. Air Force Office of Scientific Research under grant no. AFOSR-82-0271.

(1.5) $$\phi(t) = \langle\lambda,\psi^t\rangle + \langle\mu,\phi^t\rangle$$

where the inner products are taken in $M := L^2(0,\delta)$ and, for $t > \delta$ and $0 < s < \delta$,

(1.6) $$\psi^t(s) = \psi(t-\delta+s),$$
$$\phi^t(s) = \phi(t-\delta+s).$$

Here $\psi(\cdot)$ is supposed to be the available boundary observation

(1.7) $$\psi(t) := u(t,1).$$

The state space for the resulting stable semigroup will be $Y := X \times \mathbb{R} \times M^2$ with the state

(1.8) $$y(t) := [u(t,\cdot),\phi(t),\psi^t,\phi^t].$$

Applying the Gronwall inequality to a variation of parameters formula for this semigroup then shows that (1.2), (1.5) gives $u$ bounded in $X$ for arbitrary $f$ satisfying (1.3), (1.4). The construction of (1.5) proceeds by combining a full state feedback

(1.9) $$\phi(t) = \langle\kappa,u(t,\cdot)\rangle$$

(inner product in $X$) with a state estimator of the form

(1.10) $$\hat{u}(t,\cdot) = [\underset{\sim}{L}\phi^t](\delta) + \underset{\sim}{E}[\psi^t - \underset{\sim}{\tau}\underset{\sim}{L}\phi^t]$$

applying to the problem with $f \equiv 0$. (Note that it is the resulting combination (1.5) which will stabilize (1.1) even though with $f \not\equiv 0$ the formula (1.10) no longer estimates $u(t,\cdot)$.)

2. THE CONSTRUCTION In this section we consider only the linear homogeneous, autonomous equation

(2.1) $$u_t = u_{xx} + qu.$$

<u>Step 1</u>: First, using the boundary control (1.2) we consider the optimal control problem:

(2.2) $$\text{minimize } J := \int_0^\infty e^{2ct}[\phi^2(t) + \|u(t,\cdot)\|^2]\,dt$$

for some choice of $c > 0$. As in [ 8 ], [10] we have (2.1) null controllable (in arbitrary short time) by (1.2) so (a) there are controls $\phi$ making $J$ finite, (b) there is an optimal control $\phi_*$ minimizing $J$, (c) this optimal control $\phi_*$ is obtainable in feedback form

$$(2.3) \qquad \phi(t) = \langle\kappa, u(t,\cdot)\rangle = \int_0^1 \kappa(x)u(t,x)\,dx$$

for a suitably chosen $\kappa \in X$, (d) the autonomous linear system

$$(2.4) \qquad \text{(i)} \quad u_t = u_{xx} + qu$$

$$\text{(ii)} \quad u(t,0) = 0 = u_x(t,1) - \langle\kappa, u(t,\cdot)\rangle$$

obtained by using (2.3) in (2.1), (1.2) determines a holomorphic semigroup $\underset{\sim}{S}_\kappa(\cdot)$ on X. The infinitesimal generator $\underset{\sim}{A}_\kappa$ of $\underset{\sim}{S}_\kappa(\cdot)$ is given by

$$(2.5) \qquad \underset{\sim}{A}_\kappa : \omega \mapsto \omega'' + q \qquad \text{for } \omega \in \mathcal{D}(\underset{\sim}{A}_\kappa),$$

$$\mathcal{D}(\underset{\sim}{A}_\kappa) := \{\omega \in H^2(0,1) : \omega(0) = 0 = \omega'(1) - \langle\kappa,\omega\rangle\}.$$

For our present purposes it is less important that (2.3) provides the *optimal* control for (2.2) than that it is in feedback form and makes J finite for every choice of initial data $u_0 = u(0,\cdot) \in X$. It follows that

$$(2.6) \qquad \|\underset{\sim}{S}_\kappa(t)\| \leq Me^{-ct} \qquad t > 0$$

for some constant $M = M_c$.

Step 2: The feedback (2.3) is, of course, not feasible since it is assumed that we only have observations available of

$$(2.7) \qquad \psi(t) := u(t,1)$$

for the controlled system. (We do note that for any solution of (2.1), (1.2) for which control would be defined by (2.3) one would have $u(t) = \underset{\sim}{S}_\kappa(t)u_0$ so, for any $u_0 \in X$, one has $u(t) \in \mathcal{D}(\underset{\sim}{A}_\kappa)$ -- indeed, in $\cap_n \mathcal{D}(A^n)$ -- for $t > 0$ as $\underset{\sim}{S}_\kappa$ is a holomorphic semigroup; thus the measurement

$$(2.8) \qquad \underset{\sim}{T} : u \mapsto u(\cdot,1)$$

implied by (2.7) is meaningful.)

Note that we can, however, recover the state! Dual to the null controllability is the following [7], [8], [2]:

THEOREM There is a bounded linear operator $\underset{\sim}{E} : M \to X$ such that, for any v satisfying

(2.9) $$v_t = v_{xx} + qv, \qquad v(\cdot,0) = 0 = v_x(\cdot,1)$$

for $0 < t \leq \delta$, one has

(2.10) $$v(\delta,\cdot) = \underset{\sim}{E}[v(t,1) : 0 < t < \delta].$$

We also introduce the operator -- bounded by [ 5 ] --

$$\underset{\sim}{L}: M \to H^{3/2,3/4}((0,\delta)\times(0,1)): \overline{\phi} \mapsto w$$

defined by

(2.11) $$w_t = w_{xx} + qw, \qquad w(\cdot,0) = 0, \qquad w_x(\cdot,1) = \overline{\phi}, \qquad w(0,\cdot) = 0 .$$

Note that

(2.12) $$\underset{\sim}{L}_1 := \underset{\sim}{\tau}\underset{\sim}{L}: \overline{\phi} \mapsto w(\cdot,1): M \to M,$$

$$\underset{\sim}{L}_2 := [L\cdot](\delta): \overline{\phi} \mapsto w(\delta,\cdot): M \to X$$

are known to be bounded operators.

For any $t > \delta$ one can write $u(t-\delta+s,x)$ as the sum $v+w$ where $w = \underset{\sim}{L}\phi^t$ gives the effect of the control $\phi^t$ (see (1.6)) with 0 'initial data' at $t-\delta$ while the difference $v$ satisfies (2.9) with unknown data. Then

$$v(s,1) = u(t-\delta+s,1) - w(s,1) = [\psi^t - \underset{\sim}{\tau}\underset{\sim}{L}\phi^t](s)$$

for $0 < s < \delta$ so (2.10) gives $v(\delta,\cdot) = \underset{\sim}{E}[\psi^t - \underset{\sim}{\tau}\underset{\sim}{L}\phi^t]$. One thus has the formula (1.10) for $t > \delta$. Note that this depends only on (2.1), (1.2), (1.6) but not on the use of any particular choice of control $\phi(\cdot)$.

Step 3: We now substitute (1.10) into (2.3) to obtain the feedback law

(2.13) $$\phi(t) = \langle\kappa, \underset{\sim}{L}_2\phi^t + \underset{\sim}{E}[\psi^t - \underset{\sim}{L}_1\phi^t]\rangle_X$$

$$= \langle\lambda,\psi^t\rangle_M + \langle\mu,\phi^t\rangle_M$$

where $\lambda := \underset{\sim}{E}^*\kappa$ and $\mu := (\underset{\sim}{L}_2 - \underset{\sim}{E}\underset{\sim}{L}_1)^*\kappa$. Thus we are to remember the recent history intervals $\psi^t$, $\phi^t$ for the observation $\psi$ given by (2.7) and for the control used. Presumably the availability of $\psi(t)$, $\phi(t)$ for each $t$ permits recollection of $\psi^t$, $\phi^t$ so that (2.13) is an admissible feedback law.

Step 4: The combined system (still omitting f) is now given by

$$(2.14)\quad \text{(i)}\quad u_t = u_{xx} + qu \qquad 0 < x < 1,\quad t > 0$$

$$\text{with}\quad u(t,0) = 0,\quad u_x(t,1) = \phi(t) \qquad t > 0$$

$$\text{(ii)}\quad \phi(t) := <\lambda,\psi^t> + <\mu,\phi^t> \qquad t > 0$$

$$\text{(iii)}\quad \psi^t(s) := u(t-\delta+s,1) \qquad 0 < s < \delta,\quad 0 < t-\delta+s$$

$$\text{(iv)}\quad \phi^t(s) := \phi(t-\delta+s) \qquad 0 < s < \delta,\quad 0 < t-\delta+s$$

together with specification in $Y := X \times \mathbb{R} \times M^2$ of an initial state $y(0) = y_0$. (In application, the $u_0$ component of $y_0$ is unknown and $\psi^0$, $\phi^0$ are chosen arbitrarily -- say, 0 -- so one would not expect (1.10) for $t < \delta$ except by coincidence.) One easily sees that (2.14) is solvable for each $y_0 \in Y$ so (2.14) defines a $C_0$ semigroup $\underset{\sim}{T}(\cdot)$ on $Y$.

3. STABILIZATION  The construction of $\underset{\sim}{T}(\cdot)$ in (2.14) is such that (1.10) holds for $t > \delta$ so the u-component of $y$ coincides with that given (starting at $u(\delta)$) by (2.4):

$$\pi_{\underset{\sim}{u}}\underset{\sim}{T}(t)y_0 = \underset{\sim}{S}_{\kappa}(t-\delta)\pi_{\underset{\sim}{u}}\underset{\sim}{T}(\delta)y_0$$

for any $y_0 \in Y$. It follows that

$$(3.1)\qquad \|\underset{\sim}{T}(t)\| \leq \bar{M}e^{-ct}$$

as in (2.6) so (2.14) is stable.

Returning to the original equation (1.1), (1.2) and using (1.5), (1.6) as in (2.14), one obtains

$$(3.2)\quad \text{(i)}\quad u_t = u_{xx} + qu + f_0(u) + f_1$$

$$\text{with}\quad u(t,0) = 0,\quad u_*(t,1) = \phi(t)$$

$$\text{(ii)}\quad \phi(t) = <\lambda,\psi^t> + <\mu,\phi^t> + \omega(t)$$

$$\text{with}\quad \psi^t,\ \phi^t\quad \text{as in (2.14iii,iv)}$$

where, besides the perturbation $f = f_0(u) + f,$ in the equation, we have introduced a perturbation $\omega$ in the boundary condition at $x = 1$ in addition to the intended control defined by the feedback (1.5). For $t > \delta$ one then has the representation

$$(3.3) \qquad y(t) = \underset{\sim}{T}(t-\delta)y(\delta) + \int_\delta^t T(t-s)[f,0,0,0]^* \, ds + \int_\delta^t \underset{\sim}{\hat{A}}^\alpha \underset{\sim}{T}(t-s)\hat{\eta}\omega(s)\, ds$$

where $f = [f_0(u) + f_1](s)$ and, with $1/4 < \alpha < 1$, one takes $\underset{\sim}{\hat{A}} := \mathrm{diag}(\underset{\sim}{A}_\kappa, 1, 1, 1)$ where $\underset{\sim}{A}_\kappa$ is the generator of $\underset{\sim}{S}_\kappa$ and $\hat{\eta} := [\underset{\sim}{A}_\kappa^{1-\alpha}\eta, 0, 0, 0]^*$ with $\eta$ the solution of

$$(3.4) \qquad \eta'' + q\eta = 0, \qquad \eta(0) = 0, \qquad \eta'(1) = \langle \kappa, \eta \rangle$$

(well-defined because (2.4) is stable). Since $\underset{\sim}{S}_\kappa(\cdot)$ is a holomorphic semigroup, (2.6) gives

$$\|\underset{\sim}{A}_\kappa^\alpha \underset{\sim}{S}_\kappa(t)\| \leq Mt^{-\alpha}e^{-ct}$$

from which it follows that

$$\|\underset{\sim}{\hat{A}}^\alpha \underset{\sim}{T}(t-s)\| \leq \bar{M}(t-s)^{-\alpha}e^{-c(t-s)};$$

one also has $\eta$ well-defined. A standard application of Gronwall's inequality then shows $y$ is bounded in $t$ under the assumptions (1.3), (1.4) on $f$ and corresponding assumptions on $\omega$. Thus, (3.2) is stable.

This form of BIBO stability — that all solutions are bounded — is somewhat weak but is obviously the best one could hope for if persistent disturbances are permitted as in (1.4). On the other hand, if one were to have

$$(3.5) \qquad \limsup_{t\to\infty} \int_0^\varepsilon |\omega(t+s)|^p \, ds \to 0 \quad \text{as } \varepsilon \to 0 \text{ with } p > 1/(1-\alpha), \ \omega \in L^1(\mathbb{R}^+),$$

$$(3.5) \qquad |f(t,x,r)| \leq a(t,x) + b(t)r \quad \text{with } a(\cdot\cdot) \in L^1(\mathbb{R}^+ \to L^2(0,1))$$

and either $b \in L^1(\mathbb{R}^+)$ or $\limsup_{t\to\infty} b \leq \bar{\beta}$ with $M_c\bar{\beta} < c$ for for some $c > 0$,

then one would have asymptotic stability: all solutions of (3.2) would go to $0$ as $t \to \infty$. For more detail, see [11].

4. REMARKS AND FURTHER DISCUSSION The perturbation $\omega$ was introduced to permit a modification of the feedback law in which the definition of $\phi(t)$ would utilize recollection (not of the functions $\psi^t$ and $\phi^t$ but) of approximations corresponding to time-sampled observation — thus requiring only finite memory in the stronger sense of

needing only a finite number of numbers. In this case one would also re-compute the control only at corresponding intervals; the form of the control law, analogous to (3.14ii), permits this computation to be pipelined. Details of this modified construction will appear later.

It is interesting to compare this construction with some others which have appeared in the literature. One knows, following Fujii's example, that no conventional feedback with boundary observation and boundary control can stabilize (2.1) for general $q$ (although this has been shown possible [4] for the higher dimensional case). On the other hand, stabilization is possible using a dynamic compensator which can be taken to be of finite order [1], [6]. The feedback (1.5), (1.6) is, of course, not of finite order (The extended space required by the controller to retain $\psi^t$, $\phi^t$ is infinite dimensional even though the uncontrolled system has only finitely many unstable modes.) but essentially becomes so when modified as indicated above: say, giving (1.5) by a numerical quadrature rule with error $\omega$ so the number of quadrature points (sampled data!) is analogous to the order. In [1], [6] the order increases as the number of unstable modes of (2.9) increases. Here, the appearance of the scheme remains unchanged but an approximation error of the form $|\omega| \leq \varepsilon \|u\|$ would require smaller $\varepsilon$ (i.e., more accuracy, requiring more quadrature points) to achieve stabilization with $\overline{M}$ larger, as would be expected when the number of unstable modes of (2.9) increases.

We note that the presentation here depends essentially on the boundedness of the functional $u \mapsto \langle \kappa, u(t,\cdot)\rangle =: \phi(t)$ in (2.3), giving $\kappa \in X$, and we briefly sketch an argument for this. On any finite interval $[0,T]$ consider the optimal $\phi^\mu$, $u^\mu$ minimizing

$$J_\mu := \int_0^T e^{2ct}[\phi^2 + \|u\|^2]\, dt + \mu|u(t) - u^*(T)|^2$$

($u^*$ from the original problem) and note that $\phi^\mu \to \phi$, $u^\mu \to u^*$ (weak convergence). One has $\phi^\mu = w^\mu(\cdot,1)$ with $w^\mu$ satisfying an adjoint equation. Bounding $w^\mu(t,\cdot)$ for each $t < T - \delta$ using observability dual to the nullcontrollability, one can show $\phi^* = w^*(\cdot,1)$ in the limit. An argument, using the regularity results of [5] along the lines of [9], then shows $\phi^*$ is continuous so one has continuity for $\kappa\colon u(t,\cdot) \mapsto \phi(t)$ as desired.

Finally, we refer to [3] and its references for related material.

## REFERENCES


[1] R. CURTAIN, Finite dimensional compensators for parabolic distributed systems with unbounded control and observation, TW-234, Rijksuniversitat Groningen, 1982.

[2] S. DOLECKI and D. RUSSELL, A general theory of observation and control, SIAM J. Contr. Opt. 15(1977), pp. 185-220.

[3] A. ICHIKAWA, Quadratic control of evolution equations with delays in control, SIAM J. Contr. Opt. 20 (1982), pp. 645-668.

[4] I. LASIECKA and R. TRIGGIANI, Feedback semigroups and cosine operators for boundary feedback parabolic and hyperbolic equations, JDE 47(1983), pp. 246-272.

[5] J.-L. LIONS and E. MAGENES, Non-Homogeneous Boundary Value Problems, v. 2, Springer-Verlag, Berlin, 1972.

[6] J.M. SCHUMACHER, A direct approach to compensator design for distributed parameter systems, SIAM J. Contr. Opt.

[7] T.I. SEIDMAN, Problems of boundary control and observation for diffusion processes, MRR 73-10, UMBC, 1973.

[8] ___________, Observation and prediction for one-dimensional diffusion equations, JMAA 51(1975), pp. 165-175.

[9] ___________, Regularity of optimal boundary controls for parabolic equations, in Analysis and Optimization of Systems (edit. A. Bensoussan and J.-L. Lions; Lecture Notes Cont. and Inf. Sci. #28), Springer Verlag, Berlin, 1980, pp. 536-550.

[10] ___________, Regularity of optimal boundary controls for parabolic equations, I: analyticity, SIAM J. Cont. Opt. 20(1982), pp. 428-453.

[11] ___________, Construction of stabilizing control laws, in preparation.

IMPEDANCE D'UN FOUR A INDUCTION : DEFINITION, THEORIE ET CALCUL

A. BOSSAVIT
EDF, Etudes et Recherches
1, Avenue du Général de Gaulle
92141 CLAMART

Summary. The problem evoked here originates in concerns about optimal control of induction heating devices. There seems to be two separated topics there : analysis of the furnace system proper (leading to the knowledge of how its impedance changes in time) and control of the alimentation system (using such an impedance characteristic as data). This conceptual separation of the problem into two parts has obvious advantages, but depends on a workable definition of the concept of impedance. The aim of this paper is thus to give a variational formulation of the eddy-currents computation problem, from which such a definition can be derived. Next we examine how the computation of the impedance can be simplified in the case of actual induction heating devices, where skin-effect is often present.

## INTRODUCTION

La conduite optimale des dispositifs de chauffage par induction est un problème important en métallurgie. Il soulève des difficultés de tous ordres, qu'il serait naïf de tenter de décrire par une formulation mathématique unique. Par contre, on peut envisager, en vue d'une approche pluridisciplinaire, de séparer ces difficultés. En particulier, l'intervention des mathématiques appliquées semble pouvoir se développer selon deux axes : d'une part l'analyse (par le calcul numérique principalement) du four proprement dit, d'autre part le contrôle du système global constitué par le four plus son alimentation.

Un système de chauffage par induction comporte en effet deux sous-systèmes de natures assez différentes, l'alimentation d'une part, le four lui-même d'autre part. L'alimentation est un circuit électrique complexe, qu'on peut modéliser par un système différentiel non-linéaire. Le four (qui à son tour se décompose en deux parties : l'inducteur et la charge) doit plutôt être vu comme un système à paramètres distribués (les valeurs du champ magnétique en tout point). Le problème consiste à amener la charge, en temps minimal, dans un certain état thermique (caractéristique de l'opération qu'on veut effectuer : réchauffage avant laminage, trempe, recuit, etc.). Il y a bien entendu des contraintes à respecter, tant du côté de la charge (éviter des

(ce qui exclut le cas des matériaux ferro-magnétiques, pour lesquels $\mu$ dépend précisément de h).

Soit maintenant v(t) une fonction du temps donnée, régulière, périodique de période T. On montre sans difficulté qu'il existe une fonction $t \to h(t) \in \mathcal{H}$, régulière, T-périodique, vérifiant l'équation variationnelle suivante (linéaire, d'après les hypothèses faites) :

$$\begin{cases} d/dt \left[\int_{\mathbb{R}^3} \mu\, h.h'\right] + \int_{\Omega \cup \Omega_I} \rho \operatorname{rot} h . \operatorname{rot} h' = v(t)\, F(h') \quad \forall\, h' \in \mathcal{H} \\ \\ \operatorname{div}(\mu h) = 0 \end{cases} \tag{5}$$

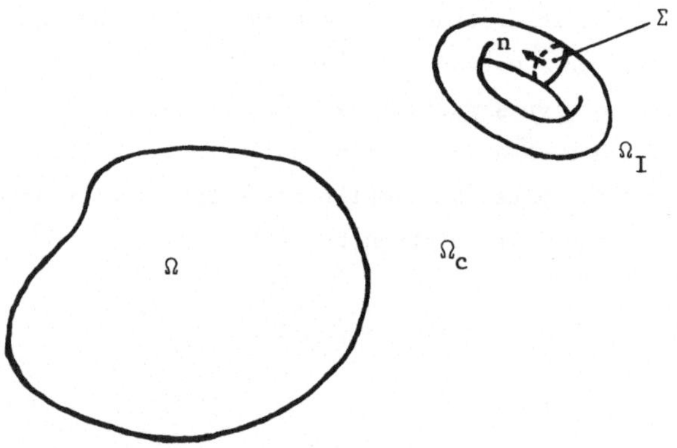


Figure 1

Dans cette équation, v(t) doit s'interpréter comme une force électromotrice présente dans $\Omega_I$. Alors, d'après la formule de Green, h vérifie

$$\frac{\partial h}{\partial t} + \operatorname{rot}(\rho \operatorname{rot} h) = 0 \tag{6}$$

dans $\Omega \cup \Omega_I$, ce qui est la loi de Faraday. Le vecteur $e = \rho \operatorname{rot} h$ est le champ électrique. La quantité

$$j(t) = F(h(t))$$

1. LES EQUATIONS DU FOUR A INDUCTION

Soit $\Omega$ un domaine de $\mathbb{R}^3$, de frontière $\Gamma$, représentant la pièce à chauffer, et soit $\Omega_c$ la région complémentaire, $\Omega_c = \mathbb{R}^3 - \bar{\Omega}$. Un sous-domaine $\Omega_I$, disjoint de $\Omega$, représentera l'inducteur (Fig. 1). Pour simplifier (mais ce n'est pas essentiel), on supposera $\Omega$ simplement connexe et $\Omega_I$ homéomorphe à un tore plein (circuit "à une seule boucle", selon la terminologie de /1, 2/). En réalité, ce circuit est ouvert et branché sur l'alimentation, avec à ses bornes une différence de potentiel v(t). Il est équivalent, du point de vue qui nous occupe, de le supposer fermé et soumis à une f.é.m. v(t), selon la théorie développée en /1, 2/, qu'on va rappeler.

Soit $\mathcal{H}$ l'espace vectoriel réel

$$\mathcal{H} = \{h \in \mathbb{L}^2(\mathbb{R}^3) \mid \text{rot } h \in \mathbb{L}^2(\mathbb{R}^3), \text{rot } h = 0 \text{ dans } \Omega_c - \bar{\Omega}_I\} \tag{1}$$

muni du produit scalaire

$$(h, h') = \int_{\mathbb{R}^3} h.h' + \int_{\Omega \cup \Omega_I} \text{rot } h .\text{rot } h' \quad . \tag{2}$$

Puisque $\Omega_I$ est l'image continue d'un tore, il existe une surface $\Sigma$ dans $\Omega_I$, dont le bord s'appuie sur $\Gamma_I$, telle que $\Omega_I - \Sigma$ soit simplement connexe. (Par exemple, si le tore est engendré par la rotation autour de l'axe des z du disque $D = \{(x, y) \mid (x - r)^2 + y^2 \leq a^2\}$, prendre pour $\Sigma$ l'image de D.) Orientons $\Sigma$ par un champ de normales n, et considérons la fonctionnelle sur $\mathcal{H}$

$$F(h) = \int_\Sigma \text{rot } h. \, n \quad . \tag{3}$$

Comme div(rot h) = 0, n.rot h est dans $H^{-1/2}(\Sigma)$, et donc F est continue sur $\mathcal{H}$.

Nous interprétons les éléments de $\mathcal{H}$ comme des champs magnétiques. Donc rot h sera (théorème d'Ampère) la densité du courant électrique, présent dans $\Omega \cup \Omega_I$, qui engendre ce champ. (Noter que la composante normale de rot h est bien nulle sur $\Gamma$ et $\Gamma_I$.) Quant à F(h), c'est l'<u>intensité</u> du courant parcourant l'inducteur.

On se donne deux fonctions de x, la perméabilité $\mu \in L^\infty(\mathbb{R}^3)$ , $\mu(x) \geq \mu_0 > 0$ p.p. et $\mu = \mu_0$ dans $\Omega_c$, et la résistivité $\rho \in L^\infty(\Omega \cup \Omega_I)$, $\rho(x) \geq \rho_0 > 0$ p.p. L'une et l'autre peuvent dépendre de facteurs externes tels que la température, mais pas de h

gradients de température trop importants, ou ne chauffer que certaines parties) que de l'alimentation.

Quant aux paramètres de commande, ils sont à l'entrée du système d'alimentation et correspondent très concrètement aux manettes dont l'agent de conduite dispose.

Or l'interface entre ces deux sous-systèmes est assez étroite, dans la mesure où les phénomènes électriques dans le circuit d'alimentation et dans le four s'influencent peu les uns les autres à distance. Elle se caractérise par deux paramètres seulement, l'intensité et la tension appliquée aux bornes du four (dans le cas d'une alimentation monophasée). La relation entre ces deux grandeurs est donnée par l'impédance du four. Celui-ci se comporte comme un transformateur, dont la charge est le secondaire. Perméabilité et résistivité de la charge varient beaucoup avec son état thermique, de sorte que l'impédance change au cours de la montée en température. Du point de vue du contrôle du four dans son ensemble, il suffirait de connaître cette "impédance du four, vue des bornes", comme disent les techniciens, pour être à même de résoudre le problème par les méthodes de l'automatique.

Nous nous intéressons donc au problème du calcul de cette impédance. Cette notion est familière en Electricité, mais dans le cadre de l'analyse des circuits. A notre connaissance, et si étonnant que cela puisse paraître, aucun traité classique ne donne de l'impédance une définition utilisable dans le cas de systèmes à un nombre infini de degrés de liberté.

L'objet de la première partie de cet article est de proposer une telle définition. Plus précisément, on introduira une formulation variationnelle d'un certain problème aux limites, qui s'avère correspondre aux équations classiques des courants de Foucault (théorème d'Ampère et loi de Faraday), et on définira l'impédance comme une intégrale de la solution. La solution peut elle-même être approchée par éléments finis, selon la méthode que nous avons exposée dans /1, 2, 3/.

La deuxième partie examinera quelques cas où ce calcul peut être réduit par une modélisation appropriée : fours à traitement de surface (trempe, etc.), où l'effet de peau est prononcé. Par des techniques de perturbation, le calcul de l'impédance se découple en un problème unidimensionnel simple d'une part, un problème de type intégral sur la surface des conducteurs d'autre part.

En conclusion, on aborde le problème de l'extension de la notion d'impédance au cas d'une caractéristique b-h non linéaire.

est l'intensité parcourant l'inducteur. Faisant h' = h dans (5), on obtient un bilan énergétique :

$$\frac{d}{dt}\left[\frac{1}{2}\int_{\mathbb{R}^3}\mu|h|^2\right] + \int_{\Omega\cup\Omega_I}\rho|\mathrm{rot}\ h|^2 = v(t)\ j(t) \tag{7}$$

$(\frac{d}{dt}$ *énergie du champ + pertes Joule = puissance apportée*)

L'équation (5), ou "équation de l'électricité", a donc bien pour solution le champ magnétique h qui s'établit lorsqu'on alimente l'inducteur avec la différence de potentiel v(t).

Remarque. On peut objecter que la loi de Faraday (6) n'est vérifiée, comme conséquence de (5), que dans les conducteurs, et non dans tout l'espace. Effectivement, la relation rot e = $-\ \mu\ \partial h/\partial t$ en dehors des conducteurs n'est pas conséquence de (5). Tout au contraire, elle doit être utilisée pour déterminer e. Ce point est traité en détail dans /4/.

## 2. NOTION D'IMPEDANCE

Introduisons maintenant une simplification essentielle, due à ce que la répartition des courants est connue (à une constante multiplicative près, qui est j(t)), dans $\Omega_I$. C'est légitime, car les inducteurs sont conçus, comme tous les enroulements conducteurs de machines électriques, de manière à minimiser les pertes Joule, en particulier les pertes supplémentaires que provoquerait l'effet de peau s'il était sensible. C'est pourquoi le diamètre des conducteurs est petit devant la profondeur de pénétration, de sorte que la répartition du courant y soit à peu près la même qu'en continu, du moins aux basses fréquences que nous considérons ici (50 Hz à quelques dizaines de k Hz).

On peut donc admettre que la densité de courant est $j(t)\ j_I(x)$ dans $\Omega_I$, où $j_I$ est la densité qui s'établirait en continu dans l'inducteur pour une intensité globale égale à 1. Appelons $h_I$ un champ, élément de $\mathcal{H}$, vérifiant la relation

$$\mathrm{rot}\ h_I = j_I \quad \text{dans } \Omega_I\ , \tag{8}$$

et à cela près quelconque. (Par exemple, $h_I$ peut être le champ qui existerait en l'absence de tout autre conducteur que $\Omega_I$ lui-même.)

Cherchons alors le champ physique sous la forme $k + j(t)\ h_I$. Alors rot k = 0 dans $\Omega_I$. Donc on cherche k dans le sous-espace suivant, fermé dans $\mathcal{H}$ :

$$\mathcal{K} = \{k \in \mathbb{L}^2(\mathbb{R}^3) \mid \text{rot } k \in \mathbb{L}^2(\mathbb{R}^3),\ \text{rot } k = 0 \text{ dans } \Omega_c\}. \tag{9}$$

Par restriction de (5) aux seuls h' appartenant à $\mathcal{K}$, on obtient la formulation variationnelle réduite :

$$\frac{d}{dt}\int_{\mathbb{R}^3} \mu(k + j(t)\ h_I).k' + \int_{\Omega} \rho \text{ rot } k.\ \text{rot } k' = 0 \qquad \forall\ k' \in \mathcal{K}, \tag{10}$$

$$k \text{ T-périodique}, \quad \text{div}(\mu(k + j(t)\ h_I)) = 0.$$

L'équation (10) détermine k si j(t) est connu. Pour trouver une équation liant j et v, faisons maintenant $h' = h_I$ dans (5). Il vient :

$$\frac{d}{dt}\int_{\mathbb{R}^3} \mu(k + j(t)\ h_I).\ h_I + j(t)\int_{\Omega_I} \rho\ |\text{rot } h_I|^2 = v(t). \tag{11}$$

On a donc deux équations couplées. L'avantage de (10) (11) par rapport à (5) est que le calcul du champ de réaction k n'a plus à être fait que dans la charge $\Omega$ et non dans l'inducteur $\Omega_I$.

Il suffit donc de tirer k de (10) et de le porter dans (11) pour obtenir la relation entre j et v qui est notre objectif. Comme (10) et (11) sont linéaires, ce programme peut être mené à bien séparément sur les différents harmoniques de j(t) et v(t), de fréquences multiples de 1/T. Donc nous allons supposer désormais que v(t) est de la forme

$$v(t) = \text{Re}\ [V \exp(i\,\omega\, t)] \tag{12}$$

avec $\omega = 2\pi/T$. On cherchera de même j et k sous la forme

$$j(t) = \text{Re}\ [J \exp(i\,\omega\, t)], \qquad k(t) = \text{Re}[J\ K \exp(i\,\omega\, t)] \tag{13}$$

(noter la "mise à l'échelle" constituée par le facteur J). Quant à $h_I$, bien que ce soit un champ réel, nous écrirons $h_I = H_I$, en n'excluant pas que $H_I$ puisse être complexe, pour faciliter la généralisation de ce qui suit à des inducteurs polyphasés. Il vient :

$$\frac{dk}{dt} = \text{Re}\ [i\omega\ J\,K\ e^{i\,\omega\, t}],$$

donc, portant dans (10),

$$\mathrm{Re}[e^{i\omega t}(\int_{\mathbb{R}^3} i\,\omega\mu(K+H_I).k' + \int_\Omega \rho \,\mathrm{rot}\, K\,.\,\mathrm{rot}\, k')] = 0 \quad \forall\, k' \in \mathcal{K}, \tag{14}$$

et pour tout t, donc en particulier si exp(i ωt) = 1 ou i, donc pour tous les K' du complexifié $\mathbb{K}$ de $\mathcal{K}$, d'où finalement

$$i\,\omega \int_{\mathbb{R}^3} \mu\,(K+H_I).K' + \int_\Omega \rho\,\mathrm{rot}\,K\,.\mathrm{rot}\,K' = 0 \quad \forall\, K' \in \mathbb{K}. \tag{15}$$

De même, (11) devient

$$[i\,\omega \int_{\mathbb{R}^3} \mu\,(K+H_I).\overline{H_I} + \int_{\Omega_I} \rho|\mathrm{rot}\,H_I|^2]\,J = V. \tag{16}$$

Faisons $K' = J\bar{K}$ dans (15) et ajoutons à (16). Il vient

$$[i\,\omega \int_{\mathbb{R}^3} \mu|(K+H_I)|^2 + \int_{\Omega_I} \rho|\mathrm{rot}\,H_I|^2 + \int_\Omega \rho|\mathrm{rot}\,K|^2]\,J = V,$$

soit V = J Z, où Z est l'impédance qu'on s'était proposé de définir.

Comme l'expression complexe du champ physique est $H = K + H_I$, que rot K est nul dans $\Omega_I$ et rot $H_I$ dans $\Omega$, on a au bout du compte

$$Z = i\,\omega \int_{\mathbb{R}^3}\mu\,|H|^2 + \int_{\Omega\cup\Omega_I} \rho|\mathrm{rot}\,H|^2, \tag{17}$$

soit $Z = i\omega L + R$, où l'interprétation de L et R est fort simple. En effet, l'énergie instantanée d'un champ h = Re [H exp(iω t)] est

$$E(h) = \frac{1}{2}\int_{\mathbb{R}^3}\mu|h(t)|^2$$

donc son énergie moyenne est

$$\frac{1}{2T}\int_0^T dt \int_{\mathbb{R}^3} \mu\,|h(t)|^2 = \frac{1}{4}\int_{\mathbb{R}^3}\mu\,|H|^2$$

L'inductance L est donc égale à quatre fois l'énergie moyenne du champ qui s'établit pour une intensité "de crête" 1. De même, R est le double de la perte Joule moyenne. Il peut être utile de remarquer que j et v obéissent à l'équation

$$L \frac{dj}{dt} + R\, j(t) = v(t). \tag{18}$$

On a donc bien trouvé une définition de l'impédance du four vue des bornes, et une formule pour la calculer : il suffit de résoudre l'équation (15), après avoir calculé $H_I$. On obtient $H_I$ explicitement par la loi de Biot et Savart, connaissant rot $H_I$ :

$$H_I(x) = \mathrm{rot}\left(\frac{1}{4\pi}\int_{\Omega_I} \frac{\mathrm{rot}\, H_I(y)}{|x - y|}\, dy\right)$$

La méthode de résolution de (15) est exposée dans /2, 3/. Un code de calcul, nommé "Trifou", a été développé à cet effet à la Direction des Etudes et Recherches d'Electricité de France /5/.

## 3. CAS OU LA PROFONDEUR DE PENETRATION EST FAIBLE

Puisque l'impédance, d'après (17), se calcule comme une intégrale étendue à tout l'espace, on peut calculer séparément, puis sommer, les impédances relatives à des sous-régions. C'est l'avantage essentiel de (17).

Supposons pour simplifier $\rho$ et $\mu$ constants dans $\Omega$. Soit $l$ une unité de longueur caractéristique de $\Omega$ (épaisseur, diamètre, ...) et supposons que la longueur $\delta$, définie par

$$\delta = \sqrt{2\rho/\omega\mu} \tag{19}$$

soit <u>très petite devant</u> $l$. Soit $\varepsilon$ le "petit paramètre" $(\delta/l)^2$.

Nous considérons maintenant (15) comme un problème parmi toute une famille, dépendant des paramètres $\rho$, T et $\mu$, et nous cherchons comment se comporte la solution (15) lorsque $\rho$, T et $\mu$ varient de manière à ce que $\delta$ ci-dessus (ou $\varepsilon$) tende vers zéro. Ce qui se passe est familier aux électriciens ("effet de peau" ou "effet Kelvin") : le champ H décroît très rapidement dans $\Omega$ en fonction de la distance à $\Gamma$. Du point de vue mathématique, c'est une situation du type "perturbations singulières" où (en un mot) on ne <u>peut pas</u> faire tout simplement $\varepsilon = 0$ dans les équations.

La méthode correcte, qu'on va décrire sans justifications (elles suivraient de près /6/, Chap. 1), consiste à substituer à (15) une équation analogue, mais où $\mathbb{K}$ est remplacé par un nouvel espace $\mathbb{K}_\varepsilon$ (formé, pour fixer les idées, des seuls champs de $\mathbb{K}$ qui "présentent l'effet de peau"). De même, alors que le champ physique $K + H_I$ de (15) était dans le complexifié $\mathbb{H}$ de $\mathcal{H}$, il sera maintenant dans $\mathbb{H}_\varepsilon$, dont les élé- sont construits comme suit.

Soit H un élément de $\mathbb{H}$ et $H_\Gamma$ sa trace tangentielle sur $\Gamma$. A partir de $H_\Gamma$, on prolonge H vers l'intérieur de $\Omega$ (d'où $H_\varepsilon$), à l'aide du "correcteur" (cf. /6/) ainsi défini. Soit $d > 0$ fixé une longueur choisie petite devant le plus petit rayon de courbure de $\Gamma$ et grande par rapport à $\delta$ (ce qui suppose évidemment $\Gamma$ régulière et $\delta$ assez petit). Soit $\Gamma_\eta = \{x \in \Omega |\ d(x,\Gamma) = \eta\delta\}$, où $d(x,\Gamma)$ est la distance de x à $\Gamma$ et $\mathcal{O} = \{x|\ d(x,\Gamma) < \delta\}$. A tout point x de $\Omega - \mathcal{O}$ on fait correspondre le pied de la normale qui passe par x, soit $\xi \in \Gamma$, et la profondeur relative $\eta$. A tout $\xi$ de $\Gamma$ correspond une fonction $a_\xi(\eta)$ dépendant des rayons de courbure de $\Gamma$ en $\xi$, qui mesure le rapport des éléments de surface homologues (cf. Fig 2) en x et $\xi$ :

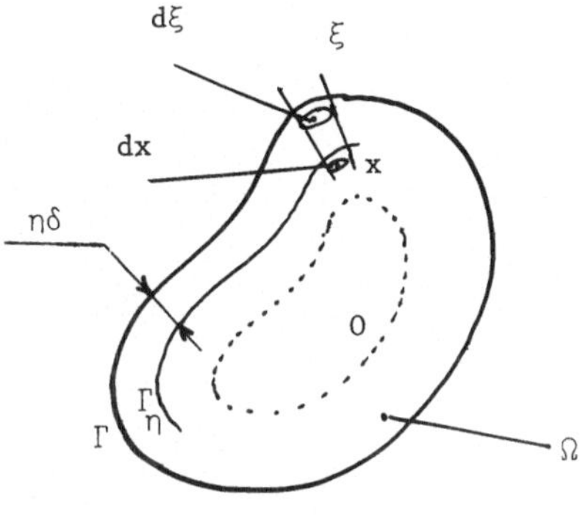


Figure 2
$dx = a_\xi(\eta) d\xi$

$$a_\xi(\eta) = 1 - \left(\frac{1}{R_1(\xi)} + \frac{1}{R_2(\xi)}\right)\eta\delta \tag{20}$$

Pour prolonger H, on procède ainsi :

1) Résoudre l'équation unidimensionnelle

$$2i\ a_\xi(\eta)h_\xi - \frac{\partial}{\partial\eta}\left(a_\xi(\eta)\frac{\partial}{\partial\eta}h_\xi\right) = 0,\ 0 < \eta < \frac{d}{\delta}, \tag{21}$$

$$h_\xi(0) = 1, \qquad h_\xi\left(\frac{d}{\delta}\right) = 0\ , \tag{22}$$

($\xi$ est un paramètre), avec h fonction complexe de $\eta$.

2) Poser $H_\varepsilon(x) = h_\xi(\eta)H_\Gamma(\xi)$ pour $x \in \Omega - \mathcal{O}$ et 0 pour $x \in \mathcal{O}$.

Remarque. On obtient un autre correcteur, plus simple (et dont on voit mieux la signification physique), en substituant h, solution de

$$2i\ h - \frac{\partial^2}{\partial\eta^2}h = 0\ , \quad 0 < \eta < \infty\ , \quad h(0) = 1,\ h \in L^2(0, \infty), \tag{23}$$

à $h_\xi$. Ces deux correcteurs sont asymptotiquement équivalents, mais le premier est plus précis pour $\varepsilon$ petit mais non nul.

Dans (15), substituons $H_\varepsilon$ à $K + H_I$ et restreignons les $K'$ à $\mathbb{K}_\varepsilon$. Compte tenu de l'expression des correcteurs dans $\Omega$, il vient

$$i\omega \int_{\Omega_c} \mu_o H_\varepsilon . K' + \int_\Gamma (H_\varepsilon)_\Gamma . K'_\Gamma [\int_0^{d/\delta} d\eta \, a_\xi(\eta)(i\omega\mu\delta |h_\xi|^2 + \ldots$$

$$\ldots + \frac{\rho}{\delta} |\frac{\partial h_\xi}{\partial \eta}|^2 )] = 0 \qquad \forall K' \in \mathbb{K}_\varepsilon$$

donc, en intégrant par parties et grâce à (21) (22) :

$$i\omega \int_{\Omega_c} \mu_o H_\varepsilon . K' - \int_\Gamma (H_\varepsilon)_\Gamma . K'_\Gamma \frac{\rho}{\delta} \frac{\partial h_\xi}{\partial \eta} (0) = 0 \qquad \forall K' \in \mathbb{K}_\varepsilon . \tag{25}$$

On voit que la quantité $\partial h_\xi / \partial \eta$, pour $\eta = 0$, ne dépend que de la <u>forme</u> du domaine $\Omega$, et du paramètre $\delta$. Donc seules les valeurs du champ dans $\Omega_c$ et sur $\Gamma$ interviennent dans (25), ce qui va simplifier beaucoup le calcul de l'impédance pour $\delta$ petit. Posons

$$r(\xi) = - \frac{\rho}{\delta} \frac{\partial h_\xi}{\partial \eta} (0) . \tag{26}$$

Nous appellerons <u>impédance surfacique</u> cette quantité.

<u>Remarque</u>. On a une bonne approximation de $r(\xi)$ à partir de (23), pour laquelle le calcul se fait "à la main" et donne :

$$r(\xi) \simeq r \equiv (1 + i) \frac{\rho}{\delta} \tag{27}$$

(qui a bien la dimension d'une impédance surfacique).

Par sa construction même, $\mathbb{K}_\varepsilon$ est isomorphe à l'espace des restrictions à $\overline{\Omega}_c$ des $K$ de $\mathbb{K}$. Or, d'après (9), il existe un potentiel $\Phi$ tel que $K = \text{grad } \Phi$ dans $\Omega_c$. Donc $K_\varepsilon \simeq \text{grad}[BL(\Omega_c)]$, où $BL(\Omega_c)$ est l'espace de Beppo-Lévi /7/, complété de $\mathcal{D}(\overline{\Omega}_c)$ pour la norme de Dirichlet :

$$BL(\Omega_c) = \{\text{grad } \Phi \mid \Phi \in \mathcal{D}(\overline{\Omega}_c), \int_{\Omega_c} |\text{grad}^2 \Phi| < \infty\}^\wedge . \tag{28}$$

(L'isomorphisme, bien entendu, dépend de $\delta$). On peut maintenant substituer à (15) l'équation variationnelle suivante :

$$i\omega \int_{\Omega_c} \mu_o (H_I + \text{grad } \Phi) . \text{grad } \Phi' + \int_\Gamma r (H_I + \text{grad } \Phi)_\Gamma . \text{grad}_\Gamma \Phi' = 0 \quad \forall \Phi' \in BL(\Omega_c) \tag{29}$$

L'impédance, par un calcul analogue à celui du point 2, est maintenant, au lieu de (17),

$$Z = i\omega \int_{\Omega_c} \mu_o |H|^2 + r \int_\Gamma |\mathrm{grad}_\Gamma \psi|^2 + \int_{\Omega_I} \rho |\mathrm{rot}\ H|^2 \tag{30}$$

(où cette fois $\psi$ est le potentiel du champ H). L'interprétation énergétique est la même.

Le calcul de $\phi$, solution de (29), est évidemment beaucoup plus simple que la résolution de (15) ! Il reste à voir dans quels cas cette simplification est possible. La résistivité $\rho$ est de l'ordre de $10^{-6}$. Pour une pièce à chauffer dont les dimensions sont de l'ordre du centimètre (cas des traitements de surface de pièces mécaniques) et pour $\mu = \mu_o$, on a à peu près $\delta = 1/2\sqrt{f}$, où f est la fréquence. Pour 100 kHz, par exemple, l'approximation est tout à fait acceptable. Si $\mu >> \mu_o$, elle l'est à des fréquences beaucoup plus basses.encore.

## 4. ETUDE DE L'EQUATION (29)

Comme on a supposé $\Omega$ simplement connexe, la composante tangentielle de $H_I$ sur $\Gamma$ est le gradient d'une certaine fonction sur $\Gamma$, notée ci-dessous $\phi_I$.

On peut voir (29) comme un problème de Poisson (par rapport à $\Phi$) dans $\Omega_c$ avec un terme de surface additionnel. Comme $\Omega_c$ est non-borné, on traitera ce problème par une méthode intégrale. Soit R l'opérateur de $\mathcal{L}(H^{1/2}(\Gamma)\ ;\ H^{-1/2}(\Gamma))$ ainsi défini : Etant donné $\phi \in H^{1/2}(\Gamma)$, résoudre

$$\Delta\Phi = 0 \quad \text{dans } \Omega_c, \quad \Phi = \phi \quad \text{sur } \Gamma, \tag{31}$$

et poser alors

$$R\phi = \partial\Phi/\partial n \tag{32}$$

où n est la normale vers l'intérieur de $\Omega$. Par la formule de Green dans $\Omega_c$, on voit que (29) équivaut à

$$i\omega\mu_0 \int_\Gamma (R\phi + n.H_I)\,\phi' + \int_\Gamma r\ \mathrm{grad}(\phi + \phi_I).\mathrm{grad}\,\phi' = 0 \quad \forall\ \phi' \in H^{1/2}(\Gamma) \tag{33}$$

Il est <u>faux</u> que $n.H_I$ soit égal à $R\phi_I$. En effet $H_I$ n'est pas un gradient dans <u>tout</u> $\Omega_c$. Par hypothèse, on sait calculer la différence entre les deux, soit $g_I$. Rebaptisons $\phi$

la somme $\phi + \phi_I$, de sorte que $\phi$ est maintenant le potentiel dont dérive le champ tangentiel total sur $\Gamma$. Alors (33) se simplifie un peu :

$$i\omega\mu_0 \int_\Gamma (R\phi + g_I)\phi' + \int_\Gamma r \text{ grad } \phi . \text{grad } \phi' = 0 \qquad \forall\ \phi' \in H^{1/2}(\Gamma) \qquad (34)$$

ce qu'on peut écrire aussi, sous forme "forte" :

$$i\omega\mu_0 R\phi - \text{div}(r \text{ grad } \phi) = - i\omega\mu_0 g_I \quad . \qquad (35)$$

Deux opérateurs très différents interviennent dans (35), R et l'opérateur de Laplace-Beltrami sur $\Gamma$. Ce dernier se discrétise simplement, et il lui correspond une matrice creuse. Par contre, R (qui s'approche par une méthode de potentiel, cf. /3/), donne une matrice pleine.

Assez souvent, inducteur et induit sont enfermés dans une boîte aux parois ferro-magnétiques, pour éviter la dispersion du champ. Cette boîte peut être considérée comme une deuxième composante connexe de $\Omega$. Le champ étant nul à l'extérieur, on peut redéfinir $\Omega_c$ comme son intérieur. Alors $BL(\Omega_c)$ est simplement $H^1(\Omega_c)$. Il se pourra (selon la forme des pièces) que le problème (29) soit plus facile à discrétiser (par éléments finis tridimensionnels dans $\Omega_c$ et bidimensionnels sur $\Gamma$) que (34)(35) en pareil cas.

Examinons maintenant l'importance relative des deux opérateurs dans (35). On prend pour r l'expression (27). Posons

$$\delta_0 = (2\rho/\omega\mu_0)^{1/2} \ ,$$

quantité qui a la dimension d'une longueur. Soit $\ell$ une dimension caractéristique de $\Gamma$, que l'on prend comme nouvelle unité de longueur. Comme $\ell$ vient en dénominateur une fois dans le premier terme de (35) et deux fois dans le second, on a, avec la nouvelle unité, et en normalisant le second membre :

$$R\phi + \frac{1+i}{2} \frac{\delta_0^2}{\delta\ell} \Delta\phi = f_I \quad . \qquad (36)$$

L'analyse qui nous a conduits à (29), donc à (36), supposait $\delta$ petit devant $\ell$. Donc si $\mu$ et $\mu_0$, et par suite $\delta$ et $\delta_0$, sont du même ordre de grandeur, c'est l'opérateur R qui domine dans (36), $\Delta$ jouant le rôle d'une perturbation.

On comprendra mieux la signification physique de cette perturbation en revenant à (29), où le premier terme est donc alors dominant. Si on néglige le terme de perturbation (l'intégrale sur $\Gamma$), on voit que $n.(H_I + \text{grad }\phi)$, c'est-à-dire la composante normale du champ magnétique sur $\Gamma$, est nul. La composante tangentielle ne l'est pas.

Le champ étant nul à une faible profondeur, il y a une variation rapide de la composante tangentielle du champ, assimilable à un <u>saut</u> de celle-ci au passage à travers $\Gamma$. Il y a donc une densité de courant surfacique sur $\Gamma$ (qui est $\mathrm{rot}_\Gamma \phi$, $\phi$ solution de (36)).

Quand au calcul de l'impédance, on peut donc négliger dans ce cas le premier des trois termes de (30).

Dans le cas maintenant où $\mu$ est très grand devant $\mu_0$ (fer doux), il peut arriver que $\delta \ell$ soit petit devant $\delta_0^2$. C'est alors R qui joue le rôle d'une perturbation, et (36) se réduit, à l'ordre zéro, à $\Delta\phi = 0$. Donc la composante tangentielle du champ magnétique est nulle (le champ est normal à $\Gamma$). Il ne reste plus qu'à calculer le champ dans l'entrefer, ce qui se fait toujours en ne conservant de (29) que le premier terme, mais cette fois-ci avec une condition de Dirichlet non-homogène sur $\Phi$. L'impédance est toujours donnée par (30), mais en négligeant cette fois le deuxième terme.

Si nécessaire, on peut sans difficulté pousser plus loin le calcul par perturbations.

## CONCLUSION

Avec (30), nous avons pu pettre en évidence trois contributions à l'impédance, celle de l'entrefer, celle de la couche de peau, celle de l'inducteur. On a aussi relevé l'existence de deux cas extrêmes, où l'influence de la couche de peau est soit dominante, soit négligeable.

Il reste à savoir ce qui peut subsister d'une telle analyse dans le cas où la caractéristique magnétique du matériau de $\Omega$ est non-linéaire. La formulation donnée dans la première partie reste valable : il suffit de remplacer dans (5) $\mu h$ par $b(h)$, avec b non-linéaire en h. En toute rigueur, "impédance" n'a plus de sens dans ce cas. Il semble pourtant exister une généralisation de ce que nous avons fait, comme suit.

Soit v la f.é.m., fonction T-périodique du temps, et j l'intensité aux bornes. Définissons (classiquement) la <u>puissance active</u> par

$$P_a = \frac{1}{T}\int_0^T v\, j \tag{37}$$

et la <u>puissance réactive</u> par

$$P_r = \frac{1}{T}\int_0^T v\, \frac{dj}{dt} \quad . \tag{38}$$

En linéaire, on vérifie que $P_a = R\, J^2$ (où J est la moyenne quadratique de j) et que

$P_r = \omega L J^2$. Nous proposons donc de prendre comme définition de R et L les quantités issues de ces formules, et d'assimiler le four à un circuit dont l'impédance serait $i\omega L + R$. La raison de (38) est que, par un calcul simple,

$$P_r = \frac{1}{T}\int_0^T \int_{\mathbb{R}^3} \frac{d}{dt} b(h) \frac{dh}{dt} , \qquad P_a = \frac{1}{T}\int_0^T \int_\Omega |\operatorname{rot} h|^2 \tag{39}$$

d'où une expression intégrale pour la pseudo-impédance, que l'on peut donc calculer par par sous-domaines. Tout ce que nous avons fait (y compris le calcul du correcteur, solution d'une équation semblable à (23), mais avec un terme d'ordre zéro non-linéaire) peut être repris dans ce contexte.

REFERENCES

/1/ A. Bossavit : "Finite Elements for the Electricity Equation", in The Mathematics of Finite Elements and Applications IV, (J.R. Whiteman, ed.), Academic Press (London), 1982, pp. 85-92.

/2/ A. Bossavit, J.C. Vérité : "A Mixed FEM-BIEM Method to Solve 3-D Eddy-Current Problems", IEEE Trans. on Magnetism, MAG-18, 2 (1982), pp 431-35.

/3/ A. Bossavit, J.C. Vérité : "The "Trifou" code : Solving the 3-D Eddy-Currents Problem by Using H as State Variable", IEEE Trans. on Magnetism, MAG-19, 6 (1984), pp. 2465-70.

/4/ A. Bossavit : "Eddy-currents in a System of Moving Conductors", in The Mechanical Behavior of Electromagnetic Solid Continua (G.A. Maugin, ed.), North-Holland (Amsterdam), 1984, pp. 345-50.

/5/ J.C. Vérité : "'Trifou' : Un code de calcul tridimensionnel des courants de Foucault", Bull. DER-EDF, Série C, 2 (1983), pp. 79-92.

/6/ J.L. Lions : Perturbations Singulières dans les Problèmes aux Limites et en Contrôle Optimal, Springer-Verlag (Berlin), 1973.

# OPTIMAL ACTUATOR LOCATION IN A DIFFUSION PROCESS

A. EL JAI and A. NAJEM
Département de Mathématiques
Faculté des Sciences
BP 1014. Rabat (MAROC)

Abstract. In this paper we present an original way for solving the problem of optimal location of a zone actuator for a class of distributed parameter systems. The control problem for which this optimization is done is a minimum energy final value one. Semi-group theory is used to solve the control problem and the optimal location is based on the use of optimum design techniques. A significant application is given and is illustrated in an example

## INTRODUCTION. STATEMENT OF THE PROBLEM

The problem of optimal location of controllers has been studied for a long time and the usual approach is based on the numerical optimization of criteria derived from the underlying control problem. The development of optimum design techniques, as shape derivative, leads to a more rigorous methodology for solving all the problems where the unknown is a domain.

The minimum energy final value control problem is considered for parabolic systems. The control is applied on a set $\Omega \subset R^2$. The aim of this paper is to show how one can find numerically an optimal location $\Omega^*$ by the use of some modern mathematical tools. Among the numerous applications, all those of finding the optimal geometry for an actuator in a diffusion process. In the second paragraph, the control problem is solved by semi group approach. The purpose of the following paragraphs is to give techniques of shape optimization which are applied to the considered problem. An example is developed in the case where $\Omega$ is a disc.

We consider the parabolic evolution equation

$$\begin{cases} \dfrac{dy}{dt}(x,t) = My(x,t) + \chi_\Omega(x)\; g(x)\; u(t) & (x,t)\in \mathcal{D}\times]0,T[ \\ y(x,0) = y^0(x) \\ L(y) = 0 \quad \text{in } \partial\mathcal{D} \text{ the boundary of } \mathcal{D} \end{cases} \qquad (I.1)$$

where $\mathcal{D} \subset R^2$ is an open bounded set. The control $u \in L^2(0,T)$, $\Omega \subset \mathcal{D}$ is the part of $\mathcal{D}$ in which the control is applied and g defines the

spatial distribution of the control in $\Omega$.
We suppose that M is generator of a strongly continuous semi group $(S(t))_{t\geq 0}$ with $\mathcal{D}(M) = \{y \in L^2(0,T;L^2(\mathcal{D}))$ / $y$ and $\frac{dy}{dt} \in L^2(0,T;L^2(\mathcal{D}))$ with $Ly = 0$ in $\partial\mathcal{D}\}$

The solution of (I.1) is then given by:

$$y(x,t) = S(t)\, y^0(x) + \int_0^t S(t-\tau)\chi_\Omega(x)g(x)u(\tau)d\tau \qquad \text{(I.2)}$$

Let $y^d$ a desired state in $L^2(\mathcal{D})$, the minimum energy final value problem can be formulated as:

$$\begin{cases} \min_u h_\Omega(u) = \min_u ||u||^2_{L^2(0,T)} \\ \text{with } y(T) = y^d \end{cases} \qquad (P_0)$$

We shall see later that $(P_0)$ has a unique solution $u^*$ which can be given in an explicit form; then $h_\Omega(u^*) \leq h_\Omega(u)$ $\forall u \in L^2(0,T)$.
The purpose of this paper is to develop a numerical way to solve the problem

$$\min_{\Omega \subset \mathcal{D}} \left( h_\Omega(u^*).||g||^2_{L^2(\Omega)} \right) \qquad (P)$$

by the use of optimum design techniques [4,5,10,11,12] . First we give the solution of $(P_0)$ for fixed $\Omega$ and then we solve (P).

## SOLUTION OF $(P_0)$

Let us suppose that M, with the boundary conditions L have an orthonormal set of eigenfunctions $(\phi_i)_{i=1}^\infty$ . Then (I.1) can be represented in $L^2(0,T;\ell_2)$ by:

$$\begin{cases} \frac{da(t)}{dt} = A\, a(t) + B(\Omega)\, u(t) \\ a(0) = a^0 \end{cases} \qquad \text{(II.1)}$$

Where $a(t) = (a_1(t), a_2(t), \dots)^T$ with $a_i(t) = \langle y(t),\psi_i\rangle_{L^2(\mathcal{D})}$
$A = \text{diag}(\lambda_1,\lambda_2, \dots)$, $\lambda_i$ is the eigenvalue corresponding to $\phi_i$
and $B(\Omega) = (b_1(\Omega),b_2(\Omega), \dots)$ with $b_i(\Omega) = \langle\chi_\Omega g,\psi_i\rangle_{L^2(\mathcal{D})}$
$a^0 = (a_1^0,a_2^0, \dots)^T$ with $a_i^0 = \langle y^0,\psi_i\rangle_{L^2(\mathcal{D})}$
$(\psi_i)_{i=1}^\infty$ is the set of eigenfunctions of $M^*$, the adjoint of M.
The solution of (II.1) is

$$a(t) = \Phi(t)\, a^0 + \int_0^t \Phi(t-\tau)B(\Omega)u(\tau)d\tau \qquad \text{(II.2)}$$

with $\Phi(s) = (\Phi_{ij}(s))_{i,j} = (\langle S(s)\phi_i,\psi_j\rangle_{L^2(\mathcal{D})})_{i,j}$
and $(P_0)$ becomes:

$$\begin{cases} \min_u ||u||^2_{L^2(0,T)} \\ \text{with } a(T) = a^d \end{cases} \qquad (P_0')$$

with $a^d = (a_1^d, a_2^d, \ldots\ldots)$ , $a_i^d = \langle y^d, \psi_i \rangle_{L^2(\mathfrak{D})}$

It is known that for distributed systems the exact controllability is too strong and never verified when the action is not distributed in $\mathfrak{D}$; so, we shall solve a weaker problem:

$$\begin{cases} \min_u ||u||^2_{L^2(0,T)} \\ \text{with } ||a(T) - a^d||^2 \text{ minimum} \end{cases} \qquad (P_0'')$$

Let H: $L^2(0,T) \to \ell_2$ defined by:

$$Hu = \int_0^T \Phi(T-\tau)B(\Omega)u(\tau)\, d\tau$$

then $H^* = B^*(\Omega)\Phi^*$

We recall the result [1,2,6,7]

*If the system is weakly controllable ( $\overline{Im(H)} = \ell_2 \Longleftrightarrow (HH^*)^{-1}$ exists) and if $(a^d - \Phi(T)a^0) \in Im(HH^*)$ then $(P_0'')$ has a unique solution $u^*$ given by*

$$u^*(t) = H^*(HH^*)^{-1}\{a^d - \Phi(T)a^0\} \qquad (II.3)$$

$$= B^*(\Omega)\Phi^*(T-t)\{\int_0^T \Phi(T-\tau)B(\Omega)B^*(\Omega)\Phi^*(T-\tau)d\tau\}^{-1}(a^d - \Phi(T)a^0)$$

*moreover*

$$||u||^2_{L^2(0,T)} = \langle a^d - \Phi(T)a^0, (HH^*)^{-1}(a^d - \Phi(T)a^0)\rangle_{\ell_2} \qquad (II.4)$$

The computation of the control $u^*$ requires the inversion of an infinite dimensional operator. To overcome this difficulty, we project the system on a finite dimensional subspace.

We obtain:

$$\begin{cases} \dot{a}^N(t) = A_N\, a^N(t) + B_N(\Omega)\, u(t) \\ a^N(0) = a^{N0} \end{cases} \qquad (II.5)$$

where

$$a^N(t) = (a_1(t), \ldots, a_N(t))^T$$
$$A_N = \mathrm{diag}\, (\lambda_1, \ldots, \lambda_N)$$
$$B_N(\Omega) = (b_1(\Omega), \ldots, b_N(\Omega))$$
$$a^{N0} = (a_1^0, \ldots, a_N^0)$$

and the minimum energy final value problem for (II.5) leads to a suboptimal control given by:

$$u_N^*(t) = B_N^*(\Omega)\Phi_N^*(T-t)\{\int_0^T \Phi_N(T-\tau)B_N(\Omega)B_N^*(\Omega)\Phi_N^*(T-\tau)d\tau\}^{-1} .$$

$$. \{a^{Nd} - \Phi_N(T)a^{N0}\} \qquad \text{(II.6)}$$

It was shown [1,3,6,7] that $u_N^* \to u^*$ as $N \to \infty$. But the application of the control (II.6) to the system (I.1) leads to an error at T given by $||a(T) - a^d||^2 = ||a^r(T) - a^{rd}||^2$ where

$$a(t) = (a^N(t),a^r(t))^T \quad \text{and} \quad a^d = (a^{Nd},a^{rd})^T.$$

This final error depends on the spatial location of the actuator; therefore we introduce an optimality criterion which takes in account this error, and our objective becomes to select $\Omega^*$ which minimizes the functional

$$J(\Omega) = J_1(\Omega) + \alpha J_2(\Omega)$$

where

$$J_1(\Omega) = ||u_N^*||^2_{L^2(0,T)} \cdot ||g||^2_{L^2(\Omega)}$$

$$J_2(\Omega) = ||a(T) - a^d||^2_{\ell_2} = ||a^r(T) - a^{rd}||^2_{\ell_2}$$

$\Omega \subset \mathcal{D}$ so that the system is weakly controllable [8].
For the location of $\Omega^*$, we shall use an algorithm which uses the gradient of J. In the next paragraph we show how to compute this gradient.

## OPTIMUM DESIGN TECHNIQUE

Let V be a regular two-dimensional vector field defined on $[0,1]\times\mathcal{U}$ where $\mathcal{U}$ is an open neighborhood of the domain $\Omega$ . Any point X in $\Omega$ is transformed by V into a point $x(s,X)$ solution of the differential equation:

$$\begin{cases} \frac{d}{ds} x(s,X) = V(s,x(s,X)) \\ x(0,X) = X \end{cases} \qquad \text{(III.1)}$$

This equation defines a transformation $T_s(V)$: $X \to x(s,X)$ which changes the domain $\Omega$ into a new domain

$$\Omega_s = \{x/\ x = x(s,X),\ X \in \Omega\} = T_s(V)(\Omega) \qquad \text{(III.2)}$$

Let $J(\Omega)$ be a real number associated to any domain $\Omega$, we define (whenever it exists) the eulerian derivative $dJ(\Omega,V)$ at $\Omega$ in the direction of the field V as

$$dJ(\Omega,V) = \lim_{s\to 0} \frac{J(\Omega_s) - J(\Omega)}{s} \qquad \text{(III.3)}$$

In this study, we are interested by the case where

$$J(\Omega) = \int_{\Omega} f_{\Omega}(x)\, dx$$

It was shown that, in this case:

$$dJ(\Omega,V) = \int_{\Omega} \partial_{\Omega} f_{\Omega}(x)dx + \int_{\Gamma=\partial\Omega} f_{\Omega} V.n\, d\sigma \tag{III.4}$$

where $\partial_{\Omega} f_{\Omega}(x) = \lim_{s\to 0} \frac{f_{\Omega s}(x) - f_{\Omega}(x)}{s}$ is the shape derivative of the function $f_{\Omega}$ [11,12].

*1. Application to the case of* $J_1$

We have $J_1(\Omega) = ||u_N||^2 \int_{\Omega} g^2(x)\, dx$ where $u_N$ is given by (II.6).

Let $\quad U_N(\Omega) = \int_0^T \Phi_N(T-\tau)B_N(\Omega)B_N^*(\Omega)\Phi_N^*(T-\tau)d\tau$

and $\quad \hat{a}^N = a^{Nd} - \Phi_N(T)\, a^{N0}$

using the fact that $U_N(\Omega)$ is self adjoint, we obtain

$$J_1(\Omega) = (\hat{a}^N)^* U_N(\Omega)^{-1}\, \hat{a}^N \int_{\Omega} g^2(x)\, dx \tag{III.5}$$

then using (III.4)

$$dJ_1(\Omega,V) = -(\hat{a}^N)^* U_N(\Omega)^{-1} \mathcal{U}_N(\Omega) U_N(\Omega)^{-1}\, \hat{a}^N \int_{\Omega} g^2(x)dx + (\hat{a}^N)^* U_N(\Omega)^{-1}\, \hat{a}^N \{\int_{\Omega} 2g(x)\partial_{\Omega} g(x)dx + \int_{\Gamma} g^2 V.n\, d\sigma\}$$

where $\mathcal{U}_N(\Omega)$ is the shape derivative of the matrix $U_N(\Omega)$:

$$\mathcal{U}_N(\Omega) = \int_0^T \Phi_N(T-\tau)\{\mathcal{B}_N(\Omega)B_N^*(\Omega) + B_N(\Omega)\mathcal{B}_N^*(\Omega)\}\Phi_N^*(T-\tau)d\tau \tag{III.6}$$

with $\mathcal{B}_N(\Omega) = (\mathcal{B}_1(\Omega), \ldots, \mathcal{B}_N(\Omega))^T$ and $\mathcal{B}_j(\Omega) = db_j(\Omega,V)$

then

$$\mathcal{B}_i(\Omega) = \int_{\Omega} \partial_{\Omega} g(x)\psi_i(x)dx + \int_{\Gamma} g(\sigma)\,\psi_i(\sigma)V(0).n\, d\sigma \tag{III.7}$$

Let Y be the solution of the linear system $U_N(\Omega)\, Y = \hat{a}^N$ ; then

$$dJ_1(\Omega,V) = -Y^* \mathcal{U}_N(\Omega)\, Y||g||^2 + Y^* U_N(\Omega)Y\{\int_{\Omega} 2g\partial_{\Omega} g dx + \int_{\Gamma} g^2\, V.n\, d\sigma\}$$

and the matrix $\mathcal{U}_N(\Omega)$ consists of

$$(\mathcal{U}_N(\Omega))_{i,j} = \frac{e^{(\lambda_i+\lambda_j)T} - 1}{\lambda_i+\lambda_j} \{\mathcal{B}_i(\Omega)b_j(\Omega) + b_i(\Omega)\mathcal{B}_j(\Omega)\}$$

so that we obtain

$$dJ_1(\Omega,V) = \sum_{i=1}^{N}\sum_{j=1}^{N} Y_i Y_j \frac{e^{(\lambda_i+\lambda_j)T} - 1}{\lambda_i+\lambda_j} b_j(\Omega)\Big\{\int_{\Omega} g\,\psi_i dx\{2\int_{\Omega} g\partial_{\Omega} g dx + \int_{\Gamma} g^2 V(0)n d\sigma\} - 2\int_{\Omega} g^2 dx\{\int_{\Omega} \partial_{\Omega} g\,\psi_i dx + \int_{\Gamma} g\,\psi_i V(0).n d\sigma\}\Big\} \tag{III.8}$$

In the particular case of a function g which does not depend on $\Omega$, we obtain:

$$dJ_1(\Omega,V) = \int_\Gamma \sum_{i,j=1}^{N} Y_i Y_j \frac{e^{(\lambda_i+\lambda_j)T} - 1}{\lambda_i+\lambda_j} b_j(\Omega)\{g^2\int_\Omega g\,\psi_i dx - 2g\psi_i\int_\Omega g^2 dx\}.$$
$$.V(0).n\, d\sigma \qquad \text{(III.9)}$$

## 2. *Application to the case of* $J_2$

We have $J_2(\Omega) = ||a(T) - a^d||^2 = ||a^r(T) - a^{rd}||^2$

It is easy to see that:

$$a^r(T) = \int_0^T \Phi_r(T-\tau)B_r(\Omega)B_N^*(\Omega)\Phi_N^*(T-\tau)d\tau\ U_N^{-1}(\Omega)\hat{a}^N + \Phi_r(T)a^{r0}$$

Let $W(\Omega) = \int_0^T \Phi_r(T-\tau)B_r(\Omega)B_N^*(\Omega)\Phi_N^*(T-\tau)d\tau$ and $Y = U_N(\Omega)^{-1}\hat{a}^N$

$$(W_{ij}(\Omega))_{ij} = \frac{e^{(\lambda_{N+i}+\lambda_j)T} - 1}{\lambda_{N+i}+\lambda_j} b_{N+i}b_j \ , \ 1\leqslant i\leqslant\infty,\ 1\leqslant j\leqslant N$$

The expression of the terminal error is then:

$$J_2(\Omega) = \sum_{i=1}^{\infty}\{\hat{a}_{N+i}(T) - \sum_{j=1}^{N}\frac{e^{(\lambda_{N+i}+\lambda_j)T}-1}{\lambda_{N+i}+\lambda_j} b_{N+i}b_jY_j\}^2 \qquad \text{(III.10)}$$

Let $F(\Omega) = \hat{a}^r - W(\Omega)U_N(\Omega)^{-1}\hat{a}^N$, $J_2(\Omega) = ||F(\Omega)||^2_{\ell_2}$ then

$$dJ_2(\Omega,V) = 2\langle\mathfrak{F}(\Omega),F(\Omega)\rangle \quad \text{with } \mathfrak{F}(\Omega) = -\mathcal{W}U_N^{-1}\hat{a}^N + WU_N^{-1}\mathcal{U}_NU_N^{-1}\hat{a}^N$$

where $\mathcal{W}$ and $\mathcal{U}_N$ are the shape derivative of W and $U_N$

$$(\mathcal{W}(\Omega))_{ij} = \frac{e^{(\lambda_{N+i}+\lambda_j)T} - 1}{\lambda_{N+i}+\lambda_j}\{\mathcal{B}_{N+i}b_j + b_{N+i}\mathcal{B}_j\};\ 1\leqslant i\leqslant\infty,\ 1\leqslant j\leqslant N$$

Using the fact that $U_N(\Omega)$ is self adjoint, we have:

$$dJ_2(\Omega,V) = 2\{\mathcal{U}_NU_N(\Omega)^{-1}\hat{a}^N\}\ U_N(\Omega)^{-1}\ W^*(\Omega)\ F(\Omega)$$
$$- 2(\mathcal{W}(\Omega)U_N(\Omega)^{-1}\hat{a}^N)^*F(\Omega)$$

Let Y and Z be the solutions of $U_N(\Omega)Y = \hat{a}^N$ and

$$U_N(\Omega)Z = W^*(\Omega)(\hat{a}^r - W(\Omega)Y) = W^*(\Omega)\ F(\Omega)$$

we obtain

$$dJ_2(\Omega,V) = -2\sum_{i=1}^{\infty}(\sum_{j=1}^{N}\mathcal{W}_{ij}(\Omega)\ Y_j)\ F_i(\Omega) + 2\sum_{k=1}^{N}Z_k(\sum_{p=1}^{N}\mathcal{U}_{kp}\ Y_p)$$

*3. The case where Ω is a disc*

We are interested in the next section by yhe case where Ω is a disc $B(x, R)$; $x = (x_1,x_2)$. This changes the problem of finding $\Omega^*$ which minimizes $J(\Omega)$ into the problem of selecting $x_1^*$, $x_2^*$, $R^*$ which minimize $j(x_1, x_2, R) = J(B(x_1, x_2, R))$.

We need to compute the gradient of j. To do this, we use the expression of $dJ(\Omega,V)$ with a particular deformation speed V.

We have:
$$\frac{\partial j}{\partial x_1}(x_1,x_2,R) = \lim_{s\to 0} \frac{j(x_1+s,x_2,R) - j(x_1,x_2,R)}{s} = dJ(B,Vx_1)$$

and
$$\frac{\partial j}{\partial x_2}(x_1,x_2,R) = dJ(B,Vx_2) \;;\; \frac{\partial j}{\partial R}(x_1,x_2,R) = dJ(B,V_R)$$

where $Vx_1$ is the field which transforms $B(x_1,x_2,R) = \Omega$ into $B(x_1+s,x_2,R) = \Omega_s$. Using the differential equation (III.1), it is easy to see that:

$$Vx_1 = (1,0)\ ,\ Vx_2 = (0,1) \text{ and } V_R = n \text{ the outward normal field}$$

so that, if $dJ(\Omega,V) = \int_\Gamma G(x)\, V.n\, d\sigma$ , we obtain

$$\frac{\partial j}{\partial x_1}(x_1,x_2,R) = \int_0^{2\pi} G(x_1+R\cos\theta,x_2+R\sin\theta)\cos\theta d\theta$$

$$\frac{\partial j}{\partial x_2}(x_1,x_2,R) = \int_0^{2\pi} G(x_1+R\cos\theta,x_2+R\sin\theta)\sin\theta d\theta$$

$$\frac{\partial j}{\partial R}(x_1,x_2,R) = \int_0^{2\pi} G(x_1+R\cos\theta,x_2+R\sin\theta)\, d\theta \ .$$

EXAMPLE - NUMERICAL RESULTS

Consider the two-dimensional diffusion process described by the equation

$$\begin{cases} \frac{\partial y}{\partial t}(x,t) = \frac{\partial^2 y}{\partial x_1^2}(x,t) + \frac{\partial^2 y}{\partial x_2^2}(x,t) + \chi_\Omega(x)u(t) \\ x = (x_1,x_2) \in ]0,a[x]0,b[ = \mathcal{D} \text{ and } t\in ]0,T[ \\ y(x,0) = 0 \text{ and } y = 0 \text{ on } \partial\mathcal{D}x]0,T[ \end{cases} \qquad \text{(IV.1)}$$

The associated eigenvalue problem leads to

$$\phi_{mn}(x) = \frac{2}{\sqrt{ab}} \sin(\frac{m\pi}{a}x_1)\sin(\frac{n\pi}{b}x_2) \quad \text{as eigenfunctions}$$

and
$$\lambda_{mn} = -\pi^2(\frac{m^2}{a^2} + \frac{n^2}{b^2}) \quad \text{the corresponding eigenvalues.}$$

To ensure the controllability of (IV.1) one needs $(a^2/b^2) \notin Q$ [8].
We consider the equivalent formulation in $L^2(0,T;\ell_2)$ as in (II.1)
We truncate the system as in (II.5) to the order N=3, and the expression of $J_2$ to the order M=6. J is considered as in the previous section with the particular case of $\Omega = B(x_1,x_2,R)$.
We choose the following values a=1. , b=1.1 , T=0.25 and the desired state is $y^d(x) = x_1^2\, x_2^2\, (a-x_1)(b-x_2)$.
The optimization algorithm used is based on [13,14] and takes in account the geometric constraint $\Omega \subset \mathcal{D}$.
We obtain the following results:

| iterations | $x_1$ | $x_2$ | R | $J(\Omega)$ |
|---|---|---|---|---|
| 1 | 0.55 | 0.63 | 0.1 | 1107.74 |
| 2 | 0.57 | 0.62 | 0.11 | 121.72 |
| 3 | 0.65 | 0.82 | 0.277 | 26.51 |
| 4 | 0.652 | 0.766 | 0.275 | 10.085 |
| 5 | 0.650 | 0.763 | 0.276 | 10.048 |
| 6 | 0.648 | 0.747 | 0.278 | 6.779 |
| 7 | 0.646 | 0.732 | 0.281 | 4.505 |
| 8 | 0.643 | 0.718 | 0.284 | 2.823 |
| 9 | 0.639 | 0.707 | 0.287 | 1.735 |
| 10 | 0.636 | 0.698 | 0.290 | 1.096 |
| 11 | 0.634 | 0.692 | 0.293 | 0.741 |
| 12 | 0.632 | 0.688 | 0.295 | 0.521 |
| 13 | 0.630 | 0.684 | 0.296 | 0.382 |
| 14 | 0.616 | 0.657 | 0.310 | 0.296 |
| 15 | 0.625 | 0.674 | 0.301 | 0.083 |
| 16 | 0.624 | 0.671 | 0.303 | 0.050 |
| 17 | 0.6231 | 0.6691 | 0.3034 | 0.0473 |
| 18 | 0.6232 | 0.6694 | 0.3032 | 0.0472 |
| 19 | 0.6235 | 0.6697 | 0.3029 | 0.0471 |
| 20 | 0.6249 | 0.6711 | 0.3015 | 0.0465 |
| 21 | 0.627 | 0.6735 | 0.299 | 0.0458 |
| 22 | 0.631 | 0.6766 | 0.2955 | 0.0449 |
| 23 | 0.6336 | 0.6788 | 0.2929 | 0.04434 |
| 24 | 0.6345 | 0.6794 | 0.2919 | 0.04410 |
| 25 | 0.6346 | 0.6794 | 0.2919 | 0.04409 |

Evolution of the parameters which define
the control support $\Omega$

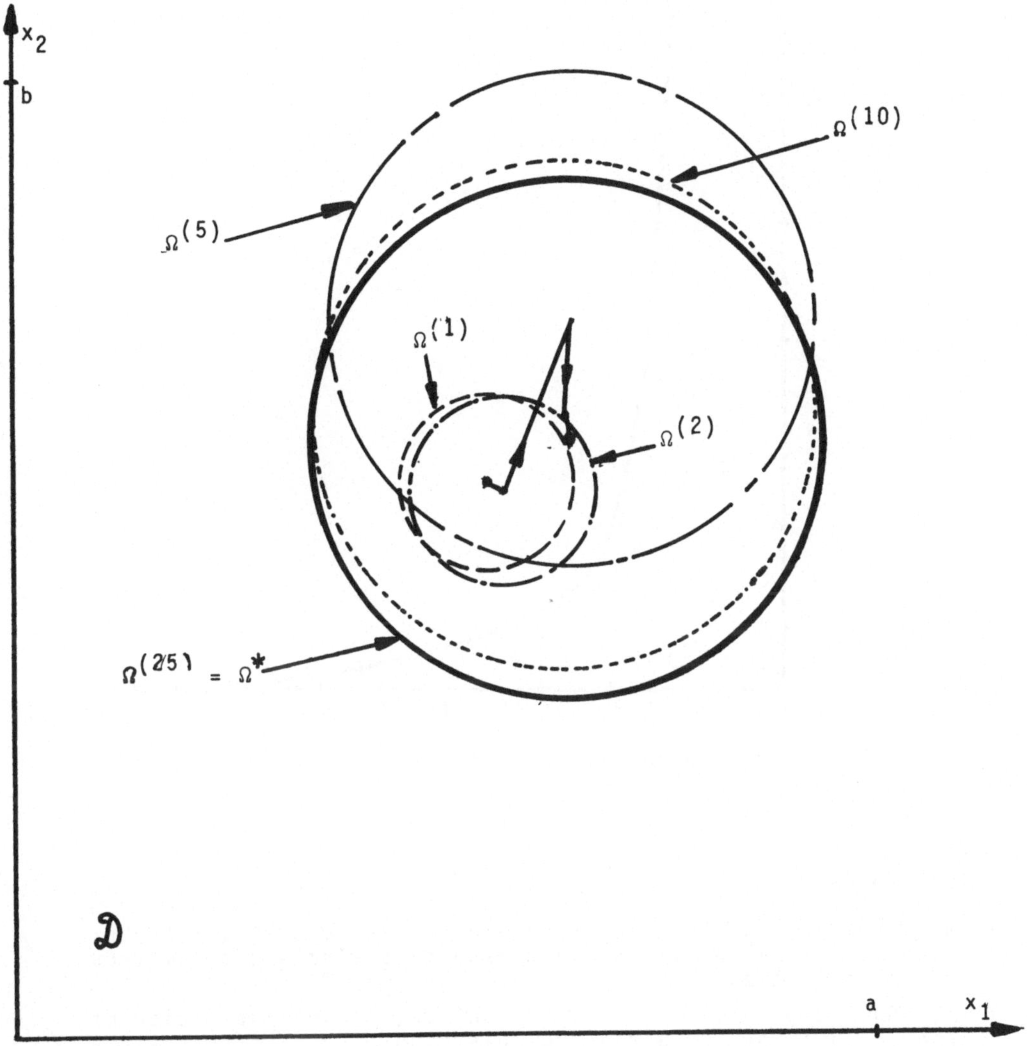


Evolution of the control support Ω

This example shows how it is important to give an optimal location of the control.
The case of two actuators ( or more ) can be considered in the same way ; but one must introduce constraints which let the different supports disjoint.
In the case of general Ω , methods using the finite elements are under consideration now.

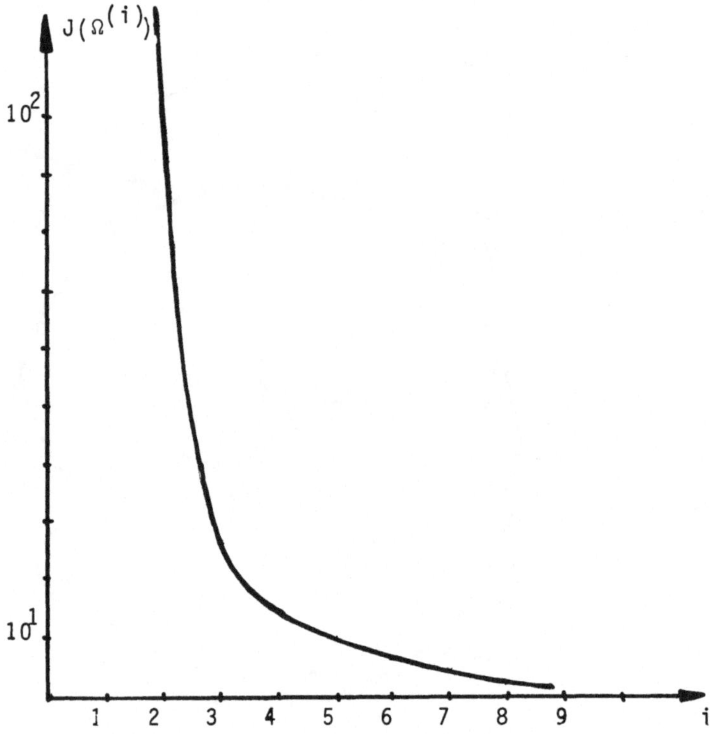


Evolution of the cost function J(Ω)

REFERENCES


(1) M. AMOUROUX, Localisation optimale de capteurs et actionneurs pour la commandabilité d'une classe de systèmes à paramètres répartis. Thèse d'Etat. Toulouse, fevrier 1977.

(2) M. AMOUROUX, G. DI PILLO and L. GRIPPO, Optimal selection of sensor location for a class of distributed parameter systems. Ricerche di Automatica. Vol.7, N°1, july 1976.

(3) M.AMOUROUX, J.P. BABARY, Détermination d'une zône d'action quasi-optimale pour une classe de systèmes à paramètres répartis. C.R.A.S. t. 282, 12 janv. 1976.

(4) J. CEA, Une méthode numérique pour la recherche d'un domaine optimal. Publication IMAP, Université de Nice. 1976.

(5) M. DELFOUR, G. PAYRE and J.P. ZOLESIO, Optimal design of a minimum weight thermal diffuser with constraint on the output yhermal flow. Applied Mathematics and optimization. Vol 9, pp225-262.

(6) A. EL JAI, Algorithmes pour la commande optimale de systèmes à paramètres répartis de type parabolique. Thèse d'Etat. Toulouse, juillet 1978.

(7) A. EL JAI, Sur la commande avec estimation de l'état initial d'une classe de systèmes à paramètres répartis. Considérations pratiques. RAIRO, Vol. 11, N° 4, 1977.

(8) A. EL JAI, L. BERRAHMOUNE, Localisation d'actionneurs zones pour la controlabilité de systèmes paraboliques. C.R.A.S. séance du 28 nov. 1983 (à paraître).

(9) J.L. LIONS, Contrôle optimal des systèmes gouvernés par des équations aux dérivées partielles. Dunod 1968.

(10) O. PIRONNEAU; Sur les problèmes d'optimisation de structure en mécanique des fluides. Thèse d'Etat. Paris VI. Mai 1976.

(11) J.P. ZOLESIO, Identification de domaines par déformation. Thèse d'Etat. Nice. 1979.

(12) J.P. ZOLESIO, The material derivative. Optimization of distributed parameter structures. Vol.II Sijthoff and Nordhoff Alphen aan den Rijn, pp1089-1153. 1981.

(13) A. BUCKLEY, an alternative implementation of Goldfarbs algorithm. AERE Harwell Report T.P. 1973.

(14) R. FLETCHER, A new approach to variable matric algorithms,Computer Journal, Vol. 13, N° 3, 1970.

Session 8

# IDENTIFICATION AND DETECTION

# IDENTIFICATION ET DÉTECTION

# PERFORMANCE EVALUATION OF MODELS, IDENTIFIED BY THE LEAST SQUARES METHOD

L. Ljung
Division of Automatic Control
Department of Electrical Engineering
Linköping University
S-581 83 Linköping, Sweden

Abstract

The least squares method is widely used for the identification of linear, dynamical systems. Here we investigate how the variances of the estimated models affect their performances in various applications, such as transfer function accuracy, prediction, minimum variance control and pole placement servo control design. Some newly developed asymptotic expressions for the variances are applied. Also, some consequences for the choice of input spectrum and model orders are drawn.

## 1. INTRODUCTION

Consider a linear system described by

$$y(t)=G_0(q^{-1})u(t)+v(t) \tag{1.1}$$

where $q^{-1}$ is the delay operator

$$q^{-1}u(t)=u(t-1)$$

and

$$G_0(q^{-1})=\sum_{k=1}^{\infty} g_k^0\, q^{-k} \tag{1.2}$$

The additive disturbance v(t) is regarded as a stationary stochastic process with zero mean value and spectral density

$$\Phi_v(\omega) \tag{1.3}$$

Many different methods for the identification of the transfer function of the system (1.1) and its noise properties have been developed. The most used and best known among these, is probably the least squares (LS) method. In this method a model is posed as

$$A_n(q^{-1})y(t)=B_m(q^{-1})u(t)+e(t) \tag{1.4}$$

where

$$A_n(q^{-1})=1+a_1q^{-1}+\ldots+a_nq^{-n} \tag{1.5a}$$

$$B_m(q^{-1})=b_1q^{-1}+\ldots+b_mq^{-m} \tag{1.5b}$$

When data: $y(t),u(t),t=1,\ldots,N$ have been measured, the parameters $a_i$ and $b_i$ are estimated as

$$\{a_i(N),b_i(N)\}=\arg\min_{a_i,b_i} \frac{1}{N}\sum_{t=1}^{N}\left[A_n(q^{-1})y(t)-B_m(q^{-1})u(t)\right]^2 \tag{1.6}$$

From these estimates we form the model

$$y(t)=\frac{\hat{B}_N(q^{-1})}{\hat{A}_N(q^{-1})}u(t)+\frac{1}{\hat{A}_N(q^{-1})}e(t) \tag{1.7}$$

where

$$\begin{aligned}&\hat{A}_N(q^{-1})=1+\hat{a}_1(N)q^{-1}+\ldots+\hat{a}_n(N)q^{-n}\\ &\hat{B}_N(q^{-1})\ \hat{b}_1(N)q^{-1}+\ldots+\hat{b}_m(N)q^{-m}\end{aligned} \tag{1.8}$$

(Here we suppress the order indices in A and B: the order of A is always n and the order of B is always m)

In the model (1.7) the sequence $\{e(t)\}$ will be thought of as white noise with variance

$$\hat{\sigma}_N=\frac{1}{N}\sum_{t=1}^{N}\left[\hat{A}_N(q^{-1})y(t)-\hat{B}_N(q^{-1})u(t)\right]^2 \tag{1.9}$$

The estimated spectral density of the output disturbance in (1.7) is thus

$$\hat{\Phi}_N^v(\omega)=\hat{\sigma}_N\cdot/|\hat{A}_N(e^{i\omega})|^2 \tag{1.10}$$

Now, the model (1.7) can be used for several purposes. We may apply it for e.g. prediction of the output, for control design like minimum varaince control or poleplacement, servo control. The success of these application will depend on the quality of the model (1.7). We shall in this paper investigate the performance degradation in such applications, arising from the fact that (1.7) is an identified model, whose elements are random variables, with certain variances.

## 2. SOME BASIC ASYMPTOTIC RESULTS

The asymptotic properties of the estimates in (1.7), when the model orders ( n and m) are allowed to increase when the number of observed data points (N) increases, are analysed in Ljung and Yuan (1983) (case n=0) and Ljung (1984a) (general case). These results hold under non-restrictive conditions and will be quoted in this section.

First introduce the parameters that minimize the expected least squares criterion (cf (1.6)):

$$\{\bar{a}_i,\bar{b}_i\}=\arg\min_{a_i\ b_i} E[A_n(q^{-1})y(t)-B_m(q^{-1})u(t)]^2 \tag{2.1}$$

and (cf(1.8)

$$\begin{aligned}\bar{A}_M(q^{-1})&=1+\bar{q}_1q^{-1}+\ldots+\bar{a}_nq^{-n}\\ \bar{B}_m(q^{-1})&=\bar{b}_1q^{-1}+\ldots+\bar{b}_mq^{-m}\end{aligned} \tag{2.2}$$

Let us also represent the true system (1.1)-(1.3) as

$$\Phi_v(\omega)=\sigma_0\cdot/|A_0(e^{i\omega})|^2 \tag{2.3}$$

$$A_0(q^{-1})=1+\sum_{k=1}^{\infty} a_k^0q^{-k} \tag{2.4}$$

$$G_0(q^{-1})=\frac{B_0(q^{-1})}{A_0(q^{-I})} \tag{2.5}$$

$$B_0(q^{-1})=\sum_{k=1}^{\infty} b_k^0 q^{-k} \tag{2.6}$$

We shall generally suppose that the input used during the identification experiments is stationary, independent of $\{v(t)\}$, and has spectral density $\Phi_u(\omega)$ We also assume that

$$\Phi_u(\omega)>\delta>0 \quad \forall \omega \tag{2.7}$$

[In fact, it is sufficient to consider

$$\Phi_u(\omega)=\sum_{k=-m}^{m} \hat{r}_N(k)e^{-ik\omega} \tag{2.8}$$

$$\hat{r}_N(k)=\frac{1}{N}\sum_{t=k}^{N} u(t-k)u(t)]$$

We now have the following results (Ljung and Yuan (1983), Ljung (1984a)): (Not all regularity conditions will be listed here)

Consistency. Suppose that the model orders n and m are functions of N: n(N),m(N) and that

$$\lim_{N\to\infty}[n(N)+m(N)][\sum_{k=n(N)}^{\infty}|a_k^0|+\sum_{k=n(N)}^{\infty}|b_k^0|]=0 \tag{2.9}$$

Then, as $N\to\infty$, with probability 1 and uniformity in $\omega$

$$\frac{\hat{B}_n(e^{i\omega})}{\hat{A}_N(e^{i\omega})} \to G_0(e^{i\omega}) \tag{2.10}$$

$$\hat{\sigma}_N/|\hat{A}_N(e^{i\omega})|^2 \to \Phi_v(\omega) \tag{2.11}$$

Variance: Suppose that (2.9) holds, that $m(N)\to\infty$ and that $(n(N)+m(N)^4/N\to 0$. Then

$$\sqrt{\frac{N}{n(N)+m(N)}}\begin{pmatrix}\hat{A}_N(e^{i\omega}) & \bar{A}_{n(N)}(e^{i\omega})\\ \hat{B}_N(e^{i\omega}) & -\ \bar{B}_{m(N)}(e^{i\omega})\end{pmatrix} \in AsN(0,P(\alpha,\omega)) \tag{2.12}$$

This expression means that the random variable on the left convergens in distribution to the normal distribution with zero mean and covariance matrix $P(\alpha,\omega)$. Here

$$\alpha = \lim_{N\to\infty} \frac{m(N)}{n(N)+m(N)} \tag{2.13}$$

and

$$P(\alpha,\omega) = \sigma_0 \begin{pmatrix} (1-\alpha)/\Phi_v(\omega) & (1-\alpha)G_0(e^{i\omega})/\Phi_v(\omega) \\ (1-\alpha)G_0(e^{-i\omega})/\Phi_v(\omega) & \alpha/\Phi_u(\omega)+(1-\alpha)|G_0(e^{i\omega})|^2/\Phi_u(\omega) \end{pmatrix} \tag{2.14a}$$

if $\alpha \geq 1/2$

and

$$P(\alpha,\omega) = \sigma_0 \begin{bmatrix} \alpha/\Phi_v(\omega)+ \dfrac{1-2\alpha}{|G_0(e^{i\omega})|^2\Phi_u(\omega)+\Phi_v(\omega)} & \alpha G_0(e^{i\omega})/\Phi_v(\omega) \\ \alpha G_0(e^{-i\omega})/\Phi_v(\omega) & \alpha[1/\Phi_u(\omega)+|G_0(e^{i\omega})|^2/\Phi_u(\omega)] \end{bmatrix} \tag{2.14b}$$

if $\alpha \leq 1/2$.

Moreover, estimates at different frequencies are asymptotically independent. It can also be shown that the actual covariance of the left hand side of (2.12) exists and converges to $P(\alpha,\omega)$ provided reasonable regularization is applied (i.e. measures to avoid division by zero when forming the estimates).

For practical purposes we introduce the signal-to-noise ratio:

$$SNR(\omega) = \Gamma(\omega) = |G_0(e^{i\omega})|^2 \Phi_u(\omega)/\Phi_v(\omega) \tag{2.15}$$

and rewrite (2.12), conceptually, as

$m \geq n$:

$$\operatorname{cov}\begin{pmatrix} \hat{A}_N(e^{i\omega}) \\ \hat{B}_N(e^{i\omega}) \end{pmatrix} \sim \frac{1}{N}\frac{\sigma_0}{\Phi_v(\omega)} \begin{pmatrix} n & n\,G_0(e^{i\omega}) \\ nG_0(e^{-i\omega}) & m\,\dfrac{\Phi_v(\omega)}{\Phi_u(\omega)} + |G_0(e^{i\omega})|^2 \end{pmatrix} \tag{2.16a}$$

$m \leq n$:

$$\operatorname{cov}\begin{pmatrix}\hat{A}_N(e^{i\omega})\\ \hat{B}_N(e^{i\omega})\end{pmatrix} \sim \frac{1}{N}\frac{\sigma_0}{\Phi_v(\omega)}\begin{pmatrix} m+\dfrac{n-m}{\Gamma(\omega)+1} & mG_0(e^{i\omega}) \\ mG_0(e^{i\omega}) & m|G_0(e^{i\omega})|^2[1+1/\Gamma(\omega)]\end{pmatrix} \quad (2.16b)$$

Generally, speaking, we see that the variance of the estimates is proportional to the orders n or m. We thus have higher variances when more parameters are estimated. On the other hand, the bias

$$\begin{aligned}&\bar{A}_n(e^{i\omega})-A_0(e^{i\omega})\\ &\bar{B}_m(e^{i\omega})-B_0(e^{i\omega})\end{aligned} \quad (2.17)$$

will of course decrease as n and m are increased (unless $A_0$ and $B_0$ are of finite orders). The best order selection is thus a trade-off between variance and bias.

We conclude this section with a result that will prove useful when discussing optimal input spectral densities.

Lemma. Let $\phi(\omega)$ be a given positive function and consider

$$\min_{\Phi(\omega)} \int_{-\pi}^{\pi} \phi(\omega)/\Phi(\omega)d\omega \quad (2.18)$$

subject to the constraint

$$\int_{-\pi}^{\pi} \Phi(\omega)d\omega \leq 1 \quad (2.19)$$

Then the minimizing function $\Phi(\omega)$ is given by

$$\Phi_{opt}(\omega)=\beta\cdot\sqrt{\phi(\omega)} \quad (2.20)$$

where $\beta$ is adjusted so that the constraint is met.

Proof.

$$(\int\sqrt{\phi}d\omega)^2=(\int\sqrt{\frac{\phi}{\Phi}}\cdot\sqrt{\Phi}d\omega)^2\leq\int\phi/\Phi d\omega\cdot\int\Phi d\omega\leq\int\phi/\Phi d\omega$$

with equalities when (2.19) and (2.20) hold according to Schwartz' inequality. □

## 3. ESTIMATION OF TRANSFERFUNCTIONS

Consider the estimate

$$\hat{G}_N(e^{i\omega}) = \frac{\hat{B}_N(e^{i\omega})}{\hat{A}_N(e^{i\omega})}. \tag{3.1}$$

Let

$$\bar{G}_{n,m}(e^{i\omega}) = \frac{\bar{B}_m(e^{i\omega})}{\bar{A}_n(e^{i\omega})}. \tag{3.2}$$

Then

$$E|\hat{G}_N(e^{i\omega})-\bar{G}_{n,m}(e^{i\omega})|^2 \approx \frac{1}{|\bar{A}_0(e^{i\omega})|^4}\{|\bar{A}(e^{i\omega})|^2 E|\hat{B}_N(e^{i\omega})-\bar{B}(e^{i\omega})|^2+$$

$$+|\bar{B}(e^{i\omega})|^2 E|\hat{A}_N(e^{i\omega})-\bar{A}(e^{i\omega})|^2-$$

$$-2\mathrm{Re}[\bar{A}(e^{i\omega})\cdot\bar{B}(e^{-i\omega})E(\hat{B}_N(e^{i\omega})-\bar{B}(e^{i\omega}))\cdot(\hat{A}_N(e^{i\omega})-\bar{A}(e^{-i\omega}))]\} \tag{3.3}$$

assuming $\hat{A}_N$ and $\bar{A}$ to be close. (for a formal justification of this, see Ljung (1984a).) Inserting (2.16) into (3.3) gives, after some calculations (also replacing $\bar{A}$ with $A_0$)

$$\mathrm{Var}\ \hat{G}_N(e^{i\omega}) \sim \frac{m}{N}\frac{\Phi_v(\omega)}{\Phi_u(\omega)} \quad \text{if } m \geq n \tag{3.4}$$

$$\mathrm{Var}\ \hat{G}_N(e^{i\omega}) \sim \frac{1}{N}\frac{\Phi_v(\omega)}{\Phi_u(\omega)}\left[m+(n-m)\frac{\Gamma(\omega)}{1+\Gamma(\omega)}\right] \quad \text{if } m \leq n \tag{3.5}$$

From these expressions we may make several non-trivial observations.

Recall, though, that (3.4) and (3.5) are valid as asymptotic expressions in N <u>and</u> n and m.

- o When $n<m$, the variance depends only on m. Hence we may increase n up to m at no cost in increased varaince, but most probably at a gain in decreased bias. <u>When the prime objective is to estimate the transferfunction there is thus never any reason to use $n<m$.</u>

- The expressions (3.4) and (3.5) reveal the relative cost in variance of increasing m and/or n and how this cost depends on the SNR $\Gamma(\omega)$.

- Suppose we would like to choose an input spectrum so as to minimize a weighted norm

$$\min_{\Phi_u(\omega)} \int_{-\pi}^{\pi} Q(\omega)\cdot \mathrm{Var}\ \hat{G}_N(e^{i\omega})d\omega \tag{3.6}$$

subject to constrained input variance. Then the optimal choice is

$$\Phi_u(\omega)=\beta\cdot\sqrt{Q(\omega)\cdot\Phi_v(\omega)} \qquad \text{if } m\geq n \tag{3.7}$$

according to the lemma of Section 2.

Notice, though that this choice of input spectrum does not necessarily minimize the bias $\bar{G}_{n,m}(e^{i\omega})$. See Yuan and Ljung (1983) for a more detailed discussion of this problem.

## 4. PREDICTION

The one-step ahead prediction for the model (1.7) is given by

$$\hat{y}_N(t)=(1-\hat{A}_N(q^{-1}))y(t)+\hat{B}_N(q^{-1})u(t) \tag{4.1}$$

The prediction based on the "expected model" (2.2) denoted by $\bar{y}(t)$ is analogous. For the difference (which is also the difference in the corresponding prediction errors) we obtain

$$\tilde{\varepsilon}_N(t)=\hat{y}_N(t)=(\bar{A}_n(q^{-1})-\hat{A}_N(q^{-1}))y(t)+(\hat{B}_N(q^{-1})-\bar{B}_m(q^{-1}))u(t) \tag{4.2}$$

The spectral density for this difference is

$$\Phi_{\tilde{\varepsilon}}(\omega)=|\bar{A}_N(e^{i\omega})-\hat{A}_N(e^{i\omega})|^2\Phi_y(\omega)+|\hat{B}_N(e^{i\omega})-\bar{B}_m(e^{i\omega})|^2\Phi_u^*(\omega)-$$

$$-2\mathrm{Re}[(\hat{A}_N(e^{i\omega})-\bar{A}_N(e^{i\omega}))(\hat{B}_N(e^{-i\omega})-\bar{B}_m(e^{-i\omega}))\cdot\Phi_{yu}(\omega)] \tag{4.3}$$

Here

$$\Phi_y(\omega)=\left|G_0(e\quad)\right| \; \Phi_u^*(\omega)+\Phi_u(\omega)$$
$$\Phi_{yu}(\omega)=G_0(e^{i\omega})\Phi_u^*(\omega) \tag{4.4}$$

where the input spectra $\Phi_u^*(\omega)$ is for the input to which the prediction is applied. This does not have to be the same as the input spectrum $\Phi_u(\omega)$ used in the identification experiment.

Taking expectations in (4.3) with respect to the estimates A and B gives, using (2.16)

$$E\Phi_{\tilde{\varepsilon}}(\omega)\sim \frac{\sigma_0}{N}\left[n+m\cdot\frac{\Phi_u^*(\omega)}{\Phi_u(\omega)}\right] \quad m\geq n \tag{4.5}$$

and

$$E\Phi_{\tilde{\varepsilon}}(\omega)\sim \frac{\sigma_0}{N}\left[m\left(1+\frac{\Phi_u^*(\omega)}{\Phi_u(\omega)}\right)+(n-m)\frac{1+\Gamma(\omega)\cdot\Phi_u^*(\omega)/\Phi_u(\omega)}{1+\Gamma(\omega)}\right] \quad m<n \tag{4.6}$$

Notice that the varaince of $\tilde{\varepsilon}_N(t)$ is

$$E\tilde{\varepsilon}_N^2(\tau)=\frac{1}{2\pi}\int_{-\pi}^{\pi}\Phi_{\tilde{\varepsilon}}(\omega)d\omega \tag{4.7}$$

and that this represents a <u>direct increase in prediction error variance</u> if the prediction $\bar{y}(t)$ is (close to ) optimal.

From this expressions we note the following:

o If $\Phi_u^*(\omega)=\Phi_u(\omega)$, then

$$E\Phi_{\tilde{\varepsilon}}(\omega)\sim \frac{\sigma_0}{N}\left[n+m\right] \quad \forall\, n,m \tag{4.8}$$

That is, the spectrum of $\tilde{\varepsilon}$ is flat, and depends only on the total number of estimated parameters. In particular we then have

$$E\,\tilde{\varepsilon}^2_N(t)\sim \frac{\sigma_0}{N}\left[n+m\right] \tag{4.9}$$

which is Akaike's (1970) classical final prediction error (FPE) result.

o Suppose we would like to choose $\Phi_u(\omega)$ so that the varaince $E\tilde{\varepsilon}_N^2(t)$ is minimized. For $m \geq n$ this gives

$$\Phi_u^{opt}(\omega)=\beta\cdot\sqrt{\Phi_u^*(\omega)} \tag{4.10}$$

The optimal input for spectrum identification is thus, somewhat surprisingly proportional to the square root of the input spectrum for which the predictor is to be applied. (Notice that it makes sense to minimize $E\,\tilde{\varepsilon}^2_N(t)$ essentially only if $\bar{y}(t)$ is close to the true prediction).

## 5. MINIMUM VARIANCE CONTROL

The problem to design a regulator for (1.7), such that the output variance is minimized is called minimum variance control. In case there is no extra time delay in the system, such control makes the closed loop system equal to

$$y(t)=e(t)$$

(see, e.g. Åström (1970).) Clearly no smaller variance of the output can be obtained.

We shall here discuss a somewhat more general problem, that of making the closed loop system equal to

$$D(q^{-1})y(t)=e(t) \tag{5.1}$$

where

$$D(q^{-1})=1+d_1q^{-1}+\ldots+d_rq^{-r}.$$

In that way the effect of highfrequency components in the output can be reduced, at the price of a higher total variance.

A regulator that, when used in conjunction with the model (1.7), gives (5.1) is:

$$\hat{B}_N(q^{-1})u(t)=(\hat{A}_N(q^{-1})-D(q^{-1}))y(t) \tag{5.2}$$

Inserted into the true system description (1.1)-(1.3) plus (2.3)-(2.6) this regulator gives the closed loop system

$$A_0(q^{-1})y(t)= \frac{B_0(q^{-1})}{\hat{B}_N(q^{-1})} (\hat{A}_N(q^{-1})-D(q^{-1}))y(t)+e(t)$$

or, suppressing arguments,

$$y= \frac{B_0+(\hat{B}-B_0)}{A_0(\hat{B}-B_0)-B_0(\hat{A}-A_0)+B_0D} e \tag{5.3}$$

Now suppose that the orders m and n are chosen so that

$$\begin{aligned} \bar{A}_n(q^{-1}) &\approx A_0(q^{-1}) \\ \bar{B}_m(q^{-1}) &\approx B_0(q^{-1}) \end{aligned} \tag{5.4}$$

so that the bias can be neglected in comparison with the variance contribution. This means that the asymptotic result (2.16) can be applied to

$$E|\hat{A}_N(e^{i\omega})-A_0(e^{i\omega})|^2$$

etc.

When these expressions are small we may expand (5.3) as

$$y= \frac{1}{D} e+ [\frac{\Delta B}{B_0 D} + \frac{1}{D} [- \frac{A_0}{B_0 D} \Delta B + \frac{1}{D} \Delta A]]e \tag{5.5}$$

The performance degradation due to the errors $\Delta A$ and $\Delta B$ is thus

$$y- \frac{1}{D} e \triangleq \tilde{\varepsilon}= \frac{1}{D^2} [ \frac{D-A_0}{B_0} \Delta B + \Delta A]e \tag{5.6}$$

with spectral density

$$\Phi_{\tilde{\varepsilon}}(\omega)= \frac{\sigma_0}{|D(e^{i\omega})|^4} \left\{ \frac{|D(e^{i\omega})A_0(e^{i\omega})|^2}{|B_0(e^{i\omega})} |\Delta B(e^{i\omega})|^2+ \right.$$

$$\left. +|\Delta A(e^{i\omega})|^2+2\mathrm{Re} \left[\frac{D(e^{i\omega})-A_0(e^{i\omega})|^2}{B_0(e^{i\omega})} \Delta B(e^{i\omega})\Delta A(e^{-i\omega})\right]\right\} \tag{5.7}$$

Taking expectation with respect to $\hat{A}_N$ and $\hat{B}_N$ using (2.16) now gives:

(arguments suppressed)

$m \geq n$

$$E\Phi_{\tilde{\varepsilon}}(\omega) \sim \frac{1}{N} \frac{|A_0|^2}{|D|^4} \left\{ \frac{|D-A_0|^2}{|B_0|^2} \quad m \cdot \left[ \frac{G}{\Phi_u \cdot |A_0|^2} + n \left| \frac{B_0}{A_0} \right|^2 \right] \right.$$

$$\left. +n+n \cdot 2\mathrm{Re}\left[ \frac{D-A_0}{B_0} \cdot \frac{B_0}{A_0} \right] \right\} \cdot \sigma$$

$$E\Phi_{\tilde{\varepsilon}}(\omega) \sim \frac{\sigma}{N} \left[ \frac{m \cdot |D(e^{i\omega})-A_0(e^{i\omega})|^2 \sigma}{|D(e^{i\omega})|^4 |B_0(e^{i\omega})|^2 \Phi_u(\omega)} + \frac{n}{|D(e^{i\omega})|^2} \right] \quad m \geq n \qquad (5.8)$$

Similarly we have for $n \geq m$

$$E\Phi_{\tilde{\varepsilon}}(\omega) \sim \frac{\sigma}{N} \left[ \frac{m}{|D(e^{i\omega})|^2} + \frac{mG|D(e^{i\omega})-A_0(e^{i\omega})|^2}{|D(e^{i\omega})|^4 |B_0(e^{i\omega})|^2 \Phi_u(\omega)} + \right.$$

$$\left. + \frac{(n-m)\sigma \cdot |A_0(e^{iW})|^2}{|D(e^{i\omega})|^4 [|B_0(e^{i\omega})|^2 \Phi_u(\omega)+G]} \right] \qquad (5.9)$$

From these expressions we note the following

- If the input energy $\Phi_u(\omega)$ at the identification experiment is small is a certain frequency band, and the closed loop system is not intended to be well damped at those frequencies ($|D(e^{i\omega})|$ not large), then the error will be large at those frequencies (unless the desired denominator D and the given one A are close at those frequencies).

- To put it another way: If the identification was performed with an input that did not excite certain frequency bands very much, then it is wise to select $|D(e^{i\omega})|$ large (to let the closed loop system be well damped) at those frequencies.

- The input spectrum that minimizes the variance $E\tilde{\varepsilon}^2(t)$ in case $m \geq n$ is, according to lemma 1,

$$\Phi_u(\omega)=\beta \cdot \frac{|D(e^{i\omega})-A_0(e^{i\omega})|}{|B_0(e^{i\omega})|} \cdot \frac{1}{|D(e^{i\omega})|^2} \tag{5.10}$$

6. POLE ASSIGNMENT

Now consider the problem of selecting a regulator

$$R(q^{-1})u(t)=-S(q^{-1})y(t)+T(q^{-1})r(t) \tag{6.1}$$

so that the closed loop system becomes

$$y(t)=\frac{N(q^{-1})}{D(q^{-1})}\, r(t) + \text{noise} \tag{6.2}$$

Here r(t) is a reference input (a setpoint).

If we choose

$$T(q^{-1})=N(q^{-1}); \quad R(q^{-1})=\hat{B}_N(q^{-1});$$

$$S(q^{-1})=\hat{A}_N(q^{-1})-D(q^{-1}) \tag{6.3}$$

we see that (6.1) together with the model (2.7) gives (6.2).

When (6.1),(6.3) is inserted into the true system description (1.1)-(1.3) plus (2.3)-(2.6) we obtain (see also (5.3)):

$$y = \frac{B_0 T\, r + \hat{B} e}{A_0(\hat{B}-B_0)-B_0(\hat{A}-A_0)+B_0 D} \tag{6.4}$$

If the errors are small this can be rewritten as in (5.5):

$$y \approx \frac{T}{D}\, r\, \frac{1}{D}\, e + \frac{T}{D^2}\left[-\frac{A_0}{B_0}\,\Delta B+\Delta A\right] r + \frac{1}{D^2}\left[\frac{D-A_0}{B_0}\,\Delta B+\Delta A\right] e \tag{6.5}$$

The performance degradation here is

$$\tilde{\varepsilon}=y-\frac{T}{D}r-\frac{1}{D}e.$$

The contribution to this error from e was analysed in the previous section. Let us here study the effect of r:

$$E\Phi_{\tilde{\varepsilon}}(\omega)\sim\frac{m\cdot\sigma}{N}\cdot\frac{|T(e^{i\omega})|^2}{|D(e^{i\omega})|^4}\cdot\frac{|A_0(e^{i\omega})|^2}{|B_0(e^{i\omega})|^2}\cdot\frac{\Phi_r(\omega)}{\Phi_u(\omega)}\qquad m\geq n \tag{6.6}$$

Here we used (5.4) and (2.17). Also $\Phi_r(\omega)$ denotes the spectral density of the reference signal r(t).

Similarly:

$$E\Phi_{\tilde{\varepsilon}}(\omega)\sim\frac{\sigma}{N}\cdot\frac{|T(e^{i\omega})|^2\cdot|A_0(e^{i\omega})|^2\,[m\sigma+n|B_0(e^{i\omega})|^2\Phi_u(\omega)]}{|D(e^{i\omega})|^4|B_0(e^{i\omega})|^2[\sigma+|B_0(e^{i\omega})|^2\Phi_u(\omega)]\Phi_u(\omega)}\,\Phi_r(\omega)\quad n\geq m \tag{6.7}$$

From these expressions we note:

- o It is important that the input spectrum $\Phi_u(\omega)$ is large at those frequencies where the closed loop system T/D has higher gain than the open loop one $B_0/A_0$, unless the reference input r has little energy there.

- o In case $m\geq n$ the input spectrum (subject to constrained variance) that minimizes the r-component of the variance $E\tilde{\varepsilon}^2(t)$ is given by

$$\Phi_u(\omega)=\beta\cdot\frac{|T(e^{i\omega})|}{|D(e^{i\omega})|^2}\,\frac{|A_0(e^{i\omega})|}{|B_0(e^{i\omega})|}\cdot\sqrt{\Phi_r(\omega)} \tag{6.8}$$

## 7. CONCLUSIONS

When an estimated model is used for some design purpose, the performance will be affected by the fact that the model is a random variable with a certain variance. We have here studied these effects for some common design purposes. The character of the results are not unexpected, such as: use more input energy for the identification at frequencies where a good model is more critical because, e.g. the closed loop gain will be increased there. However, the quantitative implications are not trivial.

When using the conclusions drawn at the end of each section it must be kept in mind that they are based on expressions that are asymptotic in the model orders. Also, we have mostly concentrated on the variance effects. When the model orders are small, so that bias rather than varaince may dominate the model error, then other effects of input may predominate.

A related study for more general transfer function models is given in Ljung (1984b).

REFERENCES

H. Akaike (1970):"Statistical Predictor Identification". Ann. Inst. Statist. Math. Vol. 22 pp 202-217.

K. J. Åström (1970):"Introduction to Stochastic Control", Academic Press, N.Y.

L. Ljung (1984a):"Asymptotic Properties of the Least Squares Method for Estimating Transfer Functions and Disturbance Spectra". Report, Dept. of Electrical Engineering, Linköping University, Linköping, Sweden.

L. Ljung (1984b):"Asymptotic variance expressions for identified black-box transfer function models". Report, Dept. of Electrical Engineering, Linköping University, Linköping Sweden.

L. Ljung and Z. D. Yuan (1983):"Properties on Non-Parametric Time-Domain Methods for Estimating Transfer Functions", Report, LiTH-ISY-I-570, Linköping University, Linköping, Sweden.

Z. D. Yuan and L. Ljung (1983):"Unprejudiced Optimal Input Design for Identification of Transfer Functions". Report, LiTH-ISY-I-0622, Linköping University, Linköping, Sweden.

# THE WEAK STOCHASTIC REALIZATION PROBLEM FOR DISCRETE-TIME COUNTING PROCESSES

J.H. van Schuppen
Centre for Mathematics and Computer Science
P.O. Box 4079
1009 AB Amsterdam
The Netherlands

Abstract. The weak stochastic realization problem is considered for discrete-time stationary counting processes. Such processes take values in the countable infinite set $N = \{0,1,2,\ldots\}$. A stochastic realization is sought in the class of stochastic systems specified by a conditional distribution for the output given the state of Poisson type, and by a finite valued state process. In the paper a necessary and sufficient condition is derived for the existence of a stochastic realization in the above specified class.

## 1. INTRODUCTION

The purpose of this paper is to present a result for the weak stochastic realization of a discrete-time counting process and to indicate the major open questions.

The weak stochastic realization problem to be considered is given a discrete-time counting process to show existence of and to classify all minimal Poisson-finite-state stochastic systems whose output equals the given process in distribution. The class of Poisson-finite-state stochastic systems is specified by a conditional distribution for the output given the state of Poisson type, and by a finite valued state process.

The motivation of this problem is the area of control and prediction for systems with point process observations. Examples of practical problems in this area are the control of queues, the prediction of traffic intensities, the estimation of software reliability, and the estimation of certain biomedical signals. The prediction and control problems for this class of systems, under the assumption that the parameter values are known, have been considered. Practical application of these results demands the solution of the system identification problem and the stochastic realization problem for the class of Poisson-finite-state systems.

The stochastic realization problem for Gaussian processes has received quite some attention the past fifteen years [2,3,6]. Both the weak and the strong version of the problem have been investigated. A considerable body of results is available for this problem. The corresponding problem for finite valued processes for which a realization is sought in the class of stochastic systems with a finite state process has also received consideration [4,5,8]. However, little progress has been made on this problem as far as a realization algorithm and the characterization of minimal realizations is concerned. The major bottle neck is a factorization question for nonnegative matrices [5].

In this paper attention is focused on the weak stochastic realization problem for stochastic processes taking values in the positive integers. This problem should be distinguished from the finite stochastic realization problem for processes taking values in a finite set. A weak stochastic realization is sought in the class of Poisson-finite-state stochastic systems described above. A necessary and sufficient condition will be stated for a discrete-time counting process to have a realization in this class. Open questions will be mentioned.

A summary of the paper follows. The problem formulation is given in section 2, while in section 3 a condition for existence of a weak stochastic realization is derived.

## 2. PROBLEM FORMULATION

Below a definition is given of a Poisson-finite-state stochastic system and the corresponding weak stochastic realization problem is formulated.

Notation and terminology that will be used in the paper, will be defined. Let $\{\Omega,F,P\}$ be a complete probability space and $T = Z$ be the time index set. The conditional independence relation for a triple of $\sigma$-algebra's $F_1,F_2,G$ is defined by the condition that

$$E[x_1x_2|G] = E[x_1|G]E[x_2|G]$$

for all $x_1 \in L^+(F_1)$ and $x_2 \in L^+(F_2)$; notation $(F_1,G,F_2) \in CI$. Here $L^+(F_1)$ is the set of all positive $F_1$ measurable random variables. The smallest $\sigma$-algebra with respect to which a random variable x is measurable is denoted by $F^x$, and that containing the $\sigma$-algebra's G and H by $G \vee H$. The set of positive integers is denoted by $N = \{0,1,2,\ldots\}$, while that of strictly positive integers by $Z_+ = \{1,2,3,\ldots\}$. For $n \in Z_+$ is $Z_n = \{1,2,\ldots,n\}$. The set of nonnegative matrices is denoted by $R_+^{n\times n}$. For material on this set see [1].

<u>2.1. DEFINITION</u>. A *Poisson-finite-state stochastic system* is a collection

$$\sigma = \{\Omega,F,P,T,N,B_N,X,B_X,n,\lambda\}$$

where $\{\Omega,F,P\}$ is a complete probability space, $T = Z$, $N = \{0,1,2,\ldots\}$, $X = \{c_1,c_2,\ldots,c_n\} \subset (0,\infty)$ for some $n \in Z_+$, $B_N,B_X$ are $\sigma$-algebra's on N and X generated by all subsets of N and X, $n\colon \Omega \times T \to X$, $\lambda\colon \Omega\times T \to X$ are stochastic processes called respectively the *output process* and the *state process*, such that for all $t \in T$, $k \in N$

$$E[I_{(n_t=k)}|F_{t-1}^{n-} \vee F_\infty^\lambda] = (\lambda_t)^k\exp(-\lambda_t)/k!$$

and $(\lambda_t, F_{t-1}^{n-} \vee F_t^{\lambda-}, t \in T)$ is a stationary finite-state Markov process. Here $F_t^{n-} = \sigma(\{n_s, \forall s \leq t\})$, $F_\infty^{\lambda} = \sigma(\{\lambda_s, \forall s \in T\})$.
Notation: $\sigma \in$ PFSΣ.

In a stochastic system one exhibits, besides the externally available output process, the underlying state process. The state process is of crucial importance for the solution of prediction and control problems. The above defined stochastic system is called Poisson-finite-state because the conditional distribution of the output process given the past and the state process is of Poisson type, and because the state process is a finite-state Markov process.

In the following a stochastic process taking values in N will be called a discrete-time counting process. The output of a Poisson-finite-state stochastic system is a discrete-time counting process.

An abstract definition of a stochastic system can also be given [4,5,8]. It can then be shown that the above defined Poisson-finite-state stochastic system satisfies this abstraction definition. For the sake of completeness this result is put on record.

2.2 DEFINITION. A (discrete-time) *stochastic system* is a collection

$$\sigma = \{\Omega, F, P, T, Y, B_Y, X, B_X, y, x\}$$

where $\{\Omega, F, P\}$ is a complete probability space, $T = Z$, Y, X are sets and $B_Y, B_X$ σ-algebra's on Y respectively X, $y: \Omega\times T \to Y$, $x: \Omega\times T \to X$ are stochastic processes called respectively the *output process* and the *state process*, such that for all $t \in T$

$$\left(F_t^{y+} \vee F_t^{x+},\ F^{x_t},\ F_t^{x-} \vee F_{t-1}^{y-}\right) \in CI,$$

where

$$F_t^{y+} = \sigma(\{y_s, \forall s \geq t\}).$$

2.3 PROPOSITION. *A Poisson-finite-state stochastic system as defined in* 2.1 *is a stochastic system as defined in* 2.2.

PROOF. Let $t \in T$, $k \in N$, $i \in Z_n$. Then

$$E\left[I_{(n_t=k)}\, I_{(\lambda_{t+1}=c_i)} \,|\, F_{t-1}^{n-} \vee F_t^{\lambda-}\right]$$

$$= E\left[I_{(\lambda_{t+1}=c_i)} E\left[I_{(n_t=k)} \,|\, F_{t-1}^{n-} \vee F_\infty^{\lambda-}\right] \,|\, F_{t-1}^{n-} \vee F_t^{\lambda-}\right]$$

$$= E\left[I_{(\lambda_{t+1}=c_i)} (\lambda_t)^k \exp(-\lambda_t)/k! \,|\, F_{t-1}^{n-} \vee F_t^{\lambda-}\right]$$

$$= E\left[I_{(\lambda_{t+1}=c_i)} \,|\, F^{\lambda_t}\right] (\lambda_t)^k \exp(-\lambda_t)/k!$$

by $(\lambda_t, F^{n-1}_{t-1} \vee F^{\lambda-}_t, t \in T)$ a Markov process,

$$= E\left[ I_{(n_t=k)} I_{(\lambda_{t+1}=c_i)} | F^{\lambda_t} \right].$$

A monotone class argument then gives that $(F^{n_t} \vee F^{\lambda_{t+1}}, F^{\lambda_t}, F^{n-}_{t-1} \vee F^{\lambda-}_t) \in CI$. An induction procedure and another monotone class argument then yields that

$$(F^{n+}_t \vee F^{\lambda+}_{t+1}, F^{\lambda t}, F^{n-}_{t-1} \vee F^{\lambda-}_t) \in CI,$$

from which the result is easily deduced.

For future use a dynamic representation of a Poisson-finite-state stochastic system is derived. Define $x: \Omega \times T \to R^n$ by $x_{it} = I_{(\lambda_t=c_i)}$, and $c \in R^n$ by

$$c^T = (c_1 \dots c_n).$$

For $c \in R^n$ define the diagonal matrix

$$D(c) = \text{diag}(c_1,\dots,c_n) \in R^{n\times n}$$

with on the diagonal entries of the vector c. Let $b \in R^n$, $b_i = \exp(-c_i)$.
Then

$$\begin{aligned}
&(\lambda_t)^k \exp(-\lambda_t)/k! \\
&= \sum_{i=1}^{n} \exp(-c_i)(c_i)^k I_{(\lambda_t=c_i)}/k! \\
&= b^T D(c)^k x_t/k!
\end{aligned}$$

Let $A \in R^{n\times n}$ be the transition matrix of the stationary finite-state Markov process $\lambda$; thus

$$A_{ij} = P(\{x_{i,t+1}=1\} \cap \{x_{jt}=1\})/P(\{x_{jt}=1\})$$

if well defined and zero otherwise. Then

$$E[x_{t+1}|F^x_t] = Ax_t.$$

Define

$$\Delta m_{1t} = x_{t+1} - Ax_t$$

$$\Delta m_{2kt} = I_{(n_t=k)} - b^T D(c)^k x_t/k!$$

Then $\Delta m_{1t}, \Delta m_{2kt}$ are martingale increments:

$$E[\Delta m_{1t}|F^{n-}_{t-1} \vee F^x_t] = 0,$$

$$E[\Delta m_{2kt} | F^{n-}_{t-1} \vee F^{x}_{t}] = 0.$$

One obtains thus the representation

$$\begin{cases} x_{t+1} = Ax_t + \Delta m_{1t}, \\ I_{(n_t=k)} = b^T D(c)^k x_t/k! + \Delta m_{2kt}. \end{cases}$$

2.4 PROBLEM. The *Poisson-finite-state weak stochastic realization problem* is, given a stationary discrete-time counting process on T=Z, to solve the following subproblems:

a. To give necessary and sufficient conditions for the existence of a Poisson-finite-state stochastic system $\sigma$ such that the output process of this system equals the given process in distribution; if such a system exists then it is called a *weak stochastic realization* of the given process;
b. to classify all minimal weak stochastic realizations, where minimal refers to the number of elements in the state space.

One may pose the question why for discrete-time counting processes attention is restricted to the class of Poisson-finite-state stochastic systems? The answer is that for systems in this class the stochastic filtering problem can easily be solved. Such systems may therefore be used in applications. The system identification problem then demands the estimation of the parameters of the filter representation. To answer questions about the identifiability of the parameters, the weak stochastic realization problem must be resolved.

For the sake of reference the solution to the stochastic filtering problem for a Poisson-finite-state stochastic system is stated below. No reference in the literature is known for this result but its proof is elementary.

2.5 PROPOSITION. *Assume given a Poisson-finite-state stochastic system with the representation*

$$x_{t+1} = Ax_t + \Delta m_{1t},$$

$$I_{(n_t=k)} = b^T D(c)^k x_t/k! + \Delta m_{2kt},$$

*as described above. The solution of the stochastic filtering problem for this system is given by*

$$\hat{x}_t = E[x_t | F^{n-}_{t-1}],$$

$$\hat{x}_{t+1} = A\hat{x}_t + \sum_{k=0}^{\infty} A[D(\hat{x}_t) - \hat{x}_t\hat{x}_t^T]$$

$$(D(c)^k b/k!)[b^T D(c)^k \hat{x}_t/k!]^{-1} I_{(n_t=k)}$$

$$= \sum_{k=0}^{\infty} [AD(\hat{x}_t)D(c)^k b/k!][b^T D(c)^k \hat{x}_t/k!]^{-1} I_{(n_t=k)}.$$

PROOF. Omitted. □

The solution of the above filtering problem is readily implemented. If $b_k \in R^n_+$, $k \in N$, is defined as $b_k = D(c)^k b/k!$ then one has the recursion

$$b_{k+1} = D(c)b_k/(k+1),\ b_0 = b.$$

## 3. THE RESULT

Below a necessary and sufficient condition is given for a discrete-time counting process to have a weak stochastic realization in the class of Poisson-finite-state stochastic systems.

Some remarks on notation follow. The family of finite dimensional distributions of a stationary counting process n is denoted by, for any $m \in Z_+$,

$$p_m(t_1,\ldots,t_m,k_1,\ldots,k_m) = P(\{n_{t_1} = k_1,\ldots,n_{t_m} = k_m\})$$

where $t_1,\ldots,t_m \in T$, $t_m \le t_{m-1} \le \ldots \le t_1$, and $k_1,\ldots,k_m \in N$. Because the process is stationary $p_m$ is dependent on the $t_i$'s only through $t_1-t_2, t_2-t_3,\ldots,t_{m-1}-t_m$.

If $c, b \in R^n_+$ then $D(c)D(b) = D(b)d(c)$, while $D(c)b = D(b)c$. Let $u \in R^n, u^T=(1\ 1 \ldots 1)$. A stochastic matrix is an element $A \in R^{n\times n}_+$ such that $u^T A = u^T$. Note that if $x\colon \Omega\times T \to R^n$ is defined as in section 2 by $x_{it} = I_{(\lambda_t = c_i)}$, that then $(x_{it})^2 = x_{it}$, while for $i \ne j$, $x_{it}x_{jt} = 0$.

3.1 THEOREM. *Assume given a stationary discrete-time counting process on* $T = Z$, *say with finite-dimensional distribution, for* $m \in Z_+$,

$$p_m(t_1,\ldots,t_m,k_1,\ldots,t_m).$$

*There exists a weak stochastic realization of this process in the class of Poisson-finite-state stochastic systems iff there exists a* $n \in Z_+$, *a stochastic matrix* $A \in R^{n\times n}_+$, *and* $r,c \in (0,\infty)^n$, *such that if* $b \in (0,\infty)^n$, $b_i = \exp(-c_i)$, *then for any* $m \in Z_+$, $t_1,\ldots,t_m \in T$, $t_m < t_{m-1} < \ldots < t_1, k_1,\ldots,k_m \in N$ *one has*

$$\begin{aligned} &p_m(t_1,\ldots,t_m,k_1,\ldots,k_m)\\ &= u\, D(b)D(c)^{k_1}A^{t_1-t_2}D(b)D(c)^{k_2}A^{t_2-t_3}\ldots\\ &\qquad \ldots D(b)D(c)^{k_m}r/k_1!k_2!\ldots k_m! \end{aligned}$$

The above existence criterion is analogous to that of the existence of a finite stochastic realization as given in [4]. However, there conditional distributions are used, as where here unconditional distributions are preferred. Remarks on a realization algorithm are given below the proof.

PROOF. a ⇒ Assume there exists a weak stochastic realization say specified by the representation

$$x_{t+1} = Ax_t + \Delta m_t,$$

$$I_{(n_t=k)} = b^T D(c)^k c_t/k! + \Delta m_{2kt},$$

as discussed in section 2. Let $r = E(x_t)$. Then for $t_1 < t_2$

$$E[x_{t_2} I_{(n_{t_1}=k)} | F^{n-}_{t_1-1} \vee F^x_{t_1}]$$
$$= E[x_{t_2} E[I_{(n_{t_1}=k)} | F^{n-}_{t_1-1} \vee F^{x-}_{\infty}] | F^{n-}_{t_1-1} \vee F^x_{t_1}]$$
$$= E[x_{t_2} x^T_{t_1} D(c)^k b/k! | F^{n-}_{t_1-1} \vee F^x_{t_1}]$$
$$= A^{t_2-t_1} x_{t_1} x^T_{t_1} D(c)^k b/k!$$
$$= A^{t_2-t_1} D(x_{t_1}) D(c)^k D(b) u/k!$$
$$= A^{t_2-t_1} D(b) D(c)^k x_t/k!,$$

$$E[x_{t_2} I_{(n_{t_1}=k)}]$$
$$= A^{t_2-t_1} D(b) D(c)^k r/k!,$$

$$p_1(t_1,k) = E[I_{(n_{t_1}=k)}] = u^T D(b) D(c)^k r/k!$$

It will be shown by induction that for

$$t_m < t_{m-1} < \dots < t_2 < t_1 < t_0,\ k_1,\dots,k_m \in N$$

$$E[x_{t_0} I_{(n_{t_1}=k_1)} \dots I_{(n_{t_m}=k_m)}]$$
$$= A^{t_0-t_1} D(b) D(c)^{k_1} \dots A^{t_m-t_{m-1}} D(b) D(c)^{k_m} r/k_1! \dots k_m!\ .$$

By the above this holds for m = 1. Suppose it is true for m − 1. Then

$$E[x_{t_0} I_{(n_{t_1}=k_1)} \dots I_{(n_{t_m}=k_m)}]$$
$$= E[E[x_{t_0} I_{(n_{t_1}=k_1)} | F^{n-}_{t_1-1} \vee F^{x-}_{t_1}] \dots I_{(n_{t_m}=k_m)}]$$
$$= A^{t_0-t_1} D(b) D(c)^{k_1}/k_1! E[x_{t_1} E[x_{t_1} I_{(n_{t_2}=k_2)} \dots I_{(n_{t_m}=k_m)}]$$
$$= A^{t_0-t_1} D(b) D(c)^{k_1} \dots D(b) D(c)^{k_m} r/k_1! \dots k_m!,$$

$$p_m(t_1,\dots,t_m,k_1,\dots,k_m)$$

$$= E[I_{(n_{t_1}=k_1)} \dots I_{(n_{t_m}=k_m)}]$$

$$= u^T D(b)D(c)^{k_1} \ldots A^{t_{m-1}-t_m} D(b)D(c)^{k_m} r/k_1! \ldots k_m!$$

b. ⇐ If the indicated factorization exists then one has $n \in Z_+$, $A \in R_+^{n\times n}$ a stochastic matrix, and $c \in (0,\infty)^n$. One can then construct a probability space and a Poisson-finite-state stochastic system on it and part a. of the proof then shows that

$$E[I_{(n_{t_1}=k_1)} \ldots I_{(n_{t_m}=k_m)}]$$
$$= u^T D(b)D(c)^{k_1} A^{t_1-t_2} \ldots D(b)D(c)^{k_m} r/k_1!k_2! \ldots k_m!$$
$$= p_m(t_1,\ldots,t_m,k_1,\ldots,k_m).$$ □

A major unsolved question for the stochastic realization problem under discussion is the construction of a realization algorithm. The following heuristic procedure may be considered.

1. Assume that the function $k!p_1(t,k)$, as function of $k \in N$, is a positive Bohl function meaning that there exists a $n \in Z_+$, $h,g \in R_+^n$, $F \in R_+^{n\times n}$ such that

$$k!p_1(t,k) = h^T F^k g.$$

Assume further that F can be chosen diagonal, say $F = D(c)$ with $c \in R_+^n$. Define $b,d \in R_+^n$ as $b_i = \exp(-c_i)$, $d_i = \exp(c_i)$. Then

$$k!p_1(t,k) = h^T D(c)^k g = u^T D(h)D(c)^k g$$
$$= u^T D(b)D(d)D(h)D(c)^k g$$
$$= u^T D(b)D(c)^k D(d)D(h)g$$
$$= b^T D(c)^k r,$$
$$1 = \Sigma_{k=0}^{\infty} p_1(t,k) = u^T r.$$

2. Determine a stochastic matrix $A \in R_+^{n\times n}$ such that for all $t_1,t_2 \in T$, $t_2 < t_1$, $k_1,k_2 \in N$,

$$k_1!k_2!p_2(t_1,k_2,k_1,k_2) = u^T D(b)D(c)^{k_1} A^{t_1-t_2} D(b)D(c)^{k_2} r.$$

Step 1 and 2 determine $n \in Z_+$, $c \in (0,\infty)^n$, $A \in R_+^{n\times n}$.

3. Check whether the condition of theorem 3.1 holds for any $m \in Z_+$.

A major difficulty with the above algorithm is that nothing is known about factorization of positive functions as in step 1 above. In addition little is known about the factorization in step 2 of positive functions with more then one countable infinite index. Analogous difficulties occur in the finite stochastic realization problem [4,5].

Another major unsolved question is the characterization of minimal realizations.

It seems that this question is also analogous to that of the finite stochastic realization problem, see [5]. There it is shown that this question leads to a factorization problem for nonnegative matrices. The latter problem is unsolved.

REFERENCES

[1] BERMAN, A. & R.J. PLEMMONS, *Nonnegative matrices in the mathematical sciences*, Academic Press, New York, 1979.

[2] FAURRE, P., M. CLERGET & F. GERMAIN, *Opérateurs rationnels positifs*, Dunod, Paris, 1979.

[3] LINDQUIST, A. & G. PICCI, *On the stochastic realization problem*, SIAM J. Control Optim., 17 (1979), pp. 365-389.

[4] PICCI, G., *On the internal structure of finite-state stochastic processes*, in: *Recent Developments in Variable Structure Systems, Economics, and Biology*, Proc. of a U.S. -Italy Seminar, Taormina, Sicily, 1977, Lecture Notes in Econ. and Mathematical Systems, volume 162, Springer-Verlag, Berlin, 1978, pp. 288-304.

[5] PICCI, G. & J.H. VAN SCHUPPEN, *On the weak finite stochastic realization problem*, Proc. Colloque ENST-CNET: *Développements récents dans le filtrage et le contrôle des processes aléatoires*, to appear; also report BW 184/83, Centre of Mathematics and Computer Science, Amsterdam, 1983.

[6] RUCKEBUSCH, G., *A state space approach to the stochastic realization problem*, Proc. 1978 Int. Symp. on Circuits and Systems, New York, IEEE, New York, 1978, pp. 972-977.

[7] SNYDER, D.L., *Random point processes*, J. Wiley & Sons, New York, 1975.

[8] VAN SCHUPPEN, J.H., *The strong finite stochastic realization problem-preliminary results*, in: *Analysis and Optimization of Systems*, A. Bensoussan, J.L. Lions (eds.), Lecture Notes in Control and Info. Sci., volume 44, Springer-Verlag, Berlin, 1982, pp. 179-190.

# LINEAR STATISTICAL MODELS AND STOCHASTIC REALIZATION THEORY

Lorenzo Finesso
LADSEB-C.N.R.
Corso Stati Uniti 4 - 35100 PADOVA Italy

Giorgio Picci
Istituto di Elettrotecnica ed Elettronica
Via Gradenigo 6A - 35131 PADOVA Italy

ABSTRACT

The problem of representing a given gaussian zero mean random vector y by linear statistical models is considered. This is a concrete formulation of a simple stochastic realization problem. Let $y = [y_1', y_2']'$ be any partition of y into two disjoint subvectors $y_1$, $y_2$. It is shown that to every random vector x, making $y_1$ and $y_2$ conditionally independent given x there corresponds an (essentially unique) model of y of the form

$$\begin{aligned} y_1 &= H_1 x + n_1 \\ y_2 &= H_2 x + n_2 \end{aligned} \tag{0}$$

where $H_1$ and $H_2$ are deterministic matrices, $n_1$ and $n_2$ are mutually independent noise terms and each $n_i$ (i=1,2) is independent of x. The family of all realizations of y of the form (0) is analyzed both probabilistically and from the point of view of explicit computation of the parameters. Possible applications especially to the theory of Factor Analysis are discussed.

## 1. INTRODUCTION

Modelling problems are becoming more and more important in modern engineering sciences and econometrics. In many instances one wants to describe mathematically the behaviour of processes (industrial or economic etc.) where the underlying "physics" is either poorly known or too complex and unreliable to be of practical use. In these cases a mathematical description of the system should come from processing observed data. This is generally referred to as identification. What is commonly called identification theory, however, mainly consists of a bunch of algorithms or statistical procedures to do parameter estimation. We believe that this view has led to neglecting some important modelling aspects of the problem; more research is needed to understand what class of models one should use to describe the data. The need for a "theory of modelling" has recently been emphasized by KALMAN [6], [7]. (In this respect see also the provocative paper by J.C. WILLEMS [20])

To describe the framework which we are referring to, let us represent our industrial or economic process by a black box which produces "unpredictable" outputs $\eta$ (we disregard exogenous or input variables). The output $\eta$ is "unpredictable" or "uncertain" just because the physical laws governing the generation of $\eta$ and schematically represented by the black box, are either unknown or too complicated to account for. For example a realistic description of the measurement mechanism might bring in hundreds of external variables which one is not willing to introduce into a model.

The following is a first basic conceptual step in the modelling process which, even if often is only implicitely made, lies at the ground of all statistical identification procedures.

One decides to describe the data-generating mechanism (i.e. the black box) by a probability space $\{\Omega, \mathscr{A}, P\}$ and the measurement $\eta$ (say an m-dimensional real vector) as the sample value taken by a random vector $y(\cdot): \Omega \to R^m$ at some "state" $\bar{\omega} \in \Omega$, randomly chosen by "nature" at the moment of performing the experiment, i.e.

$$\eta = y(\bar{\omega})$$

As one can always take a canonical sample space $\Omega = R^m$ (= space of possible values of $\eta$), the random vector y is completely described by the probability measure P. [In more realistic situations $\eta$ needs to be thought of as a time series and then the probabilistic model of the data becomes a (discrete time) stochastic process $y \equiv \{y_t\}$. The data are then forced to be interpreted as a chunk of a trajectory $\{y_t(\bar{\omega})\}_{t \in \mathbb{Z}}$ of the process]. Notice that the model of the data is now a probability distribution, P. How to determine this unknown probability law P, starting from the given observation $\eta$, is then a problem which falls into the domain of statistics and should therefore be solved by statistical means. One selects a reasonable class of probability laws $\{P_\theta;\ \theta \in \Theta \subset R^p\}$ based on a priori information and uses inference procedures to assign a reasonable P starting from the observed sample $\eta$ etc.....

The above philosophy is of course questionable and has recently been subjected to much criticism ([20]). It has however undubious merits and chiefly that of incorporating the sound idea that a model resulting from an identification method should describe a family of possible observations (in principle all possible sample values of y) and not just the particular observation which was used for its calibration. This is important for the very obvious reason that a model is useful in as much as it helps to predict or "describe" data which have not yet been observed.

Let us notice now that the probabilistic model P (say a gaussian m-dimensional distribution) is merely a phenomenological description of the data-generating mechanism. For various reasons econometricians and engineers are not happy with this kind of description of the data. A first superficial argument is that P may be too complex an object; think of a stochastic process for example, where P is actually an infinite family of distributions. A more substantial argument is that an "external" model like P does not provide any "explanation" of the data. Indeed, a basic ingredient of what is commonly conceived as a "model" is a mathematical relation whereby the random variable y is expressed as a function of simpler random quantities (of smaller dimension or with particularly simple correlation properties like white noise). Examples of such "explanatory models" which we shall call for the time being internal, are the Factor Analysis model

$$y = Hx + \varepsilon \tag{1.1}$$

where H is a deterministic matrix, x and ε are random vectors, the first with dimension smaller than y and the second with uncorrelated (or independent) components, or, in the dynamic case, the well known Gauss-Markov model

$$\begin{aligned} x(t+1) &= Fx(t) + v(t) \\ y(t) &= Hx(t) + w(t) \quad . \end{aligned} \tag{1.2}$$

What makes internal models appealing and actually much more useful to solve prediction and decision problems is some kind of data reduction mechanism they incorporate: a "long" vector y is produced as a deterministic function of a short one (x) plus "noise". This property is perhaps more transparent in the deterministic case and we shall spend a few lines to draw the parallel.

Observe that, in a deterministic framework, a list of the data, say $\{\eta(1),\ldots,\eta(m)\}$ is a bona fide external model since it trivially is a mathematical representation of the data. (The stochastic analog of this is the probability distribution P). Notice on the other hand that a scheme of the form

$$x(t+1) = f(x(t))$$
$$\eta(t) = h(x(t)) \qquad (1.3)$$

which "explains" how $\eta$ is generated, permits, once f and h are known, to condensate the whole string of data $\{\eta(t)\}$ into the single parameter x(1). The deep feature of the (internal) model (1.3) is thus the relevant data compression it allows for: a long string $\{\eta(t)\}$ can be encoded by assigning two functions f, h and the initial state x(1). This, on the other hand, exactly amounts to saying that x(1) contains all the relevant information which is needed to produce the string of future outputs $\{\eta(t)\}$. We might call x(1) a "sufficient statistic" for $\eta$. The moral of the story is that explanatory (or internal) models, data compression, and existence of a sufficient statistic are all faces of the same medal and in fact equivalent concepts. We shall prove that this equivalence continues to hold in the stochastic (gaussian) case in the next chapter.

As one would expect, all identification schemes reported in the literature are invariably algorithms for parameter estimation of internal models of the type (1.2) or equivalent ARMA schemes. But, while the description of the data by a probability measure is, at least at the abstract level, an inherently univocal process (as one random variable is, by definition, described by just one probability distribution function), the description by internal models is not. This is the basic message of stochastic realization theory. There are in general infinitely many probabilistically non equivalent internal models of given structure which "realize" the same probability distribution. This non uniqueness can be source of troubles (see the vigorous arguments reported in [6] relative to the KOOPMANS REJERSØL "errors in variables" model) and in any case it raises the problem of model choice (i.e. modelling) in identification.

It is interesting to examine how the modelling problem is solved in current identification theory and the motivations which lead engineers to describe almost everything by ARMAX models. The basic postulate is the existence of a "true" physical (deterministic) system described by a (deterministic) difference equation, subject to random disturbances which are eventually added up to the "deterministic" output. (Within this logic the user is sometimes led to believe that the "randomness" of the disturbances must be a physical characteristic of the plant and wonders whether the disturbances acting on his own plant are "stochastic" enough to apply the proposed algorithms). Now, even in the rare situations where a "true system" can be unambiguously recognized, the results of identification experiments tend (as they should) much more to describe the data rather than the "true system" (compare the "biasedness" problem in least squares methods). "Physical" motivations as a basis for modelling (which we perceive as a mathematical problem) are often shaky. Our impression is that ARMAX models are so fashionable nowadays just because they are about the only class of internal models for which the non uniqueness problem has been solved (by choosing the so called "innovations representative"). Indeed, one might wonder why models of the type

$$\begin{aligned} & y(t) + \sum_{1}^{n} a_i y(t-i) = \sum_{1}^{m} b_i u(t-i) \\ & z(t) = y(t) + n_1(t) \qquad (1.4) \\ & v(t) = u(t) + n_2(t) \end{aligned}$$

where the observed quantities are $\{z(t)\}$ and $\{v(t)\}$ = "true" output and input signals + uncorrelated observation noises $\{n_1(t)\}$ and $\{n_2(t)\}$, which also do have a legitimate "physical" motivation, have never been seriously considered in the identification literature. (The answer is likely to be that models of the type (1.4) have "identifiability" problems ([14], [18]). Actually as KALMAN [6] stresses, identifiability is a mere question of coordinatization; the problem here is that there are infinitely many descriptions of the observed variables $\{z(t)\}$ and $\{v(t)\}$ of the form (1.4) which (are minimal in an appropriate sense and) realize the joint distribution of $\{z(t)\}$ and $\{v(t)\}$. One of them must be chosen (and the theory should tell which one is best adapted to the available data structure). Only after the choice has been made, it makes sense to talk about parametrization, identifiability and parameter estimation.

After this long philosophical introduction we shall say what the aim of this paper is. We want to study in some detail the simplest possible modelling problem: that relative to a single gaussian (zero mean) random vector y. In order to do this we first need a rigorous definition of the concept of internal model then we need to investigate the mathematical structure of these objects, the concept of minimality, and how minimal models are related to each other. Also we need to study how we can compute the different models starting from the external description and finally discuss their probabilistic structure and how it relates to practical problems of data representations.

As we shall see the problem is not entirely trivial even at this elementary level. This should come as no surprise to people familiar with stochastic realization theory. The set of all minimal internal models, for example, turns out to be parametrized by the solutions of a certain quadratic matrix inequality which seems to appear in every problem of this kind (see e.g. [1], [5]). The results show, among other things, that stochastic realization methods should have a major impact in solving an ever standing problem in multivariate analysis, the characterization of Factor-Analysis models of y ([2], [8]).

Finally, we should like to acknowledge some related work, especially that of VAN-PUTTEN - VAN-SCHUPPEN, [19] which although written from a different perspective, deals with some of the specific issues raised in this paper.

## 2. A REPRESENTATION THEOREM FOR STOCHASTIC SYSTEMS

Let y be an m-dimensional zero mean real Gaussian vector and let $\underline{H}$ be a gaussian space ([15]) of real random variables on some underlying probability space, containing $Y := \text{span}\{y(k);\ k=1...m\}$, the (gaussian) subspace generated by y. The notation $\text{span}\{y(k);\ k=1...m\}$ or simply span{y}, denotes the (closed) subspace of all linear combinations of components, y(k), of y. As it is well known, $\underline{H}$ is a real Hilbert space with scalar product $\langle\xi,\eta\rangle = E(\xi\eta)$ for $\xi,\eta \in \underline{H}$. For any random vector z, we shall use the shorthand $z \in \underline{H}$, to mean that all scalar components z(k) of z belong to $\underline{H}$. Similarly, given a subspace $K \subset \underline{H}$ the symbol $E(z|K)$ or $E^K z$ will denote the vector of conditional expectations of the components of z, given the minimal σ-algebra σ(K) with respect to which all $\xi \in K$ are measurable. It is well known ([15]) that the conditional expectation operator given σ(K), coincides, in $\underline{H}$, with the orthogonal projection on to the subspace K. In what follows, bar will denote closure in $\underline{H}$ and other selfevident vector notations will be used without further comments.

If $Z_1$, $Z_2$ and X are subspaces of $\underline{H}$, we shall say that $Z_1$ and $Z_2$ are conditionally orthogonal given X and write $Z_1 \perp Z_2 \mid X$, if

$$\langle E^X\eta_1, E^X\eta_2\rangle = \langle \eta_1, \eta_2\rangle \tag{2.1}$$

for all (scalar) $\eta_1 \in Z_1$, $\eta_2 \in Z_2$. By the remark above, conditional orthogonality is actually equivalent to conditional independence of $\sigma(Z_1)$ and $\sigma(Z_2)$ given $\sigma(X)$. If $Z_1$, $Z_2$ are conditionally orthogonal given X, we shall often also say that X is a <u>splitting subspace</u> for $Z_1$, $Z_2$. The notion of conditional orthogonality also applies in an obvious way to random vectors.

DEFINITION 2.1'

A "stochastic system" is a triple of random vectors $\{z_1,z_2,x\}$ in $\underline{\underline{H}}$ such that $Z_i := \text{span}\{z_i\}$, $i = 1,2$, are conditionally orthogonal given $X := \text{span}\{x\}$. The compound random vector $z := \begin{bmatrix} z_1 \\ z_2 \end{bmatrix}$ will be called the "output" of the system, x will be referred to as the "state" and X as the "state space". The dimension of X as a vector space will be called the dimension of the system.

The concept of splitting subspace and of stochastic system is central in stochastic realization theory ([10-13]). It is a generalization of the idea of sufficient statistic (or sufficient $\sigma$-algebra), at least in the gaussian case. In fact, by definition of conditional orthogonality we have ($\vee$ denotes vector sum)

$$E(z_1|Z_2 \vee X) = E(z_1|X) \tag{2.2}$$

or, equivalently

$$E(z_2|Z_1 \vee X) = E(z_2|X) \tag{2.3}$$

and these relations, for instance (2.2), tell that what is relevant in $z_2$ for predicting $z_1$ is already contained in X so that, if we have both $z_2$ (or $Z_2$) <u>and</u> X we can disregard $z_2$ completely. The symmetric interpretation of course holds for (2.3). Similar to the idea of sufficient statistic, splitting is of interest only if it corresponds to effective data reduction. Therefore the notion of <u>minimality</u> is of central importance.

A splitting subspace X for $Z_1$, $Z_2$ is <u>minimal</u> if there are no <u>proper</u> subspaces $X' \subset X$ which are also splitting. A <u>stochastic system</u> will correspondingly be called <u>minimal</u> if $X = \text{span}\{x\}$ is a minimal splitting subspace for $Z_i = \text{span}\{z_i\}$ $i = 1,2$.

LEMMA 2.1 ([17], [13])

<u>The subspace</u> X <u>is minimal splitting for</u> $Z_1$, $Z_2$ <u>if and only if the following two conditions hold,</u>

$$\bar{E}^{X} Z_1 = X \qquad \bar{E}^{X} Z_2 = X \quad . \tag{2.4}$$

The conditions (2.4) have been introduced by RUCKEBUSCH ([17]) in a dynamic context and named <u>Observability</u> and <u>Reconstructability</u>.

It can be shown that <u>the so called predictor spaces</u>, $\bar{E}^{Z_2} Z_1$ and $\bar{E}^{Z_1} Z_2$, <u>are always minimal splitting</u> but in general there are many others and the central problem in stochastic realization theory is to find and classify all of them. At this purpose the following Lemma plays an important rôle.

LEMMA 2.2 ([10], [11])

<u>All minimal splitting subspaces for</u> $(Z_1, Z_2)$ <u>which are contained in</u> $Z_1 \vee Z_2$ <u>are subspaces of the "frame space"</u> $\bar{E}^{Z_2} Z_1 \vee \bar{E}^{Z_1} Z_2$.

The next concept is of fundamental importance for the classification of splitting subspaces.

We shall say that two subspaces $S_1$, $S_2$ of $\underline{\underline{H}}$ <u>intersect perpendicularly</u> ([11]) if

$$\bar{E}^{S_2} S_1 = S_1 \cap S_2 = \bar{E}^{S_1} S_2 \tag{2.5}$$

This notion has an intuitive geometrical meaning. Let A,B be subspaces of $\underline{\underline{H}}$, $B^{\perp}$ be the orthogonal complement of B in $\underline{\underline{H}}$ and $\oplus$ denote orthogonal direct sum. It follows from the identity

$$A = \bar{E}^{A} B \oplus (A \cap B^{\perp}) \tag{2.6}$$

(for a proof see e.g. [16]) that $S_1$ and $S_2$ intersect perpendiculary if an only if the vector sum $S_1 \vee S_2$ admits the orthogonal decomposition

$$S_1 \vee S_2 = N_1 \oplus S_1 \cap S_2 \oplus N_2 \tag{2.7}$$

where

$$N_1 = S_1 \cap S_2^{\perp} \quad , \qquad N_2 = S_2 \cap S_1^{\perp} \quad .$$

It is an easy matter to check that whenever $S_1$ and $S_2$ intersect perpendicularly there is just <u>one</u> minimal splitting subspace for $(S_1, S_2)$, namely, their intersection $X := S_1 \cap S_2$. (This follows from the fact that any (A,B)-splitting subspace X must contain the intersection $A \cap B$. In this case the intersection is splitting because the left and right members in the equality (2.5) are). This fact makes the following Lemma a very useful technical tool.

LEMMA 2.3

Let $X$, $Z_1$, $Z_2$ be subspaces of $\underline{H}$. Then $X$ is splitting for $(Z_1, Z_2)$ if and only if there is a pair of perpendicularly intersecting subspaces $(S_1, S_2)$ containing (respectively) $Z_1$ and $Z_2$, for which,

$$X = S_1 \cap S_2 \tag{2.9}$$

All such pairs $(S_1, S_2)$ are described by the formula

$$S_i = (Z_i \vee X) \oplus V_i \qquad i = 1,2 \tag{2.10}$$

where the subspaces $V_1$, $V_2$ and $Z_1 \vee X \vee Z_2$ are pairwise orthogonal i.e.

$$V_1 \perp (Z_1 \vee X \vee Z_2) \perp V_2 \quad . \tag{2.11}$$

Proof:

(if). Let $X$ be given by (2.9). Then, since $(S_1, S_2)$ intersect perpendicularly $S_1 \perp S_2 | X$; but $S_i \supset Z_i$ $i = 1,2$ implies that $Z_1 \perp Z_2 | X$.

(only if). Assume $Z_1 \perp Z_2 | X$; it is not hard to show that this relation implies $Z_1 \vee X \perp Z_2 \vee X | X$. Define then $S_i := Z_i \vee X$, $i = 1,2$. We show first that $S_1$ and $S_2$ intersect at $X$. For, since $X$ is splitting for $(S_1, S_2)$, $S_1 \cap S_2 \subset X$. On the other hand both $S_1$ and $S_2$ contain $X$ and therefore $S_1 \cap S_2 \supset X$. Thus (2.9) holds. It follows then that $S_1 \perp S_2 | S_1 \cap S_2$. Since the predictor space $E^{-S_2} S_1$ is minimal splitting for $(S_1, S_2)$, it must necessarily coincide with $S_1 \cap S_2$ and likewise for $E^{-S_1} S_2$. This shows that $S_1$ and $S_2$ intersect perpendicularly.

The above concludes the proof of the first statement in the Lemma. Let now $S_i = Z_i \vee X$ $i = 1,2$, as before and let $(\bar{S}_1, \bar{S}_2)$ be any pair of subspaces such that i) $(\bar{S}_1, \bar{S}_2)$ intersect perpendicularly, ii) $\bar{S}_i \supset Z_i$, $i = 1,2$, and iii) $X = \bar{S}_1 \cap \bar{S}_2$. It follows from iii) that $\bar{S}_i \supset X$, $i = 1,2$ and hence, by ii), $\bar{S}_i \supset S_i = Z_i \vee X$. Let $V_i$ be the orthogonal complement of $S_i$ in $\bar{S}_i$. Since $V_i \subset \bar{S}_i$ and $S_i \subset \bar{S}_i$ we have $V_1 \perp S_2 | X$ and $V_2 \perp S_1 | X$. These relations are in turn equivalent to

$V_1 \perp (S_2 \ominus X)$ , $V_2 \perp (S_1 \ominus X)$

Now, $V_1 \perp S_1$, together with the first orthogonality relation implies that $V_1 \perp [S_1 \vee \vee (S_2 \ominus X)]$ and, as $S_1$ and $S_2$ intersect perpendicularly, the vector sum between square bracket is actually an orthogonal direct sum (compare (2.7)) equal to $S_1 \vee S_2$. By the same argument we show that $V_2 \perp S_1 \vee S_2$. Finally, $V_1 \perp V_2$ follows from $V_i \perp X$ and $V_1 \perp V_2 | X$, just by recalling the definition of conditional orthogonality (2.1).

We are now ready for the main representation Theorem.

THEOREM 2.4

A triple of random vectors $\{z_1, z_2, x\}$ in $\underline{H}$ (of respective dimensions, $m_1$, $m_2$, n) is a stochastic system if and only if it admits a representation of the form

$$z_1 = H_1 x + w_1$$
$$z_2 = H_2 x + w_2 \tag{2.12}$$

where $H_i \in R^{m_i \times n}$ are constant matrices and $w_i$ are uniquely determined $m_i$-dimensional random vectors such that

$$w_1 \perp x \perp w_2 \tag{2.13}$$

If $\{x(k), k=1..n\}$ form a basis for the state space X, then the matrices $H_i$ are also uniquely determined by the triple.

The system is minimal if and only if

$$\text{rank } H_i = \dim X \qquad i = 1,2 \tag{2.14}$$

both hold.

Proof:

By definition 2.1, $\{z_1, z_2, x\}$ is a stochastic system iff X is a splitting subspace for $(Z_1, Z_2)$. This in turn happens, by Lemma 2.3, iff there are perpendicularly intersecting pairs $(S_1, S_2)$ with $S_i \supset Z_i$, representing X as $S_1 \cap S_2$. Let us choose any $(S_1, S_2)$ with the above properties. By using the orthogonal decomposition (2.7), we get

$$
\begin{aligned}
z_1 &= E^{S_1} z_1 = E^X z_1 + E^{N_1} z_1 \ := H_1 x + w_1 \\
z_2 &= E^{S_2} z_2 = E^X z_2 + E^{N_2} z_2 \ := H_2 x + w_2
\end{aligned}
\tag{2.15}
$$

Clearly the orthogonality condition (2.13) follows immediately from (2.7). It is also trivial that $E^X z_i$ will be uniquely expressible in terms of x only when x is a minimal set of generators for X.

Note that any choice of the representing pair $(S_1, S_2)$ will yield the same decomposition (2.15) because of the orthogonality relation (2.11). In fact,

$$E(z_i | S_i) = E[z_i | (Z_i \vee X) \oplus V_i] = E[z_i | Z_i \vee X] \qquad i = 1,2$$

for any $(S_1, S_2)$.

The state space X is minimal splitting iff conditions (2.4) hold. These conditions say that the projection operators $E^X : Z_i \to X$, $i = 1,2$ must be onto (here of course X and $Z_i$ are finite dimensional). Clearly this happens only if

$$\dim X \leq \min_i \{\dim Z_i\} .$$

and hence we have minimality whenever (2.14) holds.

## 3. THE MODELLING PROBLEM

Suppose y is a given m dimensional zero mean gaussian random vector. In this section we shall formulate our basic modelling problem. Somewhat roughly stated, it is concerned with the following question: when is it possible to generate y by a "linear statistical model" of the form

$$y = H x + w \tag{3.1}$$

where x is some random vector whose dimension n is (as small as possible and in particular), smaller than m; w (the "noise" term) is independent of x and has a covariance matrix of pre-assigned block-diagonal structure?

Models of the type (3.1) are encountered in the statistical literature and commonly referred to as Factor Analysis models ([2], [8]). They are extremely useful in various inference and testing procedures. As yet, however, many basic questions about the representation (3.1) like existence, bounds on the dimension of x, computation of the parameters of the model starting from the covariance data of y, but especially the inherent non uniqueness of the representation, seem to be very poorly understood.

Note that the fundamental requirement which makes the problem non trivial is the (blockwise) independence of prespecified components of the noise term. In Factor Analysis one actually often requires the covariance matrix of w to be diagonal. In this paper we shall, for the sake of simplicity, just look at the simplest situation, namely the case in which w is partitioned into two random subvectors $w = [w_1' \; w_2']'$ of respective dimensions $m_1$, $m_2 (\geq 1)$ and $(w_1, w_2)$ are required to be independent. Of course this is equivalent to describing y, conformably partitioned as $y = [y_1', y_2']'$, by a model of the form

$$
\begin{aligned}
y_1 &= H_1 x + w_1 \\
y_2 &= H_2 x + w_2
\end{aligned}
\tag{3.2}
$$

where $w_1 \perp x \perp w_2$.

Motivated by the above informal discussion, we shall, from now on, assume that, together with y, there is assigned a partitioning

$$
y = \begin{bmatrix} y_1 \\ y_2 \end{bmatrix}
\tag{3.3}
$$

with $y_1$ and $y_2$ of respective (fixed) dimensions $m_1$, $m_2$.

A precise statement of the modelling problem can now be given. It is based on the equivalence between models of the form (3.2) and the notion of stochastic system established by Theorem 2.4.

PROBLEM P.1

Given the m-dimensional zero mean random vector y partitioned as in (3.3) and a Gaussian space $\underline{\underline{H}}$ containing $Y = \text{span}\{y\}$, find a stochastic system $\{z_1, z_2, x\}$ in $\underline{\underline{H}}$, such that the output vector $z = [z_1', z_2']'$ equals y almost surely.

The above is the so called strong stochastic realization problem for y. The attribute "strong" is given because a system $\{z_1, z_2, x\}$ which "realizes" y (also called a realization of y) is required to live in the pre-specified space $\underline{\underline{H}}$. The latter object, in a sense, specifies what "source of randomness" is available to build the random variables defining the realization. In this respect, the most natural situation arises when $\underline{\underline{H}} = Y$ i.e. the observed vector y is the only available "source of randomness" which can be used to construct a probabilistic model. In this case the realizations we are looking for will be called internal or output-induced. (Note that the term internal has now a different meaning than in sect. 1). These will form the main object of our study.

Of course the most interesting realizations are the minimal ones and the main objective in solving problem P.1 will be to find and classify all of them.

There is a weaker version of the stochastic realization problem where the gaussian space $\underline{\underline{H}}$ is not a priori specified and a realization can be constructed on an arbitrary probability space. In this case the equality y=z can only be understood in the sense of equality between probability laws (i.e. covariances).

PROBLEM P.2

Given the $m \times m$ covariance matrix $\Lambda$ of an m-dimensional zero mean random vector y, partitioned as in (3.3), find a stochastic system $\{z_1, z_2, x\}$ with $\dim z_i = \dim y_i$ $i = 1,2$ such that the covariance matrix of the output z equals the assigned matrix $\Lambda$.

Obviously any strong realization necessarily satisfies the requirements of problem P.2. The converse however is clearly false as a "weak realization" (i.e. a solution to problem P.2) will not in general provide output sample values equal to y. Note also that problem P.2 can be stated purely in terms of covariances. In this respect it is best to think of a (weak) realization as a model of the form (2.12) or, even better, merely as a 5-tuple of matrices $\{H_1, H_2, Q, R_1, R_2\}$ where

$$Q = E\, xx' \quad , \qquad R_i = E\, w_i w_i' \qquad i = 1,2 \quad . \tag{3.4}$$

Clearly any 5-tuple of this kind defines completely the joint statistics of the system $\{z_1, z_2, x\}$.

Problem P.2 can, in this setting, be restated in the following way

PROBLEM P.2'

Given an $m \times m$ covariance matrix $\Lambda$, partitioned in the form

$$\Lambda = \begin{bmatrix} \Lambda_1 & \Lambda_{12} \\ \Lambda_{21} & \Lambda_2 \end{bmatrix} \quad , \qquad \Lambda_{21} = \Lambda_{12}' \tag{3.5}$$

*where $\Lambda_i$ are of dimension $m_i \times m_i$, $i=1,2$, find all 5-tuples of matrices $\{H_1, H_2, Q, R_1, R_2\}$ with $H_i$ of dimension $m_i \times n$, $Q$ of dimension $n \times n$, $R_i$ of dimension $m_i \times m_i$, such that,*

i) *The following relations hold,*

$$\begin{aligned} \Lambda_1 &= H_1 Q H_1' + R_1 \\ \Lambda_{12} &= H_1 Q H_2' \\ \Lambda_2 &= H_2 Q H_2' + R_2 \quad . \end{aligned} \tag{3.6}$$

ii) *The* $(m+n) \times (m+n)$ *matrix,*

$$\tilde{\Lambda} = \begin{bmatrix} \Lambda_1 & \Lambda_{12} & H_1 Q \\ \Lambda_{21} & \Lambda_2 & H_2 Q \\ Q H_1' & Q H_2' & Q \end{bmatrix} \tag{3.7}$$

*is symmetric and nonnegative definite (i.e. a covariance matrix).*

REMARKS

Condition (i) imposes the equality of the given covariances $Ey_i y_j'$, $ij=1,2$ with the output covariances $Ez_i z_j'$ of the model (2.12). Notice that the dimension n (of the state vector x) is an unknown of the problem and is to be determined (actually the interesting question is to determine the *smallest possible* n). Moreover the solutions to the set of algebraic equations (3.6) should of course provide matrices $Q$, $R_1$, $R_2$ which are *covariance* matrices. *This constraint is actually included in condition* (ii). Note in fact that $\tilde{\Lambda}$ in (3.7) is the joint covariance of $(z_1, z_2, x)$ and the statement of Problem P.2' demands nothing else but the fact that the (Gaussian) probability law of $(y_1, y_2)$ should actually be obtained as the *marginal* of the joint law of $(z_1, z_2, x)$.

For brevity, we shall refer to condition (ii) as the *positivity condition*.

## 4. STRUCTURE OF MINIMAL WEAK REALIZATIONS

In this and in the following section we shall describe the solution sets to problems P.2 and P.1. We shall worry only about *minimal* realizations. In addition we shall, from now on, adopt the convention of taking the state x as a *basis* in the minimal $(Z_1,Z_2)$ - splitting subspace (the state space) X.

In order to avoid uninteresting algebraic complications the following assumption will be made.

ASSUMPTION 4.1

The covariance matrix, $\Lambda$, of y is positive definite.

We shall start with the following result.

LEMMA 4.2

All minimal realizations have the same dimension $n = \text{rank}\ \Lambda_{12}$.

Proof:

Let $\{z_1, z_2, x\}$ be a minimal weak realization of dimension $n_x$. By (2.14) of Thm 2.4 the matrices $H_1$ and $H_2$ in the representation (2.12) are of full rank $n_x$. By definition we have

$$\langle E^X z_1(k),\ E^X z_2(j)\rangle = \langle z_1(k),\ z_2(j)\rangle = \langle y_1(k),\ y_2(j)\rangle$$

for all $k=1\ldots m_1$, $j=1\ldots m_2$. But this is clearly equivalent to

$$H_1 Q H_2' = \Lambda_{12} \tag{4.1}$$

where $Q = Exx'$, x being a basis in X. Since rank $H_1 Q H_2' = n_x = \text{rank}\Lambda_{12}$ for any minimal realization, the conclusion follows.

Clearly, by the Lemma above, all minimal strong realizations (irrespective of what $\underline{\underline{H}}$ is) also have the same dimension $n = \text{rank}\ \Lambda_{12}$.

Let us fix once and for all a rank factorization of $\Lambda_{12}$,

$$\Lambda_{12} = HG'$$

with H and G of respective dimensions $m_1 \times n$, $m_2 \times n$ and rank $H = \text{rank}\ G = \text{rank}\ \Lambda_{12} = n$. From now on the matrices H and G will be considered as a part of the problem data.

LEMMA 4.3

Let $\{z_1, z_2, \tilde{x}\}$ be any minimal weak realization of y and assume $\Lambda_{12}$ is factored as in (4.2). Then there is a change of basis $\tilde{x} = Tx$ such that in the representing model (2.12),

$$H_1 = H \qquad H_2 = G\,P^{-1} \tag{4.3}$$

where $P = Exx'$. Similarly we can always choose a new basis $\bar{x}$ in such a way that,

$$H_1 = H\,\bar{P}^{-1} \qquad H_2 = G$$

where $\bar{P} = E\bar{x}\bar{x}'$.

Proof:

In fact, if we start from

$$z_1 = \tilde{H}_1\tilde{x} + w_1$$

$$z_2 = \tilde{H}_2\tilde{x} + w_2$$

and introduce the basis change $\tilde{x} = Tx$ with T such that $\tilde{H}_1 T = H$ (note that such nonsingular T always exists as $\tilde{H}_1$ and H are of full rank n), we get

$$HG' = Ez_1z_2' = HPT'\tilde{H}_2'$$

from which,

$$\tilde{H}_2 = GP^{-1}\,T^{-1}$$

with $P = Exx'$. This proves the Lemma.

Note that any minimal realization can then be written in any one of the following two "canonical" forms

$$\begin{aligned} z_1 &= Hx + w_1 & \qquad z_1 &= H\bar{P}^{-1}\bar{x} + w_1 \\ z_2 &= GP^{-1}x + w_2 & \qquad z_2 &= G\,\bar{x} + w_2 \end{aligned} \tag{4.5}$$

which are related by the transformation $\bar{x} = P^{-1}x$. We shall call "type 1" and "type 2" canonical forms the first and second kind of representations (4.5) respectively. Till now we have not been worrying about showing that (minimal) realizations of y exist at all. Indeed there are plenty. We shall now explicitly construct two strong realizations which will play an important rôle in the following.

Let us choose $\underline{\underline{H}} = Y_1 \vee Y_2$, where $Y_i = \text{span}\{y_i\}$ $i = 1,2$ and consider the predictor spaces

$$X_1 := E^{Y_1} Y_2 \quad , \qquad X_2 := E^{Y_2} Y_1 \tag{4.6}$$

which are minimal splitting for $(Y_1,Y_2)$.

Let $N_1$ be the orthogonal complement of $X_1$ in $Y_1$ and $N_2$ the orthogonal complement of $X_2$ in $Y_2$. From the orthogonal direct sum decompositions

$$H = N_1 \oplus X_1 \oplus Y_1^{\perp} = Y_2^{\perp} \oplus X_2 \oplus N_2$$

it is not difficult to show (see e.g. [12]) that $X_1$ and $X_2$ can be represented, in the sense of Lemma 2.3, in terms of the perpendicularly intersecting subspaces $(Y_1,N_1^{\perp})$ and $(N_2^{\perp},Y_2)$, i.e.

$$X_1 = Y_1 \cap N_1^{\perp} \quad , \qquad X_2 = N_2^{\perp} \cap Y_2 \tag{4.8}$$

where the orthogonal complements are taken w.r. to $\underline{\underline{H}}$.

We now proceed to choose a convenient basis in the splitting subspaces $X_1$ and $X_2$. From

$$E(y_2|y_1) = \Lambda_{21}\Lambda_1^{-1} y_1 = GH'\Lambda_1^{-1} y_1 \tag{4.9a}$$

$$E(y_1|y_2) = \Lambda_{12}\Lambda_2^{-1} y_2 = HG'\Lambda_2^{-1} y_2 \quad , \tag{4.9b}$$

we see that we can choose

$$\bar{x}_1 := H'\Lambda_1^{-1} y_1 \quad , \qquad x_2 := G'\Lambda_2^{-1} y_2 \tag{4.10}$$

as (n-dimensional) basis vectors in $X_1$ and $X_2$ respectvely. Their covariances are

$$\bar{P}_1 = E\bar{x}_1\bar{x}_1' = H'\Lambda_1^{-1} H \quad , \qquad P_2 = Ex_2x_2' = G'\Lambda_2^{-1} G \tag{4.11}$$

The linear model representations of the strong realizations $\{y_1,y_2,\bar{x}_1\}$ and $\{y_1,y_2,x_2\}$ can at this point be obtained by the computations sketched in the proof of Theorem 2.4 (formulas (2.15)). In the given bases we have,

$$E^{X_1} y_1 = E(y_1|\bar{x}_1) = E\, y_1\bar{x}_1' \cdot \bar{P}_1^{-1} \bar{x}_1 = H\, \bar{P}_1^{-1} \bar{x}_1$$

$$E^{X_1} y_2 = E(y_2|\bar{x}_1) = E\, y_2\bar{x}_1' \cdot \bar{P}_1^{-1} \bar{x}_1 = G\, \bar{x}_1$$

and, by a similar computation,

$$E^{X_2} y_1 = E(y_1|x_2) = H\, x_2$$

$$E^{X_2} y_2 = E(y_2|x_2) = G\, P_2^{-1} x_2 \quad .$$

PROPOSITION 4.4

The random vector y admits minimal internal realizations with state spaces $X_1$ and $X_2$ defined by (4.6). The corresponding linear models are

$$\begin{aligned} y_1 &= H\bar{P}_1^{-1}\bar{x}_1 + v_1 \\ y_2 &= G\,\bar{x}_1 + v_2 \end{aligned} \qquad (4.12)$$

and

$$\begin{aligned} y_1 &= H\, x_2 + n_1 \\ y_2 &= GP_2^{-1}\, x_2 + n_2 \end{aligned} \qquad (4.13)$$

where $\bar{x}_1$, $x_2$, $\bar{P}_1$ and $P_2$ are given by (4.10) and (4.11). The noise terms $v_1$ and $v_2$ in (4.12) belong to $N_1$ and $Y_1^{\perp}$ respectively and are given by

$$v_1 = (I-\Pi_H)y_1 \quad , \qquad v_2 = y_2-E(y_2|y_1) \qquad (4.14)$$

where $\Pi_H := H(H'\Lambda_1^{-1}H)^{-1}H'\Lambda_1^{-1}$ is a projection operator onto the column space of the matrix H.

Likewise, the noises $n_1$ and $n_2$ belong to $Y_2^{\perp}$ and $N_2$ respectively and are given by

$$n_1 = y_1 - E(y_1|y_2) \quad , \qquad n_2 = (I-\Pi_G)y_2 \qquad (4.15)$$

where $\Pi_G := G(G'\Lambda_2^{-1}G)^{-1}G'\Lambda_2^{-1}$ is a projection operator onto the column space of G.

REMARK

Note that the model (4.12) is in canonical form 2 while model (4.13) is in canonical form 1. By a trivial change of basis they can be brought to whatever canonical form one likes.

An interesting fact emerges from the structure of the first equation in (4.12) and the second equation in (4.13). Taking into account the equality $H\bar{P}_1^{-1}\bar{x}_1 = \Pi_H y_1$, the first can for instance be rewritten as

$$y_1 = \Pi_H y_1 + (I-\Pi_H) y_1$$

which is indeed an orthogonal decomposition as $\Pi_H \Lambda_1 (I-\Pi_H)' = 0$ (i.e. $\Pi_H$ is a "$\Lambda_1$-orthogonal" projector). The above formula leads to an interpretation of $\bar{x}_1$ as a "Fisher estimate" of the deterministic parameter $\theta$ in the linear model $y_1 = H\theta + \varepsilon_1$.

At this point we are ready to study the set of all minimal weak realizations. We shall agree to choose the basis in the state space in such a way that either (4.3) or (4.4) hold i.e. we shall take the linear models either in canonical form 1 or 2. Note that this choice parametrizes each 5-tuple $\{H_1, H_2, Q, R_1, R_2\}$ uniquely in terms of the state covariance matrix. For example, in canonical form 1, we can express $H_1, H_2$, $R_1, R_2$ in terms of the state covariance P and the fixed matrices $H_1, G, \Lambda_1, \Lambda_2$ by the relations (4.3) and (3.6). The crucial condition which then determines the solution set to problem P.2' is the positivity condition (ii).

Let us consider the matrix $\tilde{\Lambda}$ in (3.7). By a standard block diagonalization procedure, it is easy to see that the positivity condition reduces to the set of matrix inequalities

$$\begin{aligned} &\Lambda_1 \geq 0 \\ &\tilde{\Lambda}_2 := \Lambda_2 - \Lambda_{21}\Lambda_1^{-1}\Lambda_{12} \geq 0 \\ &Q-QH_1'\Lambda_1^{-1}H_1Q - (QH_2'-QH_1'\Lambda_1^{-1}H_1QH_2')\tilde{\Lambda}_2^{-1} \cdot (QH_2'-QH_1'\Lambda_1^{-1}H_1QH_2')' \geq 0 \end{aligned} \tag{4.16}$$

of which the first two are trivially satisfied. Note in fact that $\tilde{\Lambda}_2$ is the covariance of $y_2 - E(y_2|y_1)$ which, given the stated assumptions on $\Lambda$, is strictly positive definite. It will turn out useful to rewrite the positivity condition in an equivalent form, where instead of $\tilde{\Lambda}$ we consider the joint covariance matrix of the same vectors but with $z_2$ in place of $z_1$, i.e. exchanging everywhere in $\tilde{\Lambda}$ the indices 1 and 2. The resulting inequalities are obtained from (4.16) just by exchanging indices; in particular we get

$$Q-QH_2'\Lambda_2^{-1}H_2Q-(QH_1'-QH_2'\Lambda_2^{-1}H_2QH_1')\tilde{\Lambda}_1^{-1} \cdot (QH_1'-QH_2'\Lambda_2^{-1}H_2QH_1')' \geq 0 \tag{4.17}$$

where,

$$\tilde{\Lambda}_1 = \Lambda_1 - \Lambda_{12}\Lambda_2^{-1}\Lambda_{21} \quad .$$

The matrix $\tilde{\Lambda}_1$ is also strictly positive definite.

THEOREM 4.5

All minimal weak realizations in canonical form 1, $\{H, GP^{-1}, P, \Lambda_1 - HPH', \Lambda_2 - GP^{-1}G'\}$, are parametrized by the symmetric solutions of the algebraic quadratic inequality,

$$P - P_2 - (P-P_2)\left[\bar{P}_1^{-1} - P_2\right]^{-1}(P-P_2)' \geq 0 \tag{4.18}$$

where $\bar{P}_1$ and $P_2$ are given in (4.11).
Dually, all minimal weak realizations in canonical form 2, $\{H\bar{P}^{-1}, G, \bar{P}, \Lambda_1 - H\bar{P}^{-1}H', \Lambda_2 - G\bar{P}G'\}$, are parametrized by the symmetric solutions of the algebraic quadratic inequality,

$$\bar{P} - \bar{P}_1 - (\bar{P}-\bar{P}_1)\left[P_2^{-1} - \bar{P}_1\right]^{-1}(\bar{P}-\bar{P}_1)' \geq 0 \quad . \tag{4.19}$$

The symmetric matrix $P$ is a solution to (4.18) if and only if $P^{-1}$ solves (4.19). Similarly for $\bar{P}$. Finally, all solutions $P$ ($\bar{P}$) to (4.18) (resp to (4.19)) admit an upper and lower bound, in fact,

$$\bar{P}_1^{-1} \geq P \geq P_2 \quad , \qquad P_2^{-1} \geq \bar{P} \geq \bar{P}_1 \quad . \tag{4.20}$$

Proof:

Recall first that $\bar{P}_1$ and $P_2$ are invertible by minimality. Likewise $G'\tilde{\Lambda}_2^{-1}G$ and $H'\tilde{\Lambda}_1^{-1}H$ are also invertible.

Let us take any minimal weak realization in canonical form 1. Clearly this realization will be representable by a 5-tuple of the form $\{H, GP^{-1}, P, \Lambda_1 - HPH', \Lambda_2 - GP^{-1}G'\}$ as stated in the lemma. Notice that this 5-tuple automatically satisfies condition (i) of problem P.2'. By definition, it also has to satisfy the positivity condition. The latter, by using the form (4.17), can be rewritten as

$$P - P_2 - (P-P_2)H'\tilde{\Lambda}_1^{-1}H(P-P_2)' \geq 0 \quad ,$$

with,

$$H'\tilde{\Lambda}_1^{-1}H = H'\left[\Lambda_1 - HP_2H'\right]^{-1}H = \left[\bar{P}_1^{-1} - P_2\right]^{-1}$$

by the matrix inversion Lemma. Hence P satisfies (4.18).

Viceversa, assume P is symmetric and satisfies (4.18). By isolating $P-P_2$ on the left of the inequality sign it is seen that $P \geq P_2$ (i.e. <u>all solutions to</u> (4.18) <u>admit</u> $P_2$ <u>as a lower bound</u>) and thus P is necessarily positive definite. Introduce now, $H_1 := H$, $H_2 := GP^{-1}$, $Q := P$, $R_1 := \Lambda_1 - HPH'$, $R_2 := \Lambda_2 - GP^{-1}G'$. This clearly is a weak realization (i.e. a solution to problem P.2') as it satisfies the positivity condition (which has already been shown to be equivalent to (4.18)) and the algebraic relations (3.6). Since $\text{rank}H_1 = \text{rank}H_2 = \text{rank}\Lambda_{12}$ it is also minimal. This argument can be repeated verbatim for realizations in canonical form 2, just use the form (4.16) of the positivity condition instead of (4.17) to show that all $\bar{P}$ must satisfy the inequality (4.11). Here we come to the conclusion that all solutions to (4.19) satisfy the inequality $\bar{P} \geq \bar{P}_1$.

It has already been noted before that a realization in canonical form 1 is transformed into canonical form 2 by the change of basis $\bar{x} = P^{-1}x$. In other words a symmetric $n \times n$ positive definite matrix P is the state covariance of a realization in canonical form 1 if and only if $P^{-1}$ is the state covariance matrix of a realization in canonical form 2. Thus every $\bar{P}$ can be written as $P^{-1}$ for a suitable P and viceversa. This last comment together with the inequalities $P \geq P_2$, $\bar{P} \geq \bar{P}_1$, justifies the last claim of the Theorem.

REMARK

The reader may wonder whether the noise covariances $\Lambda_1 - HPH'$, $\Lambda_2 - GP^{-1}G'$ etc. are actually positive semidefinite matrices. As pointed out in the remark after the statement of problem P.2' this is indeed so and in fact the positivity of $R_1$, $R_2$ is (alike that of the state covariance) a consequence of the general positivity condition (ii). This can be seen also from the following argument.

Compute the covariance matrix of the noise vector $v_1$ in the strong "$y_1$-measurable" realization (4.12). From (4.14) we get

$$Ev_1v_1' = \Lambda_1 - H(H'\Lambda^{-1}H)^{-1}H' = \Lambda_1 - H\bar{P}_1^{-1}H'$$

This matrix is obviously semidefinite, being a bona fide covariance. Use now the first inequality in (4.20) to check the positivity of $\Lambda_1 - HPH'$.

Theorem 4.5 provides a recipe for computing all minimal weak realizations of y in a given coordinate system. The structure of the set $\mathscr{P}$ of all symmetric solutions to an algebraic quadratic matrix inequality of the type (4.18) or (4.19) has been thouroughly investigated (see e.g. [5]) in connection with various system theoretic problems related to the idea of positvity. A general feature of $\mathcal{P}$ is that it is a bounded closed convex set with a nonempty interior. This means that, in a given fixed coordinate system (e.g. in canonical form 1), there are in general infinitely many minimal realizations of y of the form (3.2).

## 5. STRUCTURE OF THE SET OF MINIMAL INTERNAL REALIZATIONS

At this point the problem arises of classifying the various solutions described by Thm 4.5 i.e., describing in what minimal weak realizations differ one from another. The following result is a first step in this direction.

LEMMA 5.1

A minimal strong realization in canonical form 1 is internal (i.e. $x \in Y_1 \vee Y_2$) if and only if its state covariance matrix P satisfies the quadratic matrix equation

$$P-P_2-(P-P_2)\left[\bar{P}_1^{-1}-P_2\right]^{-1}(P-P_2)'=0 \tag{5.1}$$

Likewise, a minimal strong realization in canonical form 2 is internal if and only if the state covariance $\bar{P}$ satisfies,

$$\bar{P}-\bar{P}_1-(\bar{P}-\bar{P}_1)\left[P_2^{-1}-\bar{P}_1\right]^{-1}(\bar{P}-\bar{P}_1)'=0 \tag{5.2}$$

Proof:

Consider the joint covariance matrix $\hat{\Lambda}$ of $\{y_2, y_1, x\}$ and suppose $\{y_1, y_2, x\}$ is represented by a linear model in canonical form 1. As pointed out before (compare the derivation of (4.17)), $\hat{\Lambda}$ can be reduced to block diagonal form,

$$\hat{\Lambda} \sim \operatorname{diag}\{\Lambda_2, \tilde{\Lambda}_1, P-P_2-(P-P_2)\left[\bar{P}_1^{-1}-P_2\right]^{-1}(P-P_2)'\}$$

Clearly x is a linear combination of $(y_1, y_2)$ if and only if rank $\hat{\Lambda}=m_1+m_2$. This in turn happens if and only if (5.1) holds.

Notice in particular, that the state covariances $\bar{P}_1^{-1}$ and $P_2$ of the $y_1$-measurable and (resp.) $y_2$-measurable realizations (4.12), (4.13) (of course once both are written in form 1) trivially satisfy (5.1).

REMARK

Let us denote by $\mathscr{P}_o$, resp. $\bar{\mathscr{P}}_o$, the boundary of the solution set $\mathscr{P}$ ($\bar{\mathscr{P}}$) of the matrix inequalities (4.18), (4.19). Lemma 5.1 can be amplified somehow by saying that all symmetric $P \in \mathscr{P}_o$ can always be interpreted as state covariances of strong, internal realizations (in form 1). In fact, whenever $P \in \mathscr{P}_o$ we see from the argument given in the proof, that $x \in Z_1 \vee Z_2$. But then we can as well choose the space in which the realization lives equal to $Y_1 \vee Y_2$ by taking $z_i = y_i$ $i = 1,2$.

It can be shown (but we shall not do that here) that all weak realizations corresponding to symmetric P is in the interior of $\mathscr{P}$, still can be interpreted as strong minimal realizations. In this case however, it is necessary to use a larger space than $Y_1 \vee Y_2$ to build the state vector and an exogenous noise (e.g. independent of y) needs to be introduced. For a result in this vein see [19] sect. 5. The above picture agrees with very general results in the theory of stochastic equations obtained by ERSHOV ([3], [4]).

In the rest of this section we shall investigate the structure of internal minimal realizations.

LEMMA 5.2

Let x and $\bar{x}$ be the state vector of any strong minimal realization of y in canonical form 1 and 2 respectively. Then the following projection formulas hold,

$$E(x|y_2) = x_2 \quad , \qquad E(x|y_1) = P\bar{x}_1 \tag{5.3a}$$

$$E(\bar{x}|y_1) = \bar{x}_1 \quad , \qquad E(\bar{x}|y_2) = \bar{P}x_2 \tag{5.3b}$$

where $\bar{x}_1$ and $x_2$ are defined by (4.10) and P, $\bar{P}$ denote the covariance matrices of x and $\bar{x}$.

In particular, for any state vector x, $\bar{x}$ as above,

$$E(x|x_2) = x_2 \quad , \qquad E(\bar{x}|\bar{x}_1) = \bar{x}_1 \tag{5.4}$$

i.e. the projection of x onto $X_2$ and of $\bar{x}$ onto $X_1$ are invariant in the set of all minimal strong realizations.

Proof:

follows by straightforward computations and is therefore omitted.

Let us rewrite the $y_1$-measurable realization (4.12) in canonical form 1, i.e. perform the change of basis $x_1 = \bar{P}_1^{-1}\bar{x}_1$. We get,

$$y_1 = Hx_1 + v_1$$
$$y_2 = GP_1^{-1}x_1 + v_2 \quad , \tag{5.5}$$

where we have introduced the "maximal state covariance" (in canonical form 1)

$$P_1 := \bar{P}_1^{-1} = Ex_1x_1' \tag{5.6}$$

Introduce the random vector

$$\tilde{x}_1 := x_1 - E(x_1|x_2) = x_1 - x_2 \tag{5.7}$$

where in the last equality we have used the invariance relation (5.4). Note that $\tilde{x}_1$ is orthogonal to $x_2$ and thus $(\tilde{x}_1, x_2)$ <u>are orthogonal generators for the frame space</u> $X_1 \vee X_2 = \text{span}\{x_1\} \vee \text{span}\{x_2\}$.
Recalling Lemma 2.2, the state vector x of any minimal internal realization belongs to the frame space. Hence $x \in \text{span}\{\tilde{x}_1\} \oplus \text{span}\{x_2\}$ and from this relation we obtain,

$$x = Ex\tilde{x}_1' \cdot \Delta^{-1}\tilde{x}_1 + Exx_2' \cdot P_2^{-1}x_2$$

where $\Delta := E\tilde{x}_1\tilde{x}_1' = \bar{P}_1^{-1} - P_2 > 0$. By formulas (5.3) and after some simple algebra the following expression is obtained,

$$x = \Pi x_1 + (I-\Pi)x_2 \tag{5.8}$$

where

$$\Pi = (P-P_2)\Delta^{-1} \quad . \tag{5.9}$$

A completely analogous formula can be derived for the state $\bar{x}$ of any minimal internal realization (in canonical form 2) in terms of $\bar{x}_1$ and $\bar{x}_2 := P_2^{-1}x_2$.

Note now that $\Pi$ in (5.9) is a <u>projection operator</u>, actually a "$\Delta$-orthogonal" projector, as it satisfies

$$\Pi\Delta(I-\Pi)' = 0 \quad . \tag{5.10}$$

In fact from the basic quadratic equation (5.1) we get

$$(P-P_2)\Delta^{-1} = (P-P_2)\Delta^{-1} (P-P_2)'\Delta^{-1}$$

i.e. $\Pi = \Pi^2$ if P (and $P_2$) are symmetric. The $\Delta$-orthogonality condition is then exactly the relation (5.1).

THEOREM 5.3

The state vector x of any minimal internal realization (in canonical form 1) is a "convex combination" of the maximum and minimum variance state vectors $x_1$ and $x_2$, of the form,

$$x = \Pi x_1 + (I-\Pi)x_2$$

where $\Pi$ is a $\Delta$-orthogonal projection matrix in $R^n$ (i.e. $\Pi=\Pi^2$ and $\Pi$ satisfies (5.10)). A totally analogous statement holds for the state $\bar{x}$ of any minimal internal realization in canonical form 2. (Here $\bar{\Delta} := \bar{P}_2^{-1} - \bar{P}_1$ replaces $\Delta$)

Proof:

Let x be given by (5.8). Then clearly x belongs to $X_1 \vee X_2 \subset Y_1 \vee Y_2$ and its covariance is computed from the representation to be,

$$P = \Pi\Delta\Pi' + P_2$$

where $\Pi$ satisfies (5.10). It then immediately follows that P solves the algebraic quadratic equation (5.1). By Lemma 5.1 and the remark which follows x is the state of a minimal internal realization in canonical form 1. The opposite implication has already been shown to hold and the theorem is thus proved.

## 6. CONCLUSIONS

In this paper the problem of representing $y = [y_1', y_2']$ by means of Factor Analysis models of the type (3.2) has been solved completely by using techniques from stochastic realization theory. As it is shown in sections 4 and 5 there are infinitely many minimal representations which are probabilistically different. In the internal (or output induced) case the difference between various minimal models is apparent from the representation formula (5.8) whereby the state x is explicitely produced as a combination of different "portions" of $y_1$ and $y_2$.

The generalization of the present approach to deal with the more realistic case where y is partitioned in k subvectors $[y_1',\ldots,y_k']'$ is currently under investigation.

REFERENCES


[1] ANDERSON B.D.O. "The inverse problem of stationary covariance generation" J. Stat. Phys. 1, 133-147, (1969).

[2] COMREY A.L., A first Course in Factor Analysis Ac. Press, 1973.

[3] ERSHOV M.P. "Extension of measures and stochastic equations" Theory Prob. Appl. XIX, 3, 431-444, (1974).

[4] ERSHOV M.P. "Non anticipating solutions of stochastic equations" Proc. 3rd Japan-USSR Symp. on Probability Theory, Springer Lect. Notes Math 550, 655-691, (1976).

[5] FAURRE P., CLERGET M., GERMAIN F., Opérateurs rationnels positifs, Dunod, 1979.

[6] KALMAN R.E. "Identifiability and modeling in econometrics" in Developments in statistics, 4, P. Krishnaiah ed., 97-134, Ac Press, 1983.

[7] KALMAN R.E. "Theory of modeling" Proc. IBM Syst. Science Symp., Oiso Japan, Y. Nishikawa ed. 53-69, (1979).

[8] KENDALL M., STUART A., The Advanced Theory of Statistics, vol. III, Griffin 1976.

[9] LINDQUIST A., PICCI G. "On the stochastic realization problem", SIAM J. Control and Optim. 17, 365-389, (1979).

[10] LINDQUIST A., PICCI G. "Realization theory for multivariate stationary gaussian processes I: State space construction", Proc. 4th Intern. Symp. Math. Theory of Networks and Systems, July 1979, Delft, Holland, 140-148, (1979).

[11] LINDQUIST A., PICCI G. "Realization theory for multivariate stationary gaussian processes II: State space theory revisited and dynamical representations of finite dimensional state spaces", Proc. 2nd Intern. Conf. on Information Sciences and Systems, Patras, Greece, July 1979, Reidel Publ., Co., 108-129, (1979).

[12] LINDQUIST A., PICCI G. "State space models for gaussian stochastic processes", Stochastic Systems: The Mathematics of Filtering and Identification and Applications, M. Hazewinkel and J.C. Willems, Eds., Reidel Publ. Co., (1981).

[13] LINDQUIST A., PICCI G., RUCKEBUSCH G. "On minimal splitting subspaces and Markovian representations", Math. Syst. Theory 12, 271-279, (1979).

[14] MEHRA R. "Identification and estimation of the error-in-variables model (EVM) in structural form" Math. Prog. Study 5, 191-210, (1976).

[15] NEVEU J., Processus Aléatoires Gaussiens, Presses de l'Université de Montreal, 1968.

[16] PICCI G. "The stochastic realization problem" Proc. Symp. Sistemi Dinamici Stocastici, Rome June 1982, G. del Grosso ed. (1982).

[17] RÜCKEBUSCH G. "Répresentations Markoviennes de processus gaussiens stationnaires" C.R. Acad. Sc. Paris, Serie A, 282, 649-651, (1976).

[18] SÖDESTROM T. "Some methods for identification of linear systems with noisy input-output data", Proc. 5th IFAC Symp. Identif. Syst. Param. Estim., Darmstadt 1979, 1, 357-363, (1979).

[19] Van PUTTEN C., Van SCHUPPEN J.H. "The weak and strong gaussian probabilistic realization problem" J. Multivar. Anal. 13, 118-137, (1983).

[20] WILLEMS J.C. "From time series to linear systems" talk presented at the 6th Math. Th. of Networks and Systems Symp., Beer Sheva, Israel, June 1983.

# SIMULTANEOUS DETECTION AND ESTIMATION FOR DIFFUSION PROCESS SIGNALS*

John S. Baras
Electrical Engineering Department
University of Maryland
College Park, MD 20742

ABSTRACT

We consider the problem of simultaneous detection and estimation when the signals corresponding to the M different hypotheses can be modelled as outputs of M distinct stochastic dynamical systems of the Ito type. Under very mild assumptions on the models and on the cost structure we show that there exist a set of sufficient statistics for the simultaneous detection-estimation problem that can be computed recursively by linear equations. Furthermore we show that the structure of the detector and estimator is completely determined by the cost structure. The methodology used employes recent advances in nonlinear filtering and stochastic control of partially observed stochastic systems of the Ito type. Specific examples and applications in radar tracking and discrimination problems are discussed.

## INTRODUCTION

In a typical present day radar environment, the radar receiver is subjected to radiation from various sources. A very important function of the radar receiver is its ability to discriminate between the various waveforms received and select the desired one for further processing. Furthermore an equally important function of the receiver is to estimate important parameters of the radiating source from the received waveforms. Thus the receiver is required often to perform a "combined detection and estimation" function.

An abstract formulation of the combined detection and estimation problem in the language of statistical decision theory has been developed by Middleton and Esposito in [1]. They correctly point out that optimal processing in such problems often requires the mutual coupling of the detection and estimation algorithms. Although from the mathematical point of view estimation may be considered as a generalized detection problem, from an operational point of view the

*Research supported in part by ONR grant N00014-83-K-0731, by the U.S. Army contract DAAG29-81-D through Battelle Research, and by ARO contract DAAG-39-83-C-0028 at SEPI.

two procedures are different: e.g. one usually selects different cost functions for each and obtains different data processors as a result. It is then correctly argued in 1 that it is practically appropriate to retain the usual distinction between detection and estimation. There are various ways that the detector and estimator can be coupled leading to a hierarchy of complex processors. We describe here some important cases.

Detection-directed estimation

Here the detection operation is optimized with a priori knowledge of the existence of an estimator following it. The estimator is dependent on the detector's decision by being gated on only if the detector decides that the desired signal is present. Here the coupling is via cost terms that assess the performance deterioration when the estimator is turned off while the signal is present $C_{e,1}$, or the estimator is turned on when the signal is not present $C_{e,0}$. Therefore the average risks corresponding to the operations of detection and estimation can be minimized separately. This leads to a detection test that is a modified generalized likelihood test. If the cost terms $C_{e,1}$, $C_{e,0}$ are constant the coupling just reduces to a modification of the threshold [1]. Since the detector's decision rule does not depend on the estimate, the structure of the optimal estimator is not a function of the data region specified by the decision rule of the detector's operation, when the detector's decision is to accept the signal. In practical terms this means that we can choose to estimate only when the detector has decided that the desired signal is present.

Coupled detection-estimation with decision rejection

Here detection and estimation run in parallel and are followed by rejection of the estimate if the detector's decision is not to accept the signal. Here the detector's cost depends on the value of the estimate. Typically, one solves the detection problem knowing the estimator. Then a second optimization is performed over all estimators. This case usually results in relatively simple estimators and complex highly nonlinear detectors [1].

Motivation for these problems stems from distributed target problems, see in particular [2]-[7].

We concentrate in this paper on a two hypothesis detection formulation, but it is clear that the methods can be easily extended to M-ary detection problems. The two hypotheses are $H_0$ = the received signal is a process $y_{0t}$ plus noise, $H_1$ = the received signal is a process $y_{1t}$ (different from $y_{0t}$) plus noise. Both processes are modeled as outputs of stochastic dynamical systems of the diffusion type. The noise is the same in both cases. Due to this fact we can assume that noise is eliminated from the mathematical formulation of the problem of detection, while as we shall see its presence may be crucial for the estimation problem.

We did not study detectors with "learning" and we suggest this as a promising extension of the results reported here. We note however, that our formalism includes general "learning" algorithms. Most of the work on detectors with "learning" is problem specific and does not utilize dynamical system models for the signals as we do.

The major criticism for the work of Middleton and Esposito [1], is that although they used a Bayesian approach to the estimation problem, they considered nonrecursive solutions and detection was coupled to estimation through cost structure which explicitly considers coupling of the detection and estimation costs. Clearly nonrecursive solutions are not appropriate for advanced sensors employed in guided platforms. Furthermore it would be unrealistic to assume that the designer has such explicit knowledge of the functional couplings between detection and estimation costs.

Several other authors have analyzed the problem. Scharf and Lytle [13] studied detection problems involving Gaussian noise of unknown level, thus including noise parameters in the problem. As in [1], their solution is also nonrecursive, and focuses on the existence of uniformly most powerful tests. Spooner [14], [15] considered in detail unknown parameters in the noise model. Jaffer and Gupta [16], [17] consider the recursive Bayesian problem using a quadratic cost, Gauss-Markov processes and estimating only signal parameters.

Birdsall and Gobien [18] considered the problem of simultaneous detection and estimation from a Bayesian viewpoint. This work is close in spirit with our approach, although the class of problems we can analyze by our methods is significantly wider. We also follow a Bayesian methodology during the initial phase of analysis. It becomes clear that using Bayesian methods one can analyze the problems under consideration in an inherently intuitive, simple conceptually manner which can be easily obscured in highly structured methodologies utilizing specific detector structures and cost relationships. As a result one can analyze the special problems described earlier as specializations of a wider picture and framework. The results reported in [16] are limited by two important assumptions: (a) the observed data have densities that display finite dimensional sufficient statistics under both hypotheses for the unknown parameters and (b) the unknown parameters form a finite-dimensional vector. Both nonsequential and sequential problems are analyzed in [18]. The most important result of

[18] is the proof that through a Bayesian approach both estimation and detection occur simultaneously, with the detector using the a posteriori densities generated by two separate estimators, one for each hypothesis. A particularly attractive feature is that no assumptions are made on the estimation criterion and very flexible assumptions are made on the detecction criterion. When finite-dimensional sufficient statistics exist the optimum processor partitions naturally into three parts: a "primary" processor which is totally independent of a priori distributions on the parameters, a "secondary" processor which modifies the output according to the priors and solves the detection problem, and an estimator which uses the output of the other two in estimating the unknown parameters. Only the estimator structure depends on cost functionals.

Since dynamical system models are not utilized to represent signals in [18], there is great difficulty in analyzing the far more interesting sequential problem. It is for this reason that one is forced to make the limiting assumptions mentioned above. In our approach we consider diffusion type models for the signals and we utilize modern methods from nonlinear filtering and stochastic control to analyze the problem [19]-[23]. Corresponding results for Markov chain models can be easily obtained, but we only give brief comments for such problems here.

## NOMENCLATURE AND FORMULATION OF THE SEQUENTIAL PROBLEM

In this section we present a general formulation for the continuous time, sequential, simultaneous detection and estimation problem when the signals can be represented as outputs of diffusion type processes [20]. To simplify notation, terminology and subsequent computations we consider only the scalar observation case here. All results extend to vector observations in a straight-forward manner. The observed data $y(t)$ constitute therefore a real-valued scalar stochastic process.

The statistics of $y(\cdot)$ are not completely known. More specifically they depend on some parameters and some hypotheses. For simplicity we shall consider here only the binary hypotheses detection problem. Extensions to M-ary detection are trivial. We shall denote by $H_0$, $H_1$ the two mutually exclusive and exhaustive hypotheses.

Under hypothese $H_0$, the received data $y(t)$ can be represented as

$$\begin{aligned} dy(t) &= h^0(x^0(t),\theta^0)dt + dv(t) \\ dx^0(t) &= f^0(x^0(t),\theta^0)dt + g^0(x^0(t),\theta^0)dw^0(t) \end{aligned} \tag{1}$$

where $\theta^0$ is a vector-valued unknown parameter that may be assumed fixed or random throughout the problem. Here v(·), w(·) are independent, 1-dimensional and $n_0$-dimensional respectively standard Wiener processes [20]. In other words when hypothesis $H_0$ is true the received data can be thought of as the output of a stochastic dynamical system, corrupted by white Gaussian noise. $h^0$, $f^0$, $g^0$, $\theta^0$ parameterize the nonlinear stochastic system.
Similarly when hypothesis $H_1$ is true, the received data y(t) can be modeled as

$$dy(t) = h^1(x^1(t),\theta^1)dt + dv(t)$$
$$dx^1(t) = f^1(t),\theta^1)dt + g^1(x^1(t),\theta^1)dw^1(t) \tag{2}$$

where now $x^1$ is $n_1$-dimensional. The vector parameters $\theta^0,\theta^1$ may have common components. For instance, in the classical "noise or signal-plus-noise" problem any noise parameters clearly appear in both hypotheses and would thus be common to $\theta^0,\theta^1$.
We note that we have the same "observation noise" v(·) under both hypotheses. This is clearly the case in radar applications (see [6]). On the other hand when one is faced with state and parameter dependent observation noises, a simple transformation translates the two models in the form (1) (2). We shall assume that $h^i,f^i,g^i$, i=0,1, have sufficient properties to guarantee existence and uniqueness of probability distribution functions for y(·) under either hypothesis. As a minimal hypothesis we assume that the martingale problems for (1) and (2) are well posed [24] for all values of $\theta^0,\theta^1$ in appropriate compact sets $\Theta^0,\Theta^1$ respectively. Furthermore neither (1) nor (2) exhibit esplosions [24] for any value of the parameters. Often we shall make stronger assumptions such as existence of strong solutions to (1) (2), or smoothness of $f^i,g^i,h^i$, i=0,1, or existence of classical probability densities for $y$ under either hypothesis.
We shall denote by $p_y^i(\cdot,t|\theta^i)$, i=0,1, the probability density of y(t) under hypothesis $H^i$ and when the parameter obtains the value $\theta^i$, i=0,1. We shall denote the probability measures corresponding to y under $H^0$ or $H^1$ by $\mu_y^0$ and $\mu_y^1$ respectively. As is well known these are measures on the space of continuous functions [24].
Finally we note that although we have assumed time invariant stochastic models in (1), (2) the results extend easily to the time varying case.
Following a Bayesian approach we assume a priori densities for the two parameters $\theta^0,\theta^1$ which will be denoted by $p_\theta^i(\cdot,0)$, i=0,1 respectively.

Similarly initial densities for $x^0(0)$ and $x^1(0)$ are assumed known and independent of $\theta^0, \theta^1$ respectively. They will be denoted by $p_x^i(\cdot,0)$. The choice of these a priori densities, is frequently a very interesting problem in applications, as they represent the designer's a priori knowledge about the models used.

With these preliminaries we can now formulate the problem. Let $y^t$ denote as usual the portion of the observed sample path "up to time t", i.e. $y^t = \{y(s), s \le t\}$. Given the observed data $y^t$, we wish to design a processor which at time t will optimally select simultaneously which of the two hypotheses $H_0$ or $H_1$ is true, and optimal estimates for the parameters $\theta^0$ and $\theta^1$. Moreover the processor should operate recursively so as to permit real-time implementation.

To complete the problem formulation we need to specify costs for detection and estimation. Let $c_i(\hat{\theta}^i(t), \theta^i)$, i=0,1 be the penalty for "estimating" $\theta^i$, by $\hat{\theta}^i(t)$ at time t. If $c_i$ is quadratic we have the well known minimum variance estimates. Similarly let $\gamma(t)$ denote the decision, at time t, of whether we declare hypothesis $H_0$ or $H_1$ to hold. Then $k(\gamma(t), i)$, i=0,1 will denote the penalty when the true hypothesis is $H_i$ and we decide $\gamma(t)$, at time t. Obviously there are infinitely many variations on the possible choice for a cost function. We shall consider only two possibilities in this report. Finite time average integral cost

$$J_f = E\{\int_0^T \lambda_e [c_0(\hat{\theta}^0(t), \theta^0) X\{t, \gamma(t)=0\} + c_1(\hat{\theta}^1(t), \theta^1) X\{t, \gamma(t)=1\} dt + \lambda_d k(\gamma(t), i) dt\} \tag{3}$$

and infinite time average discounted cost.

$$J_d = E\{\int_0^\infty C(\gamma, \hat{\theta}^0, \hat{\theta}^1, x) e^{-\alpha t} dt\} \tag{4}$$

where $C(\gamma, \hat{\theta}^0, \hat{\theta}^1, x)$ is the integrand in (3) and $\alpha$ the discount rate. $\lambda_e, \lambda_d$ are weights. The reasons for the characteristic functions appearing in (3), (4) are rather obvious. The estimator will contribute cost only when utilized, and it will be utilized for $\theta^0$ only when $\gamma(t)=0$. We would like to point out that this does not preclude both estimators from running continuously. This scheme is used only to assess costs properly.

The appropriate formulation of the problem is as a partially observable stochastic control problem. The admissible controls are

$$\gamma : R \to \{0,1\}$$

$$\hat{\theta}^0 : R \to \Theta^0 \tag{5}$$

$$\hat{\theta}^1 : R \to \Theta^1$$

where all functions are nonanticipative with respect to y; i.e. measurable w.r. to $F_t^y$:

$$\gamma(\cdot),\hat{\theta}^0(\cdot),\hat{\theta}^1(\cdot)\varepsilon F_t^y \tag{6}$$

The cost is either (3) or (4). For the system dynamics we proceed as follows. The state equations are mixed consisting of the continuous components

$$dx^0(t) = f^0(x^0(t),\theta^0(t))dt + g^0(x^0(t),\theta^0(t))dw^0(t)$$

$$dx^1(t) = f^1(x^1(t),\theta^1(t))dt + g^1(x^1(t),\theta^1(t))dw^1(t)$$

$$d\theta^0(t) = 0 \tag{7}$$

$$d\theta^1(t) = 0$$

and the discrete component z(t) which can take only the values 0 or and is constant. The initial densities for $x^0,x^1,\theta^0,\theta^1$ have already been described. The initial probability vector for z(t) (which tracks which hypothesis is true) is

$$\Pr\{z(0) = 0\} = P_0, \ \Pr\{z(0) = 1\} = P_1 \tag{8}$$

The observations are

$$dy(t)=(1-z(t)h^0(x^0(t),\theta^0)dt+z(t)h^1(x^1(t),\theta^1)dt+dv(t \tag{9}$$

Since (7) are degenerate, there are some technical minor difficulties, which can be circumvented however using recent techniques. This completes the formulation of the problem.

STRUCTURE OF THE OPTIMAL PROCESSOR

Following recent results [25]-[29] in stochastic optimal control theory we have obtained first the following results that reduce the partially observed stochastic control problem described in section 2 to an equivalent, infinite dimensional fully observed problem.

Theorem 1: There exist optimal $\gamma,\hat{\theta}^0,\hat{\theta}^1$ for the stochastic optimal control problem (3)-(9).

Proof: This follows from the results of Fleming and Pardoux [27] and Bismut [29]. The only difference is that due to the structure of the dynamics here (i.e. they do not depend on the controls $\gamma,\hat{\theta}^0,\hat{\theta}^1$) we can show that optimal controls exist in the class of strict sense controls as specified in section 2 (i.e. $\gamma(t),\hat{\theta}^0(t),\hat{\theta}^1(t)$ are measurable with respect to $F_f^y$).

We then introduce as in Fleming and Pardoux [27] the associated "separated" stochastic control problem. In the separated stochastic control problem the state at time t is a measure $\Lambda_t$ on $R^N$ (where $N = n_0+n_1+2$), which is an unnormalized conditional distribution of the state $x(t)\triangleq[x_0(t),x_1(t),\theta_0(t),\theta_1(t),z(t)]^T$ of the problem formulated in section 2. The dynamics of the measure-valued process $\Lambda_t$ obey the Zakai equation of nonlinear filtering [26]-[31], and [20].

In the sequel we assume that all functions appearing in (1)-(9) are bounded and continuous and that $g^0,f^0,g^1,f^1$ are Lipschitz in $x^0,\theta^0,x^1,\theta^1$, respectively. Due to the discrete component z(t) of the state x(t) we have to consider a two dimensional measure valued process $\Lambda^0,\Lambda^1$, where $\Lambda^i$ is the unnormalized conditional distribution of the state $x(t)\triangleq[x_0(t),x_1(t),\theta_0(t),\theta_1(t)]$ (slight abuse of notation here) when hypothesis $H_i$ is true, i=0,1. We further assume that for i=0,1 the corresponding Zakai equation has a unique solution which is absolutely continuous with respect to Lebesque measure; i.e. we assume the existence of conditional unnormalized probability densities for $x(t)\varepsilon R^N$ given $y^t$. For results on this see [30], [31].

Let $u^i(x,t)$ denote the conditional probability density of x(t) given $y^t$ when hypothesis $H_i$ holds. Then $u^i(\cdot,\cdot)$ satisfies the Zakai equation

$$du^i = L_i^* u^i dt + dy(t)h^i u^i, \ i=0,1 \tag{10}$$

where $L^*$ is the formal adjoint to the infinitesimal generator of the $i^{th}$ component of (7); i.e. it has the form

$$L = \frac{1}{2}\sum_{i,j=1}^{N} a^{i}_{ij}(x)\frac{\partial^2}{x_i x_j} + \sum_{i=1}^{N} b^{i}_{i}(x)\frac{\partial}{\partial x_i} \qquad (11)$$

Here

$$a^{i} = \sigma^{i}(\sigma^{i})^{T}, \sigma^{i} = \begin{bmatrix} g^{i} & 0 \\ 0 & 0 \end{bmatrix} \quad b^{i} = \begin{bmatrix} f^{i} & 0 \\ 0 & 0 \end{bmatrix} \qquad (12)$$

To complete the description of the "separated" stochastic control problem, let $C(\gamma,\hat{\theta}^{0},\hat{\theta}^{1},x)$ denote the integrand in the cost definition (3). then if we let

$$u(x,t) = \begin{bmatrix} u^{0}(x_0,\theta_0,t) \\ u^{1}(x_1,\theta_1,t) \end{bmatrix} \qquad (13)$$

we can rewrite the cost (3) as

$$J_f(\pi) = E_y\{\int\int_0^T C(\gamma,\hat{\theta}^{0},\hat{\theta}^{1},x)[u(x,t)^{T} \begin{bmatrix} P_0 \\ P_1 \end{bmatrix}]dxdt\} \qquad (14)$$

where $\pi$ is the policy corresponding to a particular selection of $\gamma(\cdot)$, $\theta^{0}(\cdot)$, $\theta^{1}(\cdot)$, and $E_y$ is expectation with respect to y. Note that u depends explicitly on y.

The separated problem is to choose a policy $\pi$ which is a function of $u^{0}$, $u^{1}$ to minimize (14). This is a fully observed problem since $u^{0}$, $u^{1}$ satisfy (10) and enter directly into (14). We then have the following very important result:

Theorem 2: Under the above assumptions the optimal $\gamma,\hat{\theta}^{0},\hat{\theta}^{1}$ (which exist according to theorem 1) are functions of $u^{0},u^{1}$ only. That is they depend on $y^{t}$ only through the unnormalized conditional densities $u^{0},u^{1}$.

Proof: The proof follows from appropriate modifications of the results in [25]-[29] and will appear elsewhere.

The significance of the result is that it provides the basic structure of the optimal processor by identifying $u^{0},u^{1}$ as the sufficient statistics for the original problem. Furthermore the result is free from structural assumptions on the detection and estimation costs and can be established in far greater generality than the results presented here may indicate.

In figure 1 below we give a pictorial illustration of the result. We basically have to run two "filters" in parallel, one for each hypothesis. The output of each filter (which by the way is represented by the bilinear stochastic p.d.e. (10)) is the unnormalized conditional probability density of $x^0$, $\theta^0$ or

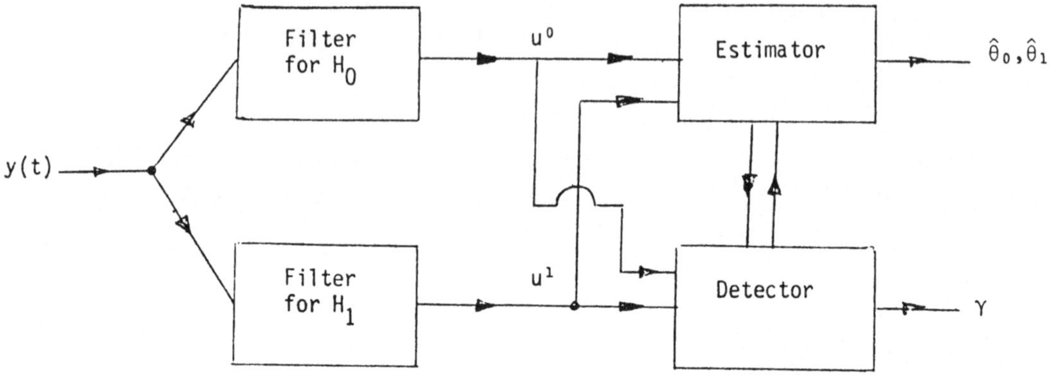


Figure 1 Illustrating the generic structure of the optimal processor $x^1$, $\theta^1$ given $H^0$ or $H^1$. Each filter is driven directly by the observations.

The estimator, detector and their coupling will depend on the explicit cost structure. They are problem dependent. Their explicit functioning can be computed as our final result indicates.

Theorem 3: The explicit dependence of $\gamma$ (which is discrete valued), $\hat{\theta}_0, \hat{\theta}_1$ on $u^0, u^1$ can be determined by solving a variational inequality on the space of solutions of (10).

Proof: The result is rather technical. A complete proof will be given elsewhere. It follows by appropriate modifications to the results of [26], [32].

This result opens the way for promising electronic implementation of the optimal processor by the following steps: (1) solve numerically the resulting variational inequality using the methods of [33], (2) implement the resulting numerical algorithm by a special purpose, multiprocessor, VLSI device along the lines of [34]. In simple cost cases explicit solutions of the variational inequality can be obtained of course.

## MOTIVATION AND EXAMPLES FROM RADAR TRACKING LOOPS

The primary motivation for the mathematical problem studied in section 3 comes from design consideration of advanced (smart) sensors in guided platforms. To be more specific let us consider radar sensors.

The radar return from a scatterer carries (depending on the radar sophistication) significant information about a scatterer. For example range, Doppler extend, shape and extend, motion, of a scatterer can be extracted from a radar return by appropriate processing. In today's dense environment a very important function of an advanced processor is classification of scatterers. This function is required for example by sensors participating in a surveillance network (since threats must be classified, so that appropriate response can be applied), in electronic warfare (since decoys and other countermeasures can be designed to emulate target characteristics) and in tracking radars (since the sensor often must develop a tracking path for a designated priority target).

A related equally important function of a radar receiver is the estimation of parameters embedded in the return signal. For example pulse length, pulse repetition frequency, amplitude scintillation spectrum, conical scan frequency, antenna pointing, surface roughness. The two problems of detection and estimation are indeed closely related, as explained earlier.

In our earlier work [2]-[5] we have developed statistical models for distributed scatterers which can represent accurately phenomena characteristic of distributed scatterer radar returns such as amplitude scintillation and angle noise or glint. In addition we have developed similar statistical models for the effects of multipath on radar returns, for sea clutter returns and for chaff cloud returns. The models developed in [2]-[5] are of the form

$$
\begin{aligned}
dx(t) &= A(t,\theta)x(t)dt + B(t,\theta)dw(t) \\
dy(t) &= h(t,x(t),\theta)dt + dv(t)
\end{aligned}
\qquad (15)
$$

Furthermore A,B,h are piecewise constant with respect to time since the models developed in [2]-[5] are piecewise stationary. For example in [2] we used models like (5) to describe the RCS scintillation for ships. The same type models can be used for other distributed targets such as tanks or armored vehicles. For example when the return appears spiky, indicating higher probability of strong return, an appropriate model is provided by a lognormal process, where $x(\cdot)$ in (15) is scalar and h is chosen to be an exponential function of x. For chaff clouds a more appropriate model is provided by a Rayleigh process, where $x(\cdot)$ is two dimensional, with the two components being identically distributed, independent Gaussian random processes and

$$h(t,x(t),\theta) = \sqrt{x_1^2(t) + x_2^2(t)}.$$

Clearly then in target discrimination problems with distributed targets of this type one encounters problems like those treated in section 3. It is important to note that since the first of (15) is linear the corresponding filtering and stochastic control problems described in section 3 are definitely more tractable. For further examples of this type we refer the reader to [2]-[5].
Further research is needed to apply the powerful results of section 3 to specific problems in order to evaluate current design principles and more importantly in order to suggest new electronic implementations capable of performing in a dense, hostile environment. In particular the methodology developed in 3 can be used to identify the cost structures that lead to the specific hierarchies suggested in the introduction.

## REFERENCES


[1] Middleton, D., and R. Esposito. (May 1968). Simultaneous optimum detection and estimation of signals in noise. IEEE Trans. on Information Theory, Vol. IT-14, No. 3, pp. 434-444.

[2] Baras, J.S. (1983 (in publication)). Sea clutter statistical models. Naval Research Laboratory Technical Report.

[3] Baras, J.S. (May 1978). Ship RCS scintillation simulation. Naval Research Laboratory Technical Report 8189.

[4] Baras, J.S. (1982). "Sea Clutter Statistical Models," Naval Research Laboratory Technical Report, (in publication).

[5] Baras, J.S. (1983). "Multipath Effects Modeling," Naval Research Laboratory Technical Report, (in preparation).

[6] Baras, J.S., A. Ephremides, and G. Panayotopoulos. (1980). "Modelling of Scattering Returns and Discrimination of Distributed Targets," Final Technical Report on ONR Contract N00014-78-C-0602.

[7] Skolnik, M.I. (1970). Radar Handbook, McGraw-Hill: New York.

[8] Agrawala, A.K. (1970). "Learning with a Probabilistic Teaching," IEEE Trans. Inform. Theory, Vol. IT-16, pp. 373-379.

[9] Cooper, D.B. and P.W. Cooper. (1964). "Adaptive Pattern Recognition and Signal Detection Without Supervision," IEEE Int. Conv. Rec., pt. 1, pp. 252-255.

[10] Katopis, A. and S.C. Schwartz. (1972). "Decision-directed Learning Using Stochastic Approximation," in Proc. Modeling and Simulation Conf., pp. 473-481.

[11] Gimlin, D.R. (September 1974). "A Parametric Procedure for Imperfectly Supervised Learning with Unknown Class Probabilities,"

IEEE Trans. on Inform. Theory, Vol. IT-20, pp. 661-663.

[12] Cooper, D.B. (November 1975). "On Some Convergence Properties of 'Learning with a Probabilistic Teacher' Algorithms," IEEE Trans. Inform. Theory, Vol. IT-21, pp. 699-702, Nov. 1975.

[13] Scharf, L., and D. Lytle. (July 1971). Signal detection in Gaussian noise of unknown level: an invariance application. IEEE Trans. Inform. Theory, Vol. IT-17, pp. 404-411.

[14] Spooner, R.L. (1968). "On the Detection of a Known Signal in a non-Gaussian Noise Process," J. Acoust. Soc. Amer., Vol. 44, pp. 141-147.

[15] Spooner, R.L. (April 1968). "The Theory of signal Detectability: Extension to the Double-Composite Hypothesis Situation," Cooley Electronics Lab., Univ. Michigan, Ann Arbor, Tech. Rep. No. TR-192, April 1968.

[16] Jaffer, A., and S. Gupta. (Sept. 1971). Recursive Bayesian estimation with uncertain observation. IEEE Trans. Inform. Theory, Vol. IT-17, pp. 614-616.

[17] Jaffer, A., and S. Gupta. (Jan. 1972). Coupled detection estimation of Gaussian processes in Gaussian noise. IEEE Trans. Inform. Theory, Vol. IT-18, pp. 106-110.

[18] Birdsall, T.G., and J.O. Gobien. (Nov. 1973). Sufficient statistics and reproducing densities in simultaneous sequential detection and estimation. IEEE Trans. on Inform. Theory, Vol. IT-19, pp. 760-768.

[19] Kailath, T. (1970). "The Innovations Approach to Detection and Estimation Theory," Proc. IEEE, 58, pp. 680-695.

[20] Liptser, R.S. and A.N. Shiryayev. (1977). Statistics of Random Processes I, General Theory; II, Applications, Springer-Verlag.

[21] Kushner, H.J. (1967). Stochastic Stability and Control, Academic Press.

[22] Kushner, H.J. (1971). Introduction to Stochastic Control, Holt, Rinehart and Winston.

[23] Fleming, W.H. and R.W. Rishel. (1975). Deterministic and Stochastic Optimal Control, Springer-Verlag.

[24] Stroock, D.W., and S.R.S. Varadhan. (1975). Multidimensional Diffusion Processes. Springer Verlag, Secaucus, NJ.

[25] Fleming, W.H. (1982). "Nonlinear Semigroup for Controlled Partially Observed Diffusions," SIAM J. Control and Optim., Vol. 20, pp. 286-301.

[26] Bensoussan, A. (1982). "Optimal Control of Partially Observed Diffusions," in Advances in Filtering and Optimal Stochastic Control, W. Fleming and L.G. Gorostiza (edts.), Lecture Notes in Control and Information Sciences, 42, Springer-Verlag.

[27] Fleming, W.H., and E. Pardoux. (1982). Optimal control for partially observed diffusions. SIAM J. Control and Optim., Vol. 20, pp. 261-286.

[28] Davis, M.H.A. (1982). "Stochastic Control with Noisy Observations," in Advances in Filtering and Optimal Stochastic Control, W. Fleming and L.G. Gorostiza (edts.), Lecture Notes in Control and Information Sciences, 42, Springer-Verlag.

[29] Bismut, Jean-Michel. (1982). Partially observed diffusions and their control. SIAM J. Control and Optim., Vol. 20, pp. 302-309.

[30] Baras, J.S., G.L. Blankenship, and W.E. Hopkins, Jr. (Feb. 1983) Existence, uniqueness, and asymptotic behavior of solutions to a class of Zakai equations with unbounded coefficients. IEEE Trans. on Autom. Control, Vol. AC-28, pp.203-214.

[31] Hopkins, Jr., W.E., J.S. Baras, and G.L. Blankenship. (1982). "Existence, Uniqueness and Tail Behavior of Solutions to Zakai Equations with Unbounded Coefficients," in Advances in Filtering and Optimal Stochastic Control, W. Fleming and L.G. Gorostiza (edts.), Lecture Notes in Control and Information Sciences, 42, Springer Verlag.

[32] Bensoussan, A., and J.L. Lions. (1982). Applications of Variational Inequalities in Stochastic Control, North-Holland.

[33] Glowinski, R., J.L. Lions, and R. Tremolieres. (1976). Analyse Numberique Des Inequations Variationnelles, Vols. 1, 2. Dunod.

[34] Baras, J.S. (1981). Approximate solutions to nonlinear filtering problems by direct implementation of the Zakai equation. Proc. of 1981 IEEE Decision and Control Conference, San Diego, pp. 309-310.

Session 9

# DETERMINISTIC CONTROL

# CONTRÔLE DÉTERMINISTE

# HEAVY VIABLE TRAJECTORIES OF A DECENTRALIZED ALLOCATION MECHANISM

Jean-Pierre Aubin
CEREMADE - Université de Paris-Dauphine
75775 - PARIS CX (16)

Abstract

We define and study the concept of heavy viable trajectories of a controlled system with feedbacks in the framework a dynamical decentralized allocation mechanism in an exchange economy.

In this framework, the controls are the prices and the states of the system the consumptions of the consumers. Consumptions of each consumer evolve according a differential equation controlled by the price. Viable trajectories are the ones which obey the scarcity constraints : the sum of the consumptions must remain in the set of available commodities. Prices regulating viable trajectories evolve according a set-valued feedback map. Heavy viable trajectories are the ones associated to prices in the feedback map which evolve as slowly as possible : at each instant, the norm of the velocity of the price is minimal among the prices regulating a viable trajectory.

In this lecture, we construct the differential equation yielding heavy viable trajectories, providing a model of how the market may govern the evolution of prices. These results were obtained in collaboration with Halina Frankowska.

Introduction

The purpose of this lecture is to illustrate the concept of heavy viable trajectories of a controlled system with feedbacks - studied in Aubin-Frankowska [1984 ] - in the framework of an economic model of resource allocation.

Economic systems, as well as ecological and biological ones, are consuming scarce resources and face many constraints.

Therefore, the state of such systems must evolve in a viability domain defined by these sacrcity constraints as well as the other constraints. Trajectories lying in the viability domain are called viable. For instance, if we have to allocate a set of available commodities among consumers, the viability domain is the set of allocations, i.e., consumptions of consumers the sum of which are available.

Now, assume that the dynamics governing the evolution of the state are controlled. For instance, in economics, prices can be regarded as such regulating controls, giving consumers an information about the market that they use in a decentralized way to change their consumptions knowing only the state of their own consumptions. (They don't need to know neither the choices of the other consumers nor the set of available resources). Viability theory provides necessary and sufficient conditions for the existence of at least one viable trajectory starting from any viable initial state. It above all provides the feedback laws (concealed in both the dynamics and the viability domain) which relate the regulating controls -the prices- to the states of the controlled system of differential equations- the allocations of scarce resources among consumers. These feedback laws are not necessarily single-valued - deterministic -. They are most often set-valued maps, associating with each allocation a set of prices. We observe that the larger these subsets of prices are, the more flexible - and thus, the more robust - the regulation of the system will be, by allowing "mistakes" to be done. If we accept this mathematical metaphor of allocation of available commodities, we may propose that the duty of the market (Adam Smith's invisible hand) or of an adequate planning bureau should be to choose at each instant a price according to the feedback law.

In this paper, we make the further assumption that prices evolve with a high inertia : the prices will change only when the viability of the system is at stake, and then, the slower the better. This is at least the case when the market sets prices : the metaphorical existence of the "market" as a decision-maker should at least assume that it is lazy. And one could say the same thing about a planning bureau ! This motivates the introduction of heavy viable trajectories, associating to an allocation a price such that at each instant, the norm of the velocity is minimal among all possible prices regulating this trajectory in a viable way. We shall provide the differential equations yielding heavy trajectories (which are also concealed in the dynamics and the viability domain) and state some existence theorems.

We observe that as long as the sum of the consúmptions lies in the interior of the set of available commodities, any regulatory price will work. Therefore, along a heavy trajectory, the system maintain the price inherited from the past (the regulatory prices remains constant, even though the consumptions may evolve quite rapidly).

When the sum of the consumptions reaches the boundary of the set of available resources, two situations may occur :

(a) If the sum of the velocities "points inward" the set of available commodities, then we can still keep the same regulatory price, which pushes the sum of consumptions into the set of commodities.

(b) If not, the prices will start to evolve as slowly as possible in order to pushes the total consumption back into the commodity set.

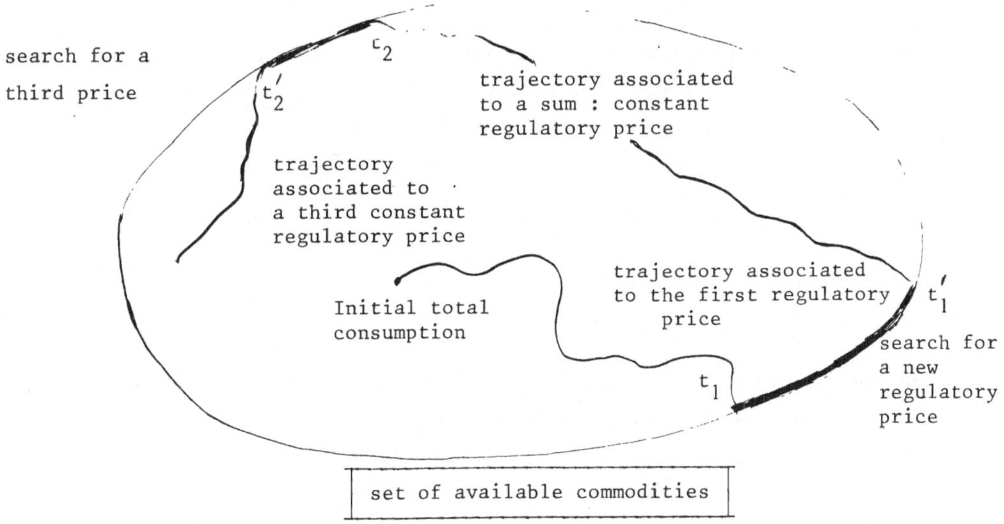

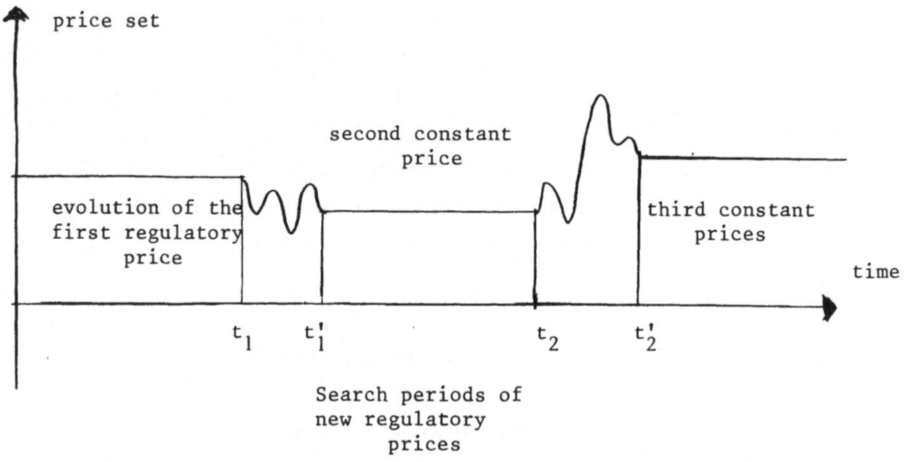


Evolution of prices

Observe that other strategies are possible, such as "enlarging" the set of available commodities instead of acting on prices. But we are far to be able to tackle all phenomena at once .

The differential equations which govern the evolution of heavy viable trajectories also reveal a division of the viability domain into "cells"; each cell is the subset of allocations which can be regulated by a constant price.

We end this informal presentation of heavy viable trajectories by mentionning that paleontological concepts of biological evolution such as punctuated equilibria proposed by Elredge and Gould are consistent with what we said.

## 1 . Description of the decentralized allocation mechanism.

We consider the decentralized dynamical allocation mechanism proposed in Aubin [ 1981 ] b) (see also Aubin-Cellina [ 1984 ] p. 245-256 and Stacchetti [ 1984 ]). We interpret $Y := \mathbb{R}^{\ell}$ as a commodity space, its dual $Y^{\star}$ as the price space. The description of the economy begins with

(1.1) the subset $M \subset Y$ of available commodities

The problem is to allocate commodities $y \in M$ among $n$ consumers $i=1,\ldots,n$ ; each consumer chooses a commodity in its <u>consumption set</u> $L_i$ .

The set $K \subset Y^n$ of <u>allocations</u> of $M$ is defined by

(1.2) $$K := \left\{ x \in \prod_{i=1}^{n} L_i \mid \sum_{i=1}^{n} x_i \in M \right\}$$

Let $P \subset Y^{\star}$ denote the set of feasible prices.

We describe the <u>behavior</u> of each consumer by <u>change functions</u> $c_i : L_i \times P \to Y$ . The decentralized allocation mechanism is described by a system of $n$ differential equations controlled by prices : For all $x_o=(x_{o_1},\ldots,x_{o_n}) \in K$ , find $T > 0$ and $n$ absolutely continuous functions $x_i(\cdot)$ satisfying,

(1.3) $$\begin{cases} \text{i) for almost all } t \in [0,T] , & x_i'(t) = c_i(x_i(t),p(t)) \\ \text{ii) for almost all } t \in [0,T] , & p(t) \in P \\ \text{iii) } x_i(0) = x_{o_i} & \end{cases}$$

which are viable in the sense that

(1.4) $$\forall\, t \in [0,T], \ \forall\, i=1,\ldots,n , \quad x_i(t) \in L_i \ \text{ and } \ \sum_{i=1}^{n} x_i(t) \in M$$

We observe that this allocation mechanism is decentralized : the actions of each consumer depend only upon his consumption and the price.

When $K$ denotes a convex subset, we recall that the tangent cone $T_K(x)$ to $K$ at $x$ defined by

(1.5) $$T_K(x) := \operatorname{cl} \left( \bigcup_{h > 0} \frac{1}{h} (K-x) \right)$$

is a closed convex cone.

We assume that

(1.6) $$\begin{cases} \text{i) the subsets } L_i \text{ and } P \text{ are closed and convex} \\ \text{ii) the subset } M \text{ is closed and convex and } M = M - \mathbb{R}^{\ell}_{+} \\ \text{iii) } 0 \in \operatorname{Int} \left( \sum_{i=1}^{n} L_i - M \right) \end{cases}$$

and that

$$(1.7)\qquad \begin{cases} \text{i)} & \text{the change functions } c_i \text{ are } C^1 \text{ around } L_i \times P \\ \text{ii)} & \forall\, x \in L_i \,,\ \forall\, p \in P \,,\quad c_i(x,p) \in T_{L_i}(x) \end{cases}$$

Assumptions (1.6) imply that

$$(1.8)\qquad T_K(x) \ := \left\{ v \in \prod_{i=1}^{n} T_{L_i}(x_i) \;\middle|\; \sum_{i=1}^{n} v_i \in T_M\left(\sum_{i=1}^{n} x_i\right)\right\}$$

(see Aubin-Ekeland [ 1984 ] p. 174).

We then define the <u>feedback map</u> $R$ from $K$ to $P$ by

$$(1.9)\qquad R(x) \ := \ \left\{ p \in P \;\middle|\; \sum_{i=1}^{n} c_i(x_i,p) \in T_M\left(\sum_{i=1}^{n} x_i\right)\right\}$$

We observe that any viable trajectory of the decentralized allocation mechanism (1.3) is a solution to the <u>feedback system</u> : $\forall\, i=1,\dots,n$ ,

$$(1.10)\qquad \begin{cases} \text{i)} & \text{for almost all } t \in [0,T] \,, \quad x_i'(t) = c_i(x_i(t),p(t)) \\ \text{ii)} & \text{for almost all } t \in [0,T] \,, \quad p(t) \in R(x(t)) \\ \text{iii)} & x_i(0) = x_{i_o} \end{cases}$$

(see Aubin [ 1981 ] b) or Aubin-Cellina [ 1984 ], p. 254). The standard viability theorem (see Haddad [ 1981 ], Aubin-Cellina [ 1984 ], p. 239-240) provides sufficient conditions for the existence of viable trajectories. Under the assumptions of this theorem, a necessary and sufficient condition for the existence of viable trajectories for all $x_o \in K$ is that

$$(1.11)\qquad \forall\, x \in K \ , \quad R(x) \neq \emptyset$$

One also shows that assumption (1.11) implies the existence of an equilibrium $(\bar{x}_1,\dots,\bar{x}_n,\bar{p})$ , a solution to

$$(1.12)\qquad \begin{cases} \text{i)} & \forall\, i=1,\dots,n \,,\quad \bar{x}_i \in L_i \,,\quad \sum_{i=1}^{n} \bar{x}_i \in M \,,\quad \bar{p} \in P \\ \text{ii)} & \forall\, i=1,\dots,n \,,\quad c_i(\bar{x}_i,\bar{p}) = 0 \end{cases}$$

Hence, in the framework of this model, at each instant $t$, the price $p(t)$ must be chosen in the subset $R(x(t))$ : it evolves according to a set-valued feedback rule.

Then the question arises whether the market - or a planning bureau - can select a price $p(t)$ in $R(x(t))$ .

We propose to answer this question by singling out heavy viable trajectory which seem to be present in the evolution of macrosystems arising in social and biological sciences (which motivated viability theory in the first place). They are trajectories which minimize at each time the norm of the velocity of the price.

The first difficulty which arises is that a solution to (1.3) is only absolutely continuous, so that the associated price is only measurable. To use the derivative in the distribution sense does not help because the concept of heavy trajectory requires the existence of the velocity of the price at almost each time. However, a straightforward strategy consists in <u>differentiating the feedback relation</u> (1.10)ii) to reveal a law relating the velocities of the prices and the consumptions.

For that purpose, we need an adequate concept of derivative of a set-valued map.

## 2 . Contingent derivative of a set-valued map.

When $K$ is a subset of a finite-dimensional space $X$, we can define many concepts of "tangent cones", among which we mention

a) the contingent cone $T_K(x)$ , defined by

$$(2.1) \qquad T_K(x) := \left\{ v \in X \mid \liminf_{h \to 0+} \frac{d_K(x+hv)}{h} = 0 \right\}$$

b) the tangent cone (introduced by Clarke [1975 ])

$$(2.2) \qquad C_K(x) := \left\{ v \in X \mid \lim_{\substack{h \to 0+ \\ y \to x \\ y \in K}} \frac{d_K(y+hv)}{h} = 0 \right\}$$

c) the Dubovickii-Miljutin [1963 ] cone

$$(2.3) \qquad D_K(x) := \{ v \in X \mid \exists\, \varepsilon > 0 \mid x + [0,\varepsilon\,](v+\varepsilon B) \subset K \}$$

We have the following relations (see Cornet [1981], Penot [1981], Aubin-Ekeland [1984] p. 409)

$$(2.4)\quad \begin{cases} \text{i)}\ C_K(x) = \liminf\limits_{\substack{y \to x \\ y \in K}} T_K(y) \subset T_K(x) \\ \text{ii)}\ \operatorname{Int} C_K(x) \subset D_K(x) \subset \operatorname{Int} T_K(x) \end{cases}$$

The tangent cone is always convex. It coïncides with the contingent cone when $K$ is a smooth manifold (tangent space) or when $K$ is convex or, more generally, when $K$ is <u>soft</u> in the sense that

$$(2.5)\quad x \to T_K(x) \quad \text{is lower semicontinuous .}$$

Consider now a set-valued map $R$ from $X$ to $Y$ and a point $(x,y)$ of its graph. The <u>contingent</u> derivative $DR(x,y)$ is the set-valued map from $X$ to $Y$ defined by

$$(2.6)\quad w \in DR(x,y)(v) \Leftrightarrow (v,w) \in T_{\text{Graph}(R)}(x,y)$$

It is equivalent to say that

$$(2.7)\quad \liminf_{\substack{h \to 0+ \\ v' \to v}} d\left(w\ ,\ \frac{R(x+hv')-y}{h}\right) = 0$$

The contingent derivative $DR(x,y)$ is a closed process (a map whose graph is a closed cone). We say that the map $R$ is <u>soft</u> if its graph is soft. Then $DR(x,y)$ is a closed convex process, because its graph is equal to the tangent cone to $\text{Graph}(R)$ at $(x,y)$.

We shall say that $R$ is <u>lower semicontinuously differentiable</u> if

$$(2.8)\quad (x,y,v) \to DR(x,y)(v) \quad \text{is lower semicontinuous}$$

We observe that in this case $DR(x,y)$ is a closed convex process because property (2.8) implies that $(x,y) \to T_{\text{Graph}(R)}(x,y)$ is lower semicontinuous, and thus, $\text{Graph}\ DR(x,y)$ is a closed convex cone.

Finally, when $K$ is a closed subset of $X$ , we denote by

$$(2.9)\quad m(K) := \left\{u \in K \mid \|u\| = \min_{v \in K} \|v\|\right\} = \pi_K(0)$$

the subset of elements of $K$ with minimal norm. If $F$ is a continuous set-valued map with closed convex images, then $m(F(x))$ is reduced to a point and the single-valued map $x \to m(F(x))$ is continuous. This is no longer the case when $F$ is only upper or lower semicontinuous (with closed convex images). However,

$$(2.10) \quad \begin{cases} \text{if } F \text{ is lower semicontinuous with closed images,} \\ \text{then } x \to d(0,F(x)) \text{ is upper semicontinuous.} \end{cases}$$

We refer to Aubin [ 1983 ] and Aubin-Ekeland, [ 1984 ] Chapter 7, Clarke [ 1983 ] for a general presentation of nonsmooth analysis relevant to this study.

## 3 . Heavy viable trajectories.

Let us consider the decentralized allocation mechanism (1.3), (1.4). We have seen that viable trajectories are solutions to the system (1.10). When the functions $x_i(\cdot)$ and $p(\cdot)$ are absolutely continuous, we deduce from the "first-order" relation (1.10)ii) the "second-order" relation

$$(3.1) \quad \begin{cases} \text{for almost all } t \in [0,T], \\ p'(t) \in DR(x(t),p(t))\,(c(x(t),p(t))) \end{cases}$$

where we set

$$(3.2) \qquad c(x,p) := (c_1(x_1,p), \ldots, c_n(x_n,p))$$

Hence we can propose a rigorous definition.

Definition 3.1

We shall say that $(x_1(\cdot),\ldots,x_n(\cdot), p(\cdot))$ is a heavy viable trajectory of the allocation mechanism (1.3), (1.4) if it is a solution to the system of differential inclusions

$$(3.3) \quad \begin{cases} \text{i)} \ \forall\, i=1,\ldots,n, \quad x_i' = c_i(x_i,p) \\ \text{ii)} \ p' \in m(DR(x,p))\,(c(x,p))) \\ \text{iii)} \ (x(0),p(0)) = (x_o,p_o) \quad \text{where } x_o \in K,\ p_o \in R(x_o) \end{cases}$$

which is viable in the sense that

$$(3.4) \qquad \forall t \geq 0 \quad , \quad p(t) \in R(x(t)) \quad .$$

▲

Remark. Viability cells.

The inverse of the feedback map $R$ associates with any price $p \in P$ the subset $R^{-1}(p)$ of allocations which can be regulated by $p$ . The viability cells $C(p)$ are the subsets (possibly empty) of $R^{-1}(p)$ defined by

$$(3.5) \qquad C(p) := \{x \in R^{-1}(p) \mid 0 \in DR(x,p)(c(x,p))\}$$

Starting with an allocation $x_o$ in a cell $c(p_o)$ in the direction $c_i(x_{o_i},p_o)$ , a heavy viable trajectory keeps the constant price $p_o$ as long as the allocation $x(t)$ remains in the state cell $C_i(p_o)$ , because in this case the system (3.3) can be written

$$(3.6) \qquad \begin{cases} \text{i)} \quad \forall\, i=1,\ldots,n \;, \quad x_i'(t) = c_i(x_i(t),p_o) \\ \text{ii)} \quad 0 \in m(DR(x(t),p_o))\,(c(x(t),p_o)) \end{cases}$$

The price system will start to evolve when the allocation leaves the viability cell $C(p_o)$ . ■

The study of the viability problem (3.3), (3.4) runs into the same difficulties that viability problems for second-order differential inclusions encounter . Therefore, we shall use the method proposed by Cornet-Haddad [ 1983 ] to overcome these difficulties. We are ready to state our main theorem.

Theorem 3.2

We posit assumptions (1.6) and (1.7) the "transversality" condition

$$(3.7) \qquad \begin{cases} \forall\, (y,z) \in Y \times Y \;,\; \exists\, \pi \in T_P(p) \;,\; \exists\, v_i \in T_{L_i}(x_i) \quad \text{such that} \\ z = \sum_{i=1}^{n} \frac{\partial}{\partial x_i} c_i(x_i,p)v_i + \sum_{i=1}^{n} \frac{\partial}{\partial p} c_i(x_i,p)\cdot\pi \\ \qquad - DT_M\left(\sum_{i=1}^{n} x_i \,,\, \sum_{i=1}^{n} c_i(x_i,p)\right)\left(\sum_{i=1}^{n} v_i - y\right) \end{cases}$$

and the regularity assumption

(3.8) the map $x \in M \rightarrow T_M(x)$ is soft

Then the contingent derivative $DR(x,p)(v)$ of the feedback map $R$ is the closed convex process from $\prod_{i=1}^{n} T_{L_i}(x_i)$ to $T_P(p)$ defined by

$$(3.9)\quad \begin{cases} DR(x,p)(v) = T_P(p) \cap \\ \left(\sum_{i=1}^{n} \frac{\partial}{\partial p} c_i(x_i,p)\right)^{-1} \left[DT_M\left(\sum_{i=1}^{n} x_i , \sum_{i=1}^{n} c_i(x_i,p)\right)\left(\sum_{i=1}^{n} v_i\right)\right. \\ \left. - \sum_{i=1}^{n} \frac{\partial}{\partial x_i} c_i(x_i,p)v_i\right] \end{cases}$$

Furthermore, let us assume that

$$(3.10)\quad \begin{cases} \text{the graphs of the set-valued maps } T_{L_i}(\cdot) \text{ and } T_P(\cdot) \\ \text{are locally compact} \end{cases}$$

and that

(3.11) the feedback map $R$ is lower semicontinuously differentiable

If both the first order condition

(3.12) $\forall\, x \in K \quad , \quad R(x) \neq \emptyset$

and the second order condition

(3.13) $\forall\, (x,p) \in \text{Graph}(R)$ , $c(x,p)$ belongs to the domain of $DR(x,p)$

hold true, then for all initial allocation $x_o \in K$ and all initial price $p_o$ satisfying either

(3.14) $p_o \in R(x_o)$ if $\text{Graph}(T_M(\cdot))$ is locally compact

or

(3.15) $\sum_{i=1}^{n} c_i\left(x_{o_i}, p_o\right) \in D_M\left(\sum_{i=1}^{n} x_{o_i}\right)$ if not ,

there exist $T > 0$ and a heavy viable trajectory of the decentralized allocation mechanism.

Remark

For checking assumption (3.15), we can assume that the set-valued map $T_M$ is lower semicontinuously differentiable and use standard theorems implying that the intersection of lower semicontinuous maps is lower semicontinuous (see Aubin-Cellina [1984] p. 49, for instance).

## 4 . Example.

We cannot have an explicit analytical expression of $m(DR(x,p)(c(x,p)))$ except in special situations. We assume, for instance, that we don't take into account the constraints on prices and consumptions of individual consumers : we take $P = Y^\star$ and $L_i = Y$ for all $i=1,\dots,n$ . Then, for all $x \in K$ ,

$$(4.1) \qquad R(x) = \left\{ p \in Y^\star \mid \sum_{i=1}^{n} c_i(x_i,p) \in T_M\left(\sum_{i=1}^{n} x_i\right)\right\}$$

The surjectivity property

$$(4.2) \qquad \forall\ (x,p) \in \mathrm{Graph}(R)\ , \quad \sum_{i=1}^{n} \frac{\partial}{\partial p}\, c_i(x_i,p) \quad \text{is surjective}$$

implies that the transversality property (3.7) is satisfied. Hence, if we assume that $T_M$ is soft, we can write

$$\begin{cases} DR(x,p)(v) = \left(\displaystyle\sum_{i=1}^{n} \frac{\partial}{\partial p}\, c_i(x,p)\right)^{-1} \cdot \\ \left[DT_M\left(\displaystyle\sum_{i=1}^{n} x_i\ ,\ \sum_{i=1}^{n} c_i(x_i,p)\right)\left(\sum_{i=1}^{n} v_i\right) - \sum_{i=1}^{n} \frac{\partial}{\partial x_i}\, c_i(x_i,p)\cdot v_i\right] \end{cases}$$

Let us set now

$$(4.3) \qquad \varphi(x,p) := \left(\sum_{i=1}^{n} \frac{\partial}{\partial p}\, c_i(x_i,p)\right)^{-1} \left(\sum_{i=1}^{n} \frac{\partial}{\partial x_i}\, c_i(x_i,p)\ c_i(x_i,p)\right)$$

We assume that the second order condition

$$(4.4) \qquad \begin{cases} \forall\ x \in K\ ,\ \forall\ p \in R(x)\ ,\ \displaystyle\sum_{i=1}^{n} c_i(x_i,p) \ \text{belongs to the domain of} \\ DT_M\left(\displaystyle\sum_{i=1}^{n} x_i, \sum_{i=1}^{n} c_i(x_i,p)\right) \end{cases}$$

holds true. By setting

$$(4.5)\quad \begin{cases} V(x,p) \;:= \\ \left(\sum_{i=1}^{n} \frac{\partial}{\partial p} c_i(x_i,p)\right)^{-1} DT_M\left(\sum_{i=1}^{n} x_i \,,\, \sum_{i=1}^{n} c_i(x_i,p)\right) \left(\sum_{i=1}^{n} c_i(x_i,p)\right) \end{cases}$$

the element of minimal norm of $DR(x,p)(c(x,p))$ can be written

$$(4.6)\qquad m(DR(x,p))(c(x,p)) \;=\; (1 - \pi_{V(x,p)})\, \varphi(x,p)$$

when $\pi_{V(x,p)}$ denotes the projection of best approximation onto the closed convex subset $V(x,p)$.

## REFERENCES


Aubin, J.P.

[1981 ]a) Contingent derivatives of set-valued maps and existence of solutions to nonlinear inclusions and differential inclusions. Advances in Mathematics. Supplementary Studies. Ed. L. Nachbin. Academic Press. 160-232.

[1981 ]b) A dynamical, pure exchange economy with feedback pricing. J. Economic Behavior and Organizations 2, 95-127.

Aubin, J.P. and A. Cellina

[1984 ] Differential inclusions. Springer-Verlag.

Aubin, J.P. and I. Ekeland

[1984 ] Applied Nonlinear Analysis. Wiley-Interscience.

Aubin, J.P. and H. Frankowska

[1984 ] Heavy viable trajectories of controlled systems.

Clarke, F.H.

[1975 ] Generalized gradients and applications. Trans. A.M.S. 205, 247-262.

[1983 ] Optimization and nonsmooth analysis. Wiley Interscience.

Cornet B.

[1981 ] Contributions à la théorie mathématique des mécanismes dynamiques d'allocation des ressources. Thèse de Doctorat d'Etat. Université de Paris-Dauphine.

Cornet, B. and G. Haddad

[1983 ] Théorèmes de viabilité pour les inclusions différentielles du second ordre. In Haddad's thesis, Université de Paris-Dauphine.

Dubovickii, A.I. and Miljutin, A.M.

[1963 ] Extremum problems with constraints. Soviet Math. 4, 452-455.

Ekeland, I.

[1979 ] Elements d'économie mathématique ; Hermann.

Haddad, G.

[1981] Monotone trajectories of differential inclusions and functional differential inclusions with memory. Israel J. Math. 39, 83-100.

Penot, J.P.

[1981] A characterization of tangential regularity. J. Nonlinear Analysis T.M.A. 5, 625-643.

Rockafellar, R.T.

[1979] Clarke's tangent cones and the boundaries of closed sets in $R^n$. Nonlinear Analysis. T.M.A. 3, 145-154.

Smale, S.

[1976] Exchange processes with price adjustments. J. Math. Econ. 3, 211-216.

Stacchetti, E.

[1984] Analysis of a dynamic, decentralized exchange economy.

# AVERAGING ET CONTROLE OPTIMAL DETERMINISTE

F. CHAPLAIS
Centre d'Automatique et d'Informatique
Ecole Nationale Supérieure des Mines de Paris
Fontainebleau - France

Résumé :

On s'intéresse ici au contrôle optimal de systèmes déterministes régis par des équations différentielles "rapidement oscillantes", et au comportement du problème d'optimisation lorsque la période tend vers zéro. On définit à partir du problème initial un nouveau problème de contrôle optimal, dit "problème moyenné".

Sous des hypothèses très générales, la fonction valeur du problème d'origine converge uniformément vers la fonction valeur du problème moyenné. Avec des hypothèses plus fortes, on montre que la commande optimale en boucle ouverte du problème moyenné engendre dans le système d'origine un coût optimal à l'ordre 2. Enfin on établit un lien entre le résultat précédent et un développement à priori de la fonction valeur.

Abstract :

The subject of this paper is the optimal control of "rapidly oscillating" deterministic systems, and the asymptotic behavior of the optimization problem as the period tends to zero. From the original problem, we define a new optimal control problem, the so-called "averaged problem".
Under weak assumptions, the value-function of the original problem tends uniformly to that of the averaged problem. Stronger assumptions ensure that the optimal open-loop control of the averaged problem induces a cost in the original system which is optimal up to the second order. Finally we relate the forementioned results to the a-priori developement of the value-function.

## INTRODUCTION :

L'approximation d'une dynamique rapidement oscillante par sa moyenne est loin d'être une idée nouvelle dans la théorie des équations différentielles ordinaires, puisqu'elle est depuis longtemps largement appliquée dans le domaine de la mécanique céleste, et qu'un long chapitre lui est consacré, par exemple, dans [1].

Néanmoins, l'utilisation théorique de cette technique dans la résolution de problèmes de contrôle optimal reste relativement récente ; on en trouve un exemple dans [5]. Notons que A. Bensoussan J.L. Lions et G. Papanicolaou ont largement traité dans [3] d'un problème voisin, puisqu'il s'agit de résoudre des équations aux dérivées partielles du second ordre, dont l'opérateur est rapidement oscillant.

A notre connaissance, la notion de problème moyenné utilisée largement ici est entièrement nouvelle. Elle permet en particulier de donner un sens à l'utilisation d'une commande en boucle ouverte extraite de ce problème, et donc d'en déduire les théorèmes d'approximation exposés ici.

Enfin précisons que les résultats énoncés dans cette étude, s'ils le sont dans un cadre relativement restreint, ouvrant la porte à de nombreuses heuristiques dans des cas plus incertains. Inversement, c'est à propos d'heuristiques développées lors d'un projet de gestion optimale de maison solaire [ 8 ], que s'est posé le problème de pourvoir une assise théorique à celles-ci, et, si possible, de leur trouver un prolongement ; c'est ainsi qu'est née cette étude.

## I - POSITION DU PROBLEME

On se donne

$$f : \mathbb{R}^n \times U^{ad} \times [0,T] \times \mathbb{R} \longrightarrow \mathbb{R}^n$$
$$(x,u,t,\theta) \longmapsto f(x,u,t,\theta)$$

et

$$L : \mathbb{R}^n \times U^{ad} \times [0,T] \longrightarrow \mathbb{R}$$
$$(x,u,t) \longmapsto L(x,u,t)$$

où $U^{ad}$ est un domaine de $\mathbb{R}^p$, contenant 0, par exemple.

On fera sur f et L les hypothèses suivantes

($\mathcal{H}_1$) il existe $k>0$ tel que :

$$\begin{cases} |f(x,u,t,\theta)| \leqslant k(1+|x|+|u|) \text{ pour tout } x,\ \theta \text{ et tout } t \text{ de } [0,T],\ u \text{ de } U^{ad} \\ |L(x,u,t)| \leqslant k(1+|x|^2+|u|^2) \text{ pour tout } x, \text{ tout } t \text{ de } [0,T],\ u \text{ de } U^{ad} \\ f \text{ et } L \text{ sont de classe } C^1 \text{ en } x \text{ et } t. \\ |\frac{\partial L}{\partial x}(x,u,t)| \leqslant k(1+|x|+|u|) \\ \frac{\partial f}{\partial x} \text{ est borné} \end{cases}$$

($\mathcal{H}_2$) $\begin{cases} U^{ad} \text{ borné ou: } L(x,u,t) \geqslant k_1|u|^2-k_2 \text{ pour tout } x, \text{ tout } t \text{ de } [0,T], \\ \text{tout } u \text{ de } U^{ad}, \text{ avec } k_1 > 0 \text{ et } k_2 > 0 \end{cases}$

($\mathcal{H}_3$) $\begin{cases} (\forall u \in U^{ad})(\forall x \in \mathbb{R}^n)(\forall t \in [0,T])(\exists\ \overline{f}_u(x,t) \in \mathbb{R}^n) \\ \frac{1}{T}\int_0^T f(x,u,t,\theta)d\theta \xrightarrow[T\to+\infty]{} \overline{f}_u(x,t) \end{cases}$

On définit le problème ($\mathcal{P}_\varepsilon$) pour $\varepsilon > 0$ comme suit :

Minimiser $\int_0^T L(x(t),u(t),t)dt$ où $\frac{dx}{dt} = f(x(t),u(t),t,\frac{t}{\varepsilon})$ sur $[0,T]$,

$x(o) = x_o$ , ceci pour u application de $[0,T]$ dans $U^{ad}$, intégrable bornée.

On voit que, sous ces hypothèses, un contrôle u meilleur, par exemple, que le contrôle identiquement nul, doit vérifier $|u|^2_{L^2[0,T]} \leqslant C(1+|x_o|^2)$ ; dans la suite de l'étude, on se restreindra à de tels contrôles. Pour des conditions initiales bornées $x_o$ , on en déduit donc que les trajectoires déterminées par de tels contrôles restent dans un domaine borné de $R^n$ , d'ailleurs estimable à partir des hyptohèses ($\mathcal{H}_1$) et ($\mathcal{H}_2$) ; le problème $\mathcal{P}_\varepsilon$ est donc inchangé si on tronque f et L de manière régulière en dehors de ce domaine. Aussi, partant du principe qu'on ne s'intéressera à $\mathcal{P}_\varepsilon$ que pour un ensemble borné de conditions initiales, <u>nous supposerons désormais f et L à support compact dans $R^n$</u>. Toutes les trajectoires sont donc bornées et $\frac{\partial L}{\partial x}(x,u(t),t)$ est donc borné dans $L^2[0,T]$ indépendamment de x pour les contrôles ci-dessus. En particulier si $V_\varepsilon$ est la fonction valeur (comme infimum) du problème $V_\varepsilon$ est globalement lipschitz en x, et indépendamment d' $\varepsilon$.

## II - PROBLEME MOYENNE

1) Soit $\mathcal{V}$ l'ensemble des applications de R dans $R^p$, bornées, localement intégrables, vérifiant :

$$(\mathcal{H}_4) \begin{cases} (\forall x \in \mathbb{R}^n)(\forall t \in [0,T])(\exists\ \overline{f}(x,v,t) \in R^n)(\exists\ \overline{L}(x,v,t) \in R) \\ \frac{1}{T}\int_o^T f(x,v(\theta),t,\theta)d\theta \xrightarrow[T\to+\infty]{} \overline{f}(x,v,t) \\ \text{et} \quad \frac{1}{T}\int_o^T L(x,v(\theta),t)d\theta \xrightarrow[T\to+\infty]{} \overline{L}(x,v,t) \end{cases}$$

et $\mathcal{V}^{ad} = \{v \in \mathcal{V}\ ,\ v(\theta) \in U^{ad}$ pour presque tout $\theta\}$

On définit $\mathcal{U}$ l'ensemble des contrôles admissibles comme étant l'ensemble des applications u de $[0,T]$ dans $\mathcal{V}^{ad}$, intégrables bornées au sens de la norme $L^\infty$ sur $\mathcal{V}$. Notons que $\mathcal{V}^{ad}$ comprend l'ensemble $\mathcal{C}$ des applications constantes à valeur dans $U^{ad}$ d'après $\mathcal{H}_3$ , et que $\mathcal{U}$ comprend les applications constantes à valeur dans $\mathcal{C}$. Enfin, si f est périodique en $\theta$ de même période pour tout x,u,t, alors les fonctions périodiques intégrables bornées sont dans $\mathcal{V}$.

2) Le problème moyenné consiste à minimiser :

$$\int_o^T \overline{L}(x(t),u(t),t)dt \quad \text{pour} \quad u \in \mathcal{U}\ ,$$

où x est défini par $\frac{dx}{dt} = \overline{f}(x(t),u(t),t)$ sur $[0,T]$, $x(o) = x_o$.

3) Validité des outils généraux du contrôle optimal

a) Equation d'Hamilton Jacobi :

On montre [ 4 ], sous les mêmes hypothèses que d'ordinaire sur f, L et $U^{ad}$ que la fonction valeur $V^o(x_o,t) = \underset{u\in\mathcal{U}}{\text{Inf}} \int_t^T \overline{L}(x(s),u(s),s)ds$ vérifie :

$$\frac{\partial V^o}{\partial x} + \underset{v\in\mathcal{V}^{ad}}{\text{Inf}} \left\{\frac{\partial V^o}{\partial x}\ \overline{f}(x,v,t) + \overline{L}(x,v,t)\right\} = 0 \quad \text{et} \quad V^o(x,T) \equiv 0.$$

On peut d'ailleurs reprendre les mêmes démonstrations.

Notons que, comme $V^\varepsilon, V^o$ est Lipschitz en x, globalement. En effet, pour $v \in \mathcal{V}$, on montre que $\overline{L}$ est de classe $C^1$, que $\frac{\partial \overline{L}}{\partial x}$ existe au sens de $(\mathcal{H}_4)$ et que $\frac{\partial \overline{L}}{\partial x} = \overline{\frac{\partial L}{\partial x}}$ ; de même pour f. En particulier $\overline{f}$ et $\overline{L}$ vérifient $(\mathcal{H}_1)$ et $(\mathcal{H}_2)$, avec les mêmes constantes, au sens de la norme $L^\infty$ pour $\mathcal{V}$. On en déduit que pour un domaine de conditions initiales identiques, on peut effectuer les mêmes troncatures que pour le problème $\mathcal{P}_\varepsilon$, et aboutir aux mêmes conclusions. De ce point de vue, on peut donc commuter troncature et moyenne. Une fois celles-ci effectuées, on peut conclure à la propriété de Lipschitz de $V^o$ .

b) Principe du minimum

Soit u controle optimal. On définit l'état adjoint par :

$$- \frac{dp}{dt} = p.\frac{\partial \overline{f}}{\partial x} (x(t),u(t),t) + \frac{\partial \overline{L}}{\partial x} (x(t),u(t),t), \quad p(T) = 0$$

Par les mêmes techniques que dans les cas classiques, on montre que u minimise $p(t) . \overline{f}(x(t),v,t) + \overline{L}(x(t),v,t)$ pour v dans $\mathcal{V}^{ad}$, ceci pour presque tout t. Noter que v est une fonction sur $\mathbb{R}_+$

## III - THEOREME LIMITE SUR LES FONCTIONS VALEUR

Nous supposons donc désormais f et L à support compact en x, indépendant de t, u et $\Theta$ et f uniformement continue en $\Theta$. Nous ferons ici les hypothèses supplémentaires suivantes :

$$(\mathcal{H}_5) \begin{cases} \text{Soit } H(p,x,t,\Theta) = \inf_{u \in U^{ad}} \left\{ p.f(x,u,t,\Theta) + L(x,u,t) \right\} \text{ pour } p \in \mathbb{R}^n \\ \text{alors : } (\forall m>o)(\exists \lambda>o)(\forall x \in R^n)(\forall t \in [0,T])(\forall \Theta \in R)(\forall (p_1,p_2) \in \mathbb{R}^n \times \mathbb{R}^n) \\ \quad (|p_1| \leq m \text{ et } |p_2| \leq m) \Rightarrow (|H(p_1,x,t,\Theta) - H(p_2,x,t,\Theta)| \leq \lambda |p_1 - p_2|) \end{cases}$$

Notons que H existe d'après $(\mathcal{H}_1)$ et $(\mathcal{H}_2)$ et que, après troncature, $(\mathcal{H}_5)$ est vérifiée dans le cas linéaire quadratique. L'hypothèse essentielle pour ce qui nous concerne est la suivante :

($\mathcal{H}_6$) il existe $\overline{H}$ : $\mathbb{R}^n \times \mathbb{R}^n \times [0,T] \to \mathbb{R}$ tel que :

$$\left\{ (\forall m>0) \sup_{\substack{x\in\mathbb{R}^n \\ t\in[0,T] \\ |p|\leq m \\ \hat{t}\geq 0}} \left| \frac{1}{T}\int_{\hat{t}}^{\hat{t}+T} H(p,x,t,\theta)d\theta - \overline{H}(p,x,t) \right| \xrightarrow[T\to\infty]{} 0 \right.$$

Remarquons que $H(p,x,t,\theta)$ est borné pour $|p|\leq m$ après troncature, d'après ($\mathcal{H}_1$) et ($\mathcal{H}_2$). En particulier, si f est périodique de période $\omega$, alors H l'est également et $\overline{H}$ est la moyenne de H sur une période.

Nous pouvons annoncer le théorème suivant [4]

Théorème 1 : sous les hypothèses ($\mathcal{H}_1$), ($\mathcal{H}_2$), ($\mathcal{H}_3$), ($\mathcal{H}_4$), ($\mathcal{H}_5$), ($\mathcal{H}_6$) pour f et L à support compact en x, f uniformément continue en $\theta$, on a :

$$\sup_{\substack{x\in R^n \\ t\in[0,T]}} \left| V^{\varepsilon}(x,t) - V^{o}(x,t) \right| \xrightarrow[\varepsilon\to 0]{} 0$$

et $V^o$ vérifie : $\frac{\partial V^o}{\partial t} + \overline{H}(\frac{\partial V^o}{\partial x},x,t) = 0$, $V^o(x,T) \equiv 0$

Plan de la démonstration :

1) On commence par établir le second point. On a en fait

$$\overline{H}(p,x,t) = \min_{v\in\mathcal{V}^{ad}} \left\{ p.\overline{f}(x,v,t) + \overline{L}(x,v,t) \right\} .$$

Le premier terme est inférieur au second car pour v dans $\mathcal{V}^{ad}$ on a :

$$p.f(x,v(\theta),t,\theta) + L(x,v(\theta),t) \geq H(p,x,t,\theta)$$

d'où l'inégalité en moyenne. Il suffit maintenant de montrer qu'il existe v dans $\mathcal{V}^{ad}$ tel que $p.\overline{f}(x,v,t) + \overline{L}(x,v,t) = \overline{H}(p,x,t)$. Pour cela on approxime $H(p,x,t,\theta)$ par $p.f(x,v(\theta),t,\theta) + L(x,v(\theta),t,\theta)$, où v est en escalier sur $R_+$, avec une erreur nulle en moyenne. Notons que dans le cas périodique on n'atteint pas toujours le Min si on se resteint à des v périodiques ; on a simplement égalité des Inf.

2) Pour démontrer le second point, on utilisera une méthode de viscosité.

En effet, on sait (cf [ 6 ]) que $V^{\varepsilon}$ peut être approché par $V_{\alpha}^{\varepsilon}$ , où $V_{\alpha}^{\varepsilon}$ est la fonction valeur d'un problème de contrôle stochastique où f est perturbé par un petit "bruit" $\sqrt{\alpha}\, dw$ (w mouvement brownien standard). L'erreur est du type $K\sqrt{\alpha}$ et en particulier K est indépendant du comportement en t de f et L, et donc d'$\varepsilon$. Il en est de même pour $V^{o}$ : il suffit de définir le problème stochastique moyenné en se restreignant aux contrôles u(x,t, ) boréliens, tels qu'ils engendrent une moyenne au sens de $(\mathcal{H}_4)$. Le théorème d'approximation par viscosité reste valide en ce qui concerne le problème moyenné. D'après ce qui précède, $V_{\alpha}^{o}$ vérifie :

$$(1)\quad \frac{\partial V_{\alpha}^{o}}{\partial t} + \alpha\, \Delta\, V_{\alpha}^{o} + \overline{H}\left(\frac{\partial V_{\alpha}}{\partial x},x,t\right) = 0 \qquad V_{\alpha}^{o}(x,T) \equiv 0 \quad \text{et on a :}$$

$$(2)\quad \frac{\partial V_{\alpha}^{\varepsilon}}{\partial t} + \alpha\, \Delta\, V_{\alpha}^{\varepsilon} + H\left(\frac{\partial V_{\alpha}^{\varepsilon}}{\partial x},x,t,\frac{t}{\varepsilon}\right) = 0 \qquad V_{\alpha}^{\varepsilon}(x,T) \equiv 0$$

$V_{\alpha}^{o}$ et $V_{\alpha}^{\varepsilon}$ sont de classe $C^{1,2}$ et sont les uniques solutions de (1) et (2). Nous allons pouvoir donc raisonner directement sur ces équations.

Plus précisément, nous allons démontrer qu'il existe $\varepsilon(\alpha,\eta)$ tel que, pour $\varepsilon < \varepsilon(\eta,\alpha)$ alors $\sup_{\substack{x\in\mathbb{R}^n \\ t\in[0,T]}} \left|V_{\alpha}^{\varepsilon}(x,t) - V^{o}(x,t)\right| \leqslant \eta$ ;

le théorème sera démontré en prenant $\varepsilon < \varepsilon\ (\eta,\frac{\eta^2}{K^2})$

La démonstration du point précédent étant assez longue, nous n'en donnerons que les principales étapes. Nous utiliserons le noyau de la chaleur et les techniques de base esposées dans [7].

a) Remarquons d'abord que $\frac{\partial V_{\alpha}^{o}}{\partial x}$ et $\frac{\partial V_{\alpha}^{\varepsilon}}{\partial x}$ sont bornés indépendamment d'$\varepsilon$. Or si W est solution de $-\frac{\partial W}{\partial t} - \alpha\, \Delta\, W = g(x,t)$ avec $W(x,T) = 0$, $t\in[0,T]$, il existe $\mu\in]0,1[$ tel que $\left|\frac{\partial W}{\partial x}(x,t) - \frac{\partial W}{\partial x}(x,t')\right| \leqslant K\ |t-t'|^{\mu}$ , où K ne dépend que de $\|g\|_{\infty}$ et T. On a donc une estimation höldérieme de $\frac{\partial V_{\alpha}^{0}}{\partial x}$ et $\frac{\partial V_{\alpha}^{\varepsilon}}{\partial x}$ indépendante d'$\varepsilon$.

b) Si $\Gamma_{\alpha}$ désigne le noyau de la chaleur utilisé, on peut ensuite estimer $\mathbb{R}_i^{\varepsilon} = \frac{\partial \Gamma_{\alpha}}{\partial x_i} *_t \left(H\left(\frac{\partial V_{\alpha}^{0}}{\partial x},x,t,\frac{t}{\varepsilon}\right) - \overline{H}\left(\frac{\partial V_{\alpha}^{0}}{\partial x},x,t\right)\right)$ pour $\varepsilon$ petit, par averaging, grâce à l'hypothèse $(\mathcal{H}_6)$ et au résultat ci-dessus ; $*_t$ désigne le produit de convolution sur $[t,T]$.

c) Posons $M(t) = \sup_{x\in\mathbb{R}^n} \left|\frac{\partial V_\alpha^\varepsilon}{\partial x} - \frac{\partial V_\alpha^0}{\partial x}\right|$. Comme H est Lipschitz en p d'après ($\mathcal{H}_5$), en utilisant $\Gamma_\alpha$ pour exprimer $\frac{\partial V_\alpha^\varepsilon}{\partial x}$ et $\frac{\partial V_\alpha^0}{\partial x}$ on a donc :

$$M(t) \leq \int_t^T M(\tau) \int_{R^n} \left|\frac{\partial \Gamma_\alpha}{\partial x_i}(x,\tau-t)\right| dx + \eta \text{ pour } \varepsilon \text{ petit ;}$$

d'où une estimation de $\left|\frac{\partial V_\alpha^\varepsilon}{\partial x} - \frac{\partial V_\alpha^0}{\partial x}\right|$ par un lemme de type Gronwall.

d) La deuxième partie consiste essentiellement a répéter la même démonstration en ramplaçant $\frac{\partial \Gamma_\alpha}{\partial x}$ par le noyau $\Gamma_\alpha$ lui-même.■

## IV - UTILISATION DU CONTROLE OPTIMAL EN BOUCLE OUVERTE DU PROBLEME MOYENNE

Le théorème présenté maintenant est à rapprocher des résultats équivalents existants en pertrubations régulières. Les hypothèses et la démonstration que nous utiliserons sont d'ailleurs très inspirées de celles utilisées par A. Bensoussan dans ce domaine [ 2 ]. On prend ici $U^{ad} = R^p$

1) Théorème 2 : Soient f et L comme dans les paragraphes I et II(*) avec de plus :

($\mathcal{H}_7$) f est périodique en $\theta$ de période $\omega$

($\mathcal{H}_8$) Le problème moyenné admet un contrôle optimal $v_o$ pour la condition initiale X ; on notera y la trajectoire optimale et q l'état adjoint et on posera $u_o(t,\theta) = [v_o(t)](\theta)$ ; on suppose $u^o \in L^2([0,T]\times[0,\omega], R^p)$

($\mathcal{H}_9$) f et L sont de classe $C^2$ en x et u, de dérivées bornées, de dérivées secondes Lipschitz en x et u

Si $h(p,x,u,t,\theta) = p.f(x,u,t,\theta) + L(x,u,t)$, alors

$$\frac{\partial^2 h}{\partial u^2}(q(t),x,v,t,\theta) \geq \beta \text{ Id}, \ \beta > 0 \text{ pour tout } x,v,t,\ \theta$$

$$\left(\frac{\partial^2 h}{\partial x^2} - \frac{\partial^2 h}{\partial x \partial v}\left(\frac{\partial^2 h}{\partial u^2}\right)^{-1} \frac{\partial^2 h}{\partial v \partial x}\right)(q(t),x,v,t,\theta) \geq 0 \text{ pour tout } x,v,t,\theta$$

($\mathcal{H}_{10}$) $f(y(t),u_o(t,\theta),t,\theta)$ et $\frac{\partial h}{\partial x}(q(t),y(t),u_o(t,\theta),t,\theta)$ sont bornés de classe $C^1$ en t, de dérivée lipschitz en t.

(*) en particulier f et L de support compact en x.

Si $g(\Theta)$ est périodique intégrable bornée, on notera $\overline{g}$ sa moyenne et $\pi(g(.))$ la primitive de moyenne nulle de $(g-\overline{g})$. ($H_9$) est identique aux hypothèses faites en perturbations régulières.
On posera $x_2(t,\Theta) = \pi(f(y(t),u_o(t,.),t,.)(\Theta)$ et $J_\varepsilon^*$ le coût optimal du problème d'origine.

On a, comme en perturbations régulières, doublement de l'ordre d'approximation. En effet, il existe $k>o$, $\varepsilon_o>o$, tels que pour $\varepsilon<\varepsilon_o$:

a) $$J_\varepsilon^* \geq \int_o^T L(y(t),u_o(t,\tfrac{t}{\varepsilon}),t)dt + \varepsilon\int_o^T \overline{\frac{\partial H}{\partial x}(q(t),y(t),t,.)x_2(t,.)}dt - \varepsilon q(o).x_2(o,o) - k\varepsilon^2$$

b) Soit $x$ défini par $\frac{dx}{dt} = f(x,u_o(t,\frac{t}{\varepsilon}),t,\frac{t}{\varepsilon})$, $x(o) = X$

Alors $$\left|\int_o^T L(x,u_o(t,\tfrac{t}{\varepsilon}),t)dt - \int_o^T L(y(t),u_o(t,\tfrac{t}{\varepsilon}),t)dt - \varepsilon\int_o^T \overline{\frac{\partial H}{\partial x}(q(t),y(t),t,.)x_2(t,.)}dt + \varepsilon q(o).x_2(o,o)\right| \leq k\varepsilon^2$$

et $u_o$ est donc optimal dans $P^\varepsilon$ à l'ordre 2.

Démonstration : Comme nous l'avons précisé, les techniques utilisées sont très proches de celles rencontrées en perturbations régulières [2]. Aussi nous ne mentionnerons ici que les principales différences :

a) Soit $p_2(t,\Theta) = -\pi(\frac{\partial H}{\partial x}(q(t),y(t),t,.))(\Theta)$

Le problème tangent à considérer est le suivant :

$$\frac{dz}{dt} = \frac{\partial f}{\partial x}(y(t),u_o(t,\tfrac{t}{\varepsilon}),t,\tfrac{t}{\varepsilon})(z+x_2(t,\tfrac{t}{\varepsilon})) + \frac{\partial f}{\partial u}(y(t),u_o(t,\tfrac{t}{\varepsilon}),t,\tfrac{t}{\varepsilon})u \quad , \quad z(o) + x_2(o,o) = o$$

$$\text{Minimiser : } \int_o^T \left\{(z',u')\begin{pmatrix} \frac{\partial^2 H}{\partial x^2} & \frac{\partial^2 H}{\partial x\partial v} \\ \frac{\partial^2 H}{\partial v\partial x} & \frac{\partial^2 H}{\partial v^2}\end{pmatrix}_{(q,y,t,\frac{t}{\varepsilon})}\begin{pmatrix} z \\ u\end{pmatrix} + p_2(t,\tfrac{t}{\varepsilon})[\frac{\partial f}{\partial x}(y,u_o,t,\tfrac{t}{\varepsilon})z+\frac{\partial f}{\partial u}(y,u_o,t,\tfrac{t}{\varepsilon})u]\right\}dt$$

On note $y_1$ la trajectoire optimale et $v_1$ le contrôle optimal.

b) On pose $r = x-y-\varepsilon x_2(t,\frac{t}{\varepsilon})-\varepsilon y_1 = \tilde{x}-\varepsilon y_1$

$$v = u-u_o(t,\frac{t}{\varepsilon})-\varepsilon v_1 \quad = \tilde{u}-\varepsilon v_1$$

$$\rho(\lambda,\mu) = (q,y+\lambda\mu(\tilde{x}+\varepsilon x_2(t,\frac{t}{\varepsilon})),u_o(t,\frac{t}{\varepsilon})+\lambda\mu\tilde{u},t,\frac{t}{\varepsilon})$$

et h désigne l'hamiltonien non minimisé, non moyenné. On rapportera toutes les estimations à :

$$Z_\varepsilon(\lambda,\mu) = v + \left|\frac{\partial^2 h}{\partial v^2}^{-1} \frac{\partial^2 h}{\partial v \partial x}\right|_{(\rho(\lambda,\mu))} (r+\varepsilon x_2)$$

et à $z_\varepsilon^2 = \int_o^1 \lambda d\lambda \int_o^1 d\mu \, |Z_\varepsilon(\lambda,\mu)|^2_{L^2(o,T)}$

A partir de là, la démonstration est assez classique, mis à part quelques simplifications d'expressions par estimations de moyennes.

Remarque : on comparera ce résultat à celui obtenu en perturbations singulières. Néanmoins, le lecteur, en raisonnant par analogie, se convaincra que minimiser la moyenne de l'hamiltonien par un contrôle en feedback sur l'état rapide est l'équivalent de la résolution à chaque instant d'un problème de couche limite, mais statique. Il n'y a donc pas à proprement parler de "contrôle lent" et, pour revenir aux perturbations singulières, on ignore donc l'ordre d'approximation intermédiaire utilisant ce "contrôle lent". Une des particularités des perturbations singulières tient à ce que, par exemple, fonctions du phénomène rapide et moyenne commutent (la moyenne étant l'état quasi-stationnaire). On peut donc considérer la moyenne du contrôle rapide, qui est le contrôle lent.

2) Conséquences numériques du théorème 2 :

On voit donc que $u_o$ est optimal à l'ordre 2. Mais que gagne-t-on à utiliser le problème moyenné plutôt que le problème d'origine ? Contrairement à ce qui se passe en perturbations singulières, on ne gagne pas en dimension d'état, bien qu'en un sens, il y ait disparition du phénomène rapide. On ne gagne pas dans la minimisation de l'Hamiltonien, puisque celle-ci se fait à l'intérieur de la moyenne ; elle nécessite donc une grille de temps fine. Le coté statique de la résolution est donc inchangé ; par contre on gagne un facteur $\frac{1}{\varepsilon}$ dans la partie dynamique. En effet la dépendance en $\frac{t}{\varepsilon}$ de f nécessite à priori une grille de temps en $\frac{\delta t}{\varepsilon}$ ; le passage à la moyenne permet de se limiter à des grilles en $\delta t$, que ce soit pour la résolution des équations d'Euler ou pour l'utilisation de la programmation dynamique. Ceci est particulièrement appréciable lorsqu'on a à raisonner sur de longues périodes de temps, ce qui est le cas lorsqu'on considère la gestion d'une maison solaire qui doit prendre en compte les variations relativement rapides des phénomènes météorologiques, alors que le problème est à horizon annuel en raison de l'importance cruciale des variations saisonnières.

Exemple :

Soit $\frac{dx}{dt} = a(t)x + u\sin\frac{t}{\varepsilon}$ , Minimiser $\int_o^T (x^2+ru^2)dt$, $r>o$, $x\in R$

Soit $P^{\varepsilon}x^2$ la fonction valeur. $P^{\varepsilon}$ vérifie :

$$\frac{dP^2}{dt} + 2aP + 1 - \frac{(P^{\varepsilon})^2}{r}\sin^2(\frac{t}{\varepsilon})=o, \quad P(T) = o,$$

d'où, à priori, l'utilisation d'une discrétisation en $\frac{\delta t}{\varepsilon}$ .

Résoudre le problème moyenne consiste à résoudre :

$$\frac{dP^o}{dt} + 2aP + 1 - \frac{(P^o)^2}{2r} = o, \quad P(T) = o,$$

d'où une grille de temps en $\delta t$. Le contrôle $u^{\varepsilon}(t) = -\frac{P^o(t)}{r}\sin\frac{t}{\varepsilon}\, y(t)$ (où $\frac{dy}{dt} = a(t)y - \frac{P^o(t)}{2r}y$, $y(o) = x(o))$ est optimal à $\varepsilon^2$ près (Dans ce cas précis, le contrôle $u^{\varepsilon}(x,t) = -\frac{P^o(t)}{r}\sin\frac{t}{\varepsilon}\, x$ est aussi optimal à l'ordre 2). On peut également procéder en boucle fermée et utiliser les mêmes résultats, puisqu'on connait la forme de feedback optimaux.

## V - CAS LINEAIRE QUADRATIQUE, PERIODIQUE

Dans le cas linéaire quadratique, tous les résultats peuvent être retrouvés par des techniques différentes, puisque l'étude se ramène à celle d'équations différentielles ordinaires (Riccati par exemple). On peut exhiber un developpement asymptotique de la fonction valeur ; il doit, à partir de l'ordre 2, dépendre nécessairement de la variable $\Theta+\frac{T-t}{\varepsilon}$ (phase finale) (cf. perturbations singulières). On généralise le doublement d'ordre d'approximation à un ordre quelconque en utilisant des feedbacks déduits du développement précédent. Notons que ceci se ramène, dans le cas linéaire quadratique, à utiliser (après transformation) des contrôles en boucle ouverte, grâce à la forme particulière des feedbacks.

## VI - DEVELOPPEMENTS A PRIORI DE LA FONCTION VALEUR

### 1) Forme du développement :

On songerait d'abord à développer la fonction valeur sous la forme :

$$V^{\varepsilon}(x,t) = V^{o}(x,t) + \Sigma\ \varepsilon^{k}\ V_{k}(x,t,\theta)\Big|_{\theta=\frac{t}{\varepsilon}}$$

Or ceci se révèle impraticable sur des cas simples à partir du troisième terme $V_2$. En effet, la partie de moyenne nulle de $V_k$ s'obtient indépendamment des conditions aux limites, et ne les vérifie en général pas(*).

Une forme plus appropriée consiste à prendre :

$$V^{\varepsilon}(x,t) = V^{o}(x,t) + \varepsilon V_{1}(x,t,\frac{t}{\varepsilon}) + \Sigma\ \varepsilon^{k}\ V_{k}(x,t,\frac{t}{\varepsilon},\frac{T}{\varepsilon})\ ;$$

elle est d'ailleurs exacte dans le cas linéaire quadratique. Il nous semble que l'apparition de $\frac{T}{\varepsilon}$ puisse être reliée à la présence d'un terme analogue à $x_2$ dans le développement de l'état adjoint. Nous allons maintenant nous limiter aux deux premiers termes du développement, et montrer la cohérence de celui-ci avec les résultats précédents.

### 2) Cohérence des développements :

#### a) Identification de $V_1$

Nous allons élargir le problème en introduisant un état supplémentaire: $\frac{dx}{dt}=f(x,u,t,\theta)$ et $\frac{d\theta}{dt}=\frac{1}{\varepsilon}$ ; $\mathcal{P}_{\varepsilon}$ correspond aux trajectoires particulières: $\theta(t)=\frac{t}{\varepsilon}$. La fonction valeur $V^{\varepsilon}(x,t,\theta)$ est périodique en $\theta$ et vérifie :

$$\frac{\partial V^{\varepsilon}}{\partial t}+\frac{1}{\varepsilon}\,\frac{\partial V^{\varepsilon}}{\partial \theta}+\underset{u}{\mathrm{Min}}\left[\frac{\partial V^{\varepsilon}}{\partial x}\ .\ f(x,u,t,\theta) + L(x,u,t)\right] = 0, \quad V^{\varepsilon}(x,t,\theta) \equiv 0$$

En supposant $\left|V^{\varepsilon} - V^{o} - \varepsilon V_{1}(x,t,\theta) - \varepsilon^{2}V_{2}(x,t,\theta,\theta+\frac{T-t}{\varepsilon})\right| \leqslant k\ \varepsilon^{2}$

(*) Si L dépend explicitement de $\theta$, on peut rencontrer ce problème dès le second terme $V_1$. C'est pourquoi nous avons supposé L indépendant de $\theta$.

et l'unicité de l'argument de minimisation, on a donc :

$$\begin{cases} \frac{\partial V^o}{\partial t} + \frac{\partial V_1}{\partial \Theta} + \underset{u}{\mathrm{Min}} \left[\frac{\partial V^o}{\partial x} f(x,u,t,\Theta) + L(x,u,t)\right] = 0 \\ \frac{\partial V_1}{\partial t} + \frac{\partial V_2}{\partial \Theta} + \frac{\partial V_1}{\partial x} f(x,v(x,t,\Theta),t,\Theta) = 0 , \end{cases}$$

où $v(x,t,\Theta)$ minimise l'hamiltonien $\frac{\partial V^o}{\partial x} f + L.$

Soit $Q_1$ la moyenne de $V_1$ et $S_1 = V_1 - Q_1$ . Alors :

$$(2) \quad S_1 = -\pi(H(\frac{\partial V_o}{\partial x},x,t,.)) = - \frac{\partial V_o}{\partial x} \pi(f(x,v(x,t,.),t,.)$$

$$-\pi(L(x,v(x,t,.),t)) ,$$ et $Q_1$ vérifie :

$$(3) \quad \begin{cases} \frac{\partial Q_1}{\partial t} + \frac{\partial Q_1}{\partial x} \overline{f(x,v(x,t,.),t,.)} + \overline{\frac{\partial S_1}{\partial x} f(x,v(x,t,.),t,.)} = 0 \\ Q_1(x,T) = 0 \end{cases}$$

b) Nous allons maintenant énoncer le

<u>Théorème 3</u> : Soient $S_1$ et $Q_1$ vérifiant (2) et (3), et $u_o$,$y$,$q$ et $x_2$ définis comme dans le théorème 2. Alors :

> a) $Q_1(X,0) = \int_o^T \overline{\frac{\partial H}{\partial x}(q,y,t,.)x_2(t,.)}dt$
>
> b) $\left|V^o(X,0)+\varepsilon S_1(X,0,0)+\varepsilon q'(o)x_2(o,o)-\int_o^T L(y,u_o(t,\frac{t}{\varepsilon}),t)dt\right| \leq k\varepsilon^2$

ce qui établit la cohérence cherchée.

<u>Démonstration</u> : Démontrons le point a). Remarquons d'abord que :

$$\overline{- \frac{\partial}{\partial x}(\pi(H(\frac{\partial V_o}{\partial x},x,t,.))f(x,v(x,t,.),t,.)} =$$

$$\overline{\frac{\partial}{\partial X}(H(\frac{\partial V_o}{\partial x},x,t,.))\pi(f(x,v(x,t,.),t,.)}$$ par intégration par parties.

D'autre part, la dynamique intervenant dans (3) est celle de y. Intégrant (3) entre 0 et T le long de y, nous obtenons donc :

$$Q_1(X,0) = \int_0^T \overline{\frac{\partial H}{\partial x}(q,y,t,.)x_2(t,.)}dt$$

Passons au point b). On sait que,

à $k\,\varepsilon^2$, $\int_0^T L(y,u_o(t,\frac{t}{\varepsilon}),t)dt$ est égal à :

$$\int_0^T \overline{L(y,u_o(t,.),t)}dt + \varepsilon\left[\pi(L(y,u_o(t,.),t)\Big|_{\theta=\frac{t}{\varepsilon}}\right]_{t=0}^{t=T}$$

Le premier terme vaut précisément $V^o(X,0)$. Quant au second, remarquons que $S_1(x,T,\theta) = -\pi(H(0,x,T,.)) = 0$ car $L$ ne dépend pas de $\theta$.

Donc $\pi(L(y(T),v(y(T),T,.),T)$ est nul et $\int_0^T L(y,u_o(t,\frac{t}{\varepsilon}),t)dt$ est égal, à $k\,\varepsilon^2$ près, à : $V^o(X,0) - \varepsilon\pi(L(X,v(X,0,.),0)\Big|_{\theta=0}$, qui vaut précisément $V^o(X,0) + \varepsilon S_1(X,0,0) + \varepsilon q'(0)x_2(o,o)$ ■

3) Conséquences envisageables :

On a montré, dans le cas linéaire quadratique, le doublement d'ordre d'approximation lorsqu'on utilise dans le système d'origine les feedbacks :

$$u = v_o + \sum_{k=1}^{i} \varepsilon^k v_k \ ;$$

où $v_j$ minimise $H(\frac{\partial V_j}{\partial x},x,t,\theta)$. Or, si on a calculé les trajectoires optimales y lors de la programmation dynamique du problème moyenné, $V_1$ est alors calculable très simplement, et permet d'espérer accéder à l'ordre 4 sur le coût optimal, ceci sans avoir recours à une grille de temps fine.

## BIBLIOGRAPHIE


[1] V. ARNOLD, "Théorie des équations différentielles ordinaires", Editions de Moscou.

[2] A. BENSOUSSAN, "Singular Perturbations in Systems and Control", Mark Ardema, ed., Springer-Verlag, 1983, pp 169-185.

[3] A. BENSOUSSAN, J.L. LIONS, G. PAPANICOLAOU,"Asymptotic Analysis for Periodic Structures", North-Holland, 1978.

[4] F. CHAPLAIS, Thèse de Docteur-Ingénieur, à paraître.

[5] F. DELEBECQUE, J.P. QUADRAT, "Utilisation d'un théorème de mélange pour le découplage gestion court terme, long terme et contrôle stochastique et application à la gestion de réservoirs", Annales des Sciences Mathématiques du Québec, 1977, Vol.2, pp. 195-205.

[6] W.H. FLEMING, R.W. RISHEL, "Deterministic and Stochastic Optimal Control", Springer-Verlag, 1975.

[7] A. FRIEDMAN, "Partial Differential Equations of Parabolic Type", Prentice-Hall, 1964.

[8] Y. LENOIR, LU RONG GUO, "Commande optimale des chauffages solaires avec appoint indépendant", Rapport à DETN - GAZ DE FRANCE - CAI ENSMP, 1982.

# THE MAXIMUM PRINCIPLE FOR A DIFFERENTIAL INCLUSION PROBLEM

Halina Frankowska

CEREMADE
Université Paris IX-Dauphine
75775 Paris CX (16) France.

The Pontriagin principle is extended to the case of minimization of solutions to differential inclusions by using a concept of derivative of set-valued maps.

## Introduction

Consider a control system with feedbacks

$$\dot{x}(t) = f(x(t),u(t)) \quad , \quad u(t) \in U(x(t)) \tag{0.1}$$

where $f : \mathbb{R}^n \times \mathbb{R}^m \to \mathbb{R}^n$ and $U : \mathbb{R}^n \rightrightarrows \mathbb{R}^m$ is a set valued map. Let $S$ be the set of all solutions to (0.1) and assume $z \in S$ solves the following problem :

$$\text{minimize } \{g(x(0),x(1)) : x \in S\}$$

$g$ being a function on $\mathbb{R}^{2n}$ taking values in $\mathbb{R} \cup \{+\infty\}$ .

If there is no feedback, i.e. if $U$ does not depend on $x$ , and the datas are smooth enough the celebrated maximum principle (see Pontriagin and others [16]) tells us that for some absolutely continuous function $q : [0,1] \to \mathbb{R}^n$ the following holds true :

$$\begin{cases} -\dot{q}(t) = [\frac{\partial f}{\partial x} (z(t),\bar{u}(t))]^{\star} q(t) \\ \\ <q(t),f(x(t),\bar{u}(t))> = \max\limits_{u \in U} <q(t),f(z(t),u)> \end{cases} \tag{0.2}$$

$$(-q(0),q(1)) = g'(z(0),z(1)) \tag{0.3}$$

where $\bar{u}$ is the corresponding control, $[\frac{\partial f}{\partial x}(z(t),\bar{u}(t))]^{\star}$ denotes the transpose of the Jacobian matrix of $f$ with respect to $x$ at $(z(t),\bar{u}(t))$, and $g'$ is the derivative of $g$.

To study the necessary conditions in a more general case we have to consider the set valued map $F : \mathbb{R}^n \rightrightarrows \mathbb{R}^n$ defined by :

$$F(x) := \{f(x,u) : u \in U(x)\}$$

and the associated differential inclusion

$$(0.1)' \qquad \dot{x} \in F(x)$$

Under some measurability assumptions on $f$ and $U$ it can be shown that the solutions to (0.1) and (0.1)' coincide.

This approach to optimal control problem was firstly proposed by Wazewski in [21] who was followed by many authors. (See for example [2], [3], [5], [6], [8], [11], [13], [14], [17], [21]).

For obtaining results similar to (0.2), (0.3) in the set valued case we need a notion generalizing the differential to a set valued map $F : \mathbb{R}^n \rightrightarrows \mathbb{R}^m$ and its transpose.

In this paper we use such a generalization, called the asymptotic differential $DF(x,y)$ and asymptotic co-differential $DF(x,y)^{\star}$ of $F$ at $(x,y) \in \text{graph}(F)$. We consider also the related notion of asymptotic gradient $\partial_a g$ of a real valued function $g$.

The necessary conditions then take the following form :

> There exists an absolutely continuous function $q : [0,1] \to \mathbb{R}^n$ satisfying the following conditions :

$$(0.2)' \qquad -\dot{q}(t) \in DF(z(t),\dot{z}(t))^{\star}(q(t))$$

$$(0.3)' \qquad (-q(0),q(1)) \in \partial_a g(z(0),z(1))$$

The outline of the paper is as follows. We devote the first section to some background definitions which we shall use. We state in section 2 the main theorem concerning the necessary conditions satisfied by an optimal solution to a differential inclusion problem. We show also how this problem can be embedded in a class of abstract optimization problems. This general problem is studied in section 3. Section 4 provides an example of application. In particular we extend in this paper to the non convex case some results obtained by Aubin-Clarke [3].

## 1. Asymptotic differential and co-differential of a set valued map.

In what follows $E$ denotes a Banach space, $\overset{\circ}{B}$ denotes the open unit ball in $E$ and $<\,,\,>$ the duality paring on $E^\star \times E$.

The tangent cone of Ursescu to a set $K \subset E$ at a point $x \in K$ is defined by

(1.1) $$I_K(x) := \bigcap_{\substack{\varepsilon > 0 \\ \delta > 0}} \bigcup_{\delta > 0} \bigcap_{h \in ]0,\delta[} \left[\frac{1}{h}(K-x) + \varepsilon\overset{\circ}{B}\right]$$

The above cone is sometimes called the intermediate tangent cone since it lies between more familiar contingent cone (of Bouligand)

$$T_K(x) := \bigcap_{\varepsilon > 0} \bigcup_{h \in ]0,\delta[} \left[\frac{1}{h}(K-x) + \varepsilon\overset{\circ}{B}\right]$$

and tangent cone (of Clarke)

$$C_K(x) := \bigcap_{\varepsilon > 0} \bigcup_{\substack{\delta > 0 \\ \rho > 0}} \bigcap_{\substack{x' \in B(x,\rho) \cap K \\ h \in ]0,\delta[}} \left[\frac{1}{h}(K-x') + \varepsilon\overset{\circ}{B}\right]$$

Indeed

$$C_K(x) \subset I_K(x) \subset T_K(x)$$

(see [4], [6] for properties of $C_K(x)$, $T_K(x)$). The cone $I_K(x)$ is less known. We only state here

(1.2) Proposition. The following statements are equivalent :

(i) $v \in I_K(x)$

(ii) For all sequence $h_n > 0$ converging to zero there exists a sequence $v_n \in E$ converging to $v$ such that $x + h_n v_n \in K$ for all $n$.

(iii) $$\lim_{h \to 0+} \frac{1}{h} d_K(x+hv) = 0$$

In the study of some nonsmooth problems we are often led to deal with convex tangent cones. We define one of them.

(1.3) Definition. The asymptotic tangent cone to a subset $K$ at $x \in K$ is given by

$$I_K^\infty(x) := \{u \in I_K(x) : u + I_K(x) \subset I_K(x)\}$$

$I_K^\infty(x)$ is closed convex cone. One can easily verify that $C_K(x) \subset I_K^\infty(x) \subset I_K(x) \subset T_K(x)$

We now define the differential and co-differential of a set valued map $F$ from $E$ to a Banach space $E_1$ .

(1.4) Definition. The asymptotic differential of $F$ at $(x,y) \in \text{graph}(F)$ is the set valued map $DF(x,y) : E \vec{\to} E_1$ defined by

$$v \in DF(x,y)(u) \quad \text{if and only if} \quad (u,v) \in I^\infty_{\text{graph}(F)}(x,y)$$

The asymptotic co-differential of $F$ at $(x,y) \in \text{graph}(F)$ is the set valued map $DF(x,y)^\star : E_1^\star \vec{\to} E^\star$ defined by

$$q \in DF(x,y)^\star(p) \quad \text{iff} \quad \langle q,u\rangle - \langle p,v\rangle \leqslant 0 \quad \text{for all} \quad v \in DF(x,y)(u)$$

(1.5) Remark. We give in [11] another characterization of $DF(x,y)^\star$ . Let us only mention that $q \in F(x,y)^\star(p)$ means that $(q,-p)$ is contained in the negative polar cone to $I^\infty_{\text{graph}(F)}(x,y)$ , the asymptotic normal cone to graph(F) at $(x,y)$ .

Let $g : E \to \mathbb{R} \cup \{+\infty\}$ , $x \in \text{Dom}(g)$ . Define

$$F(y) = \begin{cases} g(y) + \mathbb{R}_+ & \text{when} \quad y \in \text{Dom}(g) \\ \emptyset & \text{when} \quad g(y) = +\infty \end{cases}$$

Then $\text{graph}(F) = \text{Epi}(g)$ (Epigraph of $g$ ).

(1.6) Definition. The subset

$$\partial_a g(x) = DF(x,g(x))^\star(1)$$

is called the asymptotic gradient of $g$ at $x$ .

In the case when $g$ is regularly Gâteaux differentiable, i.e. it has the Gâteaux derivative $g'(x) \in E^\star$ and for all $u \in E$

$$\lim_{\substack{u' \to u \\ h \to 0_+}} \frac{g(x+hu') - g(x)}{h} = \langle g'(x), u \rangle ,$$

we have

$$\partial_a g(x) = \{g'(x)\}$$

There is also another way to introduce $\partial_a g(x)$.

Following Rockafellar [19], when a function $\Phi : U \times V \to \mathbb{R} \cup \{+\infty\}$ is given, we define

$$\limsup_{v' \to v} \inf_{u' \to u} \Phi(v',u') := \sup_{\varepsilon > 0} \inf_{\delta > 0} \sup_{v' \in B(v,\delta)} \inf_{u' \in B(u,\varepsilon)} \Phi(v',u')$$

Consider $g : E \to \mathbb{R} \cup \{+\infty\}$, $x \in \mathrm{Dom}(g)$. For all $u \in E$ set

$$i_+ g(x)(u) := \limsup_{h \to 0+} \inf_{u' \to u} \frac{g(x+hu') - g(x)}{h}$$

and

$$i_+^\infty g(x)(u) := \sup_v \left(i_+ g(x)(u+v) - i_+ g(x)(v)\right)$$

The function $i_+^\infty g(x) : E \to \mathbb{R} \cup \{+\infty\}$ is called the asymptotic derivative and enjoys the following nice properties

$$I^\infty_{\mathrm{Epi}(g)}(x, g(x)) = \mathrm{Epi}\,(i_+^\infty g(x))$$

$$\partial_a g(x) = \{q \in E^\star : \langle q,u \rangle \leqslant i_+^\infty g(x)(u) \text{ for all } u \in E\}$$

(see [11]).

## 2 . The differential inclusion problem.

Let $F : \mathbb{R}^n \rightrightarrows \mathbb{R}^n$ be a set valued map and, let $\varphi : \mathbb{R}^n \to \mathbb{R}$ be a Lipschitzean function, $g : \mathbb{R}^n \times \mathbb{R}^n \to \mathbb{R} \cup \{+\infty\}$ . We denote by $S$ the set of all solutions to the differential inclusion

$$\dot{x} \in F(x)$$

i.e.

$$S = \{x \in W^{1,1}(0,1) : \dot{x}(t) \in F(x(t)) \quad \text{a.e.}\}$$

For a function $z \in S$ the contingent cone to $S$ at $z$ is given by

$$T_S(z) = \{w \in W^{1,1}(0,1) : \text{for some sequence } h_n > 0 \text{ converging to zero there exists a sequence } w_n \in S \text{ such that } z + h_n w_n \in S , \lim_{n \to \infty} w_n = w \}$$

Assume $z \in S$ solves the following problem

$$\text{minimize} \left\{ g(x(0),x(1)) + \int_0^1 \varphi(x(t))dt : x \in S \right\}$$

In order to characterize $z$ we assume the following surjectivity hypothesis

($H$) For some $p > 1$ and all $u,e \in L^p$ there exists a solution $w \in W^{1,p}(0,1)$ to the "linearized" problem

(i) $(w(0),w(1)) \in \text{Dom } (i_+^\infty g(z(0),z(1)))$

(ii) $\dot{w}(t) \in DF(z(t),\dot{z}(t))(w(t)+u(t))+e(t)$ a.e.

and

(iii) if $u = e = 0$ then every $w$ satisfying (i), (ii) belongs to $T_S(z)$ .

Remark. The last part of the above hypothesis holds in particular when $z(t) \in \text{Int}(\text{Dom } F)$ and $F$ is Lipschitzean in Hausdorff metric. Indeed if $\dot{w}(t) \in DF(z(t),\dot{z}(t))(w(t))$ then there exists a sequence $(u_k,v_k) \in L^1$ converging to $(w,\dot{w})$ such that $[(z,\dot{z}) + \frac{1}{k}(u_k,v_k)](t) \in \text{graph}(F)$ for all $k > 0$ .

Let $y_k(t) = w(0) + \int_o^t v_k(\tau)d\tau$ and $\alpha_k(t) = u_k(t) - y_k(t)$ . Clearly $\alpha_k \to 0$ in $L^1$ when $k \to +\infty$ and

$$\text{dist}\left(\dot{z}(t) + \frac{1}{k}\dot{y}_k(t) , F\left(z(t) + \frac{1}{k}y_k(t)\right)\right) \leqslant \frac{L}{k}\alpha_k(t)$$

where L denotes the Lipschitz constant of F . Then by Corollary 2.4.1 [2] there exists a constant C and functions $x_k \in S$ such that for all $k \geqslant 1$

$$|\dot{x}_k(t) - \dot{z}(t) - \frac{1}{k}\dot{y}_k(t)| \leqslant \frac{C}{k}\left(\alpha_k(t) + \int_o^1 \alpha_k(\tau)d\tau\right)$$

$$|x_k(t) - z(t) - \frac{1}{k}y_k(t)| \leqslant \frac{C}{k}\int_o^1 \alpha_k(\tau)d\tau$$

and therefore $w \in T_S(z)$ .

(2.1) Theorem. Assume that surjectivity hypothesis (*H*) is verified. Then there exists a solution $q \in W^{1,p\star}(0,1)$ (where $\frac{1}{p} + \frac{1}{p_\star} = 1$ ) of the adjoint inclusion

$$-\dot{q}(t) \in \partial_a\varphi(z(t)) + DF(z(t),\dot{z}(t))^\star(q(t)) \qquad \text{a.e.}$$

$$(-q(0),q(1)) \in \partial_a g(z(0),z(1))$$

Proof. We first reduce the above problem to an abstract optimization problem which has many other applications. The reduction is done in two steps. Set $E = L^p(0,1;\mathbb{R}^n)$ , $W = W^{1,p}(0,1;\mathbb{R}^n)$ , $T = \mathbb{R}^n \times \mathbb{R}^n$ , $\gamma(w) = (w(0),w(1))$ , $Lw = \dot{w}$ for all $w \in W$ .

Step 1. We claim first that if $\dot{w}(t) \in DF(z(t),z(t))(w(t))$ for all $t \in [0,1]$ then

$$i_+^\infty f(z)(w) + i_+^\infty g(\gamma z)(\gamma w) \geqslant 0$$

Indeed by (*H*) there exist sequences $h_n > 0$ and $w_n \in W$ converging to zero and w respectively such that $z + h_n w_n \in S$ . Since z is a minimiser we have $f(z+h_n w_n) + g(\gamma z + h_n\gamma w_n) \geqslant f(z) + g(\gamma z)$. Thus

$$\limsup_{\substack{w' \to w \\ h \to 0+}} \frac{f(z+hw') + g(\gamma z+h\gamma w') - f(z) - g(\gamma z}{h} \geqslant 0$$

and therefore using Lipschitzeanity of $f$ we obtain

$$0 \leqslant \limsup_{h \to 0+} \inf_{w' \to w} \frac{g(\gamma z+h\gamma w) - g(\gamma z)}{h} + \limsup_{\substack{w' \to w \\ h \to 0+}} \frac{f(z+hw') - f(z)}{h}$$

$$\leqslant i_+^\infty g(\gamma z)(\gamma w) + i_+^\infty f(z)(w)$$

<u>Step 2</u>. Let $\mathcal{F} : E \vec{\to} E$ be defined by $\mathcal{F}(x) = \{y \in E : y(t) \in F(x(t)) \text{ a.e.}\}$ . Thus $z$ solves the following problem

$$\text{minimize } \{f(x)+g(\gamma x) : x \in W , Lx \in \mathcal{F}(x)\}$$

Consider the closed convex cone

$$\mathcal{C} = \{(x,y) \in E \times E : y(t) \in DF(z(t),\dot{z}(t))(x(t)) \text{ a.e.}\}$$

Using the measurable selection theorems (see for example [ 20 ]) one can verify that $\mathcal{C} \subset I_{\text{graph}(\mathcal{F})}(z,\dot{z})$ . (See [ 11 ] for the details of the proof). Let $\mathcal{C}^-$ be the negative polar to $\mathcal{C}$ . We claim that if a function $q \in W^{1,p_\star}(0,1; \mathbb{R}^n)$ satisfies the following inclusions

$$(-\dot{q},-q) \in \partial_a f(z) \times \{0\} + \mathcal{C}^-$$

$$(-q(0),q(1)) \in \partial_a g(\gamma z)$$

then $q$ satisfies also all requirement of Theorem. This can be directly proved using a contradiction argument (see [ 11 ]).

Thus to achieve the proof we have only to verify the existence of $q \in W^{1,p_\star}(0,1;\mathbb{R}^n)$ as above. This will be done in the next section where an abstract problem is treated.

## 3 . The abstract problem.

Consider reflexive Banach spaces $W,H,E,T$ where $W$ is continuously embedded into $H$ by the canonical injection $i$ . Let $L \in \mathcal{L}(W,E)$ , $\gamma \in \mathcal{L}(W,T)$ be continuous linear operators and $\gamma$ satisfies the

"trace property" $\gamma$ has a continuous right inverse and the kernel $W_o$ of $\gamma$ is dense in $H$

We denote by $i_o$ $(L_o)$ the restriction of $i$ (respectively $L$ ) to $W_o$ . Define

$$E_o^\star = \{p \in E^\star : L_o^\star p \in H^\star\}$$

Thus $L_o^\star$ maps $E_o^\star$ to $H^\star$ . (For the problem considered in § 2 $H = E$ , $E_o^\star = W^{1,p\star}(0,1;\mathbb{R}^n)$ and $L_o^\star q = -\dot{q}$ on $E_o^\star$ ). We have the following abstract Green formula (see [1]) :

There exists a unique operator $\beta^\star \in \mathcal{L}(E_o^\star,T^\star)$ such that for all $u \in W$ , $p \in E_o^\star$

$$\langle L_o^\star p,u\rangle - \langle p,Lu\rangle = \langle \beta^\star p,\gamma u\rangle$$

Let a closed convex cone $C \subset H \times E$ and functions $\pi : W \to R$ , $\psi : T \to R \cup \{+\infty\}$ be given. We assume that the epigraphs of $\pi,\psi$ are closed convex cones and define the closed convex processes $G : H \overset{\to}{\to} E$ , $G^\star : E^\star \overset{\to}{\to} H^\star$ by

$$v \in G(u) \quad \text{if and only if} \quad (u,v) \in C$$

$$r \in G^\star(q) \quad \text{if and only if} \quad (r,-q) \in C^-$$

We assume that the element $w = 0$ is a solution of the problem

$$\text{minimize} \quad \{\pi(w) + \psi(\gamma w) : Lw \in G(w)\}$$

(3.1) Theorem. Assume that the following surjectivity assumption holds true :

for all $(u,v,e) \in H \times H \times E$ there exists a solution $w \in W$ to the problem :

$$\begin{cases} \text{(i)} \quad Lw \in G(w+u) + e \\ \text{(ii)} \quad w \in \mathrm{Dom}(\pi) \quad , \quad \gamma w \in \mathrm{Dom}(\psi) \end{cases}$$

Then there exists $q \in E_o^\star$ such that

$$L_o^\star q \in \partial_a \pi(0) + G^\star(q)$$

$$- \beta^\star q \in \partial_a \psi(0)$$

Remark. For the problem considered in § 2 we have :

$$\begin{cases} \pi(w) = i_+^\infty f(z)(w) \quad ; \quad \psi(t) = i_+^\infty g(\gamma z)(t) \quad ; \quad \partial_a \pi(0) = \partial_a f(z) \quad ; \\ \partial_a \psi(0) = \partial_a g(\gamma z) \end{cases}$$

The proof of Theorem 3.1 follows immediately from the following Lemmas.

(3.2) Lemma. Under the assumptions of Theorem 3.1 the set $A$ defined by

$$A := i^\star \partial_a \pi(0) + \gamma^\star \partial_a \psi(0) + \{i^\star r - L^\star q : r \in G^\star(q)\}$$

(where $i^\star$ is the adjoint of $i$ ) is closed in $W^\star$ .

Proof. Let $a_n = i^\star \alpha_n + \gamma^\star \alpha_n' + i^\star r_n - L^\star q_n$ , where $\alpha_n \in \partial_a \pi(0)$ , $\alpha_n' \in \partial_a \psi(0)$ , $(r_n, q_n) \in C^-$ , n=1,2,... . Assume $\lim_{n \to \infty} a_n = a$ in $W^\star$ . We claim that $\{(\alpha_n, r_n, -q_n)\}_{n \geq 1}$ is bounded. This will be proved if we show that for all $(u,v,e) \in H \times H \times E$

(3.3) $$\sup_{n \geq 1} (\langle \alpha_n, v\rangle + \langle r_n, u\rangle + \langle q_n, e\rangle) < +\infty$$

Let $w$ be such that $Lw \in G(w+u) + e$ , $w \in \mathrm{Dom}(\pi)$ , $\gamma w \in \mathrm{Dom}(\psi)$ . Then $e = Lw - y$ , where $(w+u, y) \in C$ . Therefore $\langle \alpha_n, v\rangle + \langle r_n, u\rangle + \langle q_n, e\rangle = \langle \alpha_n, v\rangle + \langle r_n, u\rangle + \langle L^\star q_n, w\rangle - \langle q_n, y\rangle = \langle \alpha_n, v+w\rangle + \langle \alpha_n', \gamma w\rangle + \langle (r_n, -q_n), (u+w, y)\rangle - \langle a_n, w\rangle \leq \pi(v+w) + \psi(\gamma w) - \langle a_n, w\rangle$ and (3.3) follows. Thus by reflexivity we may assume that $(\alpha_n, r_n, q_n) \rightharpoonup (\alpha, r, q)$ weakly in $H^\star \times H^\star \times E^\star$ . By Mazur lemma [9 ] and convexity of $\partial_a \pi(0)$ , $C^-$ we have $\alpha \in \partial_a \pi(0)$ , $(r, -q) \in C^-$ . Let $\sigma$ be the continuous right inverse of $\gamma$ . Then $\alpha'_n = \sigma^{\star}\gamma^{\star}\alpha'_n = \sigma^{\star}(a_n - i^{\star}\alpha_n - i^{\star}r_n + L^{\star}q_n)$ is weakly convergent to some $\alpha' \in \partial_a\psi(0)$ . Hence $a \in A$ .

(3.4) <u>Lemma</u>. The following statements are equivalent :

(1) $\pi(w) + \psi(\gamma w) \geqslant 0$ for all $Lw \in G(w)$

(2) There is $q \in E_o^{\star}$ such that

$$L_o^{\star}q \in \partial_a\pi(0) + G^{\star}(q)$$

$$-\beta^{\star}q \in \partial_a\psi(0)$$

<u>Proof</u>. If (1) holds, then using the separation theorem we show that

$$0 \in i^{\star}\partial_a\pi(0) + \gamma^{\star}\partial_a\psi(0) + i^{\star}G^{\star}(q) - L^{\star}q$$

Let $q \in E^{\star}$ , $\alpha \in \partial_a\pi(0)$ , $\alpha' \in \partial_a\psi(0)$ , $r \in G^{\star}(q)$ be such that $i^{\star}\alpha + \gamma^{\star}\alpha' + i^{\star}r - L^{\star}q = 0$ . Thus $L_o^{\star}q = i_o^{\star}\alpha + i_o^{\star}r$ . Since $W_o$ is dense in $H$ it implies that $L_o^{\star}q \in H^{\star}$ and by consequence $q \in E_o^{\star}$ . Moreover the Green formula implies $0 = \langle\alpha,w\rangle + \langle\alpha',\gamma w\rangle + \langle(r,-q),(w,Lw)\rangle = \langle\alpha'+\beta^{\star}q,\gamma w\rangle$ for all $w \in W$ . Since $\gamma W = T$ we proved $\alpha' + \beta^{\star}q = 0$ and thus (2) .

To prove the converse, assume (2) holds. Then for some $q \in E_o^{\star}$ , $\alpha \in \partial_a\pi(0)$ , $\alpha' \in \partial_a\psi(0)$

$$L_o^{\star}q = \alpha + r \quad , \quad -\beta^{\star}q = \alpha'$$

and by Green formula $\alpha + r = \gamma^{\star}\beta^{\star}q + L^{\star}q = L^{\star}q - \gamma^{\star}\alpha'$ , $\alpha + \gamma^{\star}\alpha' = L^{\star}q - r$ . Thus if $Lw \in G(w)$ we have $\pi(w) + \psi(\gamma w) \geqslant \langle\alpha,w\rangle + \langle\alpha',\gamma w\rangle = \langle\alpha+\gamma^{\star}\alpha',w\rangle =$ $= \langle L^{\star}q-r,w\rangle = -\langle(r,-q),(w,Lw)\rangle \geqslant 0$ , which proves (1) and achieves the proof of Lemma 3.4.

Thus the proof of Theorem 3.1 is completed.

## 4 . An example.

Let $U$ be a compact subset in $\mathbb{R}^n$ , $A$ be $n \times n$ matrix, $B$ be $n \times m$ matrix and let two lipschitzean functions $\varphi : \mathbb{R}^n \to \mathbb{R}$ , $g : \mathbb{R}^n \times \mathbb{R}^n \to \mathbb{R}$ be given.

Consider the following problem :

$$(4.1) \qquad \text{minimize } \left[ g(x(0),x(1)) + \int_0^1 \varphi(x(t))dt \right]$$

over the set of solutions to the control system

$$(4.2) \qquad \dot{x}(t) = Ax(t) + Bu(t) \quad , \quad u(t) \in U$$

The corresponding differential inclusion then has the form

$$\dot{x} \in F(x) \quad , \quad F(x) = Ax + BU$$

Assume a trajectory-control pair $(z,\bar{u})$ solves (4.1), (4.2).

(4.3) Theorem. There exists an absolutely continuous function $q$ such that

$$\dot{q}(t) \in \partial_a \varphi(z(t)) - A^{\star} q(t) \qquad \text{a.e. in } [0,1]$$

$$\langle q(t),s\rangle \leqslant 0 \qquad \text{for all } s \in I_{BU}^{\infty}(B\bar{u}(t))$$

$$(-q(0),q(1)) \in \partial_a g(z(0),z(1))$$

*Proof.* To use Theorem 2.1 we verify directly that $DF(z(t),\dot{z}(t))(v) = Av + I_{BU}^{\infty}(B\bar{u}(t))$ . Fix any $s > 1$ and let $p > 1$ be defined from the equation $\frac{1}{p} + \frac{1}{s} = 1$ . Clearly for all $u,e \in L^p$ there exists $w \in W^{1,p}(0,1)$ solving the problem

$$\dot{w}(t) \in Aw(t) + Au(t) + e(t) + I_{BU}^{\infty}(B\bar{u}(t))$$

On the other hand if $w$ is such that

$$\dot{w}(t) \in Aw(t) + I_{BU}^{\infty}(B\bar{u}(t))$$

then we can find a sequence $Bu_k \in L^1$ converging to $\dot{w}(t) - Aw(t)$ such that $B\bar{u}(t) + \frac{1}{k} Bu_k(t)) \in BU$ a.e.. Let $w_k$ be defined from the equation

$$\dot{w}_k(t) = Aw_k(t) + Bu_k(t) \quad , \quad w_k(0) = w(0) \ .$$

Then $z + h_k w_k$ is a solution to (4.2) and it implies that the hypothesis (*H*) from § 2 is verified. On the other hand if $r \in DF(z(t),z(t))^\star(-\bar{q})$ then for all $v \in \mathbb{R}^n$, $s \in I^\infty_{BU}(B\bar{u}(t))$ we have $<(v,Av+s),(r,\bar{q})> \leqslant 0$ and hence $<v,r+A^\star\bar{q}> + <s,\bar{q}> \leqslant 0$ . It implies that

$$DF(z(t),\dot{z}(t))^\star(-\bar{q}) = -A^\star\bar{q} \quad ; \quad \bar{q} \in \left(I^\infty_{BU}\ (B\bar{u}(t))\right)^-$$

and by Theorem 2.1 the proof is complete.

REFERENCES :


[1] J.P. Aubin, Applied Functional Analysis, Wiley Interscience, 1979.

[2] J.P. Aubin and A. Cellina, Differential Inclusions, Springer Verlag, 1984.

[3] J.P. Aubin, F.H. Clarke, Shadow prices and duality for a class of optimal control problems, SIAM J. of Control, 17 (1979) n° 5, pp. 567-586.

[4] J.P. Aubin, I. Ekeland, Applied Nonlinear Analysis, Wiley Interscience, 1984.

[5] H. Berliocchi, J.M. Lasry, Principe de Pontriagin pour des systèmes régis par une équation différentielle multivoque, CRAS, Paris, vol. 277 (1973), 1103-1105.

[6] F.H. Clarke, Nonsmooth analysis and optimization, Wiley Interscience, 1983 .

[7 ] F.H. Clarke, The maximum principle under minimal hypothesis, SIAM J. of Control, 14 (1976), 1078-1091.

[8 ] F.H. Clarke, Optimal solutions to differential inclusions, J. Opt. Theory Appl. vol 19, n° 3 (1976), pp. 469-478.

[9 ] I. Ekeland, R. Temam, "Analyse convexe et problèmes variationels", Dunod, Paris, 1974.

[10 ] H. Frankowska, Inclusions adjointes associées aux trajectoires d'inclusions différentielles, Note C.R. Acad. Sc. Paris, t. 297 (1983), pp. 461-464.

[11 ] H. Frankowska, The adjoint differential inclusions associated to a minimal trajectory of a differential inclusion, Cahiers de CEREMADE n° 8315, 1983.

[12 ] H. Frankowska, The first order necessary conditions in nonsmooth variational and control problems, SIAM J. of Control (to appear).

[13 ] H. Frankowska, C. Olech, Boundary solutions to differential inclusions, J. Diff. Eqs. 44 (1982), pp. 156-165.

[14 ] A. Ioffe, Nonsmooth analysis : differential calculus of nondifferentiable mappings, Trans. Amer. Math. Soc., 266 (1), 1981, pp. 1-56.

[15 ] J.P. Penot, P. Terpolilli, Cônes tangents et singularités, CRAS. Paris, vol. 296 (1983), pp. 721-724.

[16 ] L. Pontriagin, V. Boltyanskii, V. Gamkrelidze, E. Mischenko, The mathematical Theory of Optimal process , Wiley Interscience Publishers, New-York, 1962.

[17 ] R.T. Rockafellar, Existence theorems for general control problems of Bolza and Lagrange. Adv. in Math. 15 (1975), 312-323.

[18 ] R.T. Rockafellar, Convex analysis , Princeton University Press, Princeton, New-Jersey, 1970.

[19 ] R.T. Rockafellar, Generalized directional derivatives and subgradients of non convex functions. Canad. J. Math., 32 (1980), 257-280.

[20] D.H. Wagner, Survey of measurable selection theorems, SIAM J. of Control, 15 (1977), 859-903.

[21] T. Ważewski, On an optimal control problem, Proc. Conference "Differential equations and their applications", Prague, 1964, pp. 229-242.

# AN EXAMPLE OF OPTIMAL CONTROL OF A SYSTEM WITH DISCONTINUOUS STATE

William S. Levine and Felix E. Zajac
Department of EE
University of Maryland
College PArk, MD 20742
USA

M.E. Dept.-Design &
Stanford University
Stanford, CA 94305
USA

R.R.&D. Center (153)
VA Med. Center
Palo Alto, CA 94304
USA

## ABSTRACT

An example of a system with discontinuities in the state vector is described. Such systems arise in manufacturing, animal and human locomotion, certain queueing problems and many other applications. Optimal control problems for such systems cannot be solved directly via dynamic programming or the maximum principle because both of these analytical tools require continuous state vectors. As a first step in the study of such problems, two alternate formulations for the dynamics of the example problem are given. Both of these formulations allow the solution of the resulting optimal control problem by elementary methods. One of the formulations produces the solution as the limit, as a parameter goes to zero, of the solutions to problems to which dynamic programming and the maximum principle apply.
The solution to the optimal control problem is given, in feedback form, throughout the state space. The optimal control includes a singular arc that is not on the state boundary. The paper concludes with a brief discussion of more realistic and practical problems.

## INTRODUCTION

For several years we have been studying the ways animals and humans control their limbs as they jump, walk and run [1], [2]. Dynamical models of locomotion have a feature, in common with models of many other physical systems, which has not, to our knowledge, received previous attention in the literature. The state of these systems cannot cross certain boundaries. For example, in walking, the foot cannot possibly pass through the surface of the ground. As we shall see shortly, the constraint imposed by the ground cannot be modeled as a state constraint. Instead, the effect of the ground, as well as similar boundaries, must be modeled as a nonlinearity in the dynamics. The fundamental feature of these problems is either the control of an impact or the control along a boundary surface. Practical examples of such problems occur continually in automated manufacturing systems.

---

Acknowledgement: This research was supported, in part, by NIH under grant NS 17622.

Any hammering or riveting operation involves control of impacts. Insertion and marking tasks are obvious examples of control along a boundary surface. Another example occurs in the control of queues when the queue can be realistically modeled by a real number rather than an integer. Then, the fact that the queue must be greater than or equal to zero represents a boundary of the type discussed here. This is not to suggest that the results described here are applicable to the practical problems mentioned above. Rather, this paper represents the very beginnings of an attempt to formulate and understand the mathematics of such problems.

There are some unusual features associated with the optimal control of systems with boundaries. In this paper we describe a very simple example of such a system, a baton propelled by an ideal torque generator. We then solve the optimal control problem of propelling the baton so it "jumps" to a maximum height. This problem was chosen because it can be solved by elementary means. The main point is not the solution. The main point is the proper mathematical model and the development of techniques whereby more realistic and complex problems might be solved.

Thus, we present two alternative mathematical formulations of the baton problem in the next section of this paper. This is followed, in section 3, by the solution. We conclude, in section 4, with some suggestions regarding extensions and further applications of these ideas.

## PROBLEM FORMULATION

The basic problem is: given an inflexible rod of length $\ell$, mass M and moment of inertia about the center of mass $I_0$. One end of the rod is resting on the ground. Cause the rod to move to the maximum possible height by applying a bounded torque to the rod. Note that the rod can "jump" if sufficient torque is applied. See Fig. 1 for notation and a picture of the physical context.

In an earlier paper [3], we analyzed a version of this problem in which the interaction with the ground was relatively simple. Thus, where the problem formulation and solution are similar to those in the earlier paper we simply state the result and assume the reader will refer to the earlier paper for details and proof.

The mathematical model of this physical optimal control problem would be straightforward except for two aspects of the interaction of the rod with the ground. First, it is possible to let the rod fall from the position shown in Figure 1 to a position where the rod lies horizontally on the ground. It is then necessary to model the effect of the impact with the ground. The natural impact models result in discontinuous phase plane trajectories (see below for details). The normal tools of optimal control do not apply when trajectories are discontinuous. Second, since the problem continues after the rod leaves the ground a natural mathematical model would change from two dimensional (while the rod is still in contact with the ground) to four dimensional (while the rod is airborne). Alternatively, one can think of the system as a four dimensional system always. Then, while the rod is on the ground its motion is along a two dimensional manifold. The choice of control then determines when the trajectory leaves the manifold, that is, jumps. Note that we will show that there are states on the manifold from which it is not possible to jump immediately. Trajectories which do leave the manifold are continuous as they do so. With the above comments in mind the next step is to write an appropriate mathematical description of the problem. There are at least two possible approaches. Consider first the dynamics when, as in Figure 1, the baton has one end resting on the ground. Then, (see Figure 1 for notation) the dynamics are

$$\ddot{\theta}(t) = \frac{u(t) - Mgx\cos\theta(t)}{I} \qquad 0<\theta(t)\leq\pi/2 \tag{1}$$

Notice that, for reasons that will become clear shortly, we are not interested in trajectories where $\theta(t)>\pi/2$.

What happens when the baton falls from a position where $\theta(t)>0$ as above to $\theta(t) = 0$? The answer depends on the details of the resulting collision. We are most interested in situations where the motion of the ground on impact is negligible. In fact, negligible ground movement and completely inelastic collision are the most common case in practice.

Thus, one natural way to augment Eq. (1) so as to account for the impact is to add an impulse as follows:

$$\ddot{\theta}(t) = \frac{u(t) - Mgx\cos\theta(t)}{I} - \alpha\dot{\theta}(t)\delta_0(\theta(t)) \qquad 0\leq\theta(t)\leq\pi/2 \tag{2}$$

where $\delta_0(x)$ is the usual impulse function and $1\leq\alpha\leq 2$

Notice that $\alpha = 1$ corresponds to a completely inelastic collision, $\alpha = 2$ corresponds to a completely elastic collision and $1<\alpha<2$ corresponds to loss of some of the energy of impact.

An alternate approach to modeling the impact is suggested by those practical situations in which the movement of the ground (or shoe cushion) is not completely negligible. Obvious examples are diving boards, trampolines and modern running tracks. In fact, McMahon and Greene [4] have estimated that the world record for the mile could be improved by as much as seven seconds by properly adjusting the track compliance. In this case, an appropriate model is

$$\ddot{\theta}(t) = \frac{u(t) - Mgx\cos\theta(t)}{I} - \delta_1(-\theta(t))[k_1\dot{\theta}(t) + k_2\theta(t)] \qquad (3)$$

where $\delta_1(x)$ is the unit step

i.e. $$\delta(x) = \begin{cases} 0 & x<0 \\ 1 & x\geq 0 \end{cases}$$

and $k_1, k_2$ are positive constants

Notice that $k_1 = 0$ corresponds to a completely elastic collision and $k_2 = 0$ corresponds to complete inelasticity.
This second model can be modified to deal with the case of negligible ground motion as follows.
Let

$$\ddot{\theta}_\varepsilon(t) = \frac{u(t) - Mgx\cos\theta_\varepsilon(t)}{I} - \frac{1}{\varepsilon}\delta_1(-\theta_\varepsilon(t))[k_1\dot{\theta}_\varepsilon(t) + k_2\theta_\varepsilon(t)] \qquad (4)$$

$$\theta(t) \stackrel{\Delta}{=} \lim_{\varepsilon\to 0} \theta_\varepsilon(t) \qquad (5)$$

Of course, one has to show that this limit exists and makes sense. This will be done shortly.
The rest of the formulation of the optimal control problem is comparatively straightforward. The control is $u(t)\varepsilon U$ where

$$U = \{u(\cdot)\varepsilon M:\ 0\leq u(t)\leq U_{max}\} \qquad (6)$$

M = set of measurable functions.

The performance criterion is

$$J = \max_{u(\cdot)\varepsilon U} \{x\sin\theta(t_f) + \frac{x^2}{2g}\cos^2\theta(t_f)\dot{\theta}^2(t_f)\} \qquad (7)$$

where $t_f$ is the instant the baton leaves the ground. The baton leaves the ground at the first instant that the vertical force, $F_v$, becomes negative. It is easy to show that

$$F_v = Mg - Mx\sin\theta(t)\dot{\theta}^2(t) + \frac{Mx}{I}[\cos\theta(t)u(t) - Mgx\cos^2\theta(t)] \qquad (8)$$

Thus, the condition for leaving the ground amounts to

$$F_v(\theta,\dot{\theta},u)<0 \qquad (9)$$

This would obviously present very serious problems. This is especially true because we will see that every interesting optimal trajectory has $F_v = 0$ for an interval of time. Fortunately, in this problem we can also show that all optimal trajectories that actually jump do so at

$$F_v(\theta,\dot{\theta},U_{max}) = 0 \qquad (10)$$

so we take Eq. (10) as the terminal condition.
We now have a completely formulated optimal control problem which we proceed to solve.

SOLUTION

In our earlier paper [3] , we have given a complete solution to the baton problem in the first quadrant of the phase plane ($\theta(t)\geq 0,\dot{\theta}(t)\geq 0$). This solution is still valid for those trajectories which never leave the first quadrant and is the basis for our solution of the more general problem. We briefly review the solution below.
As can be seen in Figure 2 the first quadrant divides into four regions. These are characterized as follows:
Region D: $F_v(\theta,\dot{\theta},U_{max})<0$ throughout region D. Thus, it is impossible for the baton to be in contact with the ground in region D. The lower boundary of region D is given by Eq. (10). All optimal trajectories that actually leave the ground do so on the boundary of region D. It should be noted that the performance, the height of the jump, is monotone decreasing with increasing $\theta$ along the boundary of region D.
Region A: It is impossible to jump if the trajectory is in region A. Furthermore, for all initial conditions in region A, the baton rotates to the vertical ($\theta = 90^\circ$) and then falls over regardless of the choice of control. Thus, all controls are optimal and the optimal performance is $V^*(\theta,\dot{\theta}) = x$.
Region B: The optimal control in region B is $u(t) = U_{max}$ regardless of the state. This optimal control is unique. Performance is monotone increasing in $\dot{\theta}$ for fixed $\theta$ throughout region B. Performance is also monotone decreasing with increasing $\theta$ for fixed $\dot{\theta}$ throughout

region B.

Region C: This is the region of most interest for this paper. In the earlier paper we showed that it is possible to reach a state where $\dot{\theta}(t)<0$ and $\theta(t)<90^{o}$ from any initial state in region C. Thus, it is possible to allow the baton to fall back to $\theta(t) = 0$ from anywhere in region C. The purpose of this paper is to describe what happens next.

Consider the dynamical model in Eq. (2) first. Suppose the baton hits the ground at $\hat{t}$. That is, $\theta(\hat{t}) = 0$. It must also be true that $\dot{\theta}(\hat{t}^{-}) = \hat{\omega}<0$. And, $\dot{\theta}(\hat{t}^{+}) = (\alpha-1)\hat{\omega}\geq 0$. Starting from the state at $\hat{t}^{+}$ the trajectory is in the first quadrant and our earlier analysis applies. Thus, since our earlier analysis showed that performance is monotone increasing with $\dot{\theta}$ for constant $\theta$ ($\theta$=0 in this case) we want to maximize $\dot{\theta}(\hat{t}^{+})$. To do this we need to minimize $\dot{\theta}(\hat{t}^{-})$.

In order to decide how to minimize $\dot{\theta}(\hat{t}^{-})$ we need to study trajectories in the fourth quadrant, $\dot{\theta}<0,\theta>0$. Consider trajectories in the fourth quadrant. From Eq. (2) we have

$$\frac{d\dot{\theta}}{d\theta} = \frac{u(t) - Mgx\cos\theta}{I\dot{\theta}} \tag{11}$$

or, holding $u(t) = u$, a constant, and letting $t_0$ denote the initial time in the fourth quadrant.

$$\dot{\theta}^2(t) - \dot{\theta}^2(t_0) = \frac{2}{I}[u(\theta(t) - \theta(t_0)) - Mgx(\sin\theta(t) - \sin\theta(t_0))] \tag{12}$$

Since $\dot{\theta}<0$ in the fourth quadrant and $\theta(t)$ is decreasing, Eq. (12) shows that $\dot{\theta}^2(t)$ is maximized ($\dot{\theta}(t)$ is minimized) over all constant u, $0\leq u\leq U_{max}$ by u=0. In particular, $\dot{\theta}^2(\hat{t}^{-})$ is maximized by setting u(t) = 0 for the entire time the trajectory is in the fourth quadrant. A simple argument shows that allowing u(t) to vary with time does not improve this result. Thus, the optimal control in the fourth quadrant is u(t) = 0 <u>provided</u> <u>that</u>

$$F_v(\theta,\dot{\theta},0)\geq 0 \tag{13}$$

However, it is not always possible to satisfy Eq. (13). To see this note that the curves of $F_v(\theta,\dot{\theta},U_{max}) = 0$ and $F_v(\theta,\dot{\theta},0) = 0$ are shown in Figure 3. We have also shown the trajectory 1-2-3-4-5-6-7 in Figure 3. This trajectory represents the boundary of the region in the phase plane from which it is possible to reach a state ($\theta$ = 0, $\dot{\theta}\leq 0$) without violating the constraint $\theta\leq 90^{o}$ or $F_v(\theta,\dot{\theta},u)\geq 0$. Notice that the segments of this trajectory from 2-3 and from 5-6 must have $0<u(t)<U_{max}$ in order to keep the baton on the ground ($F_v(\theta,\dot{\theta},u)>0$). In our previous paper we give a method to calculate the control that produces the trajectory from 2-3. The same method applies for 5-6. In essence, we set Eq. (8) to zero, solve for u in terms of $\theta$ and $\dot{\theta}$ and use this control in Eq. (2).

Now, the optimal control from any initial state should be obvious. The regions of feedback control are shown in Figure 3. Regions A,B and D are as before. In region C the optimal control is not generally unique. However, an optimal control is to set $u(t) = U_{max}$ until trajectory 1-2-3-4-5-6-7 is reached. Then set u(t) to the minimum value such that $F_v(\theta,\theta,u)>0$ until $\theta = 0$ is reached. Then, set $u(t) = U_{max}$ until the baton leaves the ground.

Notice that trajectory 1-2-3-4-5-6-7 contains singular arcs. Furthermore, the cost-to-go, as a function of states near the singular arcs is discontinuous. That is, if we imagine the cost-to-go plotted out of the plane of Figure 3 then there is a cliff at the boundary between regions B and C.

Of course, the real purpose of this research is to develop techniques for analyzing considerably more complex problems than this one. thus, it is useful to examine the solution when the model is that of Eqs. (4) and (5) instead of Eq. (2). For non-zero $\varepsilon$ trajectories can get into the second and third quadrants. Although the analysis is complicated by the $\dot{\theta}_\varepsilon$ term on the right hand side of Eq. (4) it is not hard to show that the optimal control is $u(t) = 0$ when $\dot{\theta}_\varepsilon<0$ and $\theta_\varepsilon<0$ and $u(t) = U_{max}$ when $\dot{\theta}_\varepsilon>0$ and $\theta_\varepsilon<0$. This assumes that $\varepsilon$ is small enough that $F_v>0$ with this control strategy. Furthermore, in the limit as $\varepsilon\to0$ this solution converges to the one in Figure 3.

The above result is easiest to visualize when $k_1 = 0$ in Eq. (4). Then, the system is conservative when $u(t) = 0$. Furthermore, the formal technique for deriving trajectories illustrated in Eqs. (11) and (12) works. As a result, the method of derivation used in [1] gives the optimal control cited above as well as the optimal trajectory shown in Fig. 4. As $\varepsilon\to0$, point 2 in Fig. 4 moves to the origin and point 3 moves to the mirror image of point 1. Intuitively, when the compliance is greater than zero it is possible to add a little energy to that stored in the spring to "jump" a little higher. As the compliance goes to zero it becomes harder and harder to add to this energy until, in the limit, it is impossible.

## SUGGESTIONS FOR FURTHER RESEARCH

As a practical matter the second formulation of the problem, as a limit, is somewhat better. This is because the problem can be solved for non-zero epsilon by standard techniques. In fact, the spring and dashpot introduced to deal with impacts should be smoother than in this paper if one is to apply either dynamic programming or the maximum principle. However, one can either use a cubic spring and dashpot or Frank Clarke's version of the maximum principle [5]. Then, by solving a sequence of these problems with decreasing $\varepsilon$ one can, in effect by a penalty method, solve for the optimal control.

A more important point is related to the end of the baton that rests on the ground. All of the really unusual aspects of this problem correspond to trajectories along which

$$F_v(\theta(t), \dot{\theta}(t), u(t)) = 0 \qquad (14)$$

If we think of the lower end of the baton as free to move in the vertical direction then its motion satisfies conditions similar to those for $\theta$. That is, the ground acts as a hard limit on the position of the lower end of the baton. Furthermore, trajectories for which Eq. (14) is satisfied are trajectories which slide along the hard limit. In particular, this is the singular arc in our example. For such a situation, a model with impulses is problematical. We would be interested in trajectories for which the coefficient of the impulse was zero. It is also not clear that the model using limits is appropriate since we have not yet shown that the solution in the limit as $\varepsilon \to 0$ is the solution given in this paper.

Thus, our current research related to this problem addresses three questions: How can problems in which optimal trajectories slide along the hard limits be formulated and solved? What computational methods are most effective for problems with hard limits on state variables? What is the optimal control for several practical problems of this type?

## REFERENCES

1. Zajac, F.E. and W.S. Levine, "Novel experimental and theoretical approaches to study the neural control of locomotion and jumping, in Posture and Movement, edited by R.E. Talbott and D.R. Humphrey, Raven Press: New York, pp. 259-279 (1979).

2. Levine, W.S., F.E. Zajac, M.R. Belzer and M.R. Zomlefer, "Ankle controls that produce a maximal vertical jump when other joints are locked," IEEE Trans. on AC, Vol. AC-28, No. 11, pp. 1008-1016 (1983).

3. Levine, W.S., M. Christodoulou and F.E. Zajac, "On propelling a baton to a maximum vertical or horizontal distance," Automatica Vol. 19, No. 3, pp. 321-324 (1983).

4. McMahon, T.A. and P.R. Greene, "Fast Running Tracks," Scientific American, Dec. 1978, pp. 148-163.

5. Clarke, F.H., Optimization and Nonsmooth Analysis, John Wiley & Sons, New York, 1983.

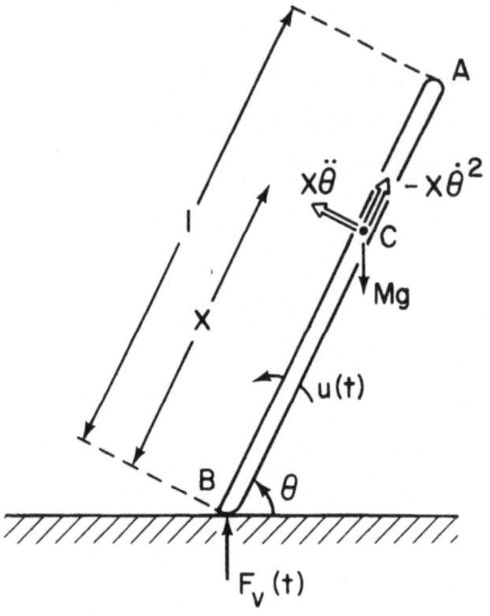


Figure 1 Baton, with forces and torques indicated

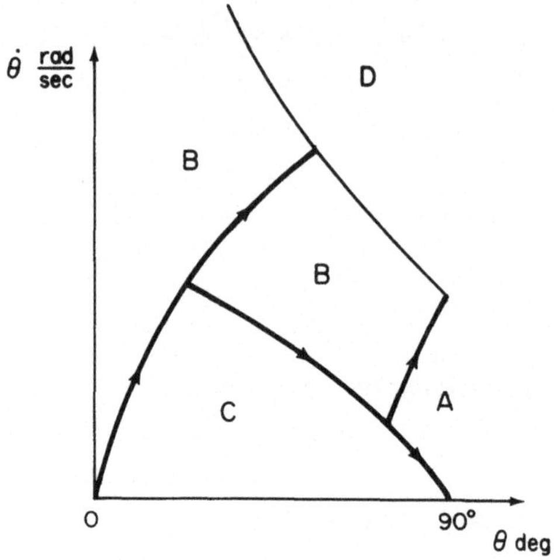


Fig. 2 Phase plane portrait of baton, showing different control regions. Lines with arrows are trajectories. $u(t) = U_{max}$ along both trajectories with arrows pointing upwards. A feedback control is used along the first part, and $u(t) = 0$ along the rest, of the trajectory with downward pointing arrows.

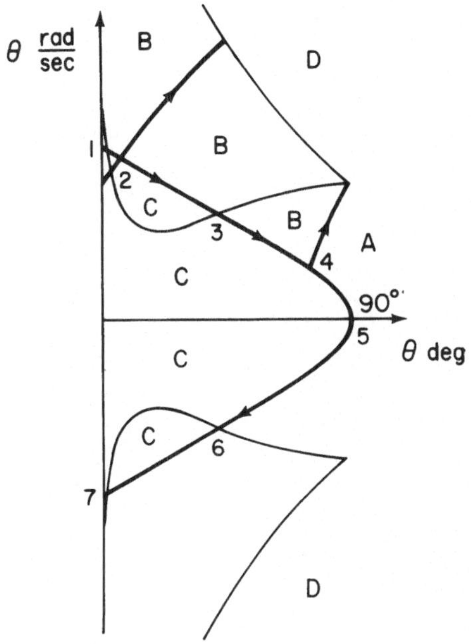


Fig. 3 Phase plane portrait of baton, showing different control regions in the first and fourth quadrants. Lines with arrows are trajectories. The outer lines without arrows correspond to $F_V(\theta,\dot{\theta},U_{max})$ = 0. The inner lines without arrows correspond to $F_V(\theta,\dot{\theta},0) = 0$. In the region between these two curves, setting u(t) = 0 causes a jump instantly while setting u(t) = $U_{max}$ keeps the baton on the ground.

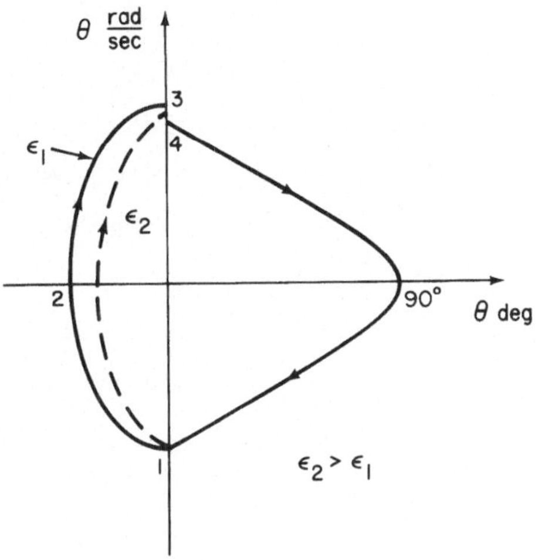


Fig. 4 Detail of phase plane portrait showing $\dot{\theta}_\varepsilon$ vs $\theta_\varepsilon$ for two values of ε.

# NON LINEAR CONTROL OF VARIABLE STRUCTURE SYSTEMS

Giorgio Bartolini
Università di Genova
Istituto di Elettrotecnica
via all'Opera Pia 11 a
16100 Genova (Italy)
Phone 010-302113

Tullio Zolezzi
Università di Genova
Istituto di Matematica
Via L.B.Alberti 4
16100 Genova (Italy)
Phone 010-515141

## ABSTRACT

Sliding modes in the control of systems described by ordinary differential equations have been investigated both from a theoretical point of view and from practical considerations, when the control variables enter linearly in the system. In this paper we extend the sliding mode control technique to fully non linear systems. We introduce a definition of equivalent control and show, under suitable conditions, the equivalence between the relevant Filippov solutions and the solutions obtained by using the equivalent control law. We discuss the physical meaning of such an equivalent control taking into account the behaviour of the corrisponding trajectories as far as approximability properties (physically relevant) are considered. We introduce a definition of approximability for nonlinear control systems with sliding surfaces, which shows the inherent limitations of the concept of equivalent control for general nonlinear dynamics. This validates some conjectures of [1] and extends basic results thereof to certain classes of fully nonlinear control systems. We show that the equivalent control may be defined and the approximability proper ty holds for control systems of the following form:

A) $z^{(n)} = g(t,z,z',\dots,z^{(n-1)},u)$ ;

B) $\dot{x} = A(t,x) + B(t,x)\,h(u)$ ,

under suitable explicit conditions about g, A, B, h and the sliding manifold.

INTRODUCTION

We consider control systems described by state equations

(1) $\dot{x} = f(t,x,u)$

where $x \in R^n$ , $u \in R^m$, $0 \leq t \leq T$,

sliding manifold S given by

(2) $s(x) = 0$ , $s = (s_1,\dots,s_m) \in R^m$ ,

and control constraints

(3) $u \in U$ a given subset of $R^m$ .

The system is controlled by using possibly discontinuous feedback control laws $u = u(t,x)$ of the following form

$$u_i(t,x) = \begin{cases} u_i^+(t,x) & \text{if } s_i(x) > 0 \\ u_i^-(t,x) & \text{if } s_i(x) < 0 \end{cases} \quad , \ i = 1,\dots,m .$$

The control methodology based on such piecewise discontinuous control laws achieves accurate tracking, stable behavior, robust control (at least in principle), whenever the state vector slides along the manifold (2) in the state space, as shown in detail in [1] if $f(t,x,\cdot)$ is an affine function of the control variable. See also [2] for a discussion of some of the drawbacks of this technique and also for applications to robotics.

The mathematical description of such control systems is based on the Filippov definition of solutions of ordinary differential equations with discontinuous right-hand side [3] .

We assume that the Filippov concept is the appropriate one. Although in principle different definitions of solution could be considered, the re sults presented in [1],[3],[4] show that the Filippov definition is significant for control systems which depend linearly upon the control variable. For a different approach see [5]. A geometrical approach has been considered in [8].

We refer the reader to [1] for motivations, applications, examples, for conditions assuring that $x(t) \in S$ if $0 \leq t \leq T$ and for a discussion of the mathematical description of control systems of the above form (1), (2), (3), (4) when the function $f(t,x,\cdot)$ is affine. Many control systems are of the form (1) with $f(t,x,\cdot)$ non affine, therefore the sliding mode control methodology should be extended to such systems (as remarked e.g. in [1] and [2]). We refer the reader to the classical paper [3] for definition and properties of Filippov's solutions.

## 2. THE EQUIVALENT CONTROL AND ITS RELATION WITH FILIPPOV SOLUTION.

We assume that f is a Carathéodory function, $s_j \in C^1(R^m)$ , $j = 1,\dots,m$ and we denote by G the mxn jacobian matrix $\partial s/\partial x$. In engineering applications (see [1] and [2]) every component $u_j$ of the control vector switches along a well defined surface $s_j(x) = 0$, so that we assume that the dimension m of the control vector is equal to the codimension m of the sliding manifold S. We suppose throughout the paper that the following basic assumption holds:

(5) for every $t \in [0,T]$ and $x \in R^n$ the map $G(x)\, f(t,x,\cdot)$ is one-to-one on U.

Denote by $u^*(t,x,v)$ the only solution (if any) in U of

$$G(x)\, f(t,x,u) = v.$$

We shall consider systems (1), (2), (3) such that with suitable control laws the corresponding state variables slide on the manifold S. Therefore we assume that $0 \in f(t,x,U)$ for all t,x of interest.

Definition 1. The equivalent control for the system (1), (2), (3) is given by the map $(t,x) \longrightarrow u^*(t,x,0)$. The above definition clearly generalizes that given in [1] , chapter II.

Such a definition should be validated by showing

(i) significant relations with the Filippov solutions of the control system;

(ii) proper (approximating) behavior of the trajectories obtained by the equivalent control with respect to perturbations acting on the system (e.g. due to uncertain parameters, delays, hysteresis, small time constants, approximate knowledge of the plant).

We shall show that condition (i) is satisfied in many nonlinear control systems, while (ii) puts severe restriction of the nonlinear system structure.

We suppose that given $x_o \in S$, a sufficiently small neighborhood of $x_o$ may be written as disjoint union of suitable portions of the surfaces $s_j(x) = 0$, $j = 1,\dots,m$, and of open connected regions $C_1,\dots,C_p$.

In contrast with the Filippov concept of solution for (1) we shall consider also classical (Carathéodory) solutions of (1), denoted by a.e. solutions.

Theorem 1. Let $u = u(t,x)$ be a feedback control law, measurable in t, continuous in x except on the surfaces $s_j(x) = 0$, $j = 1,\dots,m$. Given $x_o \in S$ denote by $u_i$ the restriction of u to the region $C_i$, $i=1,\dots,p$ (for some fixed neighborhood of $x_o$) and assume that

(6) $\quad u_i(t,x_o) = \lim_{x \to x_o} u_i(t,x)$ exists, $i = 1,\dots,p$ ;

(7) $\quad$ U is closed and f is bounded;

(8) $\quad f(t,x,U)$ is convex whenever $x \in S$.

Then if y is a Filippov solution on $[0,T]$ of (1) with $u = u(t,x)$ such that $s[y(t)] = 0$, $0 \le t \le T$, then y is an a.e. solution of (1) corresponding to the equivalent control. Conversely, if y is an a.e. solution of (1) corresponding to the equivalent control such that $s[y(0)] = 0$, then y is a Filippov solution of (1) corresponding to $u = \bar{u}(t,x)$ any feedback law satisfying (6), such that $\bar{u}$ keeps $y(t)$ on S, $0 \le t \le T$ if $y(0) \in S$, provided that $G\,\partial f/\partial u$ is non singular and $(G\,\partial f/\partial u)^{-1}$, G, f, $\partial G/\partial x$, $\partial f/\partial x$ are bounded by integrable functions of t on every set $[0,T] \times K \times U$, K compact in $R^n$.

The proof is obtained by extending to the present setting lemma 3 p. 206 of [3], so that for a.e. $t \in [0,T]$

$$\dot{y}(t) \in \mathrm{co}\,\{f(t,y(t),u_j(t,y(t))) \ : j = 1,\dots,p\} .$$

By (8) we find $u_o(t) \in U$ such that

$$\dot{y}(t) = f(t,y(t),u_o(t)) \quad \text{a.e.}$$

and then

$$G[y(t)]\ \dot{y}(t) = d/dt\ s[y(t)] = 0 = G[y(t)]\ f(t,y(t),u_o(t))$$

thus giving

$$u_o(t) = u^*(t,y(t),0).$$

This establishes the first part of the theorem. The second half of the theorem makes use of the implicit function theorem applied to

$$G(x)\quad f(t,x,u) = 0$$

which entails uniqueness for a.e. solutions of

$$\dot{y} = f(t,y,u^*(t,y,0)), \quad y(0) \text{ given.}$$

Then if y is generated by the equivalent control and z comes from some feedback u, y(0) = z(0), then y = z, and this proves the theorem. Theorem 1 generalizes the results of [1] (chapters I, II) where it is assumed that

$$f(t,x,u) = A(t,x) + B(t,x)\ u\ ,$$

G B everywhere non singular, where the equivalence is shown between states corresponding in the sense of Filippov to discontinuous feedback laws and a.e. solutions of (1) generated by the equivalent control.

From theorem 1 we see that the main properties of the equivalent control are preserved in our nonlinear setting, that is (i) the equivalent control is independent, to a large extent, upon the choice of the discontinuous feedback (4) as far as we are able to keep the state variable on the sliding manifold S; (ii) by using the equivalent control we generate, under the conditions of theorem 1, the same state variables as the control system (1). Solving the state equations (1) in the sense of Filippov can thereby avoided.

## 3. APPROXIMABILITY

Generalizing the property introduced in[1](chapter II) we introduce the following

Definition 2. The approximability property holds for the control system (1), (2), (3), if the following is true. There exist $M > 0$, $p > 1$ such that for every $a_\varepsilon \in L^p(0,T)$, such that $\| a_\varepsilon \| \leq M$ and

$\int_0^t a_\varepsilon(s)\,ds \longrightarrow 0$ uniformly in $[0,T]$ as the real parameter $\varepsilon \longrightarrow 0$, if a.e. in $[0,T]$

$$\dot{x} = f[t,x_\varepsilon, u^*(t,x_\varepsilon, a_\varepsilon(t)] , \quad s[x_\varepsilon(0)] \longrightarrow 0,$$

$$\dot{y} = f[t,y, u^*(t,y,0)] , \quad s[y(0)] = 0,$$

then $x_\varepsilon(0) \longrightarrow y(0)$ as $\varepsilon \longrightarrow 0$ imply

$$x_\varepsilon \longrightarrow y \text{ uniformly in } [0,T].$$

FOllowing the terminology of [1] the above definition means that the real states $x_\varepsilon$ of the system approximate (uniformly) the ideal ones in the sliding mode as the nonidealities disappear. The real parameter $\varepsilon$ describes such non idealities while the (essentially arbitrary) functions $a_\varepsilon$ satisfy, a.e. in $[0,T]$, the relation

$$a_\varepsilon(t) = d/dt\ s[x_\varepsilon(t)] = G(x_\varepsilon(t))\ f(t,x_\varepsilon(t), u^*_\varepsilon(t))$$

where of course

$$u^*_\varepsilon(t) = u^*(t,x_\varepsilon(t), a_\varepsilon(t)).$$

In [1] it is assumed (p.45) that all the nonidealities are described by tha single relation

$$|s[x_\varepsilon(t)]| \leq \varepsilon$$

so that by the way of the $H^{-1,\infty}(0,T)$ - convergence of $a_\varepsilon$ required above we see that definition 2 generalizes the corresponding property required in [1] , chapter II.

Example. Consider $m = 2$, $n = 3$, $T = 1$,

$$\dot{x}_1 = u_1, \quad \dot{x}_2 = u_2, \quad \dot{x}_3 = u_1 u_2, \quad |u_1| \leq 1, \quad |u_2| \leq 2,$$

$$s_1(x) = x_1, \quad s_2(x) = x_2.$$

This example has been discussed heuristically in [1] to support the belief that sliding modes may be ambiguous when u enters nonlinearly in (1). As a matter of fact, while the basic assumption (5) holds, the approximability propertiy fails, as easily seen by taking $\varepsilon = 1/k$, $k = 1,2,3,\ldots$, and considering

$$a_k(t) = 1 \quad \text{if } j/k < t < (2j+1)/k,$$

$$a_k(t) = -1 \quad \text{if } (2j+1)/k < t < (j+1)/k,$$

$0 \leq j \leq k-1$ ; $k = 1,2,3\ldots.$

Then for the corresponding state variables $x_k$, $y$, we get (for a.e. $t \in [0,1]$ )

$$\dot{y}_3 = 0, \quad \dot{x}_{3k} = 1.$$

So in this case the equivalent control has no "physical" meaning: let us remark that (8) fails too.

Definition 2 isolates a property which is true if the control variables enter linearly in the dynamics, as shown in [1] , theorem p. 45 (under some unnecessary assumptions).

The problem arises of finding significant classes of control systems (1), (2), (3), such that the approximability property holds. Such problem has interesting connections with G-convergence problems in ordinary differential equations (and also multivalued differential equations), see [6] , [7] .

By using results from [7] we prove the following

Theorem 2. The approximability property and the conclusions of theorem 1 hold for each of the following control systems A), B).

A). $z^{(n)} = g(t,z,z',z'',\ldots,z^{(n-1)},u)$,

with m = 1 (z and u scalar variables), provided that

$$s \in C^2(R^n),\ g(t,\cdot,u) \in C^2(R^n),\ \partial g/\partial u \ \text{ and } \ \partial s/\partial x_n \neq 0$$

everywhere; g, $\partial g/\partial x$ , $\partial g/\partial u$ are continuous, U is a compact interval and $|g(t,x,u)| \leq p(t) + q(t)\ |x|$ for some integrable p,q.

B). $\dot{x} = A(t,x) + B(t,x)\ h(u)$ ,

where $x \in R^n$ , $u \in R^m$, provided that

$$s \in C^2(R^n)\ ;\ A\ ,\ B\ ,\ \partial A/\partial x\ ,\ \partial B/\partial x\ ,\partial h/\partial u$$ continuous, locally bounded with respect to x; G B and G B $\partial h/\partial u$ non singular, $(G\,B\,\partial h/\partial u)^{-1}$ locally bounded in x,

$$B(\cdot,x)\ [G(x)\ B(\cdot,x)] \in L^q(0,T)\ (1/p + 1/q = 1);$$

U closed, h one-to-one with h(U) convex; $|A(t,x)| + |B(t,x)| \leq r(t) + w(t)\ |x|$ for some integrable r,w.

Part B of theorem 2 generalizes theorem p. 45 of [1] . Detailed results are given in [9] . Further results and applications will be considered elsewhere.

## REFERENCES

[1] . V.I.UTKIN. Sliding modes and their application in variable structure systems.
M.I.R. publishers, Moscow, 1978.

[2] . J.J.SLOTINE - S.S.SASTRY. Tracking control of non linear systems using sliding surfaces with application to robot manipulators. International J. of Control 38, 1983, 465-492.

[3] . A.F.FILIPPOV. Differential equations with discontinuous right-hand side.
American Mathematical Society Translations 42, 1964, 199-232.

[4] . A.F.FILIPPOV. Application of theory of differential equations with discontinuous right-hand sides to non-linear problems in automatic control.
Proceedings First IFAC International Congress, vol. I,1961,923-927.

[5] . M.A.AIZERMAN - E.S.PYATNISKII. Foundations of a theory of discontinuous systems I,II.
Avtomatika i Telemekhanika 7 (1974), 33-47; 8 (1974), 39-61.

[6] . L.C.PICCININI. Linearity and nonlinearity in the theory of G-convergence.
Recent Advances in Differential Equations, edited by R.Conti. Academic Press, 1981.

[7] . P.PATUZZO GREGO. Sulla G-convergenza delle equazioni differenziali ordinarie nel caso di domini illimitati.
Bollettino dell'Unione Matematica Italiana 16 B (1979), 466-479.

[8] . O.M.E.EL-GHEZAWI - A.S.I.ZINOBER - S.A.BILLINGS. Analysis and design of variable structure systems using a geometric approach. International Journal on Control 38 (1983), 657-671.

[9] . G.BARTOLINI - T.ZOLEZZI. Control of nonlinear variable structure systems, submitted.

Session 10

# FILTERING
# FILTRAGE

# APPROXIMATIONS OF THE NONLINEAR FILTER BY PERIODIC SAMPLING AND QUANTIZATION

H. KOREZLIOGLU and G. MAZZIOTTO

Dpt Systèmes et Communications
E.N.S.T.
46, rue Barrault
75634 PARIS - Cedex 13

PAA/TIM/MTI
C.N.E.T.
38-40, rue du G$^{al}$ Leclerc
92131 ISSY LES MOULINEAUX

Exact filtering algorithms are given and approximation rates are computed for various approximations of the nonlinear filter via periodic sampling and quantization of the observation process.

## 1. INTRODUCTION

This paper studies various approximations of the nonlinear filter by periodic sampling and quantization. The state process is a p-dimensional Markov process, strong solution of the equation

$$dX_t = a(X_t)\ dt + b(X_t)\ dB_t$$

and the observation process is a q-dimensional one defined by

$$dY_t = c(X_t)\ dt + dW_t\ ,$$

where B and W are independent Brownian motions. a, b and c are supposed to be time-independent, Lipschitzian and c is bounded. The recursive approximations considered here are the following.

1. Y is sampled with a sampling period of length h.

2. Moreover $c(X_t)$ is replaced by a process C taking the constant value $c(X_{(n-1)h})$ in the interval $[(n-1)h,nh[$.

3. After all these, X is replaced by a process $\overline{X}^h$ which is a discrete-time Markov chain approximating it at the sampling points and takes the constant value $\overline{X}^h_{(n-1)h}$ in the interval $[(n-1)h,nh[$.

4. Periodic samples of Y are quantized on a finite partition of $\mathbb{R}^q$.

In each case the corresponding recursive filtering algorithm is derived and the order of approximation is computed.

The filtering algorithm given here in Case 1 has a more explicit form than the ones obtained by LO in (7) and TAKEUCHI and AKASHI in (13). The algorithm is expressed in terms of the transition semigroup of X which is in general not easily computable. The algorithm expressed in Case 2 uses the transitions of X only on the sampling points, and Case 3 approximates these transitions by those of a discrete-time Markov process. The algorithm of this last case can be deduced from KUSHNER's work ([5]). By our approach we wanted to show that the motivation of the approximation in each case can be found in the preceding one.

If $\Pi$ denotes the normalized filter, it is proved here that for any bounded function f and t, the rate of convergence in $\mathbb{L}^1$ to $\Pi_t(f)$ of the corresponding approximated expressions is of order $\sqrt{h}$ in all three cases 1, 2 and 3.

For Case 2 a rate of convergence of order h is derived by PICARD in ([11]) under more restrictive conditions.

We would like to note that our results would still hold in case functions a, b and c depend on t. The choice of time-independent functions considerably simplifies the writing.

We have not studied here the quantization problem of X. Some results in this sense are given by DI MASI and RUNGGALDIER ([2]). We think that this problem can be considered as a source coding problem and that it needs some more detailed work.

## 2. THE MODEL

All the processes considered in this paper will be indexed on the bounded interval $[0,T]\subset\mathbb{R}_+$.

Let $B = (B_t;t \in [0,T])$ and $Y = (Y_t;t \in [0,T])$ be two independent Brownian motions with values in $\mathbb{R}^m$ and $\mathbb{R}^q$, and defined on their canonical probability spaces $(\Omega^B,\underline{\underline{B}}_T,\mathbb{P}_0^B)$ and $(\Omega^Y,\underline{\underline{Y}}_T,\mathbb{P}_0^Y)$. $\underline{\underline{F}}^B = (\underline{\underline{B}};t \in [0,T])$ and $\underline{\underline{F}}^Y= (\underline{\underline{Y}}_t;t \in [0,T])$ will denote their natural filtrations. The reference probability space on which the "state and observation" model is constructed is $(\Omega,\underline{\underline{F}}_T,\mathbb{P}_0)$ where $\Omega = \Omega^B \times \Omega^Y$, $\underline{\underline{F}}_T$ is the completion of $\underline{\underline{B}}_T \otimes \underline{\underline{Y}}_T$ with respect to $\mathbb{P}_0^B \otimes \mathbb{P}_0^Y$, and $\mathbb{P}_0$ is the extension of $\mathbb{P}_0^B \otimes \mathbb{P}_0^Y$ to $\underline{\underline{F}}_T$. $(\Omega,\underline{\underline{F}}_T,\mathbb{P}_0)$ is given the filtration $\underline{\underline{F}} = (\underline{\underline{F}}_t;t \in [0,T])$ where $\underline{\underline{F}}_t$ is the $\sigma$-algebra generated by $\underline{\underline{B}}_t\otimes\underline{\underline{Y}}_t$ and the negligible sets of $\underline{\underline{F}}_T$. We denote by $E_0$, $E_0^B$ and $E_0^Y$ the expectations on $(\Omega,\underline{\underline{F}}_T, \mathbb{P}_0)$, $(\Omega^B,\underline{\underline{F}}_T^B,\mathbb{P}_0^B)$ and $(\Omega^Y,\underline{\underline{F}}_T^Y,\mathbb{P}_0^Y)$, respectively.

Let $X = (X_t;t \in[0,T])$ be a Markov diffusion with values in $\mathbb{R}^p$, defined on $(\Omega^B,\underline{\underline{F}}_T^B,\mathbb{P}_0^B)$ (or equivalently on $(\Omega,\underline{\underline{F}}_T,\mathbb{P}_0)$) as the unique strong solution of the stochastic differential equation

$$(2.1)\quad X_t = x_0 + \int_0^t a(X_s)\, ds + \int_0^t b(X_s)\, dB_s,\quad t \in [0,T]$$

where $x_o \in \mathbb{R}^p$ and a and b are functions on $\mathbb{R}^p$ satisfying the usual Lipschitz and linear growth conditions, with values in $\mathbb{R}^p$ and the space of p x m-matrices, respectively. We denote by $(P_t \; ; \; t \in [0,T])$ the transition semigroup of X.

Let $L=(L_t \; ; \; t \in [0,T])$ be defined on $(\Omega, \underline{\underline{F}}_T, \mathbb{P}_0)$ by

$$(2.2) \quad L_t = \exp \{ \int_0^t c(X_s) \, . \, dY_s - 1/2 \int_0^t |c(X_s)|^2 \, ds \}$$

where c is a Lipschitzian $\mathbb{R}^q$ - valued bounded function on $\mathbb{R}^p$. ($|.|$ denotes the Euclidian norm on $\mathbb{R}^n$ and a.b the scalar product of two elements a, $b \in \mathbb{R}^n$).

It is known ([6]) that L is a positive $(\underline{\underline{F}}, \mathbb{P}_0)$-martingale such that $\mathbb{E}_0(L_T) = 1$, $L_T > 0$ a.s. and $L_T \in \mathbb{L}^p (\Omega, \underline{\underline{F}}_T, \mathbb{P}_0)$ for $p \in [1, \infty[$. Then $\mathbb{P} = L_T . \mathbb{P}_0$ is a probability measure on $(\Omega, \underline{\underline{F}}_T)$ equivalent to $\mathbb{P}_0$. B is also a $(\underline{\underline{F}}, \mathbb{P})$ - Brownian motion and Y has the following decomposition

$$(2.3) \quad Y_t = \int_0^t c(X_s) \, ds + W_t$$

where W is another $(\underline{\underline{F}}, \mathbb{P})$ - Brownian motion independent of B. The state and observation processes are then described on $(\Omega, \underline{\underline{F}}_T, \mathbb{P})$ by Equations (2.1) and (2.3).

The expectation on $(\Omega, \underline{\underline{F}}_T, \mathbb{P})$ will be denoted by E. $b(\mathbb{R}^n)$ will denote the space of all bounded real Borel functions on $\mathbb{R}^n$ with the sup-norm written by $\| . \|$.

The unnormalized filtering process is the process $\sigma = (\sigma_t ; t \in [0,T])$ with values in the space of positive measures on $\mathbb{R}^p$, defined by

$$(2.4) \quad \sigma_t(f)(\omega^Y) = \int_{\Omega^B} L_t(\omega^B, \omega^Y) \; f(X_t(\omega^B)) \; \mathbb{P}_0^B(d\omega^B)$$

for all $f \in b(\mathbb{R}^p)$. The filtering process is then the process $\Pi = (\Pi_t ; t \in [0,T])$ with values in the space of probability measures on $\mathbb{R}^p$ given by the Kallianpur-Striebel formula :

$$(2.5) \quad \Pi_t(f) = \sigma_t(f) \; / \; \sigma_t(1)$$

for all $f \in b(\mathbb{R}^p)$, ([4]).

3. RECURSIVE FILTERING WITH SAMPLED OBSERVATION

Suppose that the observation process Y is sampled with a constant period of length h and that the information about Y is given by the samples $Y_0 = 0, Y_h, Y_{2h}, \ldots, Y_{nh}, \ldots$ with $n \le [T/h]$. ([a/b] denotes the integer part of a/b). We denote by $\underline{\underline{Y}}_t^h$ the σ-algebra generated by $(Y_{nh}; n \le [t/h])$ and we note that it is also generated by the set $(\Delta Y_n^h; n \le [t/h])$ where

$$(3.1) \quad \Delta Y_n^h = Y_{nh} - Y_{(n-1)h} \quad \text{with } \Delta Y_0^h = Y_0 = 0.$$

Under $\mathbb{P}_0$ these random variables are mutually independent Gaussian variables with mean 0 and covariance matrix $hI_q$, where $I_q$ is the unit q x q matrix.

The filtering process $\Pi^h$ with respect to sampled observations $(Y_{nh}; n \le [T/h])$ is defined by

$$\Pi_t^h(f) = E(f(X_t) \,/\, \underline{\underline{Y}}_t^h) \text{ a.s. for all } f \in b(\mathbb{R}^p).$$

As Π in (2.5), $\Pi^h$ is given by

$$(3.2) \quad \Pi_t^h(f) = \sigma_t^h(f) \,/\, \sigma_t^h(1), \text{ for all } f \in b(\mathbb{R}^p),$$

where

$$(3.3) \quad \sigma_t^h(f) = E_0(L_t\, f(X_t) \,/\, \underline{\underline{Y}}_t^h) = E_0^Y(\sigma_t(f) \,/\, \underline{\underline{Y}}_t^h) \text{ a.s..}$$

$\sigma^h$ is the unnormalized filtering process corresponding to sampled observation $Y^h$.

<u>PROPOSITION 3.1</u> : We have for $f \in b(\mathbb{R}^p)$ and $t \in [h, [T/h]h]$

$$(3.4) \qquad \sigma_t^h(f) = E_0^B(\exp\{\sum_{k=1}^{[t/h]} (C_k^h.\Delta Y_k^h - h/2\, |C_k^h|^2)\}\, f(X_t))$$

where

$$(3.5) \qquad C_k^h = h^{-1} \int_{(k-1)h}^{kh} c(X_s)\, ds \quad .$$

<u>Proof</u>: In order to shorten the notation, we put

$$(3.6) \qquad Z_{(k-1)h}^{kh} = \exp\{\int_{(k-1)h}^{kh} c(X_s).dY_s - 1/2 \int_{(k-1)h}^{kh} |c(X_s)|^2\, ds\}$$

We first remark the following:

$$(3.7) \qquad \sigma_t^h(f)(\omega^Y) = \int_{\Omega^B} E_0^Y(L_t(\omega^B,.)\,/\,\underline{\underline{Y}}_t^h)(\omega^Y)\ f(X_t(\omega^B))\ \mathbb{P}_0^B(d\omega^B) .$$

On the other hand, since the set $((Z_{(k-1)h}^{kh}, \Delta Y_k^h)\ ;\ k = 1,2,\dots,[T/h])$ consists of mutually independent pairs of random variables, we have

$$(3.8) \qquad E_0^Y(L_t\,/\,\underline{\underline{Y}}_t^h) = \prod_{k=1}^{[t/h]} E_0^Y(Z_{(k-1)h}^{kh}\,/\,\Delta Y_k^h) \quad , \text{ for } \ t \geq h .$$

Therefore, to achieve the proof we only need to show that, for any $\alpha \in \mathbb{R}^q$,

$$(3.9) \qquad E_0^Y(Z_{(k-1)h}^{kh}\ \exp(\alpha.\Delta Y_k^h) = E_0^Y(\exp\{(C_k^h + \alpha).\Delta Y_k^h - \frac{h}{2}\ |C_k^h|^2\}\ )$$

We have

$$E_0^Y(Z_{(k-1)h}^{kh}\ \exp(\alpha.\Delta Y_k^h)) = \exp(\alpha.\int_{(k-1)h}^{kh} c(X_s)\ ds + \frac{1}{2}|\alpha|^2\ h\ )\ x$$

$$E_0^Y(\exp(\int_{(k-1)h}^{kh}(c(X_s) + \alpha)\ dY_s - \frac{1}{2}\int_{(k-1)h}^{kh}|c(X_s) + \alpha|^2\ ds)).$$

Since the process $(c(X_t) + \alpha\ ;\ t \in [0,T])$ is bounded the expectation at the right hand side is equal to 1. On the other hand, we have for the same reason

$$E_0^Y(\exp((C_k^h + \alpha).\Delta Y_k^h - \frac{h}{2}\,|\,C_k^h\,|^2)) = \exp(\alpha.C_k^h\ h + \frac{1}{2}\,|\alpha|^2 h)).$$

Therefore, equality (3.9) holds. ∎

Now let us put

$$(3.10) \qquad \hat{Z}_{(n-1)h}^{nh} = \exp(C_n^h.\Delta Y_n^h - \frac{h}{2}\ |C_n^h|^2)$$

$$(3.11) \qquad \hat{L}_{nh} = \prod_{k=1}^{n} \hat{Z}_{(k-1)h}^{kh} \quad \text{with } \hat{L}_o = 1.$$

Formula (3.7) gives for $t = nh$ :

$$(3.12) \qquad \sigma_{nh}^h(f) = E_0^B(\hat{L}_{(n-1)h}\ \hat{Z}_{(n-1)h}^{nh}\ f(X_{nh}))$$

$$= E_0^B(\hat{L}_{(n-1)h}\ E_0^B(\hat{Z}_{(n-1)h}^{nh}\ f(X_{nh})\,/\,\underline{\underline{B}}_{(n-1)h}))$$

$$= E_0^B(\hat{L}_{(n-1)h}\ E_0^B(\hat{Z}_{(n-1)h}^{nh}\ f(X_{nh})\,/\,X_{(n-1)h}))$$

where the Markovian character of X is taken into account.

One can show that there is a transition kernel $H_n$ on $\mathbb{R}^p$ such that

$$(3.13)\quad H_n(X_{(n-1)h},f) = \int_{\mathbb{R}^p} f(x)\, H_n(X_{(n-1)h},dx)$$
$$= E_0^B(\hat{Z}^{nh}_{(n-1)h}\, f(X_{nh}) \,/\, X_{(n-1)h}) \quad \text{a.s.}$$

for all $f \in b(\mathbb{R}^p)$.

For $(n-1)h < t < nh$, we have

$$(3.14)\quad \sigma_t^h(f) = E_0^B(\hat{L}_{(n-1)h}\, f(X_t)) = E_0^B(\hat{L}_{(n-1)h}\, E_0^B(f(X_t) \,/\, \underline{\underline{B}}_{(n-1)h}))$$
$$= E_0^B(\hat{L}_{(n-1)h}\, E_0^B(f(X_t) \,/\, X_{(n-1)h}))$$
$$= E_0^B(\hat{L}_{(n-1)h}\, P_{t-(n-1)h}(X_{(n-1)h},f))$$

We summarize these results in the following.

THEOREM 3.2 : For $t = nh$, $n=1,2,\dots,[T/h]$, the unnormalized filter $\sigma_t^h$ is obtained by the following recurrence formula :

$$(3.15)\quad \sigma_{nh}^h(dx) = \int_{\mathbb{R}^p} H_n(u,dx)\, \sigma^h_{(n-1)h}(du)$$

with $\sigma_o^h(dx) = \delta_{x_o}(dx)$, where $\delta_{x_o}$ is the Dirac measure at $x_o$.

For $(n-1)h < t < nh$

$$(3.16)\quad \sigma_t^h(dx) = \int_{\mathbb{R}^p} P_{t-(n-1)h}(u,dx)\, \sigma^h_{(n-1)h}(du).$$

Therefore, for these t, we have

$$(3.17)\quad \Pi_t^h(dx) = \int_{\mathbb{R}^p} P_{t-(n-1)h}(u,dx)\, \Pi^h_{(n-1)h}(du).$$

## 4. DEGREE OF THE APPROXIMATION WITH SAMPLED OBSERVATIONS

We would like to evaluate here the degree of the approximation when the filter $\Pi$ is replaced by the filter $\Pi^h$ corresponding to the sampled observation $Y^h$.

PROPOSITION 4.1 : For fixed $f \in b(\mathbb{R}^p)$ and $t \in [0,[T/h]h]$ the following assertions hold

i) The net $(\sigma_t^h(f); h > 0)$ converges to $\sigma(f)$ in $\mathbb{L}^n(\Omega^Y, \underline{\underline{Y}}, \mathbb{P}_0^Y)$, for $n \in [1,\infty[$, as h decreases to 0. The convergence is a.s. if the set of partitions $(0 < h < 2h < .. < [T/h]h; h > 0)$ is totally ordered.

ii) $E_0^Y(|\sigma_t(f) - \sigma_t^h(f)|) = O(\sqrt{h})$.

Proof : Assertion i) is a consequence of the martingale convergence theorems, (8). We have to prove only Assertion ii).

We have,

$$E_0^Y(|\sigma_t(f) - \sigma_t^h(f)|) = E_0^Y(|E_0^B(L_t\ f(X_t)) - E_0^B(E_0^Y(L_t/\underline{\underline{Y}}_t^h)\ f(X_t))|)$$
$$\leq E_0(|L_t - E_0^Y(L_t/\underline{\underline{Y}}_t^h)| \,.\, |f(X_t)|) \leq \|f\| \,.\, E_0(|L_t - E_0^Y(L_t/\underline{\underline{Y}}_t^h)|).$$

We thus have to find an upper bound of $E_0(|L_t - E_0^Y(L_t / \underline{\underline{Y}}_t^h)|)$. We write the proof for $t \geq h$, the result is trivial for $t < h$.

Taking into account the inequality $|e^x - e^y| \leq |x - y|\ (e^x + e^y)$ which holds for all $x, y \in \mathbb{R}$, and defining the process $c^h$ by $c_s^h = c_k^h$ for $s \in ](k-1)h, kh]$, $k = 1,..,[t/h]$ and $c_s^h = 0$ for $s > [t/h]h$, we get

$$|L_t - E_0^Y(L_t / \underline{\underline{Y}}_t^h)|$$
$$= \exp\{\int_0^t c(X_s).dY_s - \frac{1}{2}\int_0^t |c(X_s)|^2\, ds\} - \exp\{\sum_{k=1}^{[t/h]} (c_k^h.\Delta Y_k^h - \frac{h}{2}|c_k^h|^2)\}$$
$$\leq |\int_0^t c(X_s).dY_s - \frac{1}{2}\int_0^t |c(X_s)|^2\, ds - \int_0^t c_s^h.dY_s + \frac{1}{2}\int_0^t |c_s^h|^2\, ds$$
$$.\ (L_t + E_0^Y(L_t / \underline{\underline{Y}}_t^h))\ .$$

We get from this

$$E_0(|L_t - E_0^Y(L_t / \underline{\underline{Y}}_t^h)|)$$
$$\leq (E_0(|\int_0^t (c(X_s) - c_s^h).dY_s - \frac{1}{2}\int_0^t (|c(X_s)|^2 - |c_s^h|^2)\, ds\ |^2))^{1/2}$$
$$.\ (E_0(|L_t + E_0^Y(L_t / \underline{\underline{Y}}_t^h)|^2))^{1/2}$$
$$\leq K(E_0(|\int_0^t (c(X_s) - c_s^h).dY_s - \frac{1}{2}\int_0^t (|c(X_s)|^2 - |c_s^h|^2)\, ds\ |^2))^{1/2}$$

for $L_t \in \mathbb{L}^2(\Omega, \underline{\underline{F}}, \mathbb{P}_0)$. We then deduce

$$E_0(|L_t - E_0^Y(L_t / \underline{\underline{Y}}_t^h)|)$$
$$\leq K\ (\int_0^{[t/h]h} E_0(|c(X_s) - c_s^h|^2)\, ds\ )^{1/2} + K\int_{[t/h]h}^t E_0(|c(X_s)|^2)\, ds)^{1/2}$$

where K denotes constants independent of h. We shall go on denoting by the same symbol K various constants, independent of h, appearing in the upper bounds of various inequalities.

The last term of the above inequality is of order $\sqrt{h}$. We then need to evaluate an upper bound of

$$\int_h^{[t/h]h} E_0(|c(X_s) - c_s^h|^2)\, ds \quad .$$

We have for $s \in ](k-1)h, kh]$

$$E_0(|c(X_s) - c_s^h|^2) = E_0(|c(X_s) - \frac{1}{h}\int_{(k-1)h}^{kh} c(X_u)\, du|^2)$$

$$= \frac{1}{h^2} E_0\Big(\Big(\int_{(k-1)h}^{kh} |c(X_s) - c(X_u)|\, du\Big)^2\Big)$$

$$\leq \frac{K}{h}\Big(\int_{(k-1)h}^{kh} E_0(|X_s - X_u|^2)\, du\Big).$$

It is known ([3]) that $E_0(|X_s - X_u|^2) \leq Kh$, for all s, u such that $|s-u| < h$. Finally we have

$$E_0(|c(X_s) - c_s^h|^2) \leq Kh$$

From this we get

$$(4.1)\quad E_0(|L_t - E_0^Y(L_t/\underline{\underline{Y}}_t^h)|) \leq K\sqrt{h} \ . \ ■$$

We shall see below that inequality (4.1) will be useful for the estimation of the approximation degree on the normalized filter.

<u>PROPOSITION 4.2</u> : For fixed $f \in b(\mathbb{R}^p)$ and $t \in [0,[T/h]h]$ the following assertions hold.

i) The net $(\Pi_t^n(f); h > 0)$ converges to $\Pi_t(f)$ in $\mathbb{L}^n(\Omega^Y, \underline{\underline{Y}}_T, \mathbb{P})$ for $n \in [0,\infty[$, as h decreases to 0. The convergence is a.s. if the set of partitions $(0 < h < .. < [T/h]h; > 0)$ is totally ordered.

ii) $E^Y(|\Pi_t(f) - \Pi_t^h(f)|) = 0(\sqrt{h})$.

<u>Proof</u> : i) is deduced from the martingale convergence theorems, as in Proposition 4.2. For ii), we have

$E(|\Pi_t^h(f) - \Pi_t(f)|)$

$$= E\Big(\Big|\frac{E_0^B(f(X_t)\ E_0^Y(L_t / \underline{\underline{Y}}_t^h))}{E_0^B(E_0^Y(L_t / \underline{\underline{Y}}_t^h))} - \frac{E_0^B(f(X_t)\ L_t)}{E_0^B(L_t)}\Big|\Big)$$

$$= E_0\left(\left|\frac{E_0^B(L_t)\ E_0^B(f(X_t)\ E_0^Y(L_t / \underline{\underline{Y}}_t^h))}{E_0^B(E_0^Y(L_t / \underline{\underline{Y}}_t^h))} - E_0^B(f(X_t)\ L_t)\right|\right)$$

$$\leq E_0(|E_0^B(f(X_t)(E_0^Y(L_t / \underline{\underline{Y}}_t^h) - L_t))|) +$$

$$+ E_0\left(\left|E_0^B(f(X_t)\ E_0^Y(L_t / \underline{\underline{Y}}_t^h))\left(1 - \frac{E_0^B(L_t)}{E_0^B(E_0^Y(L_t / \underline{\underline{Y}}_t^h))}\right)\right|\right)$$

$$\leq \|f\|\ E_0(|E_0^Y(L_t / \underline{\underline{Y}}_t^h) - L_t|) + E_0(|\Pi_t^h(f)\ (E_0^B(E_0^Y(L_t / \underline{\underline{Y}}_t^h) - E_0^B(L_t))|)$$

$$\leq 2\ \|f\|\ E_0(|E_0^Y(L_t / \underline{\underline{Y}}_t^h) - L_t|)$$

Finally, inequality (4.1) provides the results. ■

## 5. SAMPLING OF THE STATE PROCESS

Within an error margin of order $\sqrt{h}$, the recursive filtering formula (3.15) replaces the well-known Zakai equation for the nonlinear filtering, ([14]). For a complete computation of $\sigma^h$ the main difficulty is then the computation of the kernels $H_n$. An approximate computation based on a modification of $c$ is proposed below.

PROPOSITION 5.1 : i) Let $\tilde{c}$ be defined by : $\tilde{c}_t = c(X_{(n-1)h})$ for $t \in [(n-1)h, nh[$, $n=1,2,\ldots,[T/h]$. Let $\tilde{\sigma}^h$ be the corresponding unnormalized filter with sampled observation, defined by

$$(5.1)\quad \tilde{\sigma}_t^h(f) = E_0(\tilde{L}_t\ f(X_t) / \underline{\underline{Y}}_t^h) \quad , \quad \forall f \in b(\mathbb{R}^p),$$

where

$$(5.2)\quad \tilde{L}_t = \exp\{\int_0^t \tilde{c}_s . dY_s - \frac{1}{2}\int_0^t |\tilde{c}_s|^2\ ds\} \quad .$$

Then, for $t = nh$, $n = 1,2,\ldots,[T/h]$ , $\sigma^h$ is given by the recurrence formula

$$(5.3)\quad \tilde{\sigma}_{nh}^h(dx) = \int_{\mathbb{R}^p} \tilde{L}(u,\Delta Y_n^h)\ P_h(u,dx)\ \tilde{\sigma}_{(n-1)h}^h(du)$$

with $\tilde{\sigma}_0^h(dx) = \delta_{x_0}(dx)$ , where

$$(5.4)\quad \tilde{L}(u,v) = \exp\{c(u).v - \frac{h}{2}|c(u)|^2\} \qquad u \in \mathbb{R}^p,\ v \in \mathbb{R}^q.$$

For $(n-1)h < t < nh$,

$$(5.5)\quad \tilde{\sigma}_t^h(dx) = \int_{\mathbb{R}^p} P_h(u,dx)\ \tilde{\sigma}_{(n-1)h}^h(du)$$

ii) For all $f \in b(\mathbb{R}^p)$ and $t \in [0,[T/h]h]$,

$$E_0^Y(|\sigma_t(f) - \tilde{\sigma}_t^h(f)|) = O(\sqrt{h})$$

$$E^Y(|\Pi_t(f) - \tilde{\Pi}_t^h(f)|) = O(\sqrt{h}) .$$

Proof : The proof of part i) is similar to that of Theorem 3.2 where L is to be replaced by $\tilde{L}$. Similarly, the proof of part ii) is identical to the proofs of Proposition 4.1 ii) and 4.2 ii).■

J. Picard obtained in (10) a bound of order $h^2$ for $E_0([\sigma_t(f) - \tilde{\sigma}_t^h(f)]^2)$ and of order h for $E(|\Pi_t(f) - \tilde{\Pi}_t^h(f)|)$, when B, X and Y are real processes, under the supplementary conditions that f should be Lipschitzian and c should be a $C^2$-function with bounded first and second derivatives. We do not know whether the same result could have been obtained under the hypotheses of the present work.

We see that without loosing the speed of convergence the filtering process $\sigma$ (resp. $\Pi$) can be replaced by $\tilde{\sigma}^h$ (resp. $\tilde{\Pi}^h$). The advantage of $\tilde{\sigma}^h$ with respect to $\sigma^h$ is that $\tilde{\sigma}^h$ can be recursively computed only on the knowledge of the transition semigroup of X.

We again see that within an error margin of order $\sqrt{h}$ the recursive filtering formulas (5.3) and (5.5) replace the Zakai equation and $\tilde{\sigma}^h$ (and hence $\tilde{\Pi}^h$) is computed by means of the transition semigroup of X. But in most of the cases this semigroup is not easy to compute . To make it easier, one naturally suggests the approximation of the Markov diffusion X by means of a discrete-time Markov process $(\overline{X}^h_{nh}; n=0,1,\dots,[T/h])$ with $\overline{X}_0^h = x_0$, such that

$$(5.6)\quad E(|X_{nh} - \overline{X}^h_{nh}|^2) = O(h)$$

and may approximate X by the process $\overline{X}^h$ defined by

$$(5.7)\quad \overline{X}_t^h = \overline{X}^t_{(n-1)h} \text{ for } (n-1)h \leq t < nh.$$

It is easily computed that $E(|X_t - X_t^h|^2) = O(h)$ for all t.

An example of such a Markov chain is given by the classical Euler scheme, (9), (11), (12) :

$$(5.8)\quad \overline{X}^h_{nh} = \overline{X}^h_{(n-1)h}) + a(\overline{X}_{(n-1)h}) h + b(\overline{X}_{(n-1)h}).\Delta B_n^h$$

where $\Delta B^h_n = B_{nh} - B_{(n-1)h}$ for $n \geq 1$, with $\bar{X}^h_0 = x_0$

We put

$$(5.9)\qquad \bar{L}^h_t = \exp\{\int_0^t c(\bar{X}^h_s).dY_s - \frac{1}{2}\int_0^t |c(\bar{X}^h_s)|^2 ds\}$$

where $\bar{X}^h$ is derived as in (5.7) from a Markov chain satisfying (5.6), and denote by $\bar{\sigma}^h$ and $\bar{\Pi}^h$ the corresponding unnormalized and normalized filters. We then have the following result.

PROPOSITION 5.2 : i) Let $\bar{P}^h$ denote the one-step transition probability of the chain $(\bar{X}^h_{nh}\,;\, n=0,\dots,[T/h])$.
Then for $t = nh$, $n = 1,\dots,[T/h]$, $\bar{\sigma}^h$ is given by

$$(5.10)\qquad \bar{\sigma}^h_{nh}(dx) = \int_{\mathbb{R}^p} \tilde{L}(u,\Delta Y^h_n)\ \bar{P}^h(u,dx)\ \bar{\sigma}^h_{(n-1)h}(du)$$

with $\bar{\sigma}^h_0(dx) = \delta_{x_0}(dx)$, where $\tilde{L}$ is defined by (5.4).
For $(n-1)h < t < nh$, we have $\bar{\sigma}^h_t(dx) = \bar{\sigma}^h_{(n-1)h}(dx)$.

ii) For a Lipschitzian $f \in b(\mathbb{R}^p)$ the following holds.

$$E_0(|\sigma_t(f) - \bar{\sigma}^h_t(f)|) = O(\sqrt{h}) \quad\text{and}\quad E(|\Pi_t(f) - \bar{\Pi}^h_t(f)|) = O(\sqrt{h}).$$

Proof: We only show the rate of convergence on $\sigma_t$ which is based on the following relations.

$$E^Y_0(|\sigma_t(f) - \bar{\sigma}^h_t(f)|) = E_0(|E^B_0(L_t\ f(X_t)) - E^B_0(\bar{L}_t\ f(\bar{X}^h_t)|)$$
$$\leq E_0(L_t\ |f(X_t) - f(\bar{X}^h_t)|) + E_0(|f(\bar{X}^h_t)|.|L_t - \bar{L}_t|)$$
$$\leq E_0(L^2_t)^{1/2}\ E_0(|f(X_t) - f(\bar{X}^h_t)|^2)^{1/2} + \|f\|\ E_0(|L_t - \bar{L}^h_t|)$$
$$\leq K\ \{(E(|X_t - \bar{X}^h_t|^2)^{1/2} + E_0(|L_t - \bar{L}_t|)\}\ .$$

We used here the fact that f is Lipschitzian. ∎

## 6. QUANTIZATION PROBLEM

The last approximation of the unnormalized filtering process $\sigma_t$ given by (5.10) needs to be numerically computed. Usually this is done by approximating the integral in (5.10) by a finite sum. This is equivalent to approximating the $\mathbb{R}^p$-valued Markov sequence $(\bar{X}^h_{nh}\,;\, n=1,..,[T/h])$ by a finite space-valued one.

Let $S = (S_0,\dots,S_n)$ be a partition of $\mathbb{R}^p$ and choose a point

$x_k$ in each $S_k$ such that the sequence $(X_{nh}^{h,S}\ ; n = 0,1,..,[T/h])$ defined by : $X_{nh}^{h,S} = x_k$ if $\bar{X}_{nh}^h \in S_k$, satisfies the condition $E_0(|\bar{X}_{nh}^h - X_{nh}^{h,S}|^2) \leq \delta$ for some small $\delta$. Then if $\bar{X}^h$ is replaced by $X^{h,S}$ in Proposition 5.2, it can be seen that the approximation rate of $\Pi$ in this proposition is of order $(h+\delta)^{1/2}$.

The best choice of the partition S and the points $(x_k)$ is a problem of coding. We will not consider it here.

But another problem of quantization may arise in the observation problems. For practical purposes it is reasonable to quantize the values of Y. We want to compute here the effect of quantizing Y on the value of the filter.

Now we consider partitions S of $\mathbb{R}^q$ defined as follows. $S = (S_i;\ i = 0,1,...,n)$ is such that for a given $R > 0$, $S_0$ is the complement of the ball $B(0,R)$ with center 0 and radius R, and $(S_i\ ;\ i = 1,...,n)$ is a partition of $B(0,R)$ into subsets whose diameters are at most equal to a given number $D > 0$ (i.e. $\forall x,y \in S_i$ : $|x-y| \leq D$). In the sequel we shall characterize the fineness of the partition S by these two numbers R and D i.e., the larger R and the smaller D are finer S becomes. We remark that the number n+1 of elements of this partition depends on R and D.

Suppose that the observation process Y is sampled with a constant period of length h, and that each variable $Y_k^h$ is quantized on the partition S. That is to say that we only know the events $(\{\Delta Y_k^h \in S_i\}\ ;\ k=0,..., T/h\ ;\ i=0,...,n)$ in order to estimate the state process X. We denote by $\underline{\underline{\Delta Y}}_k^{h,S}$ the ($\sigma$-) algebra generated by the events $(\{\Delta Y_k^{h,S} \in S_i\}\ ;\ i=0,...,n)$, and by $\underline{\underline{Y}}_t^{h,S}$ the $\sigma$-algebra generated by $(\underline{\underline{\Delta Y}}_k^{h,S}\ ;\ k=0,...,[t/h])$.

The filtering process $\Pi^{h,S}$ with respect to sampled and quantized observation is defined by

$$(6.1)\quad \Pi_t^{h,S}(f) = E(f(X_t)/\underline{\underline{Y}}_t^{h,S}),\quad \forall f \in b(\mathbb{R}^p),$$

and the associated unnormalized filtering process $\sigma^{h,S}$ is defined by

$$(6.2)\quad \sigma_t^{h,S}(f) = E_0(L_t\ f(X_t)\ /\underline{\underline{Y}}_t^{h,S}) = E_0^Y(\sigma_t^h(f)\ /\underline{\underline{Y}}_t^{h,S}).$$

By a similar proof to that of Proposition 3.1, we obtain

PROPOSITION 6.1 : We have for $f \in b(\mathbb{R}^p)$

$$(6.3)\quad \sigma_t^{h,S}(f) = E_0^B\Big(\prod_{k=1}^{[t/h]} \sum_{i=0}^{n} \frac{\mathbb{P}_0^Y(\Delta Y_k^h - hc_k^h \in S_i)}{\mathbb{P}_0^Y(\Delta Y_k^h \in S_i)}\ \mathbb{1}_{\{\Delta Y_k^h \in S_i\}} f(X_t)\Big)$$

Let us remark that Formula (6.3) leads to an algorithm similar to that of (3.15) and if the state process X is quantized then Formula (6.3) gives the classical algorithm for discrete filtering ([1]).

PROPOSITION 6.2 : Let S be a partition of the preceding type of fineness (R,D) and let h be the length of the sampling interval. If $\frac{R}{\sqrt{h}}$ is sufficiently large, more precisely greater than $\sup_{\mathbb{R}^p}|c(.)|\sqrt{h}$, and if D is such that $\frac{RD}{h} < 1/24\,(q+1)$, then for all $f \in b(\mathbb{R}^p)$ and $t \in [0,[T/h]h]$, we have

$$(6.4)\quad E_0^Y(|\sigma_t^{h,S}(f) - \sigma_t^h(f)|) \leq K\ \|f\|\ t\ (K_1\frac{RD}{h^2} + K_2\frac{R^2}{h^2}e^{-qR^2/2h})$$

where K, $K_1$ and $K_2$ are constants independent of D, R and h.

The proof lies on a lemma relative to the quantification of an $\mathbb{R}^q$-valued Gaussian normalized random variable. The proof is rather technical, and therefore is omitted.

LEMMA 6.3 : Let G be a q-dimensional Gaussian normalized random variable defined on a probability space $(\Omega,\underline{\underline{A}},\mathbb{P})$. Let C be a vector of $\mathbb{R}^q$. Let $S = (S_i\ ;\ i = 0,\dots,n)$ be a partition of $\mathbb{R}^q$ with fineness (R,D). For R greater than $|C|$ and sufficiently large, and for D such that RD is bounded by a given constant $(1/12(q+1))$, we have

$$(6.5)\ E(\sum_i \mathbb{1}_{\{G\in S_i\}}(C.G-\tfrac{1}{2}|C|^2 - \log\frac{\mathbb{P}(G - C\ \in S_i)}{\mathbb{P}(G\in\ S_i)})^2)$$
$$\leq K_1\ R^2D^2 + K_2\ R^{4-q}\ e^{-qR^2/2}$$

where $K_1$ and $K_2$ are constants independent of R and D.

Proof of Proposition 6.2: We find the upper bound for a given function $f \in b(\mathbb{R}^q)$, and for fixed $t = nh$.

$$E_0^Y(|\sigma_t^{h,S}(f) - \sigma_t^h(f)|)$$
$$\leq E_0^Y(|E_0^B(E_0^Y(L_t/\underline{\underline{Y}}_t^h)\ f(X_t)) - E_0^B(E_0^Y(L_t/\underline{\underline{Y}}_t^{h,S})\ f(X_t))|)$$
$$\leq E_0(|E_0^Y(L_t/\underline{\underline{Y}}_t^h) - E_0^Y(L_t/\underline{\underline{Y}}_t^{h,S})||f(X_t)|)$$
$$\leq \|f\|\ E_0(|E_0^Y(L_t/\underline{\underline{Y}}_t^h) - E_0^Y(L_t/\underline{\underline{Y}}_t^{h,S})|)\ .$$

Like in the above proofs, all the computations will concern the likelihood ratio $\mathbb{L}$. We use the expressions computed in Propositions 4.1 and 6.1 for the following.

$$E_0(|E_0^Y(L_t / \underline{\underline{Y}}_t^h) - E_0^Y(L_t / \underline{\underline{Y}}_t^{h,S})|)$$

$$= E_0(|\exp\{\sum_{k=1}^{n} (C_k^h.\Delta Y_k^h - \frac{h}{2}|C_k^h|^2)\} - \prod_{k=1}^{n} \sum_i \frac{\mathbb{P}_0^Y(\Delta Y_k^h - hC_k^h \in S_i)}{\mathbb{P}_0^Y(\Delta Y_k^h \in S_i)} \mathbb{1}_{\{\Delta Y_k^h \in S_i\}})$$

$$\leq E_0(|\sum_{k=1}^{n} ((C_k^h.\Delta Y_k^h - \frac{h}{2}|C_k^h|^2) - \log \sum_i \frac{\mathbb{P}_0^Y(\Delta Y_k^h - hC_k^h \in S_i)}{\mathbb{P}_0^Y(\Delta Y_k^h \in S_i)} \mathbb{1}_{\Delta Y_k^h \in S_i})|^2)^{1/2}$$

$$.4\, E_0(|L_t|^2)^{1/2} \quad .$$

$$\leq 4\, E_0(|L_t|^2)^{1/2}\, n \sup_k E_0(|(C_k^h.\Delta Y_k^h - \frac{h}{2}|C_k^h|^2) - \log \sum_i \frac{\mathbb{P}_0^Y(Y_k^h - hC_k^h \in S_i)}{\mathbb{P}_0^Y(Y_k^h \in S_i)}|^2$$

$$\mathbb{1}_{\{\Delta Y_k^h \in S_i\}})^{1/2} \quad .$$

Let us study one of the terms for k = 1,...,n :

$$E_0(|(C_k^h \Delta Y_k^h - \frac{h}{2}|C_k^h|^2) - \sum_i \log \frac{\mathbb{P}_0^Y(\Delta Y_K^h - hC_k^h \in S_i)}{\mathbb{P}_0^Y(\Delta Y_k^h \in S_i)}|^2\, \mathbb{1}_{\{\Delta Y_k^h \in S_i\}}) \quad .$$

But the term

$$E_0^Y(\sum_i \mathbb{1}_{\{\Delta Y_k^h \in S_i\}} (C_k^h.\Delta Y_k^h - \frac{h}{2}|C_k^h|^2 - \log \frac{\mathbb{P}_0^Y(\Delta Y_k^h - hC_k^h \in S_i)}{\mathbb{P}_0^Y(\Delta Y_k^h \in S_i)})^2)$$

can be majorated by a direct application of Lemma 6.3. For this purpose it is enough to notice that the random variable $\Delta Y_k^h/\sqrt{h}$ has the Normal distribution like G. The function c being supposed bounded, $\sqrt{h}\, C_k^h(\omega^B)$ is analogous to the vector C of this lemma for any fixed $\omega^B \in \Omega^B$. Similarly the partition $(S_i; i=0,..,m)$ of this lemma corresponds here to the partition $(S_i/\sqrt{h}; i=0,..,m)$, with fineness constants $R/\sqrt{h}$ and $D/\sqrt{h}$. We find then that the considered term is majorated by

$$K_1 \frac{R^2 D^2}{h^2} + K_2 (R/\sqrt{h})^{4-q}\, e^{-qR^2/2h},$$

by taking $R/\sqrt{h}$ sufficiently large, $R/\sqrt{h}$ greater than $\sqrt{h}\ C_k^h$ and D such that $RD/h < 1/24$. Then, we obtain the desired majorization by integrating on $\Omega^B$. That achieves the proof. ∎

REFERENCES


(1) K.J. ASTRÖM : Optimal control of Markov processes with incomplete state information; J. Math. Anal. and Appl. 10, 174-205 (1965).

(2) G.B. DI MASI and W.J. RUNGGALDIER : Approximations and bounds for discrete-time filtering; Preprint (1982).

(3) I.I. GIKHMAN and A.V. SKOROKHOD : Stochastic differential equations ; Springer-Verlag (1972).

(4) G. KALLIANPUR and C. STRIEBEL : Estimation of stochastic systems ; Ann. Math. Stat. 39, 785-801 (1968).

(5) H.J. KUSHNER : Probability methods for approximations in stochastic control and for elliptic equations ; Academic Press (1977).

(6) R.S. LIPTSER and A.N. SHIRYAYEV : Statistics of random processes 1 : General theory ; Springer-Verlag (1977).

(7) J.T. LO : Optimal nonlinear estimation-Part 2 : Discrete observation ; Information Sc. 7, 1-10 (1974).

(8) J. NEVEU : Martingales à temps discret ; Masson (1972).

(9) E. PARDOUX and D. TALAY : Discretization and simulation of stochastic differential equation, to be published in Acta Applicande Mathematica.

(10) J. PICARD : Approximation of nonlinear filtering problems and order of convergence ; Lect. Notes in Control and Inf. Sc. 61, Springer-Verlag (1984).

(11) E. PLATEN : Approximation of Ito integral equations ; Lect. Notes in Control and Inf. Sc. 28, 172-176, Springer-Verlag (1980).

(12) E. PLATEN : Approximation method for a class of Ito processes ; Lietuvos Mat. Rinkinys XXI-1, 121-133 (1981).

(13) Y. TAKEUCHI and H. AKASHI : Nonlinear filtering formulas for discrete time observations : SIAM J. Control and Opt. 12-2 ; 244-261 (1981).

(14) M. ZAKAI : On the optimal filtering of diffusion processes ; Z. Wahr. V. Geb. 11, 230-249 (1969).

# MARKOV JUMP-DIFFUSION MODELS AND DECISION-MAKING-FREE FILTERING

H.A.P. Blom
National Aerospace Laboratory, NLR
P.O. Box 90502, Amsterdam
The Netherlands

ABSTRACT

The problem considered is non-linear filtering of Gaussian observations of a Markov jump-diffusion with an embedded Markov chain, that is described by stochastic differential equations driven by Brownian motions and a random Poisson measure. The modelling potential of this class of Markov processes is illustrated by some simple realistic examples.
For the evolution of the conditional expectation of the Markov process decomposed representations are given. They are used as a basis to obtain approximate filtering algorithms that are free of decision making mechanisms. These algorithms are discussed for the examples given.

## 1 INTRODUCTION

Often a physical problem can be modelled realisticly by a Markov jump-diffusion with an embedded Markov chain. This is the reason that in the past a lot of attention has been paid to modelling and filtering for such Markov processes, both discrete time (Refs. 1, 2) and continuous time (Refs. 3-12). The specific problem that formed the main reason for this study is the problem of modelling and filtering trajectories of aircraft in an Air Traffic Control system.
In an Air Traffic Control system a wrong decision may be disastrous. So we wish to have a filtering algorithm that does not make use of a decision mechanism.
Existing filtering algorithms that meet this constraint have a low performance or are too complex. To gain insight into ways of bridging this complexity gap a study is started on Markov jump-diffusions that are generated by stochastic differential systems driven by a Wiener process $w_t$ and a Poisson measure, $v(dt,du)$ in U,

$$dz_t = A_t dt + B_t dw_t + \int_U C_t(u)v(dt,du) \quad , \tag{1}$$

with $A_t$, $B_t$ and $C_t$ $z_t$-adapted processes and w and v mutually independent.
Differential equations given by (1) have been studied by Gihman and Skorohod (Ref. 2). They have given sufficient conditions for z being a Markov process and representations

for the evolution of $z_t$-adapted processes.
Our interest concerns filtering Gaussian observations of a Markov process z that consists of a pair (x,m) with m a Markov chain. Further to focus the attention to the non-linearities due to the jumps, x satisfies a stochastic differential system that is linear in x.
An exact description of this Markov jump-diffusion is given in 2. To illustrate the potential of this system in modelling some examples are given. Next the problem of filtering Gaussian observations of this process is considered in 3. In the case of a finite state Markov chain m the system of equations for the conditional expectation is approximated by a finite dimensional system. This approximate filter is discussed for the examples of 2.

## 2 MODELLING OF MARKOV JUMP-DIFFUSIONS

The process considered

The most general form of the Markov jump-diffusion that will be considered consists of two components, a Markov chain $m_t$ and a semimartingale $x_t$. The Markov chain assumes values in a measurable space $M = \{m_i;\ i\in [1,N]\}$, and satisfies the stochastic differential equation,

$$dm_t = \sum_{k\in M} (k-m_t)v(dt,U_{m_t,k}) \tag{2}$$

where $v(dt,du)$ is a Poisson measure in $U\subset R^n$ with a deterministic intensity $\lambda(du)$, $\lambda(U)<\infty$ and $\sum_{k\in M} v(dt,U_{m,k}) = v(dt,U)$ for all $m\in M$. The semi-martingale x having realizations in $R^n$ satisfies the stochastic differential equation,

$$dx_t = e(m_t)dt + a(m_t)x_t dt + b(m_t)dw_t + \\ + \int_U (c(m_t,u) + d(m_t,u)x_t)v(dt,du), \tag{3}$$

where w is an l-dimensional Wiener process with mutually independent standard Brownian motion components, e and c are n-dimensional vectors, a and d are nxn-dimensional matrices and b is an nxl-dimensional matrix, the components of e, a, b, c, d are finite scalar functions on respectively M, M, M, $MxR^n$, $MxR^n$. The Wiener process w and the Poisson measure v are mutually independent. For simplicity a, b, c, d and e are time invariant.
Obviously the dependence between x and m is defined by the coefficients a, b, c, d and e in (3). When these coefficients are independent of the Markov chain the process x is a Markov process. When x is a Markov process and the components of c and d are zero outside $\bigcap_{k\in M} U_{k,k}$ then x and m are mutually independent. Our interest concerns processes x and m that are mutually dependent.

Illustrative models

As an illustration of the model that can be built with the pair (x,m) in (2) and (3) four examples are given below. In examples 1 and 2 x and m are mutually dependent Markov processes. The process x is respectively a model of Kwakernaak (Ref. 5) and a model of Au et al. (Refs. 6 and 8), while the process m is obtained in a trivial way. In example 3 x is a continuous semi-martingale and m a binary Markov chain. The model is the system with failures from Davis (Ref. 4). In example 4 x is a discontinuous semi-martingale and m a binary Markov chain. It is a simple model for suddenly accelerating aircraft.

Example 1; a river pollution model

The process m is a Poisson counting process,

$$dm_t = v(dt,U), \tag{4}$$

with $U \subset R^+$ the state space of the amount of pollutions.
The scalar process $x_t$ satisfies,

$$dx_t = ax_t dt + \int_U uv(dt,du), \quad a < 0. \tag{5}$$

Example 2; a model for the speed of a sonic wave in oil-layered earth.

A binary Markov chain m assumes values 0 and 1, 0 for earth and 1 for oil, and satisfies (2) with depth t, $U_{0,0}=U_{1,0}=U_{earth}$ and $U_{1,1}=U_{0,1}=U_{oil}$ respectively the space of earth-layers and the space of oil-layers and $\lambda(U_{earth})$, $\lambda(U_{oil})>0$.
A scalar process $x_t$ is a model for the speed of a sonic wave on depth t,

$$dx_t = \int_U (c(u)-x_t)v(dt,du), \tag{6}$$

with c a function on U and c(u) the sonic speed in a layer u.

Example 3; a linear system with failures

A binary Markov chain assumes initially the value 0 and jumps to 1 at the moment of the failure, $dm_t=(1-m_t)v(dt,U)$, with U the space of failures which contains only one element.
The scalar process x satiesfies,

$$dx_t = (a+gm_t)x_t dt + bdw_t, \; g \neq 0. \tag{7}$$

More general systems of this kind can be found easily (Refs. 4, 7, 9).

Example 4; manoeuvring aircraft

The model accounts for one axis in the airspace. A binary Markov chain m assumes 0 during uniform motion and 1 during accelerated flight; $dm_t = -m_t v(dt,U) + v(dt,U_{acc})$, with $U_{acc}$ the acceleration state space,

$$U_{acc} \subset U,\ \lambda(U) > \lambda(U_{acc}) > 0,\ \int_{U_{acc}} u\lambda(du) = 0 \text{ and}$$

$$\int_{U_{acc}} u^2 \lambda(du) = \delta_u^2\, \lambda(U_{acc}) < \infty$$

The acceleration, speed and position of the aircraft are respectively $\ddot{s}_t$, $\dot{s}_t$ and $s_t$. The process $x_t = \text{col}\{s_t, \dot{s}_t, \ddot{s}_t\}$ satisfies (3) with e=b=0, $c(m,u) = (1-m)\text{col}\{0,0,u\}$ and $d(m,u) = \text{diag}\{0,0,0\}$ for $u \in U_{acc}$, $c(m,u) = \text{col}\{0,0,0\}$ and $d(m,u) = m.\text{diag}\{0,0,-1\}$ for $u \in \bar{U}_{acc}$. Some illustrative realization of the pair (x,m) is given in figure 1.

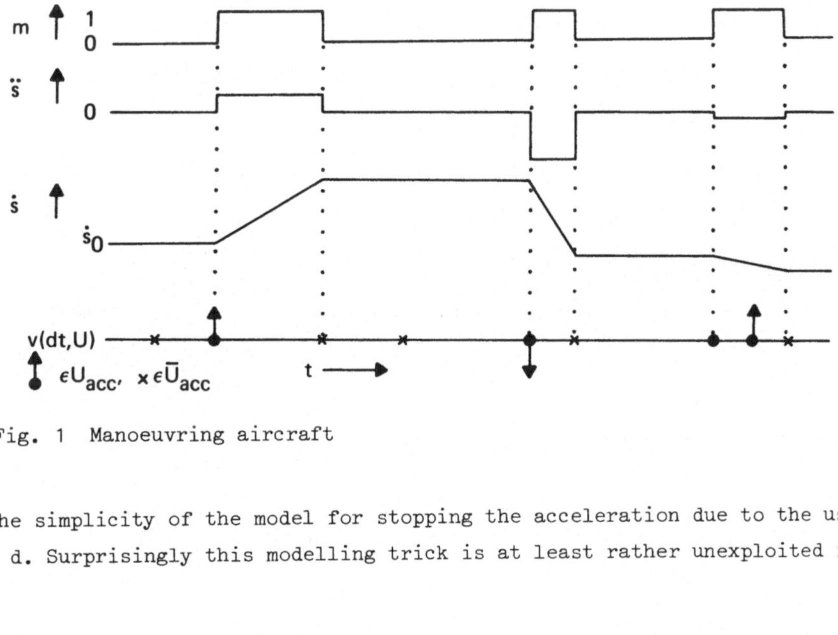


Fig. 1 Manoeuvring aircraft

Notice the simplicity of the model for stopping the acceleration due to the use of the function d. Surprisingly this modelling trick is at least rather unexploited in the past.

## 3 FILTERING OF MARKOV JUMP-DIFFUSIONS WITH AN EMBEDDED MARKOV CHAIN

### The filtering problem

Assume an increasing information field, $Y_t = \{y_s;\ s \in [0,t]\}$, of scalar observations,

$$dy_t = g(m_t)dt + h(m_t)x_t dt + db_t, \tag{8}$$

where $(x_t, m_t)$ is the pair defined in 2, $b_t$ is a standard Brownian motion, b and (x,m) are mutually independent, g is a finite real function on M and $h^T$ is an n-dimensional vector with components that are finite real functions on M.

The problem is to obtain differential equations for the conditional expectations $E^{Y_t}\{x_t\}$ and $E^{Y_t}\{m_t\}$, for which promising finite dimensional approximations exist. When $h(k)=0$ for all $k \in M$ the problem of filtering for the Markov chain m is solved by Wonham's filter (Ref. 10). When in addition x and m are mutually independent $E^{Y_t}\{x_t\}=E\{x_t\}$. When h and g are independent of m while x and m are mutually independent then $E^{Y_t}\{m_t\}=E\{m_t\}$.

For the general problem stated above exact and approximate differential equations are given in the sequel. This filtering algorithm and its relation with some previous results is discussed.

Exact filtering

Differential equations for the conditional expectation of the process (x,m) can be obtained by a repeated use of the differentiation rule for discontinuous semi-martingales and the fundamental filtering theorem for Gaussian observations of a semi-martingale (Ref. 10). A representation with potential for practical application is stated by theorems 1 and 2.

Theorem 1

For the Markov process $m_t$ in (2) a representation for the conditional expectation $E^{Y_t}\{m_t\}$ with $Y_t$ the increasing information field of observations $y_t$ in (8) is,

$$E^{Y_t}\{m_t\} = \sum_{m \in M} \hat{I}_t(m)m, \tag{9}$$

where $\hat{I}_t(m) = P\{m_t=m|Y_t\}$, $\forall m \in M$, is an $Y_t$-adapted process, which satisfies under the condition $\int_0^t E\{x_s^2\}ds < \infty$,

$$d\hat{I}_t(m) = \mathcal{L}\hat{I}_t(m)dt + \hat{I}_t(m)(\hat{H}_t(m)-\hat{H}_t)d\upsilon_t, \tag{10}$$

with:

$$\mathcal{L}i(m) = -i(m)\lambda(U) + \sum_{k \in M} i(k)\lambda(U_{k,m}) \quad , \tag{11}$$

$$\hat{H}_t(m) = g(m) + h(m)E^{Y_t}\{x_t|m_t=m\} \tag{12}$$

$$\hat{H}_t = \sum_{k \in M} \hat{I}_t(m)\hat{H}_t(m) \qquad \text{and} \tag{13}$$

$$d\upsilon_t = dy_t - \hat{H}_t dt \tag{14}$$

[ $\bar{\ }$ ]

The proof of theorem 1 is given in the Appendix.

Theorem 1 defines a finite dimensional system of stochastic differential equations with inputs $\hat{I}_0(m)$, $\hat{H}_t(m)$ and $y_t$, $\forall m \in M$, and outputs $\hat{I}_t(m)$, $E^{Y_t}\{m_t\}$, $\forall m \in M$. A representation of the inputs and outputs is given in figure 2.

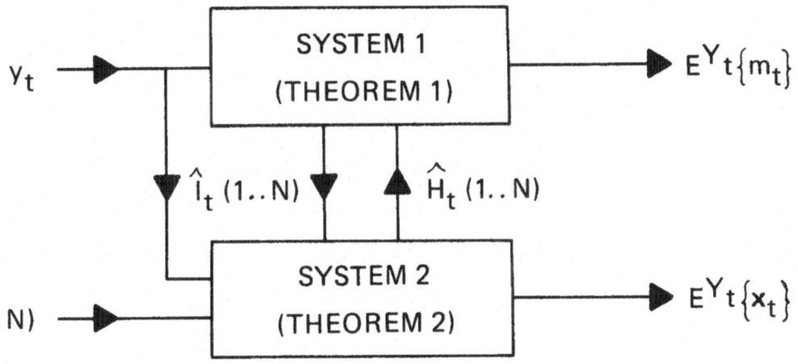


Fig. 2 The system defined by theorems 1 and 2

Verify that for $\lambda(\bigcup_{k\in M} \bar{U}_{k,k}) = 0$ theorem 1 states a version of Kailath's (Ref. 10) likelihood ratio formula. In that case system 2 will not use $\hat{I}_t(m)$ at all.

Theorem 2

For the semi-martingale $x_t$ in (3) a representation for the conditional expectation $E^{Y_t}\{x_t\}$ with $Y_t$ the increasing information field of observations $y_t$ in (8) is,

$$E^{Y_t}\{x_t\} = \sum_{m\in M} \hat{I}_t(m)\hat{x}_t(m), \tag{15}$$

where $\hat{I}_t(m)=P\{m_t = m|Y_t\}$ and $\hat{x}_t(m)=E^{Y_t}\{x_t|m_t=m\}$.

The process $\hat{x}_t(m)$ satisfies under the conditions $\int_0^t E\{x_s^T x_s\}ds<\infty$ and $\hat{I}_t(m)>0$,

$$d\hat{x}_t(m) = e(m)dt + a(m)\hat{x}_t(m)dt + \hat{r}_t(m)dt + \hat{P}_t(m)h^T(m)d\upsilon_t(m), \tag{16}$$

where: $$d\upsilon_t(m) = dy_t - \hat{H}_t(m)dt \quad , \tag{17}$$

$$\hat{P}_t(m) = E^{Y_t}\{(x_t-\hat{x}_t(m))(x_t-\hat{x}_t(m))^T|m_t=m\} \quad \text{and} \tag{18}$$

$$\hat{r}_t(m) = \hat{I}_t^{-1}(m)\sum_{k\in M}\hat{I}_t(k)(\hat{x}_t^+(k,m)-\hat{x}_t(m))\lambda(U_{k,m}), \tag{19}$$

with: $$\hat{x}_t^+(k,m) = \int_{U_{k,m}} \hat{x}_t^u(k)\lambda(du)/\lambda(U_{k,m}) \tag{20}$$

$$\hat{x}_t^u(k) = \hat{x}_t(k) + c(k,u) + d(k,u)\hat{x}_t(k) \text{ and} \tag{21}$$

$\hat{H}_t(m)$ given by (12).

In addition the process $\hat{P}_t(m)$ defined by (18) is also an $Y_t$-adapted process and satisfies

$$d\hat{P}_t(m) = a(m)\hat{P}_t(m)dt + \hat{P}_t(m)a^T(m)dt + b(m)b^T(m)dt + \hat{R}_t(m)dt +$$
$$- \hat{P}_t(m)h^T(m)h(m)\hat{P}_t(m)dt + \hat{P}_{3t}(m)h^T(m)d\upsilon_t(m) \quad , \tag{22}$$

where: $$\hat{P}_{3t}(m) = E^{Y_t}\{(x_t-\hat{x}_t(m))(x_t-\hat{x}_t(m))^T(x_t-\hat{x}_t(m))^T|m_t=m\} \quad , \tag{23}$$

$$\hat{R}_t(m) = \hat{I}_t^{-1}(m) \sum_{k\in M} \hat{I}_t(k)(\hat{P}_t^+(k,m)-\hat{P}_t(m)+(\hat{x}_t^+(k,m)-\hat{x}_t(m)).$$

$$\cdot (\hat{x}_t^+(k,m)-\hat{x}_t(m))^T\lambda(U_{k,m}) \quad , \tag{24}$$

with

$$\hat{P}_t^+(k,m) = \int_{U_{k,m}} (\hat{P}_t^u(k)+(\hat{x}_t^u(k)-\hat{x}_t^+(k,m))(\hat{x}_t^u(k)-\hat{x}_t^+(k,m))^T).\lambda(du)/\lambda(U_{k,m}) \tag{25}$$

and

$$\hat{P}_t^u(k) = (I+d(k,u))\hat{P}_t(k)(I+d(k,u))^T. \tag{26}$$

[ $\overline{\phantom{x}}$ ]

The proof of theorem 2 is given in the Appendix.

Theorem 2 defines a system of stochastic differential equations with inputs $\hat{x}_0(m)$, $\hat{P}_0(m)$, $\hat{I}_t(m)$, $\hat{P}_{3t}(m)$ and $y_t$, $\forall m\in M$, and outputs $E^Yt\{x_t\}$ and $\hat{H}_t(m)$, $\forall m\in M$. A representation of these inputs and outputs is given in figure 2.

A further interpretation of theorem 2 learns that system 2 consists of N equivalent subsystems. Each subsystem is described by the $n+n^2$ scalar differential equations of (16) and (22). Without the term $\hat{r}_t(m)dt$, $\hat{R}_t(m)dt$ and $\hat{P}_{3t}(m)h^T(m)d\upsilon_t(m)$ each subsystem would be a Kalman filter. The terms $\hat{r}_t(m)$ and $\hat{R}_t(m)$ in (19) and (24) describe the interaction between the N subsystems. This interaction is governed by the conditional probabilities of the Markov chain. It can be expected that system 2 is a very adaptive filter because of this interaction.

The terms $\hat{r}_t(m)$ and $\hat{R}_t(m)$ in (19) and (24) are rather complicated. However, for simple realistic models the combination of the Markov chain, the intensity function $\lambda$ and the coefficients c and d will have a lot of geometric structure, by which (19) and (24) can be simplified significantly.

From figure 2 it can easily be seen that to build a system with $y_t$ the only input and outputs $E^Yt\{m_t\}$ and $E^Yt\{x_t\}$, systems 1 and 2 can only be used when we have a third system with $y_t$ and the outputs of systems 1 and 2 the only inputs but $\hat{P}_{3t}(m)$, $\forall m\in M$, the output. Apart from some trivial cases such a system is infinite dimensional.

For practical use of the system in figure 2 we have to accept approximations of $\hat{P}_{3t}(m)$ and subsequently of other conditional expectations. The most simple approximation $\hat{P}_{3t}(m) = 0$ will be discussed in the sequel.

## Approximate filtering

Assume the approximation $\overline{P}_{3t}(m)$, of $\hat{P}_{3t}(m)$ is $\overline{P}_{3t}(m) = 0$, $\forall m\in M$. In that case systems 1 and 2 define a filtering algorithm that consists of $N(n^2+n+1)$ scalar differential equations. This filtering algorithms is discussed for the four examples of 2.

For simplicity the observation equation (8) has coefficients $g(m_t)=0$ and $h(m_t)=h$,

$$dy_t = hx_t dt + db_t \tag{27}$$

For examples 1 and 3 some relations with the results of respectively Kwakernaak (Ref. 5) and Davis (Ref. 4) will be given. As in examples 2, 3 and 4 the embedded Markov chain, $m_t$, is in {0,1}, $\hat{I}_t(1)$, $\hat{I}_t(0)$, $\overline{I}_t(1)$ and $\overline{I}_t(0)$ are replaced by respectively $\hat{m}_t$, $1-\hat{m}_t$, $\overline{m}_t$ and $1-\overline{m}_t$. For simplicity the initial conditions are not stated.

Example 1

Given (4), (5) and (27) the approximation proposed yields a filter consisting of a growing number (N) of interacting Kalman-like filters. For practical application some form of approximation has to be introduced to overcome the problem of growing complexity of the algorithm. Probably this can be done acceptably well by using a decision making mechanism, but in that case the resulting algorithm conflicts the constraints assumed initially.

The filtering algorithms given by Kwakernaak (Ref. 5) satisfy the constraints, but they do not give the conditional expectation of the counting process $m_t$.

When decision making methods are acceptable, for this problem good algorithms exist (Ref. 2).

Example 2

For the model of example 2 in 2 and observations according to (27) the filtering algorithm proposed consists of five scalar differential equations, for $\overline{m}_t$, $\overline{x}_t(1)$, $\overline{P}_t(1)$, $\overline{x}_t(0)$ and $\overline{P}_t(0)$:

$$d\overline{m}_t = (1-\overline{m}_t)\lambda_1 dt - \overline{m}_t\lambda_0 dt + \overline{m}_t h(\overline{x}_t(1)-\overline{x}_t)(dy_t-h\overline{x}_t dt) \tag{28}$$

with

$$\lambda_1 = \lambda(U_{oil}),\ \lambda_0 = \lambda(U_{earth}),$$
$$\overline{x}_t = \overline{x}_t(0) + \overline{m}_t(\overline{x}_t(1)-\overline{x}_t(0)), \tag{29}$$

$$d\overline{x}_t(1) = \overline{r}_t(1)dt + \overline{P}_t(1)h(dy_t-h\overline{x}_t(1)dt), \tag{30}$$

$$d\overline{P}_t(1) = \overline{R}_t(1)dt - h^2\overline{P}_t^2(1)dt \quad \text{, where} \tag{31}$$

$$\overline{r}_t(1) = \overline{m}_t^{-1}(E\{c(u)|u\in U_{oil}\}-\overline{x}_t(1))\lambda_1, \text{ and} \tag{32}$$

$$\overline{R}_t(1) = \overline{m}_t^{-1}(E\{c^2(u)-E^2\{c(u)|u\in U_{oil}\}|u\in U_{oil}\}-\overline{P}_t(1) + + (E\{c(u)|u\in U_{oil}\} -\overline{x}_t(1))^2)\lambda_1 \tag{33}$$

and similar equations for $d\overline{x}_t(0)$ and $d\overline{P}_t(0)$.

For the application to oil exploration it is necessary to extend the present results to smoothing, by which we obtain smoothed estimates of m and x. Then the final decision making procedure can be done independently of the smoothing algorithm. This is in sharp contrast with the methods developed by Au et al. (Refs. 6,8) for such a problem.

So the present result throws at least new light on the problem.

Example 3

For the model of example 3 in 2 and observations according to (27) the filtering algorithm proposed consists of five scalar differential equations for $\overline{m}_t$, $\hat{x}_t(0)$, $\hat{P}_t(0)$, $\overline{x}_t(1)$ and $\overline{P}_t(1)$. The one for $\overline{m}_t$ satisfies (28), with $\lambda_1=\lambda$ and $\lambda_0=0$, and (29). The two for $\hat{x}_t(0)$ and $P_t(0)$ are the exact Kalman equations for the system without failure. The two scalar differential equations for $\overline{x}_t(1)$ and $\overline{P}_t(1)$ satisfy,

$$d\overline{x}_t(1) = (a+g)\overline{x}_t(1)dt + \overline{r}_t(1)dt + \overline{P}_t(1)h(dy_t-h\overline{x}_t(1)dt), \quad (34)$$

$$d\overline{P}_t(1) = 2(a+g)\overline{P}_t(1)dt + b^2dt + \overline{R}_t(1)dt - h^2\overline{P}_t^2(1)dt, \quad (35)$$

with:

$$\overline{r}_t(1) = (\overline{m}_t^{-1}-1)(\hat{x}_t(0) - \overline{x}_t(1))\lambda \quad (36)$$

$$\overline{R}_t(1) = (\overline{m}_t^{-1}-1)(\hat{P}_t(0) - \overline{P}_t(1) + (\hat{x}_t(0) - \overline{x}_t(1))^2)\lambda \quad (37)$$

Notice that for this algorithm only one approximation has been introduced: $\overline{P}_{3t}(1) = 0$. For this specific example we also propose a second filtering algorithm. It is obtained by approximating $\hat{P}_t(1)$ in the exact equation for $d\hat{x}_t(1)$ by $\hat{P}_t(0)$. So it consists of four scalar differential equations for $\hat{x}_t(0)$, $\hat{P}_t(0)$, $\overline{m}_t(1)$ and $\overline{x}_t(1)$, which is one less than the first algorithm because of the delection of $d\hat{P}_t(1)$.
Finally we show the relation between the second algorithm with the one proposed by Davis (Ref. 4). His algorithm can be obtained by introducing some approximations in the exact equations for $d\hat{x}_t(0)$, $d\hat{P}_t(0)$, $d\hat{m}_t$ and $dE^Yt\{m_tx_t\}$, where $dE^Yt\{m_tx_t\}$ is given in the proof of theorem 2. The subsequent approximations are:

1 Approximate $\widehat{q_t(1)H_t}$ in $dE^Yt\{m_tx_t\}$ by $\hat{m}_t\hat{P}_t(0)h$.

2 Approximate $\hat{x}_t$ in $d\hat{m}_t$ and $dE^Yt\{m_tx_t\}$ by $\hat{x}_t(0)$.

3 Modify the filter after the detection of the failure.

By comparing these approximating steps with the approximating step for the second filtering algorithm above, step 2 appears to be unnecessary at all. It can easily be verified that this is due to the fact that Davis did not exploit the equalities $E^Yt\{m_tx_t\} = \hat{m}_t\hat{x}_t(1)$ and $E^Yt\{m_tx_t^2\} = \hat{m}_tE^Yt\{x_t^2(1)\}$, which are proven in the Appendix. Verify that the results above can be generalized to other systems with Markovian coefficients (Refs. 7, 9), certainly for systems with failures.

Example 4

For the model of example 4 in 2 and observations according to (27) the filtering algorithm proposed consists of differential equations for $\overline{m}_t$, $\overline{x}_t(0)$, $\overline{P}_t(0)$, $\overline{x}_t(1)$ and

$\bar{P}_t(1)$. The differential equation for $\bar{m}_t$ is given by (28) with $\lambda_1=\lambda(U_{acc})$ and $\lambda_0=\lambda(U)-\lambda_1$. The differential equations for $\bar{x}_t(i)$, $\bar{P}_t(i)$, $i\in\{0,1\}$ satisfy,

$$d\bar{x}_t(i) = a\bar{x}_t(i)dt + \bar{r}_t(i)dt + \bar{P}_t(i)(dy_t-h\bar{x}_t(i)dt), \quad (38)$$

$$d\bar{P}_t(i) = a\bar{P}_t(i)dt + \bar{P}_t(i)a^Tdt + \bar{R}_t(i)dt - \bar{P}_t(i)hh^T\bar{P}_t(i)dt, \quad (39)$$

with interacting terms

$$\bar{r}_t(0) = (1-\bar{m}_t)^{-1}\bar{m}_tA(\bar{x}_t(1)-\bar{x}_t(0))\lambda_0 \quad , \quad (40)$$

$$\bar{r}_t(1) = \bar{m}_t^{-1}(1-\bar{m}_t)(\bar{x}_t(0)-\bar{x}_t(1))\lambda_1 \quad , \quad (41)$$

$$\bar{R}_t(0) = (1-\bar{m}_t)^{-1}\bar{m}_tA(\bar{P}_t(1)-\bar{P}_t(0) + (\bar{x}_t(1)-\bar{x}_t(0))(\bar{x}_t(1)-\bar{x}_t(0))^T)A\lambda_0 \quad (42)$$

$$\bar{R}_t(1) = \bar{m}_t^{-1}(1-\bar{m}_t)(\bar{P}_t(0) + B-\bar{P}_t(1) + (\bar{x}_t(0)-\bar{x}_t(1)(\bar{x}_t(0)-\bar{x}_t(1))^T)\lambda_1 \quad (43)$$

with $A = \text{diag}\ \{1,1,0\}$ and $B = \text{diag}\ \{0,0,\sigma_u^2\}$.

Notice that the third component of $\bar{x}_t(0)$ and the third column of $\bar{P}_t(0)$ are zero, by which their differential equations define a two-dimensional Kalman-like filter, while the differential equations for $\bar{x}_t(1)$ and $\bar{P}_t(1)$ define a three-dimensional Kalman-like filter.

The method of modelling and filtering that is illustrated by this example is exploited very succesfully for the specific problem mentioned in the introduction (Ref. 14).

## 4 CONCLUDING REMARKS

### Modelling results

Markov jump-diffusions $(x_t,m_t)$ with an embedded Markov chain $m_t$ are realistic models for a lot of physical processes. This has been illustrated by some simple examples with $x_t$ and $m_t$ mutually dependent and $(x_t,m_t)$ described by a finite dimensional system of stochastic differential equations. In these four examples $x_t$ is respectively a Markov jump-diffusion, a Markov jump process, a continuous semi-martingale and a discontinuous semi-martingale.

The first three examples are well-known models. The last example illustrates a very effective modelling trick, that is at least rather unusual. The example accounts for the well-known problem of modelling aircraft trajectories.

### Filtering results

Given Gaussian observations of the pair $(x_t,m_t)$ filtering representations are given and an approximate filtering algorithm is proposed. The algorithm consists of a number of interacting Kalman-like filters, one for each possible state of $m_t$, and a Markov chain filter. In comparison with previous results (Refs. 4-8) the present results give a new look on the problem of filtering the pair $(x_t,m_t)$:

- The joint filtering of x and m even in some cases that x is a Markov process and the observations are independent of m.

- Decision-making-free algorithms proposed for linear systems with Markovian coefficients (Refs. 4, 7, 9).
- A decision-making-free algorithm proposed for discontinuous semi-martingales x.
- The output of the decision-making-free algorithms can be used for the detection of jumps of m.

## Implementation aspects

Both for a systematic evaluation and for the actual use of the filtering algorithm some implementation problems have to be overcome. First of all the stochastic differential equations for the conditional probabilities of the Markov chain involve multiplications of a semi-martingale with its own martingale part. By the introduction of suitable transformations the associated implementation complications can be overcome (Ref. 13). Secondly we need to develop a stable and efficient numerical integration scheme to implement the filtering algorithm in a computer. This problem can also be formulated as to develop discrete-time models and filters that are similar to the continuous time ones. It is relevant to verify that with such a result we should have improved the discrete-time Generalized Pseudo Bayesian Algorithm, (GPBA, Ref. 1):

- GPBA has the structure of the present algorithm, but
- GPBA uses at least $N^2$ Kalman-like filters instead of N.
- GPBA is for systems with Markovian coefficients only.

By now this discrete time problem is solved and the results will be given in a future publication.

## Practical application

Although the structure of the algorithm proposed is promising, the actual use of the algorithm for several kinds of problems has to be justified by a systematic evaluation. An important question thereby is for which kind of problems the decision-making-free algorithm proposed performs better than algorithms that use a decision mechanism. For the problem of filtering trajectories of an aircraft for air traffic control the algorithm has been evaluated in depth (Ref. 14), which resulted in a definite positive justification of its implementation in air traffic control systems.

# APPENDIX

## Proof of theorem 1

As the proof is straighforward only an outline of it is given. First define for $\forall m \in M$ an indicator function, $I_t(m)=1$, $m_t=m$, and $I_t(m)=0$, $m_t \neq m$. With this we have for an arbitrary information field B, $P\{m_t =m|B\}=E^B\{I_t(m)\}$, which yields (9).
The indicator function satisfies the stochastic differential equation,

$$dI_t(m) = -I_t(m)v(dt,U) + \sum_{k \in M} I_t(k)v(dt,U_{k,m}), \; \forall m \in M.$$

By applying the fundamental filtering theorem for Gaussian observations (Ref. 10) the proof of theorem 1 is completed.

Proof of theorem 2

From the definition of $I_t(m)$ in the proof of theorem 1,

$$E^{Y_t}\{x_t\} = E^{Y_t}\{\sum_{k\in M} I_t(k)x_t\} = \sum_{k\in M} E^{Y_t}\{I_t(k)x_t\},$$

by which the filtering problem has been reduced to the filtering of

$$q_t(m) \stackrel{d}{=} I_t(m)x_t,\ \forall m\in M.$$

A differential equation for $q_t(m)$ follows from the differential equations for $I_t(m)$ and $x_t$,

$$dq_t(m) = I_t(m)(e(m_t)dt + a(m_t)x_t dt + b(m_t)dw_t) - I_t(m)x_t v(dt,U) +$$

$$+ \sum_{k\in M} \int_{U_{k,m}} I_t(k)(c(m_t,u) + d(m_t,u)x_t + x_t)v(dt,du).$$

Applying the fundamental filtering theorem yields a differential equation for $\hat{q}_t(m) = E^{Y_t}\{q_t(m)\}$,

$$d\hat{q}_t(m) = \hat{I}_t(m)e(m)dt+a(m)\hat{q}_t(m)dt-\hat{q}_t(m)\lambda(U)dt+(\widehat{q_t(m)H_t}-\hat{q}_t(m)\hat{H}_t)d\upsilon_t+$$

$$+ \sum_{k\in M} \int_{U_{k,m}} (\hat{I}_t(k)c(k,u)+d(k,u)\hat{q}_t(k)+\hat{q}_t(k))\lambda(du)dt,$$

with $\hat{H}_t=E^{Y_t}\{H_t\}$, $H_t=g(m_t)+ h(m_t)x_t$ and $d\upsilon_t= dy_t-\hat{H}_t dt$.

Next we define on M $\hat{x}_t(m) = \hat{I}_t^{-1}(m)\hat{q}_t(m)$ if $\hat{I}_t(m) > 0$ and $\hat{x}_t(m)=\hat{x}_t$ if $\hat{I}_t(m)=0$, by which for $\forall m\in M$ with $\hat{I}_t(m)>0$,

$$\hat{x}_t(m) = \hat{I}_t^{-1}(m)E^{Y_t}\{I_t(m)x_t\} = \hat{I}_t^{-1}(m) \sum_{k\in M} \hat{I}_t(k)E^{T_t}\{I_t(m)x_t|m_t=k\}=$$

$$= \hat{I}_t^{-1}(m)\hat{I}_t(m)E^{Y_t}\{x_t|m_t=m\} = E^{Y_t}\{x_t|m_t=m\},$$

which yields (15).

By applying Itô's differentiation rule on $\hat{x}_t(m) = \hat{I}_t^{-1}(m)\hat{q}_t(m)$ the differential equation (16) for $\hat{x}_t(m)$ follows from $d\hat{I}_t(m)$ and $d\hat{q}_t(m)$. The proof proceeds by deriving a differential equation for $\hat{P}_t(m)$, defined in (18).

First derive a differential equation for

$$\hat{Q}_t(m) \stackrel{d}{=} E^{Y_t}\{q_t(m)q_t^T(m)\}-\hat{x}_t(m)\hat{q}_t^T(m) \text{ from } dq_t(m),\ d\hat{x}_t(m) \text{ and } d\hat{q}_t(m).$$

By its definition holds,

$$\hat{Q}_t(m) = E^{Y_t}\{I_t(m)x_t x_t^T\} - \hat{I}_t(m)\hat{x}_t(m)\hat{x}_t^T(m) =$$

$$= \sum_{k\in M} \hat{I}_t(k)E^{Y_t}\{I_T(m)x_t x_t^T|m_t=k\} - \hat{I}_t(m)\hat{x}_t(m)\hat{x}_t^T(m) =$$

$$= \hat{I}_t(m)E^{Y_t}\{x_t x_t^T|m_t=m\} - \hat{I}_t(m)\hat{x}_t(m)\hat{x}_t^T(m) =$$

$$= \hat{I}_t(m)\hat{P}_t(m).$$

Finally define on M $\hat{P}_t(m)=\hat{I}_t^{-1}(m)\hat{Q}_t(m)$ if $\hat{I}_t(m)>0$ and $\hat{P}_t(m) = \hat{P}_t$ if $\hat{I}_t(m)=0$, and apply Itô's differentiation rule on $d\hat{Q}_t(m)$ and $d\hat{I}_t(m)$ to obtain (20).
With this result the proof of theorem 2 is completed.

REFERENCES


1 Tugnait J.K., Detection and estimation for abruptly changing systems. Automatica, 18 (1982), 607-615.
2 Basseville M., Benveniste A., Design and comparative study of some sequential jump detection algorithms for digital signals. IEEE Transactions on ASSP, 31 (1983), 521-535.
3 Gihman I.I., Skohorod A.V., Stochastic differential equations, Springer-Verlag, Berling, 1972.
4 Davis M.H.A., The application of nonlinear filtering to fault detection in linear systems, IEEE Transactions on Automatic Control, 20 (1975), 257-259.
5 Kwakernaak H., Filtering for systems excited by Poisson white noise. Eds: A. Bensoussan, J.L. Lions, Control theory, numerical methods and computer systems; Springer Lecture Notes in Economics and Mathematical Systems, Vol. 107, Berlin, 1975, 468-492.
6 Au S.P., Haddad A.H., Suboptimal sequential estimation-detection scheme for Poisson driven linear systems. Information Sciences, 16 (1978), 95-113.
7 Loparo K.A., Roth Z., Suboptimal nonlinear filters for systems with random structure, IEEE Conf. on Decision and Control, 1981, 320-323.
8 Au S.P., Haddad A.H., Poor H.V., A state estimation algorithm for linear systems driven simultaneously by Wiener and Poisson processes, IEEE Trans. on Automatic Control, 27 (1982), 617-626.
9 Brockett R.W., Blankenship G.L., A representation theorem for linear differential equations with Markovian coefficients, Proc. 1977 Allerton Conf. on Circuits and Systems Theory, Urbana, Illinois, 671-679.
10 Liptser R.S., Shiryayev A.N., Statistics of random processes, Part I and II, Springer Verlag, 1977, 1978.
11 Marcus S.I., Modeling and analysis of stochastic differential equations driven by point processes, IEEE Transactions on Information Theory, 24 (1978), 164-172.
12 Snyder D.L., Random point processes, New York, Wiley, 1975.
13 Davis M.H.A., Pathwise nonlinear filtering, Eds: M. Hazewinkel, J.C. Willems, Stochactic systems: the mathematics of filtering and indentification and applications, Proc. of the NATO Advanced Study Institute, Les Arcs, 1980, D. Reidel, Dordrecht, 1981, 505-528.
14 Blom H.A.P., A sophisticated tracking algorithm for ATC surveillance radar data, Proceedings of the International Conference on Radar, Paris, may 1984, session B III.

# NONLINEAR FILTERING FOR MARKOV PROCESSES: AN $L^2$ APPROACH

A. Germani[*] and M. Piccioni[**]

* Istituto di Analisi dei Sistemi ed Informatica del C.N.R., Viale Manzoni 30, 00185 Roma, Italy

** Istituto di Matematica, Informatica e Sistemistica dell'Università di Udine, Via Mantica 3, 33100 Udine Italy

ABSTRACT

In this paper the Zakai equation of nonlinear filtering is directly derived as a mild stochastic differential equation on a Hilbert space. This is established when the state process is Markov, with a generator on some $L^2$ space, and the observation process is corrupted by white noise. The main step is the derivation of a Feynman-Kac like formula for mild stochastic differential equations. In such a way well-known results in literature are generalized and at the same time their proofs are made much more simpler; moreover the Hilbert space setting is the most appropriate for approximation procedures. A final example is given in which the Zakai equation is written for a stochastic differential system in which the state equation is linear with an arbitrary number of noises, partly overcoming one of the up to now most serious limitations of such theory.

## 1. INTRODUCTION

Recently, some general results about approximation of stochastic bilinear differential equations in a mild form have been given [1]. These results have allowed to prove the convergence of Galerkin procedures when the state is a diffusion process whose generator is strongly elliptic [2]. In the general case of Markov processes, with the observation process corrupted by additive white noise, it is required that the corresponding Zakai equation [3] evolves in a $L^2$-space of densities on the state space. In this paper it is pointed out that this is possible provided that the same property is assumed for the Fokker-Planck equation.

Under such hypothesis a Feynman-Kac like formula is readily derived for the solution of a backward mild stochastic differential equation obtained by a stochastic perturbation of the Markov process generator. For the case of diffusion processes see [4,5,6], whereas in [7] the case of Markov processes with a more general discussion is reported. Moreover the Zakai equation, obtained with a standard argument as the adjoint of the previous one, is derived directly in the mild form, so that all the existence and uniqueness problems are more easily solvable.

## 2. MILD FORMS OF ABSTRACT STOCHASTIC BILINEAR PDE's

The first section deals with a general stochastic bilinear PDE on a real separable Hilbert space H. Let A be the infinitesimal generator of a $C_o$-semigroup $\{S(t)\}$ of bounded linear operators on H and let B be a linear bounded operator on H.

Let $(\Theta, F, \mu)$ be a probability space and let $\{W_t\}$ be a standard Wiener process on it. Denote also by $F^W_{s,t}$ the $\sigma$-algebra generated by the increments $\{W_u - W_v,\ s \leq v \leq u \leq t\}$. For any fixed $x_o \in H$ let us consider the stochastic bilinear PDE in the mild form

$$X_t = S(t)x_o + \int_0^t S(t-\tau)BX_\tau dW_\tau \qquad t \in [0,T] \tag{1}$$

This equation is studied in the subspace of $F^W_{0,t}$-adapted process in $C(0,T;L^2(\Omega;H))$, which will be denoted by $Z^W_f(0,T;H)$. This is a Banach space with the norm

$$|||x|||_T = \sup_{t\in[0,T]} (E\|x_t\|^2)^{1/2} \tag{2}$$

The existence and uniqueness of the solution of (1) in $Z^W_f(0,T;H)$ is known [8] but a Volterra series expansion of it is needed [1,4,9]. First of all note that if X is in $Z^W_f(0,T;H)$ the stochastic integral remains in this space [2]; therefore equation (1) can be intended as an equality in $Z^W_f(0,T;H)$.

THEOREM 1. *The solution of equation* (1) *is unique and it is given by*

$$X_t = S(t)x_o + \sum_{k=1}^{\infty} \int_0^t \cdots \int_0^{t_{k-1}} S(t-t_1)B\ldots S(t_{k-1}-t_k)BS(t_k)x_o dW_{t_k}\ldots dW_{t_1} \tag{3}$$

For the proof see [1]. □

For the sequel the family of equations adjoint to (1) will be extensively used. They are backward stochastic PDE's, in the sense that the solution is adapted to the future increments of $\{W_t\}$. More precisely define by $Z_b^W(0,t;H)$ the closed subspace of $F_{s,t}^W$-adapted process in $C(0,t;L^2(\Omega;H))$, endowed with the norm $|||\cdot|||_t$. For processes in $Z_b^W(0,t;H)$ a backward stochastic integral can be defined. The easiest way is the following: let

$$V_s = W_t - W_{t-s}\ , \quad 0 \le s \le t,$$

then $\{V_s,\ 0 \le s \le t\}$, is a standard Wiener process. For $z \in Z_b^W(0,t;H)$

$$\int_s^t z_\sigma \oplus dW_\sigma = \int_0^{t-s} z_{t-\tau}dV_\tau \tag{4}$$

which defines, for $s \in [0,t]$, a process in $Z_b^W(0,t;H)$.

Then it is possible to consider the equation

$$z_s^t = S^*(t-s)f + \int_s^t S^*(\sigma-s)B^*z_\sigma^t \oplus dW_\sigma \tag{5}$$

in $Z_b^W(0,t;H)$, where $f \in H$ is fixed.

THEOREM 2. *The solution of equation* (5) *is unique and it is given by*

$$z_s^t = S^*(t-s)f + \sum_{k=1}^{\infty} \int_s^t \dots \int_{s_{k-1}}^t S^*(s_1-s)B^*\dots S^*(s_k-s_{k-1})B^*S^*(t-s_k)f \oplus$$
$$\oplus\, dW_{s_k} \dots \oplus dW_{s_1} \tag{6}$$

PROOF. The backward equation (5) is equivalent to the forward one

$$z_{t-\theta}^t = S^*(\theta)f + \int_0^\theta S^*(\theta-\tau)B^*z_{t-\tau}^t dV_\tau$$

which by Theorem 1 has the unique solution in $Z_b^W(0,t;H)$ given by

$$z_{t-\theta}^t = S^*(\theta)f + \sum_{k=1}^{\infty} \int_0^\theta \dots \int_0^{t_{k-1}} S^*(\theta-t_1)B^*\dots S^*(t_{k-1}-t_k)B^*S^*(t_k)f\, dV_{t_k}\dots dV_{t_1} \tag{7}$$

so that, with the position $t-\theta = s$ and applying the definition in (4), one has the equation (6). □

The sense in which equation (5) is adjoint to equation (1) is given by the following theorem, which will be used in the sequel.

THEOREM 3. *Let* $\{X_t\}$ *and* $\{Z_s^t\}$ *be as given in* (3) *and* (6) *respectively. Then for* $t \in [0,t]$

$$(X_t,f) = (x_o,Z_o^t) \quad \text{a.e.} \tag{8}$$

PROOF. By direct computation

$$(X_t,f) = (S(t)x_o,f)+\sum_{k=1}^{\infty}\int_0^t\cdots\int_0^{t_{k-1}}(S(t-t_1)B\ldots S(t_{k-1}-t_k)BS(t_k)x_o,f)\cdot$$

$$\cdot\, dW_{t_k}\ldots dW_{t_1} = (S(t)x_o,f)+\sum_{k=1}^{\infty} I_k(\chi_{0\le t_k\le t_{k-1}\le\ldots\le t_1\le t}(S(t-t_1)B\ldots$$

$$\ldots S(t_{k-1}-t_k)BS(t_k)x_o,f)) \tag{9}$$

where $I_k$ is the multiple Wiener integral of order k in $[0,t]^k$ [10] and $\chi$ is a characteristic function. Similarly

$$(x_o,Z_o^t) = (x_o,S^*(t)f) + \sum_{k=1}^{\infty}\int_0^t\cdots\int_{s_{k-1}}^t (x_o,S^*(s_1)B^*\ldots$$

$$\ldots S^*(s_k-s_{k-1})B^*S^*(t-s_k)f) \oplus dW_{s_k}\ldots \oplus dW_{s_1} =$$

$$= (S(t)x_o,f) + \sum_{k=1}^{\infty} I_k(\chi_{0\le s_1\le s_2\le\ldots\le s_k\le t}(S(t-s_k)B\ldots$$

$$\ldots S(s_2-s_1)BS(s_1)x_o,f)) \tag{10}$$

as it is easily verified following the same argument as in [10]. Finally, the theorem is proved because for each k the integrands in (9) and (10), respectively, have the same symmetrization on $[0,t]^k$ and therefore the same Wiener integral [10]. □

## 3. A FEYNMAN-KAC LIKE FORMULA

The background results of the previous section are now applied to the case of Markov semigroups. Let us suppose that $\{\xi_t,\ t\in[0,T]\}$ is a time-homogeneous regular measurable Markov process on $(\Omega^\xi,F^\xi,P^\xi)$, taking values in a Hilbert space K, endowed with a $\sigma$-finite measure m on its Borel $\sigma$-algebra. Let $\{P_{tx},\ t\in[0,T],\ x\in K\}$, be the family of transition measures of $\{\xi_t\}$ and suppose that the semigroup $\Gamma$ defined by

$$(\Gamma(t)f)(x) = \int_{\Omega\xi} f(\xi_t(\omega))dP_{ox}(\omega) = E_{ox}\{f(\xi_t)\} \tag{11}$$

is $C_o$ on $L^2(K,m)$. Let h be a bounded measurable function on K, so that the multiplication by h defines a linear bounded self-adjoint operator on $L^2(K,m)$ which will be denoted by h itself. Let $\{Y_t,\ t \in [0,T]\}$ be a Wiener process on some probability space $(\Omega^Y, F^Y, \lambda)$. Then the following equation of the type (5) is studied in $Z_b^Y(0,t;L^2(K,m))$

$$Z_s^t = \Gamma(t-s)f + \int_s^t \Gamma(\sigma-s)hZ_\sigma^t \oplus dY_\sigma \tag{12}$$

for $f \in L^2(K,m)$. For the solution of such equation the following theorem gives a representation which is similar to the Feynman-Kac formula for deterministic perturbed linear equations [11].

THEOREM 4. *The unique solution in* $Z_b^Y(0,t;L^2(K,m))$ *of equation* (12) *is given by*

$$Z_s^t(x) = E_{sx}[f(\xi_t)\exp(\int_s^t h(\xi_\tau)dY_\tau - \frac{1}{2}\int_s^t |h(\xi_\tau)|^2 d\tau)] \tag{13}$$

*where the stochastic integral is a Wiener integral in* $\Omega^Y$, *for each path of the process* $\{\xi_\tau,\ \tau \in [s,t]\}$.

PROOF. Let us compute the k-th Kernel of the Volterra series expansion for the solution of (12)

$$[\Gamma(s_1-s)h\ldots\Gamma(s_k-s_{k-1})h\Gamma(t-s_k)f](x) = \{\Gamma(s_1-s)h\ldots$$

$$\ldots\Gamma(s_k-s_{k-1})hE_{s_k\cdot}[f(\xi_t)]\}(x) =$$

$$= \{\Gamma(s_1-s)h\ldots\Gamma(s_k-s_{k-1})E_{s_k\cdot}[h(\xi_{s_k})f(\xi_t)]\}(x) =$$

$$= \{\Gamma(s_1-s)h\ldots E_{s_{k-1}\cdot}[h(\xi_{s_k})f(\xi_t)]\}(x) =$$

$$= E_{sx}[h(\xi_{s_1})\ldots h(\xi_{s_k})f(\xi_t)],\quad s \le s_1 \le \ldots \le s_k \le t, \tag{14}$$

where the definition (11), the time-homogeneity and the Markov property of $\{\xi_t\}$ have been exploited repeatedly.

Now let $\phi \in L^2(K,m)$ such that $\phi \geq 0$ a.e. and $\int_K \phi dm = 1$: then for $k = 1,2,\ldots$

$$\int_s^t \cdots \int_{s_{k-1}}^t (\Gamma(s_1-s)h\ldots\Gamma(s_k-s_{k-1})h\Gamma(t-s_k)f,\phi) \oplus dY_{s_k} \oplus \ldots \oplus dY_{s_1} =$$

$$= I_k(\chi_{s\le s_1\le\ldots\le s_k\le t} \int_K E_{sx}[h(\xi_{s_1})\ldots h(\xi_{s_k})f(\xi_t)]\phi(x)dm(x)) =$$

$$= I_k(E_\nu[\chi_{s\le s_1\le\ldots\le s_k\le t}\ h(\xi_{s_1})\ldots h(\xi_{s_k})f(\xi_t)]) =$$

$$= E_\nu(f(\xi_t)I_k(\chi_{s\le s_1\le\ldots\le s_k\le t}\ h(\xi_{s_1})\ldots h(\xi_{s_k}))) \tag{15}$$

where $\nu(A) = \int_K P_{sx}(A)\phi(x)dm(x)$ is a probability measure on $(\Sigma,\sigma\{X_\tau,$ $s \le \tau \le t\})$, and a Fubini theorem for multiple Wiener integrals is used. Therefore the unique solution $\{Z_s^t\}$ of equation (12) has the property:

$$(Z_s^t,\phi) = \lim_{n\to\infty} J_{s,t}^{(n)}, \quad t \in [0,T] \tag{16}$$

uniformly in mean square, where

$$J_{s,t}^{(n)} = E_\nu[f(\xi_t)\sum_{k=0}^n I_k(\chi_{s\le s_1\le\ldots\le s_k\le t}\ h(\xi_{s_1})\ldots h(\xi_{s_k}))],$$

where $I_o \equiv 1$. Moreover, let $\hat{Z}_s^t(x)$ be the r.h.s. of (13). The proof is accomplished once it is shown that for any $t\in[0,T]$ $\hat{Z}_s^t = Z_s^t$ a.e. $\lambda$; for this at first it is proved that $J_{s,t}^{(n)}$ converges in mean square to $(\hat{Z}_s^t,\phi)$, for any $0 \le s \le t$. In fact

$$E_\lambda|(\hat{Z}_s^t,\phi)-J_{s,t}^{(n)}|^2 = E_\lambda|E_\nu(f(\xi_t)[\exp(\int_s^t h(\xi_\tau)dY_\tau - \tfrac{1}{2}\int_s^t |h(\xi_\tau)|^2 d\tau) +$$

$$- \sum_{k=0}^n I_k(\chi_{s\le s_1\le\ldots\le s_k\le t}\ h(\xi_{s_1})\ldots h(\xi_{s_k}))])|^2 =$$

$$\le E_\lambda\{E_\nu(f(\xi_t))^2 E_\nu[\exp(\int_s^t h(\xi_\tau)dY_\tau - \tfrac{1}{2}\int_s^t |h(\xi_\tau)|^2\, d\tau) +$$

$$- \sum_{k=0}^n I_k(\chi_{s\le s_1\le\ldots\le s_k\le t}\ h(\xi_{s_1})\ldots h(\xi_{s_k}))]^2\} =$$

$$= E_\nu(f(\xi_t))^2 E_\nu E_\lambda[\exp(\int_s^t h(\xi_\tau)dY_\tau - \tfrac{1}{2}\int_s^t |h(\xi_\tau)|^2 d\tau +$$

$$- \sum_{k=0}^n I_k(\chi_{s\le s_1\le\ldots\le s_k\le t}\ h(\xi_{s_1})\ldots h(\xi_{s_k}))]^2 \tag{17}$$

Next note that for any path of the process $\{\xi_\tau , \tau \in [s,t]\}$

$$\exp\Big( \int_s^t h(\xi_\tau)dY_\tau - \frac{1}{2}\int_s^t |h(\xi_\tau)|^2 d\tau\Big)$$

is the unique solution of the stochastic differential equation in $\Omega^Y$,

$$dx_\tau = h(\xi_\tau)x_\tau dY_\tau, \; x_s = 1, \quad \tau \in [s,t] \tag{18}$$

computed at the right endpoint t, whereas

$$\sum_{k=0}^{n} I_k(\chi_{s\leq s_1 \leq \ldots \leq s_k \leq t}\, h(\xi_{s_1})\ldots h(\xi_{s_k}))$$

is just the n-th Picard iteration of (18), which converges in mean square to the solution, uniformly on the paths of $\{\xi_\tau\}$, beacause of the boundedness of h. This assures that

$$E_\lambda(Z_s^t - \hat{Z}_s^t, \phi_i)^2 = 0, \; \phi \in L^2(K,m), \; \phi \geq 0, \; \int_K \phi dm = 1.$$

The same is true for any $\phi \in L^2(K,m) \cap L^1(K,m)$, in view of the Jordan decomposition for signed measures. Finally, let $\{\phi_i\}$ be an orthonormal basis in $L^2(K,m) \cap L^1(K,m)$ (this is possible by denseness), so that

$$E_\lambda \| Z_s^t - \hat{Z}_s^t \|^2 = E_\lambda \sum_{i=1}^{\infty} (Z_s^t - \hat{Z}_s^t, \phi_i)^2 = \sum_{i=1}^{\infty} E_\lambda (Z_s^t - \hat{Z}_s^t, \phi_i)^2 = 0. \qquad \square$$

## 4. THE ZAKAI EQUATION FOR MARKOV PROCESSES

The derivation of the Zakai equation as a mild stochastic differential equation in $Z_f^Y(0,T;L^2(K,m))$ is now quite easy to obtain.Let $\{\xi_t, t \in [0,T]\}$ be the Markov process of the previous section and let $\xi_o$ have the density $p_o \in L^2(K,m)$ w.r.t. the measure m so that on the $\sigma$-algebra $F^\xi$ the probability $P^\xi$ is obtained by

$$P^\xi(A) = \int_K P_{ox}(A) p_o(x) dm(x) \tag{19}$$

Next let $\{V_t, t \in [0,T]\}$ be a standard scalar Wiener process on $(\Omega^V, F^V, P^V)$, h be a bounded measurable function and consider the equation

$$Y_t = \int_0^t h(\xi_\tau)d\tau + V_t , \quad t \in [0,T]$$

on $(\Omega,G,P) = (\Omega^{\xi}\times\Omega^{V}, F^{\xi}\times F^{V}, P^{\xi}\times P^{V})$. Let $F^{Y}_{s,t}$ be the σ-algebra generated by $\{Y_\theta - Y_\tau,\ s\le\tau\le\theta\le t\}$ and let λ be the unique probability measure on $(\Omega,G)$ such that $\{Y_t, F^{Y}_{0,t}\}$ is a Wiener process. Then [10] $P|_{F^{Y}_{0,t}} \ll \lambda$ and for any bounded measurable function f on K and $t\in[0,T]$, P-a.e.

$$E(f(\xi_t)|F^{Y}_{0,t})(\omega) = \frac{E_{P^{\xi}}[f(\xi_t)\exp(\int_0^t h(\xi_\tau)dY_\tau(\omega) - \frac{1}{2}\int_0^t |h(\xi_\tau)|^2 d\tau)]}{E_{P^{\xi}}[\exp \int_0^t h(\xi_\tau)dY_\tau(\omega) - \frac{1}{2}\int_0^t |h(\xi_\tau)|^2 dt]} \qquad (20)$$

At this point the main theorem is easily established.

THEOREM 5. *The solution of the mild stochastic bilinear differential equation in* $Z^{Y}_{f}(0,T;L^2(K,m))$

$$X_t = \Gamma^*(t)p_o + \int_0^t \Gamma^*(t-s)hX_s dY_s \quad , \ t\in[0,T] \qquad (21)$$

*is such that for any* f *bounded in* $L^2(K,m)$, P-*a.e.*

$$(X_t,f) = E_{P^{\xi}}[f(\xi_t)\exp(\int_0^t h(\xi_\tau)dY_\tau - \frac{1}{2}\int_0^t |h(\xi_\tau)|^2 d\tau)] \qquad (22)$$

PROOF. The equation (21), restricted to [0,t] is adjoint to equation (12), so that Theorem 3 applies, that is λ-a.e.

$$(X_t,f) = (p_o, Z^t_0). \qquad (23)$$

On the other hand, by Theorem 4, taking scalar products with $p_o$

$$(p_o,Z^t_o) = \int_K E_{ox}[f(\xi_t)\exp(\int_0^t h(\xi_\tau)dY_\tau - \frac{1}{2}\int_0^t |h(\xi_\tau)|^2 d\tau)]p_o(x)dm(x) =$$

$$= E_{P^{\xi}}[f(\xi_t)\exp(\int_0^t h(\xi_\tau)dY_\tau - \frac{1}{2}\int_0^t |h(\xi_\tau)|^2 d\tau)]$$

from which the equality (22) is obtained λ-a.e. But $P|_{F^{Y}_{0,T}} \ll \lambda$, so that the equality holds P-a.e. □

The equation (21), written in differential form, is usually referred to as the Zakai equation for the unnormalized conditional density. In fact it can be easily seen with a slight modification of the proof given in [6] that P-a.e.

i) $X_t(x) \geq 0$ m-a.e.,

ii) $$\int_K X_t(x)f(x)\,dm(x) = E_{P^\xi}[f(\xi_t)\exp(\int_0^t h(\xi_\tau)dW_\tau - \frac{1}{2}\int_0^t |h(\xi_\tau)|^2 d\tau)]$$

for any bounded measurable function f on K.
From ii), taking $f \equiv 1$ it is obtained that actually $X_t \in L^1(K,m)$ P-a.e., so that $\dfrac{X_t(x)}{\int_K X_t(x)\,dm(x)}$ is the conditional density with respect to the measure m. When h = 0 the equation (21) becomes the Fokker-Planck equation for the normalized unconditional density: in fact the starting hypothesis on the semigroup $\Gamma$ is equivalent to assume that such equation, governed by the $C_o$-semigroup $\Gamma^*$, evolves in $L^2(K,m)$.

Finally note that usually the state space K is a $R^n$-space with the Lebesgue measure m; but it is not difficult to give examples in which one can profitably exploit the freedom in the choice of such a measure: for example m could be an invariant measure for $\{\xi_t\}$. The following section gives a sample.

## 5. AN EXAMPLE OF APPLICATION

Let us consider the following stochastic differential system

$$d\xi_t = A\xi_t dt + BdW_t \quad , \quad \xi_t \in R^n$$

$$dY_t = h(\xi)dt + dV_t$$

$\{W_t, V_t\}$ being a standard $R^{p+1}$ valued Wiener process. Let us suppose that h is a bounded measurable function on $R^n$, A is a stable matrix and (A,B) is a controllable pair. It is well known [10] that $\{\xi_t\}$ is a Markov process with the transition probability measure on $R^n$

$$P_{tx}(E) = G_n(E; e^{At}x, Q(t))$$

where $G(\cdot;\mu,\Sigma)$ is the n-dimensional Gaussian measure with mean $\mu$ and covariance matrix $\Sigma$, and

$$Q(t) = \int_0^t e^{As}BB^* e^{A^*s}ds.$$

Under our hypotheses $Q(t)$ is positive definite and its limit for $t \to \infty$ exists and will be denoted by $\tilde{Q}$ [13].

The semigroup $\Gamma$ associated to $\{\xi_t\}$ defined by (11) is not strongly continuous on $L^2(R^n)$ with the usual Lebesgue measure, so that the Zakai equation cannot be written in such a space. However the probability measure $m = G_n(\cdot;0,\tilde{Q})$ is invariant for the process $\{\xi_t\}$ [12], so that $\Gamma$ is a contraction semigroup on $L^2(R^n,m)$ [13, p. 381]. Moreover $\Gamma$ is $C_o$, too: in fact, for any uniformly Lipschitz function f in $L^2(R^n,m)$

$$\|\Gamma(t)f-f\|^2 = \int_{R^n} \left| \int_{R^n} (f(y)-f(x))G_n(dy;e^{At}x,Q(t)) \right|^2 G_n(dx;0,\tilde{Q}) =$$

$$\leq L \int_{R^n}\int_{R^n} |y-x|^2 G_n(dy;e^{At}x,Q(t))G_n(dx;0,\tilde{Q}) =$$

$$= L \int_{R^n} |y-x|^2 G_{2n}(d(x,y);0,R(t))$$

where $R(t) = \begin{bmatrix} \tilde{Q} & \tilde{Q}e^{A^*t} \\ e^{At}\tilde{Q} & \tilde{Q} \end{bmatrix}$, so that for the last term, denoting by E expectations w.r.t. $G_{2n}$, one has

$$E(y-x)'(y-x) = trE(y-x)(y-x)' = tr(Eyy'-Exy'-Eyx'+Exx') =$$

$$= tr(2\tilde{Q} - \tilde{Q}e^{A^*t} - e^{At}\tilde{Q})$$

which goes to zero for $t \to 0$. This establishes the announced strongly continuity of $\Gamma$, because uniformly Lipschitz functions are dense in $L^2(R^n,m)$.

If the probability measure of $\xi_o$ is absolutely continuous w.r.t. $G(\cdot;0,\tilde{Q})$ with R.-N. derivative square-integrable with respect to the same measure (which in terms of the density $\pi_o$ w.r.t. the Lebesgue measure means $\int_{R^n} \pi_o^2(x)\exp(\frac{1}{2}x'\tilde{Q}^{-1}x)dx < +\infty$) the derivation of the Zakai equation is therefore possible following the analysis of the previous section.

REFERENCES


[1] A. GERMANI, M. PICCIONI: *Finite Dimensional Approximation for Stochastic Bilinear Differential Equations on Hilbert Spaces*, Report R.61, IASI-CNR, 1983.

[2] A. GERMANI, M. PICCIONI: *A Galerkin Approximation for the Zakai Equation*, Proceedings 11-th IFIP Conference on System Modelling and Optimization, Copenaghen, 1983, Springer-Verlag, to appear.

[3] M. ZAKAI: *On the Optimal Filtering of Diffusion Processes*, Z. Wahrschein. verw. Geb., 11, 1969, pp. 230-243.

[4] H. KUNITA: *Cauchy Problem for Stochastic Partial Differential Equations arising in Nonlinear Filtering Theory*, Systems & Control Letters, 1, 1981, pp. 37-41.

[5] N.V. KRYLOV, B.L. ROZOVSKII: *On the First Integral and Liouville Equations for Diffusion Processes*, in Stochastic Differential Systems, Ed. M. Arato, D. Vermes, A.V. Balakrishman,Springer-Verlag, 1981.

[6] E. PARDOUX: *Stochastic Partial Differential Equations and Filtering of Diffusion Processes*, Stochastics, 3, 1979, pp. 127-167.

[7] M.H.A. DAVIS: *On a Multiplicative Functional Transformation Arising in Nonlinear Filtering Theory*, Z. Wahrsch. Verw. Geb., 54, 1980, pp. 125-139.

[8] G. DA PRATO, M. IANNELLI, L. TUBARO: *Linear Stochastic Differential Equations in Hilbert Spaces*, Rend. Acc. Naz. Lincei, LXIV, 1978, pp. 22-29.

[9] G. KOCH: *Volterra Series Expansion for Stochastic Bilinear Systems*, Report R. 2-07 of Istituto di Automatica, University of Rome, 1972.

[10] G. KALLIANPUR: *Stochastic Filtering Theory*, Springer-Verlag, New York, 1980.

[11] M. KAC: *On Some Connections between Probability Theory and Differential and Integral Equations*, Proc. 2nd Berkeley Symp. on Math. Stat. and Prob., 1951, pp. 189-215.

[12] L. ARNOLD, W. KLIEMANN: *Qualitative Theory of Stochastic Systems*, in: Probabilistic Analysis and Related Topics, Vol. 3, Ed. by A.T. Bharucha Reid, Academic Press, New York, 1983.

[13] T. KAILATH: *Linear Systems*, Prentice-Hall, Englewood Cliffs,1980.

[14] K. YOSIDA: *Functional Analysis*, Springer-Verlag, Berlin, 1980.

# Lecture Notes in Control and Information Sciences

Edited by A. V. Balakrishnan and M. Thoma

## Lecture Notes in Control and Information Sciences

Edited by A. V. Balakrishnan and M. Thoma

Vol. 62: Analysis and Optimization
of Systems
Proceedings of the Sixth International
Conference on Analysis and Optimization
of Systems
Nice, June 19–22, 1984
Edited by A. Bensoussan, J. L. Lions
XIX, 591 pages. 1984.

Vol. 63: Analysis and Optimization
of Systems
Proceedings of the Sixth International
Conference on Analysis and Optimization
of Systems
Nice, June 19–22, 1984
Edited by A. Bensoussan, J. L. Lions
XIX, 700 pages. 1984.